IEE TELECOMMUNICATIONS SERIES 40

Series Editors: Professor Charles J. Hughes
Professor David Parsons
Professor Gerry White

SPREAD SPECTRUM IN MOBILE COMMUNICATION

Other volumes in this series:

Volume 1 **Telecommunications networks** J. E. Flood (Editor)
Volume 2 **Principles of telecommunication-traffic engineering** D. Bear
Volume 3 **Programming electronic switching systems** M. T. Hills and S. Kano
Volume 4 **Digital transmision systems** P. Bylanski and D. G. W. Ingram
Volume 5 **Angle modulation: the theory of system assessment** J. H. Roberts
Volume 6 **Signalling in telecommunications networks** S. Welch
Volume 7 **Elements of telecommunications economics** S. C. Littlechild
Volume 8 **Software design for electronic switching systems** S. Takamura, H. Kawashima, N. Nakajima
Volume 9 **Phase noise in signal sources** W. P. Robins
Volume 10 **Local telecommunications** J. M. Griffiths (Editor)
Volume 11 **Principles and practices of multi-frequency telegraphy** J. D. Ralphs
Volume 12 **Spread spectrum in communications** R. Skaug and J. F. Hjelmstad
Volume 13 **Advanced signal processing** D. J. Creasey (Editor)
Volume 14 **Land mobile radio systems** R. J. Holbeche (Editor)
Volume 15 **Radio receivers** W. Gosling (Editor)
Volume 16 **Data communications and networks** R. L. Brewster (Editor)
Volume 17 **Local telecommunications 2** J. M. Griffiths (Editor)
Volume 18 **Satellite communication systems** B. G. Evans (Editor)
Volume 19 **Telecommunications traffic, tariffs and costs** R. E. Farr
Volume 20 **An introduction to satellite communications** D. I. Dalgleish
Volume 21 **SPC digital telephone exchanges** F. J. Redmill and A. R. Valdar
Volume 22 **Data communications and networks 2nd Edn.** R. L. Brewster (Editor)
Volume 23 **Radio spectrum management** D. J. Withers
Volume 24 **Satellite communication systems 2nd Edn.** B. G. Evans (Editor)
Volume 25 **Personal & mobile radio systems** R. C. V. Macario (Editor)
Volume 26 **Common-channel signalling** R. J. Manterfield
Volume 27 **Transmission systems** J. E. Flood and P. Cochrane (Editors)
Volume 28 **VSATs: very small aperture terminals** J. L. Everett (Editor)
Volume 29 **ATM: the broadband telecommunications solution** L. G. Cuthbert and J.-C. Sapanel
Volume 30 **Telecommunication network management into the 21st century** S. Aidarous and T. Plevyak (Editors)
Volume 31 **Data communications and networks 3rd Edn.** R. L. Brewster (Editor)
Volume 32 **Analogue optical fibre communications** B. Wilson, I. Darwazeh and Z. Ghassemlooy (Editors)
Volume 33 **Modern personal radio systems** R. C. V. Macario (Editor)
Volume 34 **Digital broadcasting** P. Dambacher
Volume 35 **Principles of performance engineering for telecommunication and information systems** M. Ghanbari, C. J. Hughes, M. C. Sinclair and J. P. Eade
Volume 36 **Telecommunication networks, 2nd edition** J. E. Flood (Editor)
Volume 37 **Optical communication receiver design** S. B. Alexander
Volume 38 **Satellite communication systems, 3rd edition** B. G. Evans (Editor)
Volume 39 **Quality of service in telecommunications** A. Oodan, K. Ward and T. Mullee

SPREAD SPECTRUM IN MOBILE COMMUNICATION

Olav Berg, Tore Berg,
Svein Haavik, Jens Hjelmstad
and Reidar Skaug

The Institution of Electrical Engineers

Published by: The Institution of Electrical Engineers, London,
United Kingdom

The Institution of Electrical Engineers,
Michael Faraday House,
Six Hills Way, Stevenage,
Herts. SG1 2AY, United Kingdom

British Library Cataloguing in Publication Data

A CIP catalogue record for this book
is available from the British Library

ISBN 0 85296 935 X

Printed in England by Bookcraft, Bath

Contents

List of abbreviations

ACK	Acknowledgment
ACS	Automatic channel selection
AFD	Average fading duration
AGC	Automatic gain control
ARQ	Automatic repeat request
ASIC	Application specific integrated circuit
ATM	Asynchronous transfer mode
AWGN	Additive white Gaussian noise
BCH	Bose–Chaudhuri–Hocquenghem
BER	Bit error rate
BS	Base station
BSC	Base station controller
BTMA	Busy-tone multiple access
BTS	Base transceiver station
CCIR	International radio consultative services
CDMA	Code division multiple access
CL	Connectionless
CMOS	Complementary metal oxide semiconductor
CNR	Combat net radio
CO	Connection-oriented
CPM	Continuous phase modulation
CPU	Central processing unit
CR	Central location register
CRC	Cyclic redundancy check
CRC	Cyclic redundancy code
CS	Circuit switching
CSMA	Carrier sense multiple access
CSP	Chip size packages
CTS	Clear to send
CVSD	Continuous variable slope delta modulation

CW	Continuous wave
DCE	Data circuit terminating equipment
DPSK	Differential PSK
DS(SS)	Direct sequence (spread spectrum)
DSP	Digital signal processor
DSSS	Direct sequence spread spectrum
DT	Data
DTE	Data terminal equipment
EMC	Electromagnetic compatibility
EMI	Electromagnetic interference
EPM	Electromagnetic protective measures
EREMES	Communication system
ETSI	European Telecommunications Standard Institute
EUROCOM	Communication standard
FAR	False-alarm rate
FCS	Free channel search
FDMA	Frequency division multiple access
FEC	Forward error correction
FFT	Fast Fourier transform
FH(SS)	Frequency hopping (spread spectrum)
FIFO	First in first out
FIR	Finite impulse response
FM	Frequency modulation
FT	Fourier transform
GBn	Go back n
GIS	Geographic information system
GMSK	Gaussian minimum shift-keying
GSM	Global system for mobile communication
GT	Gabor transform
HF	High frequency (approx. 3–30 MHz)
HIPERLAN	High performance radio LAN
HLR	Home location register
HOS	Higher order statistics
ICI	Interface control information
IDU	Interface data unit
IF	Intermediate frequency
IFT	Inverse Fourier transform
IN	Integrated network
INMARSAT	International Maritime Organisation satellite (communication)
ISDN	Integrated services digital network

ISI	Inter-symbol interference
ISM	Industrial, scientific and medical
ISO	International Organisation for Standardisation
ITU	International Telecommunication Union (formerly CCITT)
IUI	Interuser interference
IWF	Interworking function
IWT	Integral wavelet transform
JAGUAR V	Military radio system
LAN	Local area network
LCR	Level crossing rate
LI	Length indicator
LLC	Logical link control
LMN	Land mobile network
LPD	Low probability of detection
LPI	Low probability of intercept
LPM	Local protocol mapping
LR	Location register
LSC	Local switching centre
MAC	Medium access control
MAI	Multiple access interference
MCM	Multi chip module
MMI	Man machine interface
MPEG	Moving pictures expert group
MRR	Multi role radio
MS	Mobile station
MSC	Mobile services switching centre
MSE	Mobile switching entity
MSK	Minimum shift keying
MT	Mobile terminal
NBSS	Narrow band spread spectrum
NMT	Nordic mobile telephone system
OMC-S/R	Operation and maintenance centre – switching/radio
OSI	Open system interconnection
PAMR	Public access mobile radio
PCI	Protocol control information
PCM	Pulse code modulation
PCS	Personal communication system (/services)
PDU	Protocol data unit
PMR	Private mobile radio
PN	Pseudorandom noise

PR	Packet radio (packet switched radio networks)
PRC-77	Military radio system
PROP	Packet radio organisation packet
PRP	Packet radio control
PS	Packet switching
PSK	Phase shift keying
PSMA	Preamble sense multiple access
PSTN	Public switched telephone network
RD-CDMA	Receiver directed CDMA
REP	Residual error probability
RF	Radio frequency
RS	Reed-Solomon
RTS	Request to send
SAP	Service access point
SDU	Service data unit
SER	Symbol-error rate
SHF	Super high frequency (approx. 3 – 30 GHz)
SINCARS	Military radio system
SNR	Signal-to-noise ratio
SRn	Selective repeat n
SRP	Source routing protocol
SS	Spread spectrum
STFT	Short time Fourier transform
SW	Stop-and-wait
SWIA	SW with immediate acknowledgement
TDMA	Time division multiple access
TETRA	Trans european trunked radio
TH	Time hopping
UHF	Ultra high frequency (approx. 0.3 – 3 GHz)
UMTS	Universal mobile telecommunication system
UTM	Universal transverse mercator, co-ordinate system
VHF	Very high frequency (approx. 30 – 300 MHz)
VLR	Visitors location register
VLSI	Very large scale integration
VRC-12	Military radio system
WSSUS	Wide sense stationary uncorrelated scatter
WT	Wavelet transform

Preface

When two of the authors were initially approached by the publishers, they were asked if it would be possible to update the work 'Spread spectrum in communication' published in 1985. As the publishers pointed out, a lot has happened recently in the field of spread-spectrum communications, not the least with respect to the use of mobile radio systems and the introduction of radio networks rather than point-to-point radio connections.

The results presented in this book show that so much has changed that it was not really a question of updating the book from 1985 but rather one of writing a completely new book. Any readers of the previous book will thus only recognise a minor re-use of text and figures. The bare fact that we are no longer two co-authors, but five, points to the great widening of the field of spread-spectrum radiocommunications.

The first book did, to a large degree, focus on technology and the implementational aspects relating to radio design and spread-spectrum modulation. To generate and demodulate the spreading codes was a major issue, as was the necessity for methods to synchronise the transmitter and receiver. Today, embedded digital signal-processing power, and the general technology available, has enabled a large variety of possibilities for practical spread-spectrum implementations. At the same time, the integration and use of data-network protocols in radios enables the development of radio systems with the kind of functionality and menu of services previously only possible for wired systems.

This book is, as the previous 'Spread spectrum in communication', based on research and development (R&D) work carried out at or on behalf of FFI (Norwegian Defence Research Establishment). Although we have increased the number of co-authors from two to five, we recognise and greatly appreciate permission to also make unrestricted use of the work of our colleagues. In particular we would like to thank Georg

Anuglen, Hanne Hodnesdal, Herman Lia, Berta Omtveit, Snorre Prytz, Bjørn Skeie, Arne Slåstad and Bjørn Solberg.

FFI is an applied research establishment which means that most of the issues and solutions discussed in this book have been tried out in the laboratory and/or in the field, and that the simulation models have been developed to evaluate alternative practical solutions. Even with all this material at hand, it has been necessary to carry out a considerable amount of supplementary work during the writing process. We wish in that respect to thank FFI and Telenor Research and Development for supporting this book project.

Although we have tried to give a coherent presentation of the field of spread-spectrum radio communication, we are well aware that the book by no means treats the subject in a complete and textbook-like manner, but rather represents a monograph of selected issues with respect to challenges and possible solutions.

Chapter 1 introduces the concepts of modern mobile communication and gives an overview of the evolution of mobile communication networks. Certain basic fundamentals for spread-spectrum modulation and the use of radio waves as information carriers are introduced as a background for reading and understanding the rest of the book.

Chapter 2 introduces mobile-radio users and their requirements, and shows how these requirements must influence the system design. The radio system itself is defined and the concept of an architecture is presented. Finally, the central issues relating to mobility, switching and security are elaborated on.

Chapter 3 presents a short introduction to the operational environment especially relevant to mobile spread-spectrum systems, focusing on the effects of the channel characteristics due to different propagational effects and the presence of noise and interference.

Chapter 4 gives an overview of the radio transmission system. The Chapter describes the radio and its different characteristics and elaborates on different forms of spread-spectrum technique. In particular, the Chapter introduces the effect of spread-spectrum modulation on radios used in packet-switched networks. The Chapter also treats the signal design issues relevant for performance analysis, including error control. Finally, a technology forecast is presented together with a description of new trends within radio design.

Chapter 5 covers a broad presentation of packet switching in radio networks. The importance of how the transmission medium is accessed and the effect of network topology are treated initially. The influence of the radio transmission system and radio-channel quality is discussed in the context of capture models. Several examples showing how different network parameters can be set to achieve satisfactory medium-access control are also given. The Chapter then moves up the network hierarchy and explains how a logical link control is introduced to provide satisfactory reliability for the delivery of data packets. In a distributed radio system the problem with nodes without direct radio contact will exist and the Chapter discusses the different issues relevant for the hidden-node problem. The Chapter finally discusses the network layer and the issues of routing through a certain network topology.

Chapter 6 presents a case study. The case exemplifies the process from the initial presentation of the user requirements through the many iterations necessary in order to match requirements to the operational environment as well as the technical and technological challenges.

To write a book is much more than thinking a book. This means that a number of people have been involved in the process between thinking and writing. We are particularly indebted to Lise Gulbrandsen as an effective overall co-ordinator, to Peter Toombs for helping with the English language, to Kari Skovli and her colleagues for the writing process and to Unni Næss and Guri Bolstad for their work on the many illustrations.

Chapter 1

Introduction

1.1 Concepts of modern mobile communication

1.1.1 Fundamental concepts

This book is dedicated to the science and engineering of mobile communication systems. Mobile communication is, at this stage, already one of the biggest segments of information technology and every technology forecaster predicts a seemingly endless growth. Why is it that mobile communication now seems to be an essential part of society, and with an equally strong market demand for both personal users and professional users?

To look into this question we must first realise that the term 'mobile communication' consists of two separate elements, 'mobile' and 'communication'. Taken separately, each one has its own history dating way back to the beginning of mankind. At this time in history both terms are epitomes of modern society; mobility is necessary for individuals to function in any modern environment and efficient ways for the communication of information in every sense of the word have always been vital to any organisation consisting of sets of individuals.

The total amount of communication spread around the globe is increasing at an exponential rate. It is happening because of a combination of technological breakthroughs improving cost effectiveness and an increasing demand. This gives a positive feedback since these factors result in a technology push and the need for ever more capacity at an increasingly competitive cost. At this stage there is little evidence that this trend will depart from its present course; however, the type of information transmitted and the associated requirements for the type of information network will change. This leads on to one of the main attributes of this book, namely that of covering a wide variety of user requirements and communication needs and presenting the rationale and reasoning behind design of mobile communication systems.

The mobility aspect is increasingly important, not only because individuals move around more but also because technology provides solutions to the engineering challenges of designing advanced mobile radio networks. Mobility may be taken to mean that one is actually moving or merely imply flexibility as to where one operates or is positioned. Both these interpretations require similar technical solutions and the majority of systems developed are capable of addressing both aspects of mobility.

Users moving from place to place may establish communication by connecting to land-based wired networks. This will satisfy the need for broadband information transfer, but this solution is rapidly becoming obsolete for a number of reasons, the main ones being its impracticality due to its inflexibility and specific problems such as the lack of guaranteed access because the user might not be 'plugged in' anywhere.

As a consequence, modern interpretations of mobility tend to be synonymous with radio mobility, which is the key issue of this book. Looking at the history of radio we find that it interacts with various aspects of mobility and communication in a very interesting way. When radio was first used it was specifically for communication purposes. The first radio telegrams were transmitted more than a century ago and the main virtue of the radio waves was their ability to reach distant stations. The reasons for the instant popularity and demand for radiocommunication was that within a few years of its invention radio telegrams could be transmitted to the most distant parts of the globe. In this way radiocommunication was an instant commercial success and, as in any industry, efforts were focused on ways of improving the technology.

Radio soon matured to the stage where the devices used for communication could in fact be moved. From this point, the mobility aspect became a major driving force for technology. The achievable communication ranges of portable units soon increased to practical distances and radio systems were instantly to be found in military forces and other high-demand agencies.

The term mobility might, furthermore, have different implications in a different context. The usual implication, and the one which usually comes to mind for the nonexpert, is terminal mobility, where the user carries a terminal which most conveniently will have radio access to a switching or relaying station. Another way of achieving mobility, however, is through a scheme which might be called user mobility. In this case the user does not carry the actual communication device but merely a proof of identity such as a card which might be inserted into the communication system. This approach is seeing a renaissance in cellular telephone systems where a person might rent a phone and identify himself for calling and billing purposes using an identity card.

Interestingly, today's communication systems exhibit a mutual dependence between mobility and the radio. System solutions for applications focusing on mobility and those focusing on communication are becoming more and more alike. One interesting observation in this context is that the transmission ranges for radiocommunication have seen a steady shift towards ever shorter distances as technology has matured. Cross-continent communications have been replaced by urban and tactical systems such as those used by municipal units or military forces. This trend is due to the development of wired ground-based communication and high-capacity switching units. Radiocommunication devices now play an important role in supplying services to wired users connected to digital computerised exchanges as well as connecting nonwired users using their own network.

Mobile radiocommunication is now inextricably intertwined with computerised information processing and data routing. Even the most basic communication requirement might be more efficiently served using computerised mobile radiocommunications systems. This is not merely the result of technological breakthroughs; operational demands and regulation requirements necessitate the integration of radio units into digital data-processing systems.

This marriage of technologies has ushered in a new era, where the radio segment is sandwiched between layers of digital signal processors and where the information transmitted is structured in layers of protocols and routing strategies. One result of this is that one protocol might be used for a wide variety of systems with widely different transmission ranges. Data-communication protocols may be the same for a satellite downlink to a mobile user as in a land-based network. Furthermore, a set of networks might share protocols or parts of protocols when operating jointly.

Data-transmission ranges with the explicit exception of satellite downlinks and crosslinks have shown a trend toward the use of smaller and smaller cells. This situation has arisen as ground-based networks increasingly use wide bandwidth fibre-optic networks connecting high-capacity switches. Mobile radio systems might be connected to a fixed network or make use of a backbone wired network for parts of the communication structure. As an alternative to a backbone wired network, several radio units might be connected in relay mode. Not only will this extend achievable ranges but it will also facilitate numerous other functions, as will be discussed in later sections of this book.

Finally, the last keyword of this book is spread spectrum. The term 'spread spectrum' refers to a class of radio-modulation formats that gives optimum performance for a wide variety of challenging applications. The obvious implication of using radio waves is that of remote access without any physical links. Radio transmission can have virtually any range: distances might range from light years to distances within a computer processing chip. In any case, radio waves are electromagnetic waves which may have superimposed information that can be stripped off after transmission along the propagation path. The science of designing the radio link is a complicated one, where numerous requirements have to be met. These are requirements for data-transmission capacity and quality, restrictions on transmitted power and bandwidth and numerous others which will be covered in this book. As will be seen, there will also be an unconditional requirement to perform in a scenario with numerous other participants that also emit radiocommunication signals.

This situation has led to the term spread-spectrum radio systems. Originally, the term spread spectrum was used on bandwidth-spreading schemes where the bandwidth was extended by superimposing a code on the data stream. This use of extra bandwidth brought the advantage that the signal-to-noise ratio needed for adequate reception was dramatically reduced. Indeed, the novel concept was that a signal with amplitude far below the noise level could be used to transmit information. By now, the term spread-spectrum modulation has come to include a large family of techniques which adapt radiocommunications to a computerised data-information processing network. As will be shown in this book, digital information may be transmitted through radio links in numerous ways, with varying degrees of sophistication. These range from traditional modulation schemes to sophisticated digital formats. Advanced mathematical methods are used to select formats that meet user requirements and radio-channel characteristics.

As is evident from Figure 1.1 which shows the number of quotes of the terms 'spread spectrum', 'radio' and 'mobile radio' in the INSPEC reference files for the years 1994 and 1995, spread-spectrum applications are on the increase and are also moving rapidly into the radio and mobile radio business segment.

This is one of the reasons why the authors felt there would be a demand for this updated book dealing with the integration of spread spectrum in mobile communication. This is a very interesting theme for engineers and scientists alike, demanding analytical and theoretical skills as well as creative and innovative engineering. It is evident that this is a technology that will continue to increase its marked shares within the field of information technology and will be more and more noticeable in everyday life for individuals and businesses.

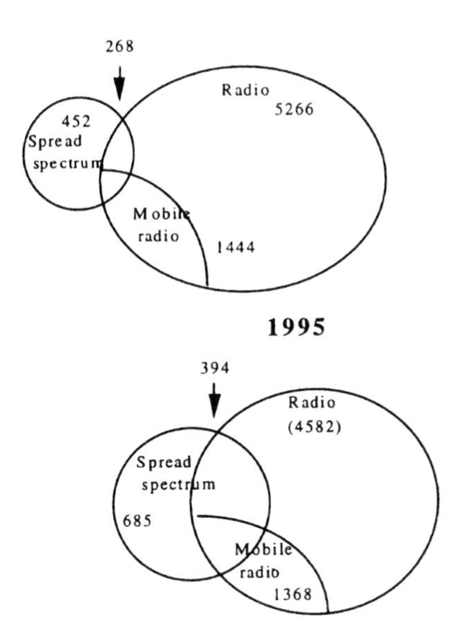

Figure 1.1 *Number of quotes of the terms 'spread spectrum', 'radio' and 'mobile radio' in the INSPEC reference files for the years 1994 and 1995. Citations for spread-spectrum applications show an upward trend and also appear increasingly in the same context as 'radio' and 'mobile radio'.*

1.1.2 Topology of radio networks

The need for mobility has necessitated changes in the design and implementation of radiocommunication systems, and has been a major driving force in developing the concept of networking. Mobility considerations have caused a change in both network structure and manifestation over the years, and different realisations of mobile radio systems have developed. The simplest communication systems use point-to-point connection, as was indeed the original implementation. This approach works perfectly with only one radio on the network at any one time. Any additional users in that area and on the same frequency slot will require some administrative procedure which would normally take the form of a centralised network control allocating the channel to users through e.g. a time sharing procedure. A typical protocol for the use of a point-to-point system is:

(i) the caller listens in to the network (traditionally by tuning in to the right frequency) and finds out if the radio channel is vacant or not;

(ii) the caller sends a calling signal to alert the person being called; he usually would identify himself in the same process;
(iii) the person being called responds; in many systems such as in traditional military radio he would also at this stage indicate the quality (strength and readability) of the signals which he has received;
(iv) the caller acknowledges the response and might indicate the quality of the signal he receives;
(v) communication commences;
(vi) when communication is complete both parties acknowledge this by a statement.

The inadequacies of a point-to-point system might be averted by introducing a base station. This approach involves relaying all radiocommunication paths through a terminal which includes a radio receiver and a radio transmitter, back to back. This configuration might be used in two different ways:

(i) the base station acts as a beacon and reradiates all the signals it receives;
(ii) the base station has the capability to handle many channels (e.g. frequency channels or code-division channels) and reradiates selectively to one particular user.

In the latter case, the base station will need to include a switching function which co-ordinates the use of the different channels and acts as a terminal for the two communication paths. Multiple communication channels might be operational simultaneously, each making use of a pair of radio channels terminated at the base station. The set-up and close-down procedures will be as for the point-to-point case, only in this case performed on each one of the twin radio paths.

There are numerous advantages of this set-up, which explains its widespread use in many commercial applications:

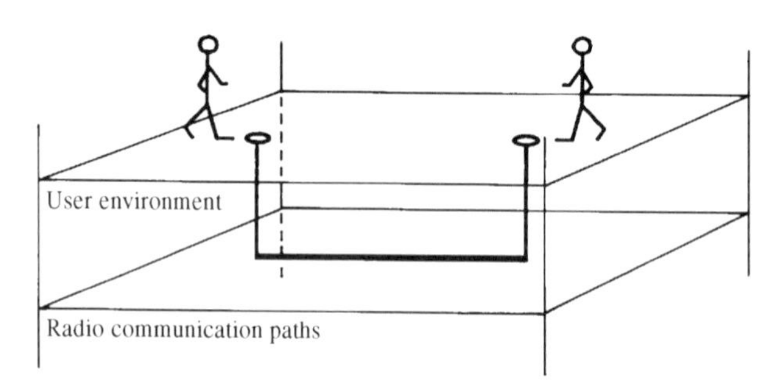

Figure 1.2 Point-to-point communications

- the radio range will be much extended, in fact by a factor much larger than two as the base station is typically located at some elevated platform to give wide coverage;
- the mobile units might be much lighter as the base station will have more transmitted power and higher-gain antennas;
- the switching functions in the base station facilitate automatic operation and advanced user services.

To further enhance the system, a number of base stations might be linked using hard wired or radio links giving a multiple base-station network.

External trunk lines can be used to connect the base stations. These lines may be ground based using wire, fibre or microwave links, or use a high-capacity satellite link. This will give a much improved capability in connecting users in different locations. The penalty is increased network complexity, requiring protocols for establishing, maintaining and disconnecting communication paths and for keeping track of users.

For many applications this method of combining radio links, switching functions and trunk lines has proved to be a reliable and useful concept. In fact, the majority of modern cellular systems use such a structure and this situation is not likely to change in the immediate future.

An alternative approach takes advantage of the processing capacity of modern processing units and distributes all switching functions in each radio unit. This is what might be described as a distributed network. In this system there are alternative routes between the base stations and each call or information package might be routed via several basestations.

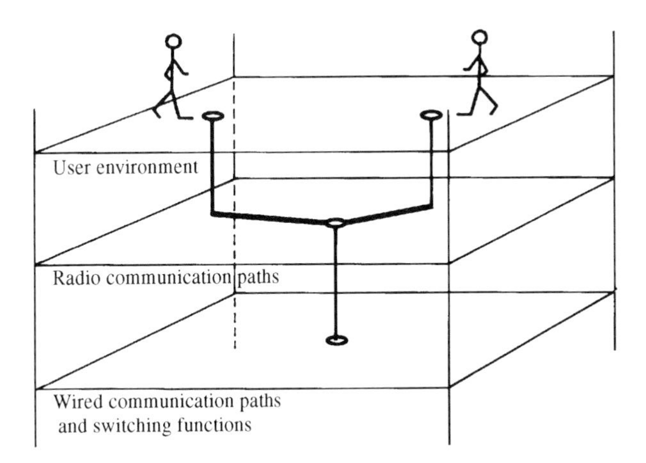

Figure 1.3 Point-to-point communications via a base station including a switching function

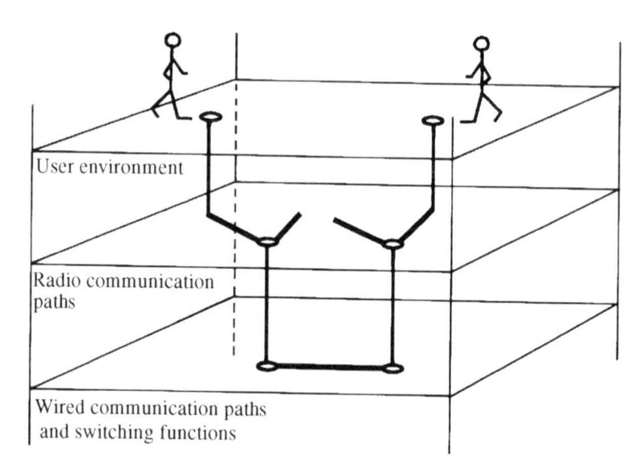

Figure 1.4 *A multiple base-station network based on base stations linked with a point-to-point or point-to-multipoint connection*

Normally there will be redundant communication paths and thus this type of system represents a more reliable way of networking a number of units.

Such a distributed network may be implemented using wired links or radio links; in the latter case the system is referred to as a distributed radio network. It is thereby assumed that each station is within radio coverage of at least one neighbour. A noticeable distinction between a network wired with cables and one connected with radio waves is that radio signals might reach more than one station and also out to stations further afield. In this way a distributed radio network has the possibility of giving a more tightly

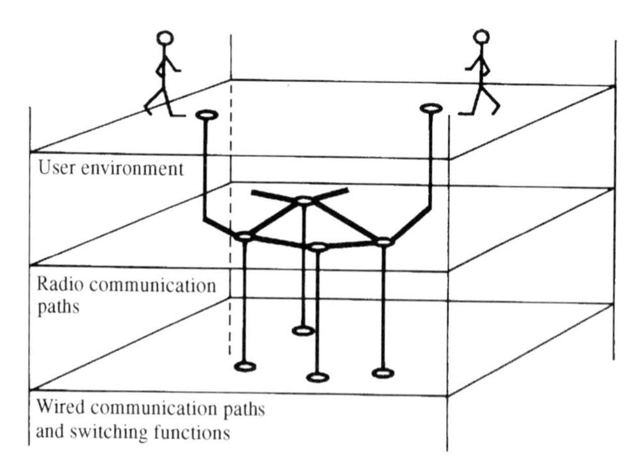

Figure 1.5 *A distributed radio network where all the nodes may act as relays*

knit network with more crosslinks than its physically-wired counterpart. The consequence is that a radio system might be more robust in challenging situations.

The relative uncertainty of the transmission range of radio signals, particularly in built-up areas, poses more stringent requirements on network control. Transmissions generated by communication nodes at large distances might be received at a certain location. It might also happen that transmissions from very local units might not be received. The challenge of designing a distributed system that allows for these characteristics is significantly different from that of designing a distributed network with fixed communication paths.

A further complication is the dynamic nature of a communication network. In a mobile system the connections will change continuously and the system will need to cope with this changing environment. This necessitates dynamic distributed network control, a task which requires a certain additional overhead in the network and which might be solved in a number of different ways. The various strategies for obtaining this function will be presented in the next Chapter and are indeed a core topic in this book.

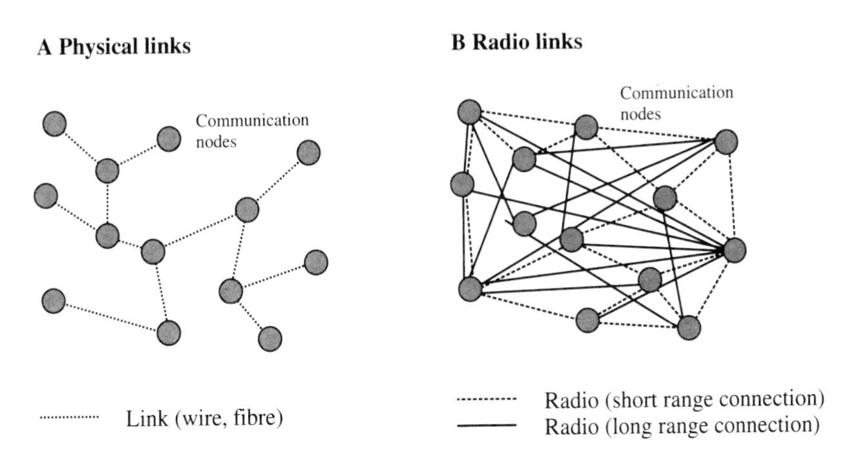

Figure 1.6 *A distributed system using wired links has wires or fibres connecting the communication nodes. A distributed radiocommunication system will have more communication paths and each transmission might be picked up by a multiplicity of other communication nodes, even those far away*

1.1.3 Packet radio or circuit switching?

In a radio network there are two fundamental approaches to connecting two terminals; these are the methods of circuit switching or packet switching.

In a circuit switched system a communication path is established and maintained throughout the duration of the call or data transfer. This is how nearly all telephone networks operate where each switch routes the call through to the next switching position and keeps the connection until the call is finished.

Advantages of this approach include:

- continuous access to the data channel;
- no latency or wait time;
- return communication on the same channel.

Disadvantages include:

- inefficient use of the radio spectrum and communication resources;
- susceptibility to dynamic changes in the communication channel;
- not flexible for roaming users.

A particular application of a line-switched network is the voice application, which has the following requirements:

- the nature of voice is that a relatively high bit error rate (BER) is acceptable (probability of bit error should be less than 10^{-3});
- the transmission delay must be low (<40 ms).

In a packet-oriented network, all data is grouped in packets and launched into the network which routes each individual packet through to its destination. An implication of this approach is that each individual packet is regarded as having its own identity and the system might choose to deal with each one differently, the choice being determined by the packet and the network's characteristics at the time of submission. A packet-oriented system can be perceived as being much like a postal package delivery system which has to cope with packets of different kinds coming and going to different locations.

Advantages of packet-oriented systems are:

- potential for efficient use of the radio spectrum;
- highly flexible for mobile users;
- might include advanced network functions.

Inherently, packet networks are more complicated as the transmission and networking system includes more advanced functions; however, for achieving the operational goal packet networks are seen as the most viable approach. Consequently, this book deals primarily with this approach and deals with applications where the forms and functions of line switching are required as a subset of packet radio.

With packet-switched services, digital data is exchanged as logical units of data. A continuous stream of data is not needed as it is with circuit switching, and the blocks of data may be separated by random idle periods. This type of service may operate with demanding delay distributions.

1.1.4 Teleservices against bearer services in a mobile radiocommunication system

In principle, a network designer is only restricted by two sets of design rules. The first set contains the restrictions introduced when using radio waves as the information carrier. The second set contains the design rules introduced by the telecommunication services which the network shall adhere to. For example, users experience bad voice quality if excessive network delay is introduced during a conversation.

The challenge for the network designer is to find efficient solutions for providing telecommunication services. In a later section we specify a layered model which facilitates a structured approach to performing this task.

Telecommunication services are provided to the user through the user interface. A common practice is to divide telecommunication services into bearer services and teleservices, see Figure 1.7. Telecommunication

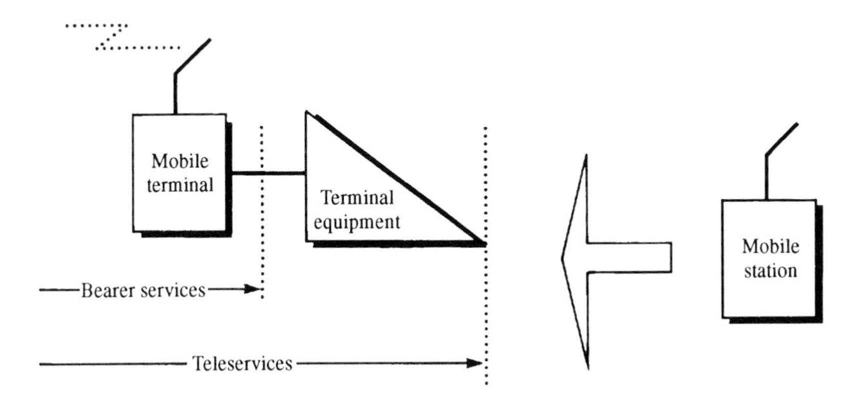

Figure 1.7 Functional split of the mobile station

services or teleservices are offered by the terminal equipment and include all capabilities for communication between two applications. Bearer services are provided by the network and include all capabilities for transmission between two or more end points. These services are used by the mobile terminal to exchange data with other mobile terminals and do not include anything concerning the terminal equipment.

It is important to know the nature of the different teleservices to understand the impact on the air interface in Figure 1.7. As we will see throughout this book, some services may cause a heavy burden on the radio channels, whereas others will only represent a negligible load to the system.

In contrast to voice, data services demand very low BER because data delivered with errors generally introduce unpredictable behaviour of the communication protocols in the end systems. If the bearer services do not provide sufficiently low BER, some entity within the end system must perform such a check and take the appropriate action when errors occur.

1.1.5 Interfaces and functional units in a mobile radio network

The *user interface* represents the connection with the user (a person or mechanical device). In the simplest form, this is the buttons for pressing numbers together with the acoustic transducers for a telephone. In more complex systems the user interface might take the form of a man–machine interface, possibly run as software on a computer.

The *air interface* represents the transition from one mobile station to another mobile or fixed terminal. This term includes all the radio modulation and demodulation operations along with the actual transmission of the radio signals.

The *mobile station* represents the actual piece of equipment that fulfils the functions of transmission and reception of information. This station will connect through its user interface to the user and through its air interface to other mobile units or fixed units (base station).

The *base station* is the functional unit that connects to the air interface (by transmission and reception of radio waves), thereby connecting several mobile users to each other and also possibly to ground-based networks.

The *switching function* represents the intelligence that connects the right terminals according to some requirement or request. In Figure 1.8 this functionality is found at the location of the base station. The switching function can, however, be included at the mobile stations as in the case of a distributed network.

Thus, the network can be classified as having one mobile part and one fixed part. The mobile part will rely on transmissions using radio waves,

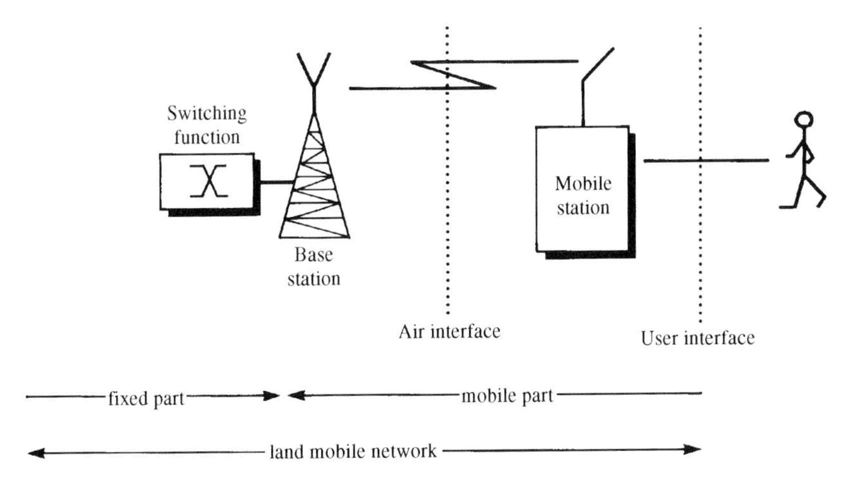

Figure 1.8 A mobile network will typically include mobile stations, base stations and switching functions. In this example the switching function is included in the base station

whereas the fixed part will use ground-based wires or fibre-optic connections. The total system is a mobile network or, for land applications, a land mobile network.

The way in which mobile stations, switching functions and possible base stations are connected is referred to as the network architecture. Figure 1.9 indicates three methods which may serve as examples of approaches that will have dramatically different operational characteristics and performance.

Figure 1.9*a* refers to the standard configuration. The switching function is included into or is attached to the base station. Mobile stations *A* and *B* are connected using two air interfaces (transmission circuits). This means that the mobile stations have a minimum of complexity; however, they are critically dependent on the base station to serve them.

In case *b* a certain amount of switching function is integrated into the mobile station. Mobile station *A* is unable to reach the base station and requests mobile station *C* to serve as an intermediate and relay his signals to the base station, thus reaching mobile station *B*. This terminal need not possess any switching capability as it connects directly to the base station. In this case three air interfaces are used, and three switching units are activated. This extra functionality as compared to case *a* greatly extends the available range and accessibility of the system.

When the switching capability is provided in all the mobile units, these may establish a communication link without connecting to the base station or the base station need not be present at all. This is indicated in case *c*

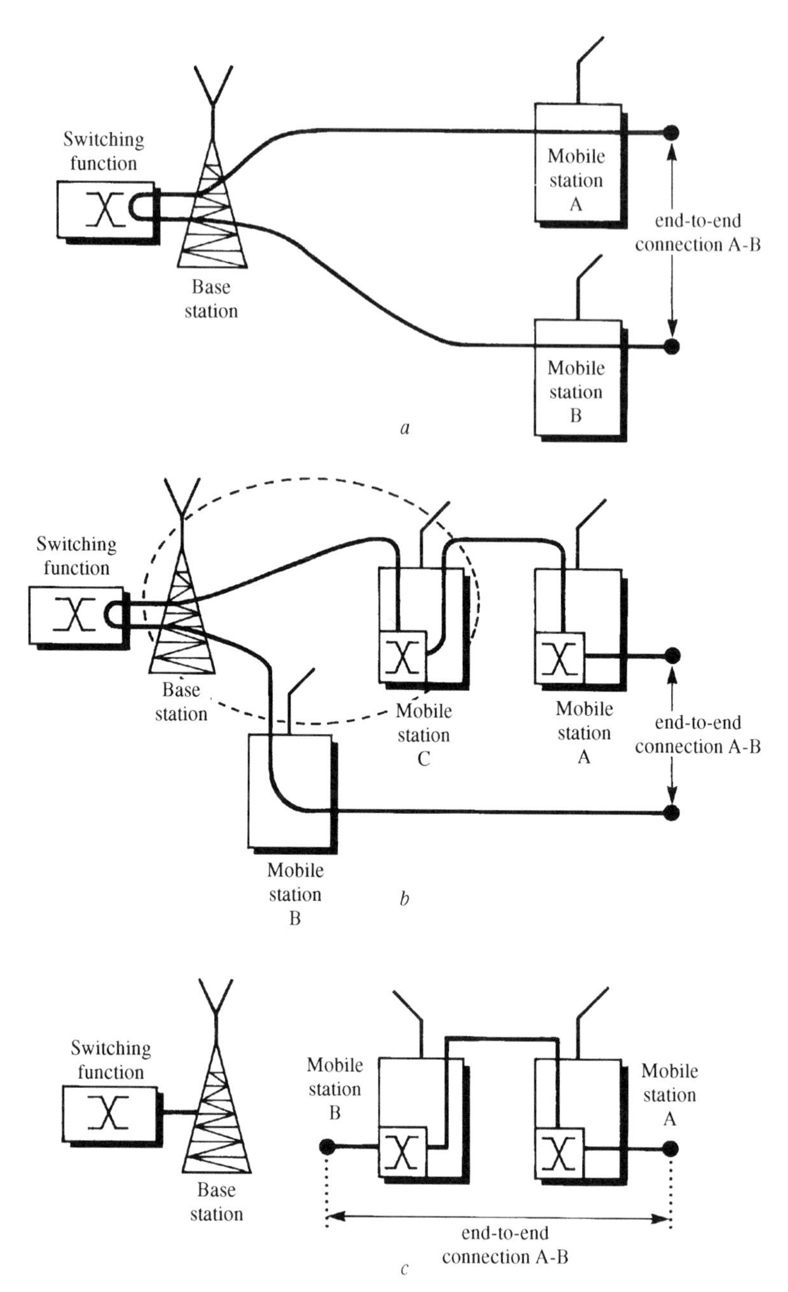

Figure 1.9 Operating mode of standard, relay and mobile-to-mobile station communications
a standard configuration
b mobile station relaying
c mobile station to mobile station

where mobile station *A* is connected directly to mobile station *B* using one air interface. This represents a more demanding approach and is possibly much more robust. When a large number of stations operate simultaneously in a given area, and each one might connect to all the other ones directly or via relays, all the stations form a highly-redundant network at the expense of an increased complexity.

1.2 The evolution of mobile communication networks

Mobile communication is currently the fastest growing sector within the telecommunications industry. The thrust of the development is seen in the public sector where mobile telephones for voice and e-mail connection are dominating the market.

A fast transition to digital systems has been seen in the 1990s. This development is driven by technology availability and also the demand for more advanced services which can only be provided by digital systems. The diverse user requirements have led to many different technologies for public/private mobile communications. This Section presents an introductory overview of the development in public and private communication systems. Public services are here defined as services open for free sale to the public and offered by a national or international operator. Private services are defined as systems not available to the public. Typical examples of private services are military systems and other systems proprietary to corporations or agencies.

1.2.1 Evolution of public communication systems

The following gives a review of current public communication systems for the benefit of understanding the landscape of evolving technologies and also to point in the direction of more universal systems. In order to understand the great variety of systems available let us first review the traditional classification of systems based on their technology and function.

Cellular networks offer an extended service area by dividing the area into a number of smaller cells, each having its own base station. As subscribers with a call connected move from one cell to another, cellular systems automatically reroute the connection to the base station in the next cell in order to enable continuous service. The general trend in cellular systems is toward microcellular systems which reduce the typical current cell size of 3 km down to 1 km or less in high density urban areas.

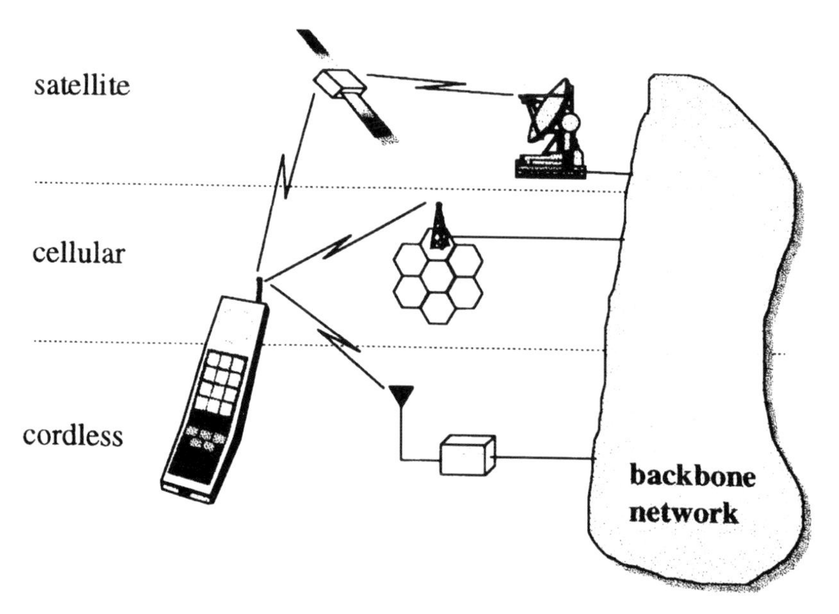

Figure 1.10 *Satellite, cellular and cordless mobile systems are examples of public systems which all connect into a backbone network*

Today, cellular systems may give the users better coverage than cordless systems, but with the introduction of intelligent network functionality together with mobility functions within the fixed network, the user may no longer experience any difference.

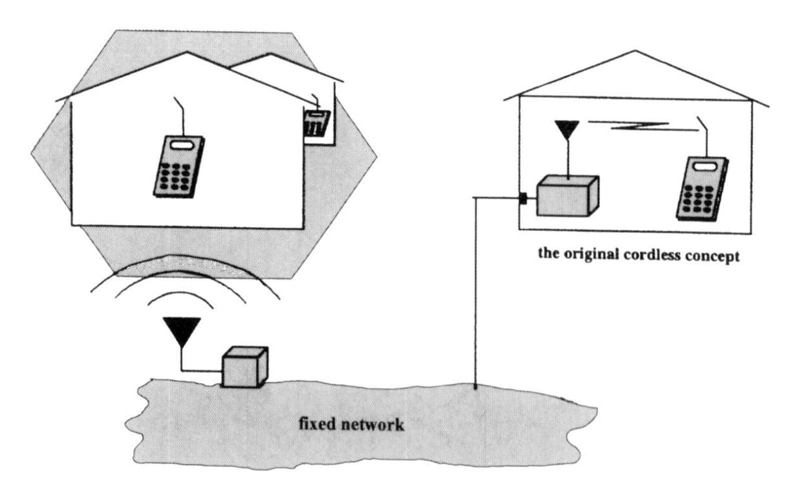

Figure 1.11 *Cordless systems, originally developed as cordless telephone extensions, are currently being integrated into cellular systems*

Systems aimed at offering mobility within limited geographic areas, e.g. around the office or home, belong to the *cordless* application domain. The original cordless usage was to provide indoor mobility for domestic users and short-range outdoor coverage (say, 50 m). By the introduction of DECT, Digital European Cordless Telecommunications, the cordless domain has been extended to the scenario where the local exchange can be enhanced by radio access. However, cordless systems provide limited mobility and therefore a terminal is denied service if it moves outside the coverage area of its local exchange.

Telepoint systems may be seen as a replacement of the conventional pay phone commonly found in public areas such as train stations, shopping centres and airports. Telepoint coverage is limited to relatively small zones, typically a few hundred metres.

Satellite systems such as INMARSAT and others have, for two decades, represented the most advanced communication concepts and have had a high cost, high reliability profile for users such as those in shipping, oil exploration and other professions operating in areas where no alternative had previously existed.

Paging systems typically provide three types of services: tone-only paging and numeric/alphanumeric paging. By the tone-only service the calling party sends different alert signals, which appear at the receiver as different acoustic tones. Numeric/alphanumeric paging facilitates the transfer of single messages. A paging terminal contains no transmitter so that the system cannot detect any packet loss.

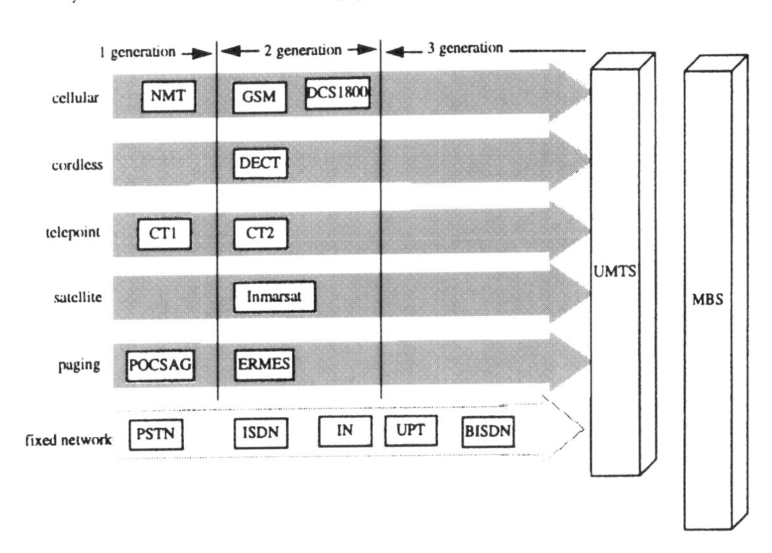

Figure 1.12 Evolution of wireless and ground-based communication systems for public services

The general evolution trend has been that the need for a new service is followed by an effort to develop a communication system which provides it. Users wanting different services are faced with the burden of using more than one mobile terminal, e.g. one for GSM and one for ERMES. Also, cellular operators may want to use different systems to have better utilisation of their frequency resources, e.g. DCS1800 with its smaller cell size may be used in areas with high population densities while GSM is used in the rural domain. Therefore, we see a common interest for multimode terminals, and multimode terminals such as DCS1800/GSM are now available. These multi-mode terminals are referred to as generation 2.5 and may support limited (PCS).

Mobile systems fulfilling the PCS concept are referred to as generation 3. This generation uses the IN architecture to allow rapid and inexpensive deployment of new services on a wide range of platforms.

The main differences between UMTS and MBS are:

- MBS is not expected to provide universal coverage and will probably be used by users which mainly require specialist services;
- MBS supports a set of widely different bit rates when compared with mobile speed scenarios;
- MBS has to tackle more demanding handover conditions due to cell size, frequency, data rates and terminal speed.

The most widespread public communication systems are listed as follows:

- cellular (GSM, DECT, DCS1800, IS-95);
- cordless (DECT);
- telepoint (CT2, DECT);
- paging (ERMES);
- mobile data (MOBITEX);
- terrestrial flight (TFTS);
- satellite (Inmarsat-P, IRIDIUM).

1.2.2 Evolution of private and military communication systems

Private communication systems including dedicated military systems have seen an accelerating development fuelled by technological improvements and user needs. This development was for a number of years concentrated within the defence industry. The recent price reduction in radio integrated circuit technology, and not least in digital processing power, has caused technology previously afforded only by military users to be available to most users.

Private mobile communication systems are operated by private organisations and are not open for public sale. Some examples of private mobile communication systems are:

- mobile radio (PMR, DSRR);
- trunked systems (TETRA);
- mobile data networks;
- wireless local area networks (LAN) (HIPERLAN);
- wireless PABX.

The evolution has been triggered by a number of user-defined requirements. Some of these concern the quality and nature of the transmission link with requirements such as:

- requirement for the data transfer of widely different data types;
- requirement for message integrity;
- requirement for more secure data and voice transmission;
- requirement for higher transmission speed.

Other requirements concern how the system will be used, such as:

- requirement to get access to outside nets and land-based systems;
- requirement for a wider range of services;
- requirement for a higher degree of automation.

Early advanced military communication systems were developed for very specific tactical or strategic purposes. Strategic plans needed a web of high reliability and high survivability communication backbones. Although mobility was not a major consideration for many of these communication lines, the vulnerability of cable-based networks made it necessary to develop radio-based systems. These included troposcatter long-range microwave links, meteoscatter links, very low-frequency broadcasting and high-frequency (HF) systems.

Tactical considerations saw the development of HF and very high-frequency (VHF) communication systems with a high degree of survivability in a battlefield environment. Digital technology saw its first uses in these environments and very soon scientists started to explore more sophisticated modulation and communication schemes. It was in this context that spread-spectrum modulation was devised with the main selling point of being a robust modulation offering better security and survivability. At this stage spread spectrum was synonymous with direct-sequence binary-frequency spreading and soon found itself facing a seemingly totally different approach, namely that of frequency hopping. To date this struggle goes on and has escalated into a commercial battle. As in most struggles, there is no right answer and it is the aim of this book to introduce modern mobile communication in a way that encompasses different implementations of the actual modulation scheme.

The need for the digital transmission of data found the mobile radio community needing to devise methods for data transfer. Packet radio was one of the propositions and has gained widespread popularity over the past few years. Multirole radio (MRR), integrating voice and data, was a later refinement and exists today in commercial versions.

The important issue for the radio designer is whether there is a trend towards an umbrella system which, within a certain span of time, would combine all types of system into a big happy universal system. This is an unresolved question, but most critics agree that continuing technological progress does not agree with such thinking and that the diversity of systems seen today is essential for the evolution of more and more advanced concepts.

Fundamental to the understanding of the growth and diversity of communication concepts is the fact that user needs are very different. As an example, Figure 1.13 illustrates that e.g. a GSM system is very mobile but provides only a very limited transmission bandwidth, whereas more modern cousins such as various forms of wireless local area network offer much more traffic and possibly the same degree of mobility.

Military radio systems have conveniently been divided into larger networks performing a backbone operation and having smaller tactical networks connected at different locations and at different levels. The larger tactical networks include systems such as EUROCOM and the future TACOMS-2000, scheduled for operation beyond year 2000. These networks work on top of and connect into other mainframe systems such as the international civilian telephone and data exchange system.

Military radio systems are conveniently divided into generations, based on technology and network complexity. The first-generation military radio system of the 1970s utilised analogue FM/AM modulation

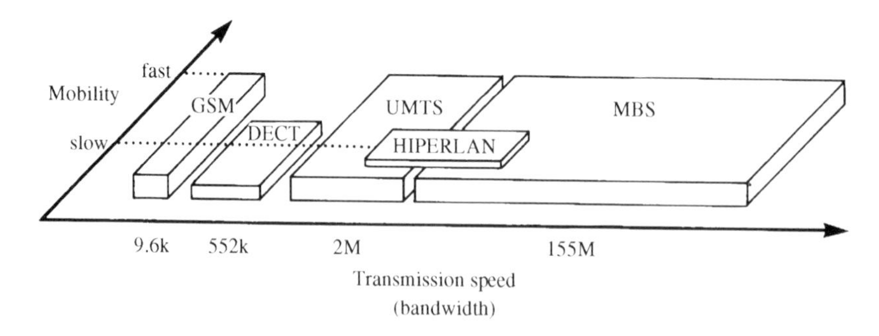

Figure 1.13 The mobility against bandwidth classification of some private and public systems such as GSM, DECT, HIPERLAN, UMTS and MBS

and did not include a centralised switching or network function. Examples of these are the VRC-12 and PRC-77 NATO systems.

The second-generation military radio system of the 1980s still used basically analogue modulation but included more advanced modes of operation such as digitally controlled frequency hopping to meet more demanding electronic warfare requirements. A leading example of such a system is JAGUAR V.

Third-generation military radio systems of the 1990s use a complete digital architecture and already include a complete decentralised network control and advanced operating modes. Typically, they make use of hybrid direct sequence/frequency hopping modulation aimed at giving the robustness required for military systems. Examples of third-generation military radio systems are:

- PRR4G;
- multirole radio (MRR);
- SINCARS;
- BOWMAN.

Military radio systems of the future are referred to as fourth-generation military radio systems. Additional features over third-generation systems are:

- software radio architecture;
- further integration of network functions and terminal functions;
- more advanced modulation formats and structures;
- much increased bandwidths.

Software radio architecture is thought of as a system where many of the analogue functions of a radio (filtering, down conversion, demodulation etc.) are replaced by operations in a digital processor.

This book aims to outline the background for potential improvements and suggest ways of approaching the technological issues that will eventually lead to improvements in system performance.

1.2.3 The designer's choice, choose off-the-shelf or make your own?

The potential developer or purchaser of a mobile radio network might be one out of a number of different categories. These categories may range in size from an international telecommunication company to a small, private organisation; in complexity the system could range from a handful of voice transmission radio units to an integrated system with a multitude of services and operating modes. Obviously, within such a diversity there will be differences in design and production strategies although fundamental components in technology and system design will be identical or related.

Despite this variety a large fraction of mobile communication developments will fall into two categories: the large national/international commercial networks and medium-sized customised networks for dedicated users within the private and government sectors. The reason for this is obviously related to the fact that the cost of purchasing or developing a standalone or integrated system is prohibitive for a number of smaller applications. These users might find that the cost of subscribing to services from a major telecommunications operator is much more affordable, at least initially, than having to invest capital in a development programme which might span years prior to getting an operational system. In order to accommodate particular requirements, most users will choose to have some sort of customised overlay on the systems, which might be in the form of software or hardware that is associated with user terminals or switching units. Typical requirements might be enhanced security, priority delivery or guaranteed access. The complexity of providing these overlays might cause unexpected surprises to the developer, both regarding time and cost. This lesson has caused a lot of potential buyers to divert to a course that involves going through a thorough specification phase, including an examination of total system cost and development time. These deliberations will generally make the developer rethink its original strategies and more carefully consider all aspects of the communication system. A possible outcome might be that even the smaller developer will enter onto the avenue of choosing its own system strategies and provide the users with a proprietary system.

Telecommunication users might include a range of small to medium-sized users such as police departments, ambulance services, transportation providers etc. All users of this category will have the option of relying on products available from big operators in the market and will have to carefully consider system cost and whether or not public systems meet their needs.

An additional complexity stems from the fact that all users will carry some traditions from the use of previous communication systems. There will be previous suppliers that hope to carry on being the provider of communication systems, a situation which is more rigid as in some cases the new system components need to be phased in gradually and thus a gradual transition to a newer generation of communication tools is needed. Sticking to a previous provider will also, in some cases, be preferred by in-house personnel and also management as this helps to keep training and other personnel costs down. Consequently, if a previous supplier has new products which are reasonably well adapted to the user's need, it will have a strong case in getting its new product sold.

A new situation in the 1990s is that a new mobile communication system will include in one packet what in the typical previous generation was split across at least three packages or units: voice communication, data communication and also in some cases administrative functions (e.g. command and control for military or paramilitary users, positioning, data logging for transportation users). Each of these functions will have previously been provided from independent sources and although many of these will be co-operating in developing joint platforms, the user will see a new set of suppliers as it tries to get its needs met.

In the process of developing a new mobile communication product, one of the most important and urgent needs is to establish a consortium or group of companies which can undertake the operation of developing, manufacturing, installing and servicing the communication network. In many cases, such as in providing communication units for national defence, national interests will require the use of domestic firms. This limits the flexibility and choice and, furthermore, there might be national requirements for several companies or consortia to take part in the development and production process. Particularly in these cases, the user will require that there will be a lead partner which has overall responsibility and acts as the prime contractor.

Given that the industrial consortium is established, the first phase is that of establishing all requirements for system performance and for the functions of the communication network. At this stage, adhering to existing systems and possibly connecting national and global telecommunication networks is a viable option. However, the specification phase will limit the choice of what may be purchased off-the-shelf and at this stage the necessary overlay might be defined and analysed. It is also of utmost importance at this stage that a thorough modelling of system capabilities for the various options is carried out. For most applications, analysis of system life-span performance is required to assess the cost as a function of time and expected lifetime of the particular system.

Concurrently with the efforts relating to establishing system specifications, most organisations will see the need to establish resources in terms of man power, equipment and development tools. This is important in order to achieve continuity in the development process and establish the best possible organisation. A particular concern for developers of mobile communication systems is to attract the right profile of personnel skills. Modern mobile communication systems need expertise in such varied fields as radio engineering, antenna design, packaging technology, radio frequency integrated circuitry, digital signal processing, data transmission, switching functions, software engineering, protocols, coding and many more. These skills need to exist within the firm or associated organisations. Increasingly, different mobile-system developers team up with specialised firms which possess niche skills that they provide for mutually competing firms. Altogether, the total sum of all contributors must have the right flavour and punch to carry the development through its various phases.

The process of making system specifications is quite tedious in that all levels of hardware, and software on both a system and unit level, need to be specified. A usual complication at this stage is that there exists an uncertainty as to what technology or principles should be used in various parts of the system. These questions will need to be resolved by a combination of feedback from the specification process and actually building and testing subunits. The risk associated with using new technologies needs to be assessed and seen in relation to the expected life span of the system. In many situations, the competitive aspects will spur the developer into pursuing very challenging technologies as it will know that it may at this stage still be months or years ahead of the product hitting the customer if it is developing a product in a competitive market.

Marketing efforts will, at this stage, give indications as to whether the product range will penetrate further than to the initial customer. Most mobile communication systems will include a range of functions enabling them to interconnect to external networks or global nets. In addition to being used in different geographic areas or countries, intelligent marketing will single out markets that will be able to take advantage of particularities of the system. In today's deregularised communication business segment, market penetration depends largely on strategic alliances in sales organisations and the interconnect facilities of communication networks.

1.3 From spread-spectrum modulation to integrated communication

1.3.1 Classical and new challenges in relation to intelligent radiocommunication

The challenges confronting the radio engineer have changed dramatically during the past decades. Traditional requirements, such as maximising operational range, have been replaced by other demands such as optimising traffic throughput per unit volume or per unit cost. System developments have made considerable use of advances in fixed-point wired communication systems. Standardised components and manufacturing techniques are emerging and hardware and software components will, in many cases, be available off-the-shelf for many applications. This may not be so for custom designed mobile communication systems. Specialised user requirements will necessitate that the system developer considers the virtues of all system components. One of these is the radio interface and the way in which information is conveyed using radio waves. Advances in this sector might possibly give system capabilities that are advantageous or essential for the communication network as a whole.

It is crucial to the understanding of the potential of radio for the communication of information to realise that radio waves are vectorial fields in a five-dimensional space; three spatial dimensions, polarisation and the time dimension. Each one of these domains has the potential for communicating information. In most practical systems the engineer does not really have the option of exploiting this multidimensional space, although some features such as directional beams, polarisation sensitive antennas and wideband complex waveforms are exploitable. Some of the underlying ideas of these concepts will be outlined in the following and also serve as an indication of those novel principles which might be used in future communication systems.

1.3.2 The use of radio waves to carry information

The aim of this Section is to provide an understanding of the concepts and principles required to describe signals at various stages in a radiocommunication link, and to indicate the degrees of freedom which are available to a designer of radio systems. Waveform coding, modulation, frequency spreading and so on should be regarded as subsets of the spectrum of techniques that can be used for the secure transmission of messages through space, even in hostile environments. All these techniques entail processing and coding of waveforms, one-dimensional or multidimensional.

In a communication link, a waveform (or carrier) is first passed through the various modulation stages in a transmitter, where information is superimposed by varying one variable or a set of variables of the waveform. Immediately, the signal is transformed into an electromagnetic wave by an antenna and it propagates in various directions in space until it is collected by another antenna and fed into a receiver. To fully describe this process, the complete representation of the electromagnetic wave has to be introduced, namely the polarisation properties and the spatial distribution of transmitting and receiving elements.

Figure 1.14 gives a schematic picture of the various stages in the communication system. The source information, I, is imposed on the

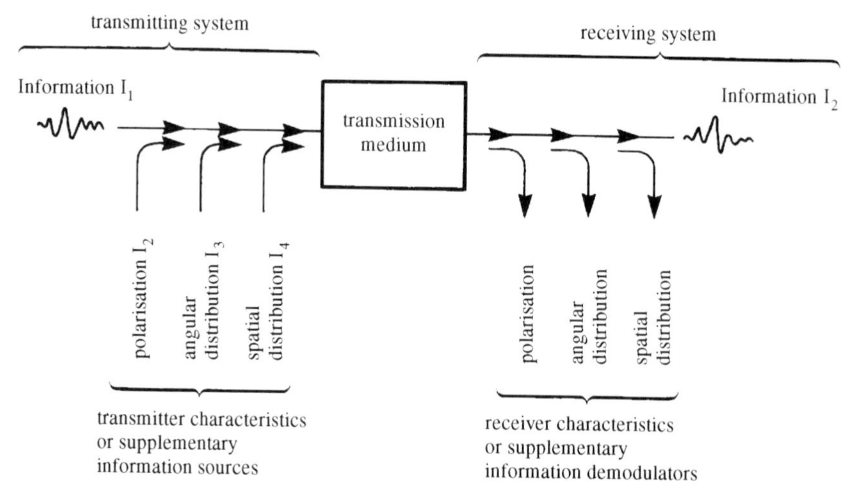

Figure 1.14 Physical properties of the electromagnetic wave introduce added information to the transmitted message

electromagnetic carrier through a modulation process and then transmitted into space by a set of radiating elements. This radiation process introduces a new set of domains in addition to the time domain. Theoretical aspects of this situation will not be covered here in detail, but some of the possibilities will be presented. A more complete analysis of communication networks should include coding in the new domains.

Any transmitter will give certain characteristics to the polarisation as well as to the angular and spatial distribution of the electromagnetic field. The receiver will, likewise, have certain characteristics which determine its sensitivity to these phenomena. These properties will have to be included as part of the total description of the system. Several very interesting features can be seen if these characteristics are allowed to be time variables.

The polarisation state of an electromagnetic wave depends on the relationship in phase and amplitude of its two transversal components. Coding methods in the polarisation domain take advantage of the transversal nature of the electromagnetic wave, whereby two independent domains are introduced. It is well known that any antenna emits an electromagnetic wave with a given polarisation state, but polarisation coding is rarely used for transmitting information. Obvious applications are propagation-path adaptation techniques and interference-rejection techniques which can take advantage of polarisation manipulation.

Similarly it can be envisaged that, when the emissions from a set of radiating elements are superimposed to give an angular variation of radiated field strength, it is possible to consider a system with individually-coded radiation elements. The traditional directional antenna could then be replaced by a set of uncorrected beams, and individual coding would give an improvement if a co-ordinated and synchronised network of receivers was used.

On a larger scale, a network of transmitting and/or receiving stations permits spatial processing and focusing. This degree of freedom is of use when it is required to tune in to a specific cell in space or time. This is the inverse of the synthesis of the coded antenna pattern: the electromagnetic field in space is probed by a set of receivers to deduce the angular distribution of radiation. For example, a jammer could be eliminated or a concealed source detected using spatial focusing. Owing to the reciprocity of electromagnetic waves, these operations can be performed equally well at the receiving station or stations.

The most important aspects of spatial coding are those associated with beam-forming antenna concepts and spatial focusing, with the polarisation offering another dimension which can be utilised for jammer avoidance etc. However, for the remainder of this book, we will be concerned with time- (and frequency-) domain coding since this represents the mainstream approach to coding and is of most common use in communication systems.

1.3.2.1 Time/frequency-domain operations

The basic modulator, as used from the dawn of radiocommunications more than a century ago, still exists relatively unchanged today. The first transmissions of information were performed with impulse modulation using spark transmitters, a method of communication that may be seen as the most basic form of transforming information by a radio wave. Later on, spark-modulation systems were replaced by technologically more advanced methods such as amplitude, frequency or phase modulation in various forms. These still very unsophisticated, although efficient, modulation techniques even today account for the vast majority of waveforms used in radio broadcast and point-to-point radiocommunications. However, for the remainder of this book emphasis will be put on time- (and frequency-) domain coding since this is of most common interest.

As stated in its title, this book will focus on spread-spectrum communications. As such, spread-spectrum modulation represents an improvement over the traditional modulation methods. This improvement has come about because of a number of factors. Although the availability of digital signal-processing hardware has been instrumental, the real forcing issue has been market pull. The demand for high quality mobile communications has caused professional users to turn to more sophisticated systems, particularly since modern spread-spectrum systems are competitive both in performance and in system or unit cost.

1.3.2.2 Transverse properties of the electromagnetic wave: polarisation

Polarisation properties of electromagnetic waves have wilfully or accidentally been used to improve the quality of radiocommunications since the very early attempts at long-distance communications. The complexity of the propagation channel has left little opportunity to use the polarisation domain for actual coding of information. Now, due to greatly improved technology and coding schemes, this option is much more feasible. In particular, modern mobile communication systems have moved up in frequency into the microwave region where the polarisation properties of the propagation medium are better behaved. Figure 1.15 indicates the transversal nature of the electromagnetic wave.

It should at this stage be noted that as the modulus of the electric-field vector is the vector sum of the horizontal- and vertical-field components, it can only be zero if both the components are zero. This implies that the transmitter can operate both polarisations simultaneously, but it cannot conceal energy in one polarisation channel by adding a random or deterministic signal in another. This is an obvious result of the independence of

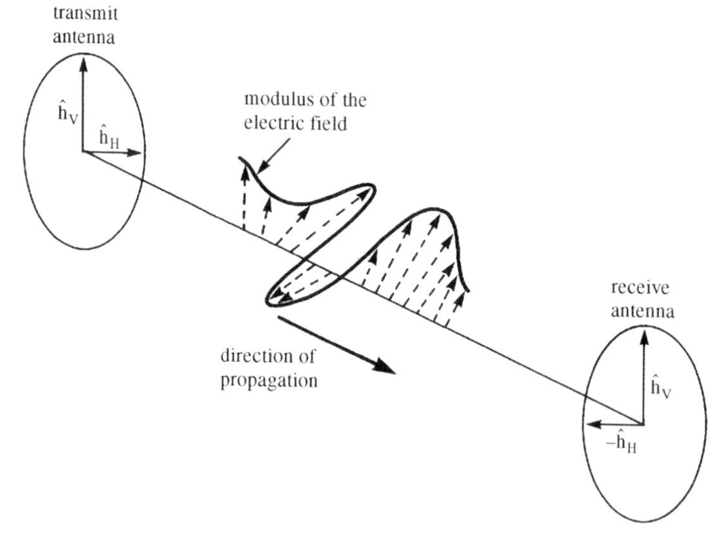

Figure 1.15 Polarisation properties of the electromagnetic wave are described by the modulus and transversal direction of the electromagnetic field. For a narrowband (sinusoidal) signal, the trajectory is ellipsoidal

the two dimensions, but should be kept in mind when the polarisation properties of a wave are exploited for anti-eavesdropping purposes.

Fundamental to the theory of polarisation is the fact that if two orthogonal polarisations are transmitted and received independently, any polarisation state can be generated synthetically making use of polarisation transformation matrices. In consequence, by applying signal processing in the receiver in a fully coherent dual-polarised system, a range of improvements to the communication path could be achieved. This could be used for cancellation of multiple-reflection paths, reduction of the interference from adjacent transmitters and so on.

1.3.2.3 Antenna beam forming and focusing

Increasingly, all modern mobile communication systems rely on the use of antenna directionality to obtain the required performance of the system.

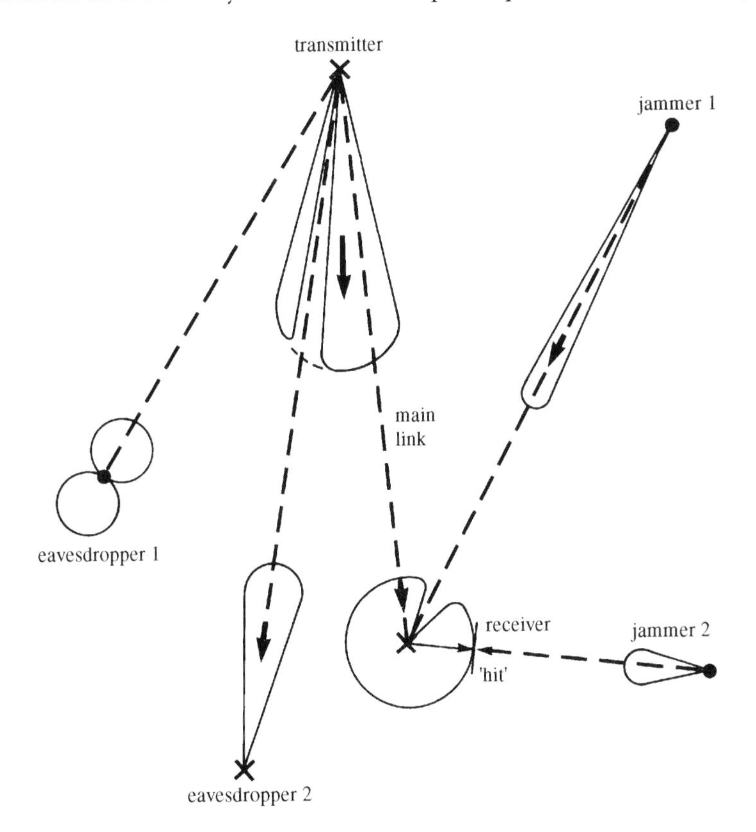

Figure 1.16 Antenna directivity allows for a geographic selection of users. In this way many users may share a frequency slot and when this is important other listeners can be denied access

Antenna directionality is used at base stations for large volume traffic such as in the GSM public system to allow for many more users to share a frequency slot or, as in the case of a spread-spectrum system, a code set.

The beamwidth of an antenna is proportional to the ratio between the radio wavelength and the size of the antenna used. A directional antenna will thus be many wavelengths in size in one or more dimensions. A reflector antenna or a phased array will be large in directions orthogonal to the propagation direction, whereas a YAGI structure or a spiral antenna will have its largest dimension along the propagation direction.

For these reasons base stations will be the obvious candidates for directional antennas. As higher frequencies come into use, directional antennas will be more and more appropriate even for smaller portable units. In particular, when directional antennas are implemented using phased arrays, allowing for rapid beam steering, directional antennas will play an important part in the design of mobile units.

1.3.2.4 Co-operative antennas/spatial focusing

When mobile communication units are too small to accommodate large antennas giving sufficient directionality, individual nodes can be connected to improve directivity and spatial selectivity.

Combining two receivers or transmitters makes it possible to generate a null in the resulting radiation pattern. This is useful both for transmitter and receiver locations. At the transmitter the null might be placed in the direction of a user that should not receive the transmission either because it is denied access or because it may be a co-operative user but will not tolerate the disturbance. At the receiver the null might be made use of in much the same way to stop interference or to suppress a co-operative transmitter in order to be able to communicate with a more distant party.

Combining three or more units makes it possible to achieve spatial focusing. Focusing implies that the electromagnetic field from a number of positions in space is concentrated in a smaller volume. This principle can be applied on reception as well as upon transmission.

The simplest form of focusing is the irradiation of a point with several independently-transmitted beams. If these beams are coherent, then the response at the illuminated spot will be proportional to the number of beams. Coherent beams require that the illuminators (transmitters or receivers) are synchronised. If they are not, then the bandwidth of the focusing signal waveform will be affected.

The relative time lag between the stations in space affects the field strength and also the phase of the wave in the focal point. The absolute time reference is of importance in most modulation schemes and the

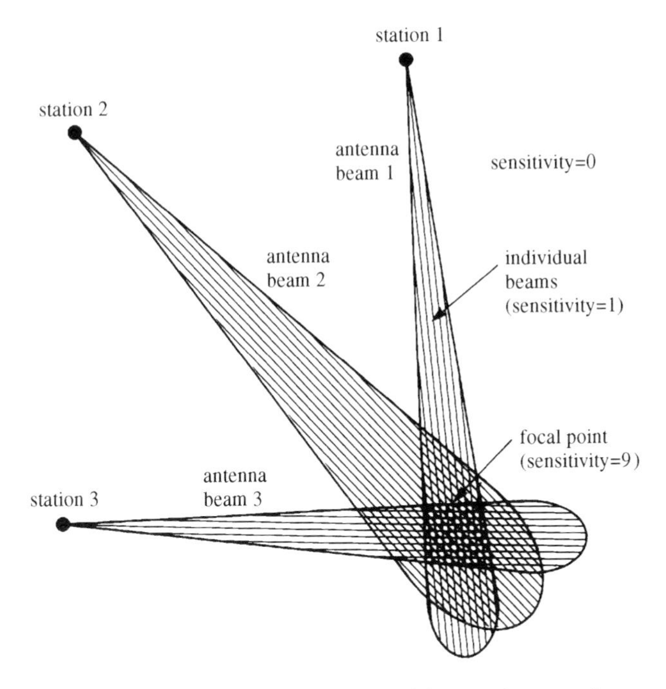

Figure 1.17 *By illuminating a segment in space with several rays and compensating for the propagation time, coherent addition of energy is achieved. This may be used for optimum detection of a transmitter concealed in a noisy environment*

receiver typically establishes an internal time reference based on the received signal or on an accurate time reference. As shown in the previous section, the exploitation of the time reference is crucial to many intelligent countermeasures. Therefore, the space-domain efficiency factor has to include the effect of spatial variability of the time reference. By proper timing, focusing of signal power can be achieved and, in addition, the time of arrival can be synchronised with the receiver.

Owing to the reciprocity of the electromagnetic waves, the same procedure can be applied equally well for reception. This can be useful for eliminating interference sources which could not be adequately suppressed by other means. Also, the position of transmitters can be found by using these procedures. By measuring the direction of the phase front at two or more positions in space, the location of the transmitter can be established.

1.3.3 Introduction to current spread-spectrum principles

1.3.3.1 Definition of spread-spectrum modulation

The evolution of spread-spectrum communication takes us as far back as the 1920s. At this stage, technology had matured to the point where more sophisticated means of making use of radio waves of more complex form were a viable option. This took designers through various stages of utilising wideband frequency-modulation schemes and eventually back to time-modulation schemes which represented the very beginning of radiocommunication.

To see the perspective it should be noted that the first mention of electrical communication over distance was in the *Scot Magazine* in 1753. In modern times it was in the period following World War II that combined time- and frequency-domain operations were utilised to design more robust communication systems.

The term spread spectrum might in the context of a communication system be defined as:

> *A spread-spectrum communication system is one in which each bit of information is represented by one out of a set of waveforms each of which has specified properties in the time- and frequency-domains. In most cases, the resulting bandwidth will be much greater than that of the modulating data stream.*

This process might be achieved by a system as illustrated in Figure 1.18. When an elementary unit of information, a data bit, is input into a spread-spectrum modulator it responds by transmitting a more complex waveform. This waveform is the one that is transmitted and upon

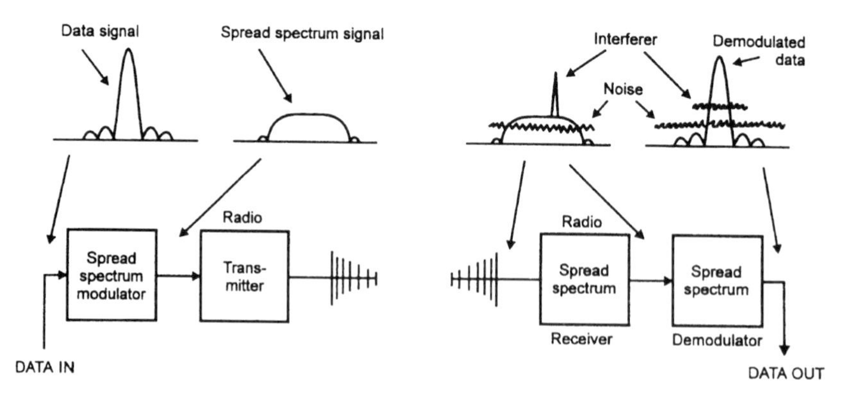

Figure 1.18 A generic spread-spectrum communication link (one way link)

reception it is correlated with a replica of the transmitted waveform to generate a decision as to the value of the data received.

In general, this process also generates frequency spreading as the bandwidth of the spread-spectrum waveform is much wider than the inverse of the duration of each individual data pulse (the bandwidth of each individual data pulse or pulse train). As will be indicated later, the use of parallel chains in the receiver chain might offset the bandwidth expansion and the resulting bandwidth could be less than that of the modulating data stream.

1.3.3.2 Generation of spread-spectrum waveforms

Spread-spectrum systems may be in the form of frequency hopping or direct sequence spread-spectrum modulation or even more complex waveforms. Although the end result may have similar characteristics, the properties are different and the technologies behind them are usually vastly different. The following definitions should be noted:

Frequency-hopping systems achieve frequency spreading and processing gain by transmitting a continuous wave waveform which more or less abruptly changes instantaneous frequency in a prearranged pattern as dictated by a code.

Direct-sequence systems, on the other hand, are basically time-domain systems which manipulate the phase of a continuous carrier to give a resulting waveform that is spread out over the frequency band.

The distinction and relative merits of these two systems and possible combinations will be explained in a later Chapter.

Spread waveforms might thus be generated in a number of different ways:

- frequency synthesisers;
- phase modulators;
- stored waveform memories.

The actual code-generation process will eventually be a computerised operation where the code is generated sample by sample from a certain equation or rule or be sequentially loaded from memory.

1.3.3.3 Definition of processing gain

In a spread-spectrum system, the spread-spectrum modulator replaces a digital symbol which might be binary (carrying the values 0 or 1) or *m*-ary (taking one of *m* possible values). This waveform might be one of those

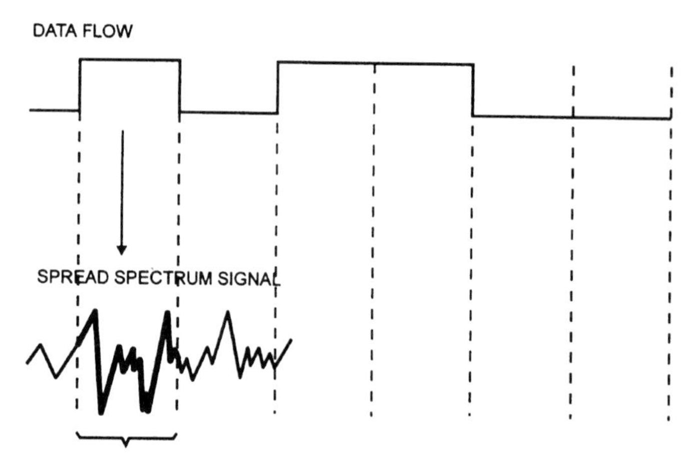

Figure 1.19 *The spread-spectrum modulator represents each element of a symbol stream with an arbitrary complex waveform. The processing gain describes the bandwidth expansion associated with this operation*

discussed previously and have an arbitrary complexity. Although the detailed structure of this waveform is of great importance to the operation of the system, the resulting bandwidth of the spread-spectrum signal is an essential parameter. In fact, the ratio between the bandwidth of the spread-spectrum signal and that of the modulating data stream relates directly to the quality of the operation. The following definition of the term 'spread-spectrum processing gain' might therefore be given:

$$\text{processing gain (PG)} = \text{output bandwidth}/\text{input bandwidth}$$

As seen in Section 1.3.3, the choice of spread-spectrum waveform might be from a selection of two (binary) or m (m-ary). In the latter case the receiver runs a set of parallel matched receivers and, as the transmission circuitry and medium might be considered as linear transmission paths, the multiplicity of the channel results in a factor we might call the m-ary gain factor:

$$m\text{-ary gain} = \text{multiplicity of codes being simultaneously used}$$

A fundamental difference between m-ary gain and spread-spectrum modulation gain is that the parallel nature of the detection process in an m-ary detector works in the same way as increasing radio output power. This means that all parameters that are dependent on output power or thermal noise conditions will be affected by the choice of m-ary gain factor.

Table 1.1 A comparison of features where the addition of a multiple symbol set gives an improvement relative to binary spread-spectrum modulation

	Spread-spectrum processing gain	Parallel signal set *m*-ary gain
Suppression of gaussian noise (wideband noise)	No	Yes
Suppression of non-gaussian noise (narrowband noise)	Yes	No
Improved security	Yes	Yes
Reduced output power requirement	No	Yes
Reduced bandwidth requirement	No	Yes

In the context of this brief introduction (the topic of spread-spectrum and *m*-ary gain is covered in more depth in Chapter 4, including a review of the essential mathematics) *m*-ary gain might be described by the following formula relating total power used in the transmission of the data as perceived by the detector, actual electric power transmitted by the radio transmitter and *m*-ary gain processing power:

$$\text{total power} = \text{electric power} + m\text{-ary gain}$$

In a channel with wideband gaussian noise such as thermal noise or interfering spread-spectrum systems, spread-spectrum modulation does not improve the transmission quality in terms of bit error rate; parallel processing giving an *m*-ary gain will however increase the performance according to the *m*-ary gain factor.

1.3.3.4 Operational advantages of spread-spectrum modulation

Low probability of intercept

This was the earliest justification of spread-spectrum modulation and the first property of the spreading technique that aroused interest in particular among military users. Low probability of intercept implies that a third party cannot easily eavesdrop on the conversation or has to utilise expensive means to accomplish this. A standard communications receiver selects one particular frequency band for reception and the operator selects a demodulation circuitry, such as an amplitude demodulator or a frequency demodulator depending on the modulation used at the transmitter. In a spread-spectrum system the receiver demodulates the transmitted energy through some correlation process and effectively combines various components within a wider bandwidth. A single-channel receiver will thus only detect a small part of the transmitted signal which will be too faint for normal detection and even if it could be magnified to a detectable level would be incomplete and thus not possible to understand.

Low probability of position fix

Among many users there is a concern about revealing their geographical position. Apart from military users, commercial units may also not want to be located. Conventional radio transmitters are easily pinpointed by simple and inexpensive direction finders. The spectral-spreading concepts make this task much more demanding as greater processing power and integration time will be needed within each resolution.

Low probability of signal exploitation

Low probability of signal exploitation refers to the possibilities which exist within a communication scenario to exploit the communication link by some manipulation of the waveforms used to convey the message. These could be:

- destruction of synchronisation messages;
- destruction or alteration of routing information;
- destruction or alteration of the message contents;
- invisible or concealed addition of data bits;
- assumption of user role.

High resistance to jamming and interference

Spread-spectrum systems are inherently more robust against jamming and interference than systems not using spreading techniques. This property of spread-spectrum modulation is used in very sophisticated ways, particularly in some military systems where the main design driver is to develop a system which is able to deliver a message through a very hostile and impenetrable ether.

Some of these scenarios will see a frequency spectrum crowded by radio networks of various kinds and also some transmitters with the sole purpose of blocking the communication path for other users. These transmitters are referred to as jammers in military jargon.

The success of one transmitter in affecting the information transfer in a communication link in the same area depends on a number of factors. These factors are:

- distance of jammer/interferer to link receiver terminals;
- antenna directivity factor at jammer/interferer;
- antenna directivity factor at link stations;
- transmitted power of jammer/interferer;
- frequency coverage of jammer/interferer in relation to data link spectral utilisation.

Finally, spread-spectrum modulation techniques provide an additional factor which is not seen in conventional systems. This is due to the fact that a code is used in the spreading process and unless the jammer/interferer manages to get hold of this code the impact of the jamming/interference is reduced by a significant amount. The actual amount may be described by a figure of merit as mentioned in subsection 1.3.3.3 above, namely the processing gain. This figure determines the handicap of the jammer in relation to how well it affects other communication channels.

There are two very important aspects of spread spectrum in respect to jamming and interference. The first relates to the actual protection provided by a spread-spectrum code. As is obvious from its block diagram, the spread-spectrum receiver provides a post-detection signal-to-noise and signal-to-interferer improvement. This means that if the spread-spectrum signal can be received with sufficient clarity and strength and without interference signals, intelligent jammers might redirect the same waveform with more power and cause problems for the detection process in the data link.

This is a practical limitation since a limited number of spreading codes will be used in a spread-spectrum modem and thus the transmission will tend to repeat the same codes a large number of times. The vulnerability lies in the fact that the code might be revealed if it sticks out clearly only once, for instance if it is received at a very short distance from the transmitter. This points at the very important requirement of spread-spectrum systems, that the spectral-spreading scheme should be sufficiently agile and use frequent change-of-code structures.

The next important aspect of spread-spectrum modulation in respect to jamming and interference is that spread-spectrum modulation provides a practical way of coping with frequency-band congestion. In most practical scenarios the requirement is for the transmitter to get the message across to the receiver without being intentionally disturbed or jammed. The rapidly expanding marked acceptance of spread-spectrum systems is probably the best proof of the practicality of spread spectrum.

High time resolution/reduction of multipath effects

Multipath effects are one of the nonavoidable effects in radiocommunication. Multipath implies that the signal reaching the receiver antenna has travelled by two or more paths. Because these routes inevitably are of different lengths, the time delays of signals that have come along the respective paths are different and the signal will fade in or out with small displacements of the transmitter and receiver (displacements of the order of half a wavelength).

For a wide bandwidth signal, such as a spread-spectrum signal, only one of the frequency components of the signal will fade out at any one time. This means that the overall effect on the signal is smaller or even negligible.

Spread-spectrum signals using advanced time-frequency modulation even have the potential of adapting to this phenomenon such that the overall performance is maintained by adapting the waveform to the propagation path. Methods exist whereby the signal is coded such that the signals reaching the receiver via different paths add up in phase at the receiver.

Adaptive methods capitalising on the wider bandwidth of spread-spectrum waveforms make it possible to make use of radiocommunication under extremely severe multipath conditions. Examples of such are wireless local area networks used inside rooms or buildings where the propagation conditions are very poor.

Frequency-band sharing/code-division multiple access

Frequency-band sharing has been one of the driving factors behind the success of spread-spectrum systems. Spread-spectrum systems may be seen as the democratic approach to ether resources; take one or very few users within a certain geographical area and allocated bandwidth and get superior performance for everybody; add on more and more users and things still work reasonably well for all parties until some limit when everybody starts experiencing reduced performance. At the limit, the system will be saturated and nobody will get any communication going.

Apart from the demographic features where everybody suffers the same and the ether has a graceful degradation when total communication demand increases, code-division multiple access or CDMA has proven to be a practical way to implement multiuser access to communication resources. In a CDMA system, each link uses its own particular waveform or waveforms and with the particulars of that or those waveforms known (or at least used) by only that particular link. In general, these waveforms will be within the same frequency band.

Cryptographic capabilities

The coding aspects of spread-spectrum modulation have implications for possible security functions in a communication link. In principle, the spreading codes can be chosen such that they serve the dual purpose of spreading the frequency spectrum of the transmitted signal and making it impossible to decipher the message. This requires the following:

- the code gives the required spectral signature (usually reasonably flat);

- the code has good anticryptographic capabilities (difficult to break);
- the code is sufficiently redundant (has many realisations).

As these criteria place rather tough criteria on the coding strategy, a more common strategy is to implement spread-spectrum coding and cryptographic coding independently and usually sequentially. This way it is possible to optimise the performance of the respective codes independently.

Spread-spectrum coding enhances the efficiency of other coding schemes through its intrinsic properties (concealment and adaptivity) while performing independently of other modes makes the system resistant to the less desirable properties of spread-spectrum modulation. In particular, the near–far problem is critical for the cryptographic capabilities of a spread-spectrum system.

1.3.4 Cochannel interference and jamming

The performance of a communication network is generally critically dependent on the environment in which it operates. That environment affects the fidelity of the information transfer process through:

- the way in which the signals are propagated through the environment, and
- the number of disturbing signals that are present at the radio receiver terminals.

The latter of these is what is generally referred to as interference or as in the case of military applications, intentional disruption of the service through jamming.

Communication networks might be seen as possessing a multilayer defensive system:

- location (distance from disturbing transmitter);
- antenna (directional characteristics);
- frequency bands (radio wavelength);
- signal format (analogue services, digital packet radio etc.);
- modulation methods (e.g. spread-spectrum modulation);
- identity parameter (time slot, code number etc.);
- error-correcting codes;
- operator or system intervention.

For a given disturbance, each of the layers in the defensive system will have a certain effectiveness, and only a residual amount of disturbance will affect the quality of the communication channel. This disturbance will generally be very small and, particularly for interference sources that have characteristics which are very dissimilar to these of the signal being transmitted, the receiver system will be able to suppress it almost completely. For signals with very similar characteristics, however, many of these measures will have small impact leaving only one or two protective measures. In particular, when the disturbing signal is a transmission from the same net, all properties might be identical except the identity parameter. From this it follows that in many cases interference from other users in the same net might be the single most important noise source in the system. An additional problem is that in many situations this signal might originate from a much closer location and thus be much stronger than the signal from the desired transmitter.

Cochannel interference between the various users on the same net is therefore a major concern when designing a communication network. Advanced coding and modulation schemes such as spread-spectrum modulation are one means of achieving sufficient discrimination between the various communication links, even in a mobile environment where the characteristics of the signals received vary with time.

1.3.5 Evolution of spread-spectrum systems

The application of spread spectrum has seen dramatic changes in the last few decades. At the same time, spread-spectrum principles have matured continuously as their systems have gained a foothold in various civilian and military applications.

Originally, spread spectrum was sold and marketed to military users on the grounds of its stealth and potential for secure communications.

Code-division multiple access was soon seen as a way of using digital technology in communications. This gave the designer a powerful concept and very soon high-performance systems could be developed for terrestrial as well as for satellite links. Spread-spectrum modulation and code-

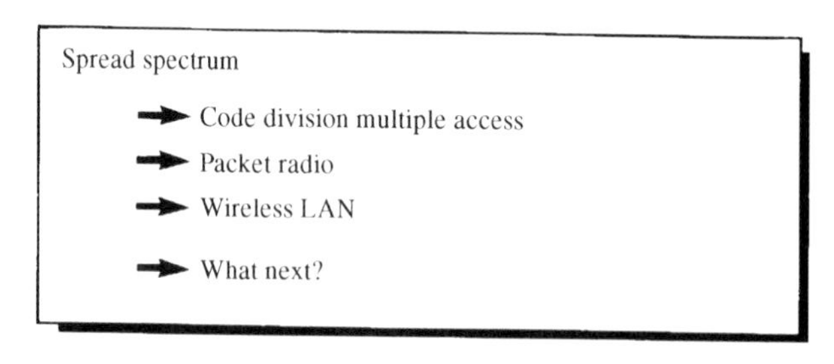

Figure 1.20 Major manifestations of spread-spectrum modulation in operational systems over the past decades

division multiple access provided the necessary flexibility for designing systems in a world of more demanding operational requirements.

With the increasing demand for the networking of computers and digital terminals, the concept of packet radio was born. Very soon the concept of spread spectrum was readapted to a new environment where information was carried in bursts (packages) of information. These systems required extensive networking control but further gained in operational flexibility.

The concept of packet radio has further been carried forward by the concept of wireless local area networks (wireless LANs). This technology is highly popular for computer networks linking computers and peripherals primarily within buildings. It is interesting to note that this represents an application for short-range communication whereas the original interest in radiocommunication was for very long-range communication and even intercontinental signalling.

It is also interesting to note how applications of spread-spectrum modulation change with time. As concepts evolve and standards are defined spread-spectrum techniques will have defined functions and will be seen as parts of larger communication structures. A significant trend in communication is towards standardisation of protocols and functions. In this picture spread-spectrum modulation emerges as a building block being integrated into vastly different systems and for different reasons. It is to be expected that most future communication systems will make use of spread-spectrum modules and it will pose an important requirement for communication engineers—to understand and make use of the virtues of spread-spectrum modulation in an efficient way.

Chapter 2

Designing systems to meet user requirements

This Chapter will take the user view as the starting point and try to follow the interaction between the user and the system designer in the process of designing a system to fulfil the user's requirements. Each user requirement will be discussed with respect to its implications for the final system. Some of these requirements are more important than others, but it is not necessarily the requirements which are considered most important to the user that are also most important for the system designer.

Designing a system requires knowledge about the following aspects:

- do you have any customers? who are they?
- what are their needs? what is their mode of operation?
- getting and understanding their requirements;
- processing the requirements with respect to system implications.

As previously stated, this book is primarily intended for designers and operators of PMR- (private mobile radio) type networks. In such cases one usually has an intended customer, and this customer may be well known. The customer has some kind of communication need, which often is not very well described. Very often his needs are described in terms of apparently very specific requirements, but unless the customer is very well informed, his requirements are often influenced by what products he might believe are available, or by his present communication solution. It is often difficult to make the customer describe his basic requirements without having an eye to a specific product or existing procedure. The first commandment of the system designer must be to thoroughly analyse the user, his operation and his basic requirements with a fresh view.

'You cannot build a system unless you know for what and how it is going to be used'

From the user's point of view the different factors which theoretically may influence the communication channel should ideally be taken care of by the system itself. The user would also prefer to specify his requirements in the simplest possible way and hope for the system to be able to adapt to his immediate needs. In other words, some user requirements may turn out to be restricted by the basic laws of physics.

In practice, such a total adaptability and match between required services and channel characteristics can only be approximated by developing realistic user ambitions based on knowledge of the radio channel and a careful design process. This Chapter takes a first look at what the user may be asking for and confronts the requirements with some of the expected characteristics of the transmission medium, both nature made and man made. The different issues are illustrated where possible by measurements and tests from past and ongoing research and development.

Basic user requirements may be divided into the following categories:

- types of teleservices;
- service coverage area;
- mobility;
- probability of intercept and interference/jamming;
- user traffic and performance measures.

This Chapter will focus on the system designer's view, trying to design a system to fulfil the user's requirements. Based on the user requirements the best way to approach a system configuration may be to create an abstract description or architecture. Such a description is the first step towards a refinement of abstraction. The next step would be to define a system specification which describes the implementation of the functions. In addition to the user requirements, the system is faced with a number of other requirements, regulations and restrictions and must fulfil boundary conditions of the following categories:

- *public restrictions and regulations*—the system is not the only one in the world, and must operate beside other systems without interfering with or being interfered with by the other systems complying to the same regulations;
- *interoperability and interconnection restrictions*—the system may be required to co-operate with other systems;
- *implementation restrictions*—when it comes to realisation and industrialisation we are limited by available technology and production costs

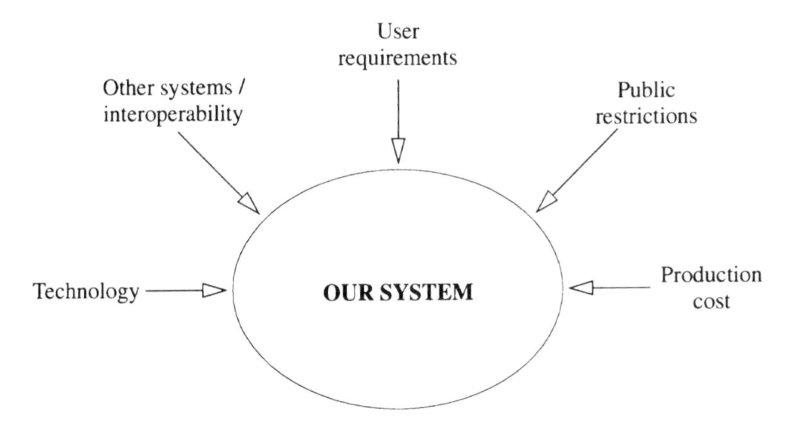

Figure 2.1 System limitations

associated with different production techniques. Also, there are requirements for the reliability and availability of the final product, as well as ease of maintenance and repair.

All these system requirements are summarised in Figure 2.1.

Before trying to describe the system, we must first analyse the implications of the different requirements.

The most important user requirements with regard to the impact on the system architecture are the following:

- autonomous operation (without infrastructure present);
- mode of operation (direct mode required or always via an infrastructure);
- service type—voice and/or data (and type of data service);
- service coverage area and mobility.

Depending on the user's requirements, different architectures are needed. The other requirements are also important to the system. But they will mainly influence the architecture in a more technical manner, which is considered not to be so important for the first approach to an architecture. We will now study these requirements and their implications to the system design in more detail.

2.1 User requirements and system aspects

User requirements refers in this case to the end-user requirements, and not to the full list of requirements as developed in a technical design process when matching the original needs to alternative technical and practical solutions and implementations. The list of requirements is thus limited to include the main issues like:

- coverage area;
- mode of operation;
- autonomy;
- services;
- mobility;
- performance;
- interoperability;
- compatibility;
- security and privacy;
- survivability;
- operations and maintenance.

The environmental influence including factors like propagation characteristics, RF spectrum utilisation and noise and interference will also strongly influence the system design. But these topics will be treated in full in the next Chapter. Here we will only briefly refer to these constraints.

2.1.1 Service-coverage area

One of the very first and most important issues for any requirements discussion is very likely to be about the service-coverage area. That is, how are the users dispersed across a certain geographic area going to be serviced by the mobile radio system? In a military context this is often referred to as the user scenario.

2.1.1.1 Radio-coverage area

For a certain transmitter–receiver distance the signal will experience a path loss and possible fading related to the terrain, frequency and any movement. In addition to these factors at propagation level, successful operation may depend on how the mobile radio receiver is able to handle problems such as capture, that is the performance when several legitimate signals try to access (capture) the receiver simultaneously. Further, the network level with its radio-access protocols and frequency-re-use strategy and resulting possible interference from surrounding users will influence coverage. Finally, factors like user-location probability distribution and user-traffic distribution at the user level will be important for coverage. Fading is more extensively discussed in Section 3.2, and its effects on performance and different ways of improving performance are treated in Section 4.3.5 and Chapter 6. The themes of noise and interference are expanded in Section 3.2, and the network and user levels are themes for Chapter 5.

Chapter 3 will dwell on propagation characteristics, but to gain a superficial understanding of propagation and radio range it might be useful to refer to Egli's formula [10]. This is a classical semi-empirical formula giving propagation attenuation for ground-based radiocommunication. It states that the median attenuation loss in a rolling-hill terrain is given by L, which may be written as:

$$L = 88\text{dB} + 40\log d + 20\log f - 20\log(h_t h_r) \qquad (2.1)$$

where the propagation loss, L, is in dB and the distance, d, between transmitter and receiver is given in km, the frequency, f, is in MHz and the effective antenna heights, h_t and h_r, of transmitter and receiver are in metres.

The power received, P_r, from a transmitter with output power P_t is given by:

$$P_r = P_t + G_t + G_r - L \qquad (2.2)$$

where all values are in dB, and G_t and G_r are the antenna gains of the transmitter and receiver, respectively. For successful reception of the signal, a minimum input power, P_{rMin}, is required at the receiver. This fact combined with eqns. 2.1 and 2.2 gives us a rough median estimate for the radio range:

$$\begin{aligned} 40\log d &= L - 88\text{dB} - 20\log f + 20\log h_t h_r \\ &= P_t - P_{rMin} + G_t + G_r - 88\text{dB} - 20\log f + 20\log h_t h_r \end{aligned}$$

$$d = 10^{\frac{P_t - P_{rMin} + G_t + G_r - 88\text{dB} - 20\log f + 20\log h_t h_r}{40}} \qquad (2.3)$$

One important fact to be read from eqn. 2.3 is that a doubling of the radio coverage area (equals 41% increase in radio range) requires an increase in output power by a factor of four or, alternatively, an improvement of both antenna gains by 100%.

The radio path-loss variability with geographic position and radio movement makes it obvious that the radio coverage can only be expressed in statistical terms. This is due to effects such as terrain shielding and multipath fading which may cause very local variations in the signal strength within the general radio coverage area. The influence of fading may be illustrated by two examples.

First, assume two radio stations located in a straight long valley with almost perfectly reflecting sides. Both transmitter and receiver are located $10 \cdot \lambda$ (λ = wavelength) from one side and $20 \cdot \lambda$ from the other. Figure 2.2 shows the variation in the received signal as a function of the distance between the two radios, assuming Egli's formula (eqn. 2.1) for propagation

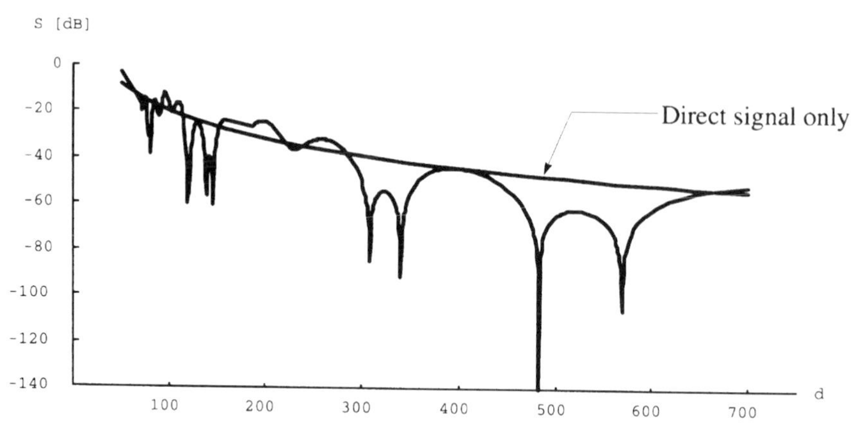

Figure 2.2 Variation in received signal as a function of distance, due to fading caused by two parallel reflecting planes (emulating a valley); distance unit is λ

loss. In this idealised example fading is rather rare. Fading caused by such a small number of reflections is most typical for low carrier frequencies (e.g. HF/VHF).

Another, more typical example for mobile radio, especially at higher frequencies (UHF), is shown in Figure 2.3. In this case the fading is caused by ten randomised reflecting points, emulating what is known as a fading Rayleigh channel. Each of the reflecting signals has ten per cent of the power of the direct signal. This results in a more dynamic signal variation, even within a displacement of a wavelength. These fading examples are valid for frequency-modulated (FM) systems only. Spread-spectrum

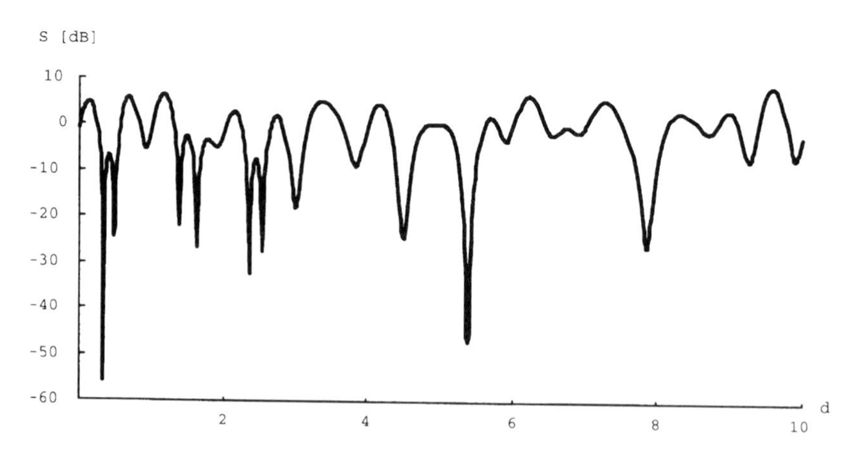

Figure 2.3 Local fading caused by ten randomised reflections with signal powers adding up to match the direct signal. Distance originates from a random receiver point; reference level is 0 dB (direct signal only)

systems may possess a certain ability to combat fading, thus reducing this problem. This is further discussed in Section 4.2.3.

2.1.1.2 *Network concepts*

For a certain signal level giving an acceptable performance, the radio source will have to transmit with a certain output power. In a mobile-radio application the radio terminals will be small in size with limited output power and antenna efficiency. In addition, they are exposed to a great variation in geographic position, both favourable and unfavourable to propagation conditions.

As a result of this situation and the factors mentioned at the radio, network and user levels, there may not be correspondence between possible transmitter–receiver distances (radio-coverage area) and the required service-coverage area that is the geographic area where the mobile radio network shall provide the services to the users. To circumvent such an unacceptable situation two basic solutions exist (also indicated in Figure 2.4):

(i) the base station concept, that is, a centralised fixed transceiver positioned propagation-wise in a favourable position, and also able to connect to other base-station transceivers by some other means of communication;
(ii) the concept of distributed switching where every mobile radio transceiver also functions as an (automatic) relay between any other transmitter and receiver.

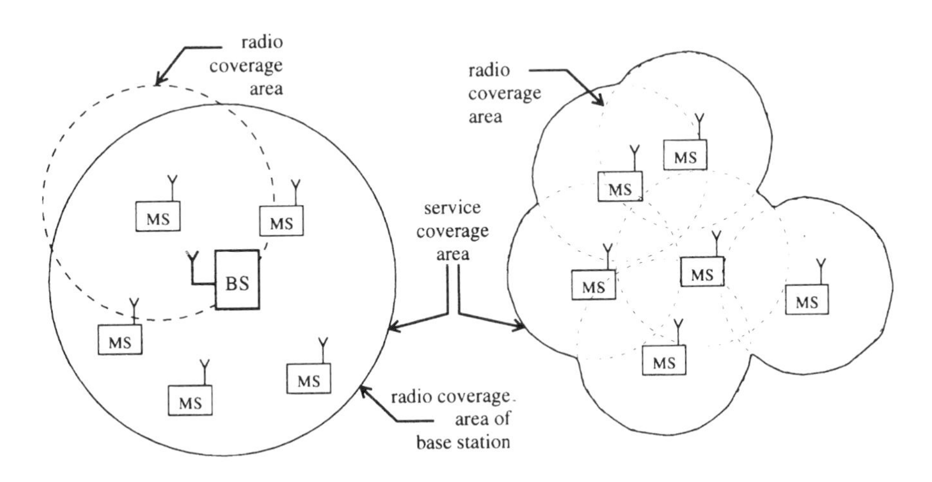

Figure 2.4 *Service-coverage area provided by*
a a base-station solution
b a distributed switched solution

Several base stations may be interconnected by trunk lines (see subsection 2.2.5.4) to obtain a larger service coverage area than that which can be obtained from a single base-station position. The hexagonal regular structure as shown in Figure 2.5 is the most efficient in a flat terrain, with respect to effective complementing but nonoverlapping radio-coverage area per base station. In practice, of course, the situation is generally not so simple. Irregular terrain and inhomogeneous traffic requirements lead to the use of sector antennas, variable cell sizes and irregular base-station positioning.

From eqn. 2.3 we see that a number of factors may be used to increase the radio coverage area of each base station. Both increased output power, antenna gain/height and improved geographic position are seen to be alternative actions. But, bear in mind that although an increase in base-station output power will increase its radio range, this will not affect the range of the mobile station. A useable connection requires two-way communication. But much can be achieved by improving base-station position, antenna height and/or gain. This effort may be taken out in the form of increased coverage area, or in the form of reduced requirements to output power or antenna size for the mobiles, thus improving user-friendly operation.

The choice between a distributed switched network or a centralised network with base stations depends on the user requirements. Generally, base-station concepts are chosen for most public applications, while distributed concepts are chosen for military and other highly specialised applications. Table 2.1 gives a short summary of factors which might influence the choice of network concept.

2.1.1.3 Interference and collocation

The received signal level itself is not sufficient to guarantee any performance, but rather the signal-to-noise and interference ratio. Thus, the coverage will also depend on interference from other users both when it comes to reduced achievable path length, that is at the point where the signal received from the wanted transmitter no longer exceeds that from the unwanted transmitter by the required and predetermined factor, and when it comes to cositing.

It may be the wish of the user to be able to collocate several transceivers. The basic philosophy for avoiding interference has been to make the signals for different destinations orthogonal to each other either in the frequency domain or in the time domain. This is easily achieved through a base-station network concept where the base station may be in full control of all channel access. However, in a distributed system time multiplexing might

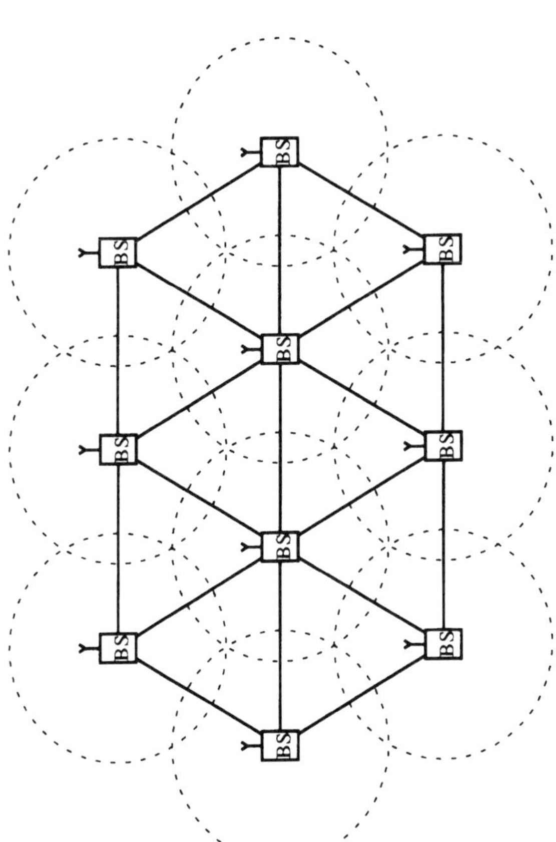

Figure 2.5 Interconnected base stations with complementing areas of radio coverage; the frequently-used hexagonal pattern is the most efficient regular structure

Table 2.1 Summary of factors influencing choice of network concept

User requirement	Base station concept	Distributed switching concept
Service coverage area	fixed geographic position, large area---expandable	flexible, moves with the users, limited area
Availability	predetermined	higher, more flexible
Network establishment	slow	rapid
Traffic capacity	large, easily scalable	limited
Spectrum efficiency	high	generally lower
Control & maintenance	centralised	distributed
Initial cost	large	small
Survivability	low---vulnerable	high, graceful degradation
Frequency planning	very easy	may be complicated
Individual users	inhomogeneous	homogeneous loyal group
Predominant traffic pattern	most externally directed	most internal within group
Mobile collocation	easily controllable	requires special attention
Automatic power control	easy	more complicated
Access to other networks	very easy	requires base station of some kind
Predominant service	circuit switching	packet switching
Effective use of (direct sequence) spread spectrum	easy due to power control, requires multiple receiver functionality in BS	power control may be problem, parallel transmissions to different destinations
Typical application	Public telecommunications	Specialised systems

not be possible and two different kinds of problem may arise relating to channel multiplexing (separation of different channels) and the duplexing method (to distinguish signal direction).

The first problem is illustrated in Figure 2.6. A transmitter is generally not able to limit its emission to the allocated channel (frequency bandwidth). Although the power emitted outside this channel is reduced with the distance to the centre frequency, there may still be a considerable skirt extending over a sufficiently large frequency spectrum to limit the sensitivity of a collocated receiving mobile radio, operating at a nearby frequency. This problem may be solved by special collocation filters, but they are both expensive and space consuming. This makes hand-held implementations difficult, especially in the lower frequency bands such as VHF. Often, such collocation filters are not implemented in the radio itself, but as a separate unit to be used by the few users which are most exposed to this problem, e.g. vehicles with a number of radio transceivers installed.

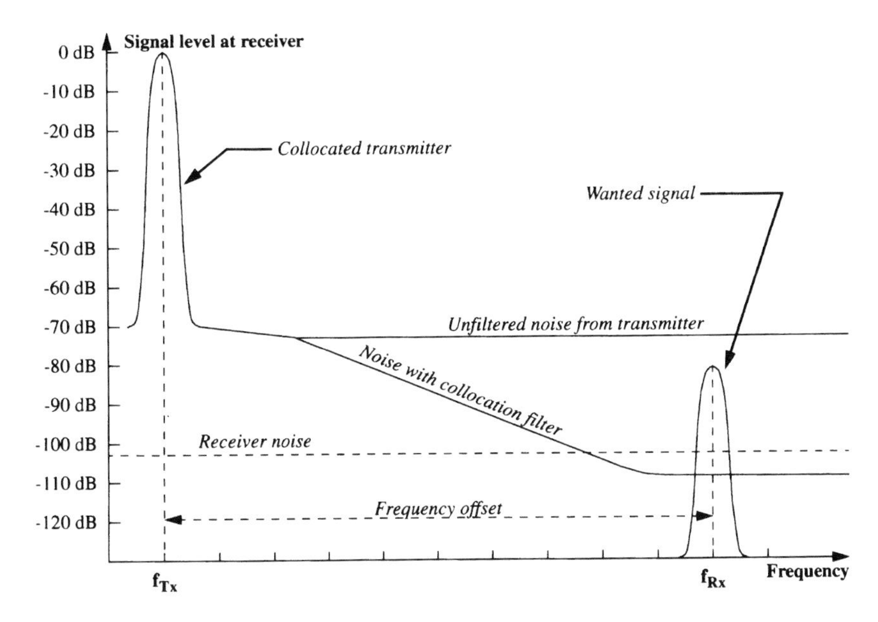

Figure 2.6 Nearby channel interference by collocated transmitter

The other problem is that of receiver filtering to handle interfering signals with frequency offset from receiver carrier. This problem also requires special filters. Many of the same challenges are found here as with the transmit filters. In addition, receiver filters should have as low an attenuation as possible, in order not to reduce the sensitivity. This may not be so vital for transmit filters as the final signal amplification often is done after filtering (when the problem of intermodulation can be ignored). The problem of collocation will be elaborated on in Chapter 6.

The problem arising in a network of radios where several transmitters may attempt to access the same receiver at the same time, often referred to as the capture problem, must also be solved. The particular solution must be matched to the receiver characteristics and access protocol used.

In spread-spectrum systems code-division multiple access (CDMA) may be used to separate the different signals and channels. However, the codes are generally not strictly orthogonal, introducing noise between the different signals. Thus one is left with the problems of capture and interference. The selection of coding technique and possible code alphabets based on various criteria is further analysed in Sections 4.2.3 and 4.3.4. An example of cross correlation between two 128-chip spreading codes is shown in Figure 2.7. These two codes are particularly suitable for a synchronised system as the cross correlation is best (lowest) when the two codes are shifted by only a few chips (around ± 128 in the Figure).

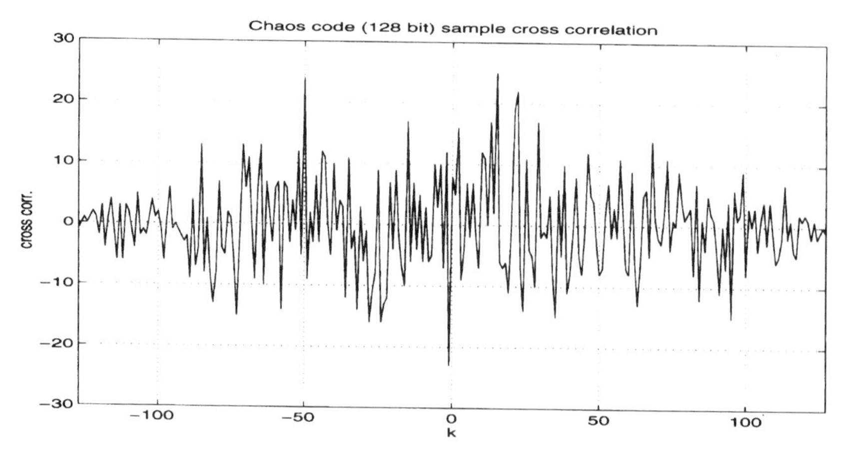

Figure 2.7 *Cross correlation of two 128-chip spreading codes*

The interference from other users will also be subject to fading. In practice, the criteria for a minimum signal level and an adequate signal-to-interference ratio must therefore be met simultaneously. In this context the probability of connectivity is expressed as the probability of meeting both criteria at the same time.

The two most important terms introduced in this section, radio- and service-coverage area, may both be considered to be functions of vital parameters or strategies/concepts. In short this can be summarised as:

$$\text{radio-coverage area} = f(\text{radio characteristics, path loss, terrain features, interference from other users}) \tag{2.4}$$

$$\text{service-coverage area} = f(\text{radio-coverage area, network concept for connecting geographic areas}) \tag{2.5}$$

2.1.2 Autonomy

Most civilian LMNs, e.g. mobile telephone systems, depend on a fixed network, an infrastructure with base stations, in order to provide service for mobile terminals. But it may be a requirement from several user groups not to be so absolutely dependent on this infrastructure. An example to illustrate this; imagine two cellular telephone subscribers located close to each other (well within practical radio range). If for some reason the base station (or another vital part of the infrastructure) is out of operation, there is no way that these two subscribers are able to communicate with each other. For many applications (e.g. military) this is not at all acceptable. That is why many land mobile radio systems also offer different levels of autonomy with variable service availability.

Several users also have requirements for service availability in remote areas, and cannot allow themselves to depend on the presence of infrastructure in all locations as this would be too expensive. This is often the case for many rescue operations as well as military operations.

For instance, in the case of a major disaster (e.g. earthquake, storm or act of war), fixed installations may be destroyed. In these situations the availability of communications is especially vital. In addition, in particular for military forces but also for civilian, the dependence on an infrastructure makes the communication system a valuable target for hostile attack, by an enemy or a terrorist group.

Thus, the possibility of autonomous operation where mobiles are able to communicate directly, in many cases proves to be of vital significance. This is why, for instance TETRA (trans European trunked radio system), the new standard for private mobile radio (PMR) and public-access mobile radio (PAMR) systems [5], also includes a direct communication mode between mobile radios as well as the ability to use mobile stations as repeaters to increase the service coverage area.

Autonomous operation is not a uniquely-defined concept. Many different levels of autonomy may be imagined, all with a different range of services available to the users. A first degradation of the network may occur if the central network control centre is out of operation, but communication between different base stations is maintained. Users may still be able to communicate relatively unnoticed, but the attempt of new users to enter the network may be inhibited. Some other, not frequently used, services may also be unavailable. But to most users this degradation may go unnoticed if the network has been properly designed for such a degradation to lower levels of autonomy.

Another lower level of autonomy occurs when users connected to one base station are cut off from the rest of the network (with other base stations). Users connected to the same base station may still be able to communicate, but obviously they are unable to reach users connected to other base stations or other networks. Also, some services may no longer be available.

The lowest level of autonomy is in the absence of any base station. Two or more radios operating on the same channel (frequency and other parameters) may still maintain communication, but many services may not be available to the communicating parties.

This is further discussed and illustrated when addressing system architecture in Section 2.2.5.

2.1.3 *Mode of operation*

It is convenient to distinguish between three different modes of operation for a mobile communication system:

(i) *Normal mode* (voice and data): this can be compared to the traditional (mobile) telephone service. The mobile unit communicates via a switched infrastructure through a base station. The infrastructure contains a number of base stations which together may give a large geographic area of radio coverage. In this mode the mobile station always communicates through a base station, even with other mobiles within direct radio range.

(ii) *Direct mode* (voice and data): mobiles communicate directly and are not dependent on any infrastructure. The use of this mode may often be attributed to the fact that the radios are outside the infrastructure area of coverage. Another reason for using this mode is limited radio resources at the base station. Two mobile stations (MS) communicating in direct mode occupy only one radio channel, compared to two when operating in the normal mode, see Figure 2.8. Also, they do not need to occupy any of the radio resources at the base station (BS), and are as such not at all dependent on the presence of a base station within radio range.

(iii) *Packet data mode* (optimised): packet switching (compared to voice- and circuit-switched data) is characterised by the fact that all transmissions are relatively short, and that there is no requirement for continuous or instantaneous radio transmission. Here, many mobile stations can share the same radio channel. Thus, this requires specialised protocols in order to obtain a reasonable utilisation of the channel resources.

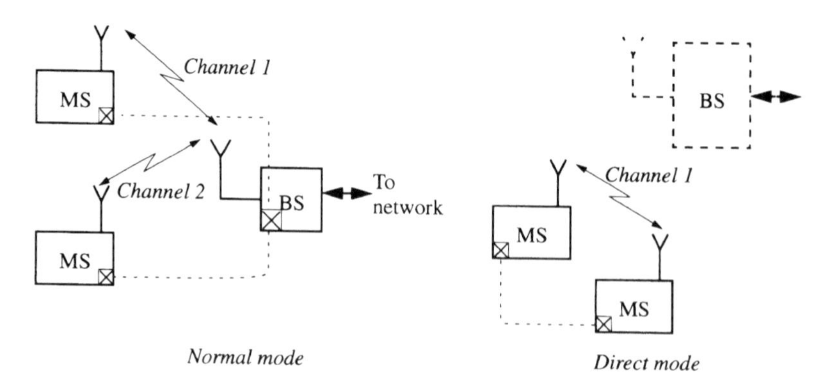

Figure 2.8 Radio resource utilisation in direct and normal mode

A mobile communication system may use one or more of these modes, depending on the services to be offered to the users and the requirements for availability, survivability and autonomy as mentioned in Section 2.1.2. A modern LMN system should be designed with all these modes in mind.

2.1.4 Services

Different applications often require different services from the telecommunication network. In the future the multimedia revolution will merge the transfer of video, voice and data, also for wireless subscribers. These different user services make different requirements of the network. Obviously, video requires a much larger capacity than voice, but capacity is not the only difference. There are also many different data applications with varying requirements.

One of the near-future switching techniques is asynchronous transfer mode (ATM). It is not necessarily so that wireless networks will be required to use this technique in the future, but, connected applications and interconnected networks will use ATM. Even if ATM is not suited to all wireless networks, future networks will be required to be interoperable with ATM networks in some way, and wireless subscribers will need to access ATM networks through wireless networks.

2.1.4.1 User requirements

For multimedia applications let us divide the user services into the following three categories: voice, video and data. Within these categories there is a whole range of subcategories with differing requirements:

Voice transmission

Traditionally voice was transferred through telecommunication networks in analogue form, e.g. within a bandwidth from 300 to 3400 Hz. Future networks will be completely digital, which requires some kind of voice coding (from analogue to digital representation). Currently, vocoders (voice coders) are found with a large variation in required data rate. Public switched telephone networks (PSTN) use 64 kbit/s PCM (pulse code modulation), tactical military networks apply 32 or 16 kbit/s delta modulation and mobile networks use a large range of low-rate vocoders (e.g. 4.8 kbit/s in INMARSAT M and 13 kbit/s in GSM). The trend towards lower bit rate coders will still continue for some time. Obviously, analogue voice and high-rate coding schemes have an inherent redundancy. This means that such voice transmissions will still be comprehensible even under conditions introducing severe noise (high bit error rates) on the channel. Low-rate vocoders are much more susceptible to channel bit errors, and are therefore accompanied by some kind of error control. This is absolutely vital for wireless networks which are more exposed to channel errors.

Voice transmission usually requires a service with fixed end-to-end delay. Actually, the requirement for fixed delay is not so important as a guarantee of a maximum delay. As seen from the user, the quality of a voice connection degrades rapidly as the delay increases. Many readers will be familiar with the annoying ~250 ms delay introduced by a (geostationary) satellite connection. In addition to the pure transfer delay many low-rate vocoders introduce an extra significant delay due to their block structure (the coder samples voice for a time period before coding and transferring to receiver). Generally, the lower the bit rate that the vocoder needs, the higher is the coding delay introduced.

Although voice services may still be the least demanding with respect to the quality of the propagation medium, in the form of distortion and interference, the service's sensitivity to delay and any communication disruption requires a close look at how the service-coverage area is implemented. In this context it may be useful to divide voice users into the following two categories:

(i) typical mobile telephone system users transferring long voice messages to or involved in a conversation with another user;
(ii) typical mobile radio system users transferring short messages within a group of users.

The first category of users with a session or hold time of up to several minutes may tolerate a set-up delay of several seconds, but the second category with a hold time of just a few seconds will require a set-up delay of less than a second. The hold time is also a critical factor when it comes to movement of the users and the need to avoid noticeable breaks in communication. This vital topic to mobile telephony is called handover and is discussed in Section 2.3 on mobility.

Video transmission

Video has many similarities to voice. It is also very susceptible to delay or loss of data as it constitutes a continuous stream of consecutive images. Traditional analogue video (TV broadcasting) requires a very large bandwidth (approximately 5 MHz), although the introduction of digital video and compression techniques has brought the requirement down to an average of three to six Mbit/s for MPEG (moving pictures expert group) video with acceptable picture quality. The bandwidth, of course, depends on the channel coding used and the actual channel quality. (MPEG is a standard for compressing both video and audio.)

As for voice, the use of compression techniques makes the signal more susceptible to bit errors, and some kind of error control is normally required for wireless transmission. In addition, the coding also introduces an inherent delay. Video has the same handover requirements as voice, only with stricter requirements related to the increased bit rate.

Data transmission

The transmission of data does not constitute a homogeneous class of service requirements. There are different applications ranging from the transfer of large data files with loose requirements for delay, to the short but urgent messages at the other end of the scale. Other applications may be the repetitive transfer of current status of a remote sensor. For file and message transfer it is of vital importance to receive this information correctly, and delay may be less important. For the repetitive sensor information there is usually no harm done if one or a few status reports are lost since fresh information soon will be sent again. But for many of these remote sensor applications the transfer delay must be minimised.

Error control

Information to be passed between two users connected to a telecommunication network is exposed to a number of potential errors: arbitrary bit errors may be introduced, parts of the information may be lost, the message may be duplicated, or it may be delivered to the wrong destination. To some applications such errors may be fatal, while other applications may have built-in error tolerance. This puts different constraints on the network concerning the required quality of service. Generally, the higher the requirements on the network for quality, the higher the communication cost in some way (e.g. effective communication capacity). It is thus very important not to put higher requirements for the network quality than are absolutely needed by the actual application in order to minimise this cost.

Delay

There are two aspects of delay worth discussing. As previously mentioned, some applications (e.g. voice) need an upper limit on the transmission delay in order to maintain the required quality of the user service.[1] For other applications the actual delay is not important, there is no strict requirement, but the application may need to know the quantity of the actual transmission delay related to every information package sent. This latter

[1]For many such applications information arriving too late will be thrown away by the application as it is no longer of any value. In such cases the network might as well not spend valuable resources attempting to deliver the information, knowing it will arrive too late.

consideration requires the network to be able to measure the transmission delay and pass this over to the receiving application.

The requirements for delay have been discussed for voice, video and data transmission. As will be discussed later, some network services are of such a nature that they give a fixed transmission delay, and others may offer a highly variable delay. One important aspect closely related to delay is error control or guarantee of delivery. It may not be obvious, but applications requiring low error rate (or allowing no errors at all) and guaranteed delivery will generally experience a longer mean transfer delay (assuming otherwise equal conditions). This is best illustrated by comparing protected and unprotected services. Protected services are used when error-free delivery is required, but delay is not as important. Unprotected services are better used when short/fixed delay is important and some errors or loss of information may be acceptable.

Unprotected services provide a constant throughput and a short delay (which for circuit switching also is constant). However, there is no guarantee regarding the error rate, as performance will depend on the bit rate and the channel conditions. For *protected* services a virtually error-free channel with highly variable throughput and delay is established. The information received is checked for errors. If errors exist the receiver requests retransmission of the data. The two types of transmission services are illustrated in Figure 2.9

Circuit and packet switching

Applications such as voice and video generate a constant stream of information and are often best served by the network service called circuit switching (CS), which is the traditional switching concept. In subsection 2.1.1.2 it was shown how it was possible to achieve a service coverage area that was larger than the radio coverage area without the need for introducing base stations. The concept was based on the ability of each radio to act as a possible relay and introduced a mechanism which can allow automatic multihop transmissions. Packet switching (PS) will enable this. One of the major differences between voice and circuit-switched data services on one hand and packet-switched data services on the other is the packing of information of the latter, allowing for transmission breaks to be taken care of at the network level. This allows the sharing of a single channel between several users without any perceived deterioration in performance until blocking occurs as the network becomes really congested.

Both circuit and packet switching are treated in more detail in subsection 2.1.4.5, while the characteristics of packet switching in radio networks are extensively treated in Chapter 5. To summarise this introduction to

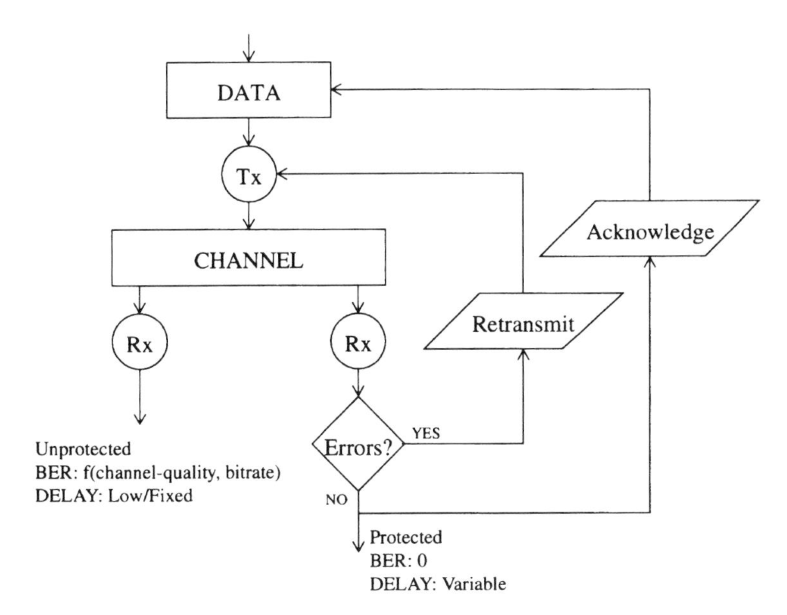

Figure 2.9 Comparing protected and unprotected services

services, we can state that what may be provided to the users depends both on the terrain and the technology available.

2.1.4.2 Network services

There are two basic different kinds of network services, as described in Figure 2.10:

(i) *Bearer services*: these are the services offered by the network (levels 1 to 3 of the open systems interconnection (OSI) model, see Section 2.2.3) to the connected (subscriber) terminal equipment. They provide communication capabilities between terminal network interfaces.

(ii) *Teleservices*: these are the services offered to the user by the subscriber equipment. They provide complete capabilities, including terminal functions, for communication between users. Obviously, these services are enhancements of the bearer services offered by the network.

In this Chapter we will focus on the bearer services; the teleservices, as seen from the user's perspective, are less important to the communication system designer.

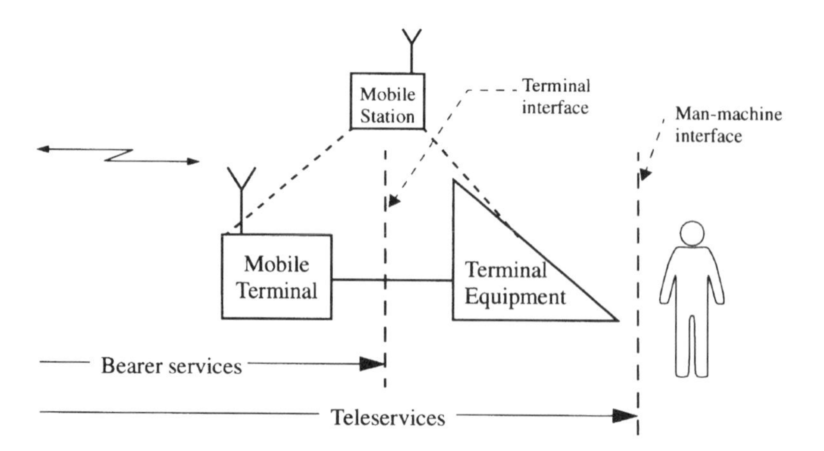

Figure 2.10 Functional decomposition of the mobile station and the services

2.1.4.3 Teleservices

The teleservices are often separated into two groups:

- basic teleservices;
- supplementary teleservices.

The basic teleservices are mandatory and must be available in all systems complying with a certain standard. They are considered to be of vital importance to all users. The supplementary teleservices modify or supplement the basic teleservices.

Basic teleservices may, for instance, be for both voice and data in clear or concealed (encrypted) mode in each of the following:

- individual call (point-to-point);
- group call (point-to-multipoint bidirectional);
- acknowledged group call;
- broadcast call (point-to-multipoint one way only).

Supplementary services may be separated into two groups, PMR type and telephone type.

PMR-type supplementary services:

- access priority, pre-emptive priority call, priority call;
- include call, transfer of control, late entry;
- call authorised by dispatcher, ambience listening, discrete listening;
- area selection;
- short number addressing;
- dynamic group number assignment.

Telephone type supplementary services:

- list search call;
- call forwarding–unconditional/busy/no reply/not reachable;
- call barring–incoming/outgoing calls;
- call report;
- call waiting;
- call hold;
- calling/connected line identity presentation;
- calling/connected line identity restriction;
- call completion to busy subscriber/on no reply;
- advice of charge;
- call retention.

These supplementary teleservices do not require the introduction of new bearer services. But most of them require additional information to be exchanged within the network. This may be done either by special information exchange, or by adding the extra information to the previously required control information.

2.1.4.4 Bearer services

There are different ways of implementing the required teleservices, using the bearer services available. To support the basic teleservices the network should offer the following bearer services: individual call, group call, acknowledged group call and broadcast call in each of the following:

- circuit-switched mode unprotected data (one or more data rates);
- circuit-switched mode protected data (one or more data rates);
- packet-switched connection-oriented data;
- packet-switched connectionless data.

Another bearer service that is special to radio networks is called semibroadcast. This service is introduced because it is very easy to implement in a radio network, and also it may, for some applications, help to reduce the traffic load and congestion in the network. This is discussed in subsection 2.1.4.7.

2.1.4.5 Switching concepts

Figure 2.11 shows the basic difference between the alternative switching concepts: circuit switching (CS) and message switching or packet switching (PS).

The traditional *circuit-switching* method implies the establishment of an end-to-end physical connection with reservation of a dedicated static

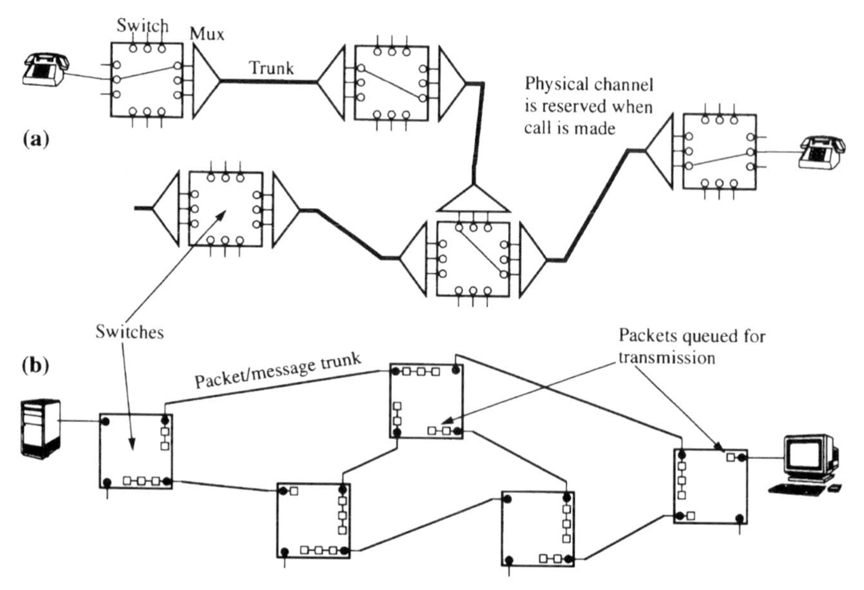

Figure 2.11 Comparing (a) circuit switching to (b) message/packet switching

transmission path occupying link channel bandwidth resources through the network before any data can be sent. The connection is then free to be used (or just to be idle) until the connection is closed. The data bits flow uninterrupted through the connection. Unless the connection with the reserved bandwidth is well utilised for the period it is established, this may represent a waste of resources. The reserved bandwidth is fixed, and may not readily be altered in the case of changing requirements during the connection lifetime.

In *message switching*[1] or *packet switching* the channel resources are shared between all connections, and are only occupied for the actual data transmission period. No dedicated physical connection is established. The information is partitioned into messages or packets. The whole packet/message is transferred over a link and stored in the network node before it proceeds on the next link on its way to the destination. Packet switching is preferred to message switching for several reasons: messages may be larger than the available storage capacity in a network node, transmission delay may in some instances be shorter when messages are sectioned into shorter packets, a large message may tie up transmission resources for a long period—blocking other (interactive) traffic. In packet switching there is a strict upper limit on the block size. Large messages then must be transmitted as a number of packets to be reconstructed at the destination.

[1]Not to be confused with message-handling systems (MHS) which are often found as enhancements to packet switching (or circuit switching).

Packet-switching services are further separated into connection-oriented (CO) and connectionless (CL) services. CO services (as well as CS services) encompass three distinct phases as seen from the user or application:

- connection-establishment phase;
- data-transfer phase;
- disconnection phase.

The connection-establishment phase is first entered when the user signals that a connection is needed. Successful termination of this phase results in a virtual connection on which the users may exchange blocks of data. When the information transfer has reached its end, the user may release the connection by initiating the disconnect procedure. CL services facilitate information transfer as self contained logical units without the prior need for establishing a connection. Thus CL packets belonging to the same message may take different routes through the network and may arrive in reverse order at the destination node (to be rearranged before being presented to the user or application).

Figure 2.12 shows the possible benefits of message and packet switching compared to circuit switching when considering network delay. The Figure applies to networks with separate links, allowing parallel transmissions. It is not necessarily applicable to radio networks. If all (or many) links share the same radio channel, unable to transmit simultaneously, the point on transfer delay does not apply. But there may be several other benefits of using packet switching compared to message or circuit switching for data transmission. For voice connections the use of circuit switching is still preferred in most networks.

One drawback (for some uses) of packet switching is related to the fact that bandwidth resources are often not reserved for the established channels,[1] and certainly not when using CL services. At instances when many terminals generate packets simultaneously, the instantaneous traffic load may exceed the capacity of the node or its links to other nodes. Then some packets must be stored for some time before being transmitted. This means that there is no guarantee on packet delay, in fact it may seem quite unpredictable. For some applications this may not be acceptable, and CS may be a better alternative as this gives a predictable, constant delay.

A new packet switching technique has been introduced in an attempt to reduce the delay problem, making packet switching suitable for applica-

[1]Packet switching allows a capacity-reservation mechanism for CO services, but this should be used with care as it contradicts one of the principles of PS — efficient resource utilisation in cases of multiple users with fluctuating needs.

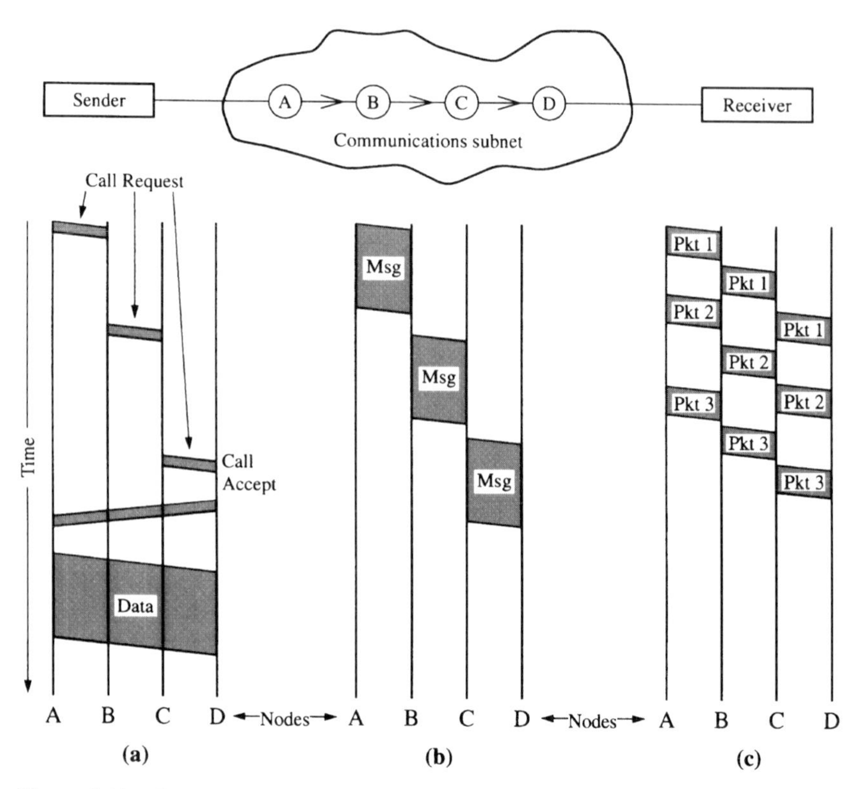

Figure 2.12 Comparing (a) circuit switching, (b) message switching and (c) packet switching

tions which traditionally have relied on circuit switching. This technique is called ATM (asynchronous transfer mode) or fast packet switching. The technique is primarily aimed at LAN (local area network) applications with its primary data rate of 155 Mbit/s, but will probably also find a use in radio networks. The reduced delay is achieved because ATM differs from traditional packet switching in many respects. First of all it uses a standard cell length of 48 bytes of user information. In addition, every cell has a simple five-byte header. The user information is not error protected in any way; this is left to the end application. When an ATM cell enters a switch its header is read, the decision on where to forward the cell is taken, a new header is calculated and the cell (with the new header) is forwarded directly to the correct output port. In this way the first part of the cell might be forwarded before the complete cell has been received. This is not possible in traditional packet switching since the complete block must be received and error checked before it may be forwarded.

Packet switching techniques applied with radio as the transmission medium are often called packet radio (PR). When speaking of packet switching in radio networks (packet radio) in this book we will generally understand a network where all the switches (nodes) use a common radio channel. This gives the extra challenge of how to find the most effective way of sharing the common channel equally and just between the network nodes. Switching in wireless networks related to system design is more extensively treated in Section 2.4.

2.1.4.6 Point-to-point addressing

Point-to-point transmission is related to the bearer service *individual call* in circuit- or packet-switched mode. The only real challenge related to the implementation of this service is relaying when the two communicating parties are out of direct radio range. For PS a relaying function may be implemented in every mobile radio by adding extra software functionality (see detailed description in Section 5.6). The packet is received and stored in the relay node until forwarding is possible. This relaying function will be based on a routing function which usually ends up as a compromise between efficiency (channel utilisation), robustness (mobility/dynamics and distress) and fairness (providing equal service level to all users).

For CS services a relaying function may be much more difficult to implement. However, in time-division multiple-access (TDMA) systems (with instantaneous channel bit rate greater than the user bit rate) this is possible by allocating another time slot for the relay transmission. The only drawback is the introduction of an additional time delay, although insignificant if the slot size is sufficiently short. For other systems CS relaying usually requires dedicated relay stations, e.g. two radios may be connected back-to-back; one radio is receiving, transferring the information to the other radio, which transmits the same information on another radio channel. This introduces very short time delays (of the same order as a synchronisation preamble).

2.1.4.7 Multipoint addressing

Let us distinguish between three different multipoint services: broadcast, multicast and semibroadcast. A true broadcast service implies that all users who happen to be located within the specified network domain (e.g. a subnetwork) shall receive a copy of the information being broadcasted. This means that delivery is guaranteed, and that there must be a stringent control of delivery to all recipients.

A multicast service implies that every member of a given group shall receive a copy of the information. The group members may be located in a number of network domains, and not all members of a domain need be members of the group. This service is usually implemented by addressing a multicast server, which in turn sends a dedicated copy to all members of the group, using point-to-point addressing for each individual member.

Semibroadcast is a special service usually found in radio-broadcast networks only, because it is so easy and natural to implement in such networks. The nature of radio broadcast (all radios communicating on a single shared channel) is such that many radios may receive the same transmission. But at the same time, there is no guarantee whatsoever that every radio in the network will be able to do so. Also, depending on the network dynamics, which radios do receive the transmission from a specific transmitter may vary with time. The semibroadcast service is performed by addressing all network members in a single call, but with no control or any guarantee of delivery.

Packet-switched multipoint addressing is fairly easy to implement in a radio system, compared with the analogue circuit-switched service. The required number of copies of the original packet are forwarded asynchronously. If not received at first attempt, the information may be retransmitted. The originator will be informed if the packet is not successfully delivered to all recipients, unless the invoked service does not support information on success of delivery. Broadcast and multicast can support such confirmation of delivery, but semibroadcast cannot.

Circuit-switched multipoint addressing is a service to be used with care as it may occupy large resources. In radio networks this service may be supported only in combination with semibroadcast. The service in combination with broadcast or multicast is not possible to implement in radio networks where all mobile stations share a single common channel, unless this channel has sufficient bandwidth to handle the required CS connections in parallel. Also, no data transmission can take place until all receivers are connected.

CS semibroadcast is the traditional service found in not-so-sophisticated closed PMR networks, especially for voice transmissions. All mobiles within the radio range of the transmitter may receive the information unless otherwise protected. Confirmation of delivery is left to the user (or application) if required. For data, this may seem like a useless service. But there are applications where speed is more important than guarantee of delivery. This may apply to cases where a sensor regularly distributes updated information, and where current state, not history, is what matters to the receivers.

2.1.4.8 Quality of service

The quality of a service may be specified with a number of parameters, e.g. transit delay, throughput, residual error probability, priority and maximum lifetime.

Transit delay
We have previously discussed the delay requirements. It was mentioned that some applications might be interested in knowing the transmission or end-to-end transit delay through the network. For a circuit-switched connection the application may obtain a good estimate (of the one-way transit delay) by measuring the round-trip delay (time from sending a signal until a response is received), subtracting the response time (delay introduced by the remote application in order to generate a response) and dividing by two. Generally, this transit delay is constant for CS connections. For packet-switched connections the transit delay may vary dynamically, in which case it makes sense to be able to measure the transit delay for each packet individually. Some networks may offer this service parameter, but not necessarily for all PS services. Transit delay may be related to maximum lifetime, as discussed below.

Throughput
It is obvious that some applications, e.g. voice, have a requirement regarding the minimum throughput available. For CS services the throughput is guaranteed since the channel bit rate is constant (as long as the channel is available). For PS services it may be more difficult both to specify and guarantee a certain throughput class due to the dynamic nature of this service. But for connection-oriented PS services it may be possible to negotiate on the logical channel throughput class. Even if the network accepts a certain channel throughput, this most often cannot be guaranteed 100% under all circumstances, but the network might attempt to reserve some capacity by not accepting too many connections. The sum of the throughput classes of all connections might thus exceed the capacity of the switch and its trunks to other switches, as some statistical traffic fluctuation is assumed in an attempt to obtain a higher utilisation of the network resources (compared to circuit switching).

Protection and residual error probability (REP)
By protection we understand the ability of a radio network to continue (more or less) undisturbed operation, even in the presence of external interference sources, whether deliberate or not. External noise will introduce bit errors into the user data being transferred over the transmission channel. Some applications may tolerate a certain error rate even unnoticed, although others may require absolute 100% error-free delivery.

Spread spectrum (SS) is one method introduced at the physical level to combat interference, forward error correction (FEC) another. But the price to pay for such protective measures is reduced channel capacity, unless the bandwidth is increased. If there is good reason to believe that the channel is undisturbed, an unprotected (i.e. less protected) transmission may be advantageous. For applications where reliable delivery is more important than channel capacity, protected transmission is preferred. Both modes may be implemented in the same system, as a service to be invoked by the user at will. This is usually done by letting a spreading or error-correction code assume one of a range of lengths, depending on the user's choice of protection level. Figure 2.13 illustrates the differences between protected and unprotected transmission.

Traditionally, the PSTN offers only unprotected transmission as this network is designed for circuit-switched voice only. Data services introduced on top of this network will usually apply their own error control. Networks providing packet-switched services must at least have error detection, and if the error rate is large, as it often is in radio networks, some kind of protection using error correction generally is required. For circuit-switched services, where low delay most often is required, retransmission of erroneous data is not acceptable and a powerful error-correction code is necessary.

Residual error probability (REP) is defined as the probability of undetected errors being introduced into the user data. This may be caused by lost, duplicated, erroneous or incorrectly delivered information. Even with the best error detecting code there is always a small possibility of not detecting certain combinations of bit errors. Obviously the REP should be low, but demanding REP ≈ 0 would require an unreasonably strong error correction and/or detection mechanism, which would reduce the effective channel throughput. Certain applications may accept a higher REP than others, which calls for at least two REP classes. Traffic which requires no stringent limit on the REP may be transferred using the cheap unprotected transmission mode, and protected transmission can be used for traffic which requires a lower REP.

As an example, let a block of 128 bytes of user information be protected by error-detecting codes of different lengths. The average probability of an undetected bit error for this block size is given in Table 2.2 for a number of channel bit error rates. The formula for calculating this upper limit of undetected error may be found in [11]. The typical channel bit error rate of a wired link may be 10^{-5} or less although a radio connection may experience much higher error rates; even BER $= 0.5$ (a fully blocked channel) may occur at the edge of the radio range or in extremely noisy situations.

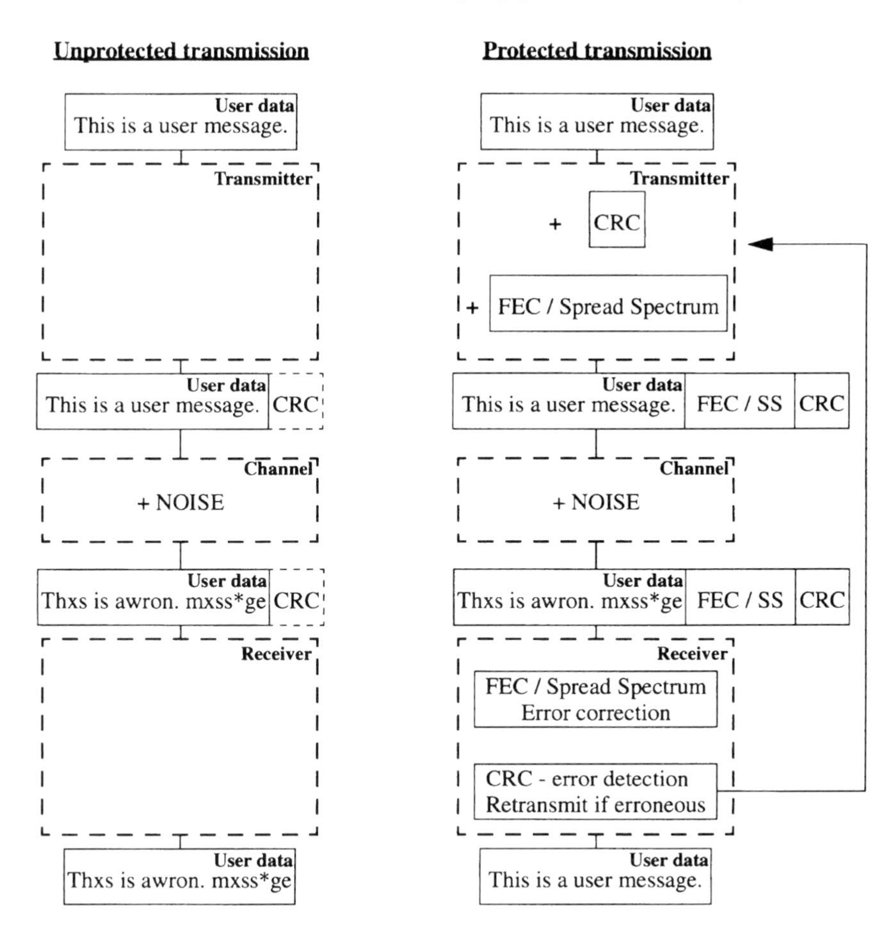

Figure 2.13 The principles of protected against unprotected transmission. User data is error controlled by e.g. a CRC code (cyclic redundancy code) and optionally error protected by an FEC (forward error correction) code for protected transmission

Table 2.2 Probability of undetected bit error in a 128 byte data block as a function of error-detecting code length and channel bit error rate

Error-detecting code length (bits)	Channel BER = 0.5	Channel BER = 0.1	Channel BER = 10^{-5}
7	7.8×10^{-3}	7.8×10^{-3}	8.0×10^{-5}
16	1.5×10^{-5}	1.5×10^{-5}	1.6×10^{-7}
27	7.5×10^{-9}	7.5×10^{-9}	7.8×10^{-11}
32	2.3×10^{-10}	2.3×10^{-10}	2.5×10^{-12}

REP may also be related to channel forming. For a direct-sequence spread-spectrum (DS SS) system it is possible to operate with a number of code sets of different length. Short codes may be used under undisturbed conditions or when the application can accept a higher (poorer) REP. The benefit will be increased channel capacity assuming a constant signal bandwidth. If the application requires low REP and/or the transmission conditions are difficult (high levels of interference), longer spreading codes may be applied at the expense of reduced channel capacity.

Priority
In a network serving a number of users with potentially different requirements and applications, there should be a network mechanism to assure a prioritised order of serving these customers. This service, rarely found in public telecommunication networks, is signalled through the service parameter called priority. This parameter may be selected by the user on a per-call basis, and is signalled to the receiver. The use of this parameter for precedence and pre-emption is further discussed in subsection 2.1.4.9.

Maximum lifetime
As previously mentioned, for some applications data arriving too late may be of no use. The best option would be to let the network abandon these data blocks instead of spending valuable resources in a vain attempt to deliver them to the receiver. This calls for the introduction of the service parameter called maximum lifetime. This enables the application to set a maximum time after which this information expires. This facility does not make sense for circuit-switched or connection-oriented packet-switched connections. But for the connectionless PS service the application may set a maximum lifetime for each packet. If the packet has not reached its destination within the specified time the network is free to discard this packet.

Maximum lifetime may be combined with transit delay. The network may use the maximum lifetime set by the application, and count it down towards zero as delay is added at each step in the network. Assuming that the receiving application knows the initial lifetime setting (through an agreement with the transmitting application) it may calculate the transit delay by subtracting the signalled remaining lifetime from the maximum lifetime:

$$\text{transit delay} = \text{maximum lifetime} - \text{remaining lifetime} \qquad (2.6)$$

The use of priority, precedence and pre-emption is more easily introduced in PMR systems compared to public systems. The reason for this is that a PMR system is usually serving a homogeneous group of users with a common superior objective. Such users may more easily accept the idea of helping each other for their common good, even if that should result in a reduced quality of service for the individual user. In public systems serving a large group of users with little in common, the individuals are more inclined to fight for themselves, thus not so willing to accept a lower priority than others.

Voice transmission
Voice is usually transmitted by the use of CS services. This is due to the fact that the throughput cannot be guaranteed and the time delay introduced by PS is variable and unpredictable. The quality of voice rapidly deteriorates under such conditions. The reason for addressing voice transmission separately is the fact that there are many different voice coding (vocoder) algorithms. Within one subnetwork usually only one common vocoder is used. But the interconnection of subnetworks with different vocoders introduces a necessity for transcoding. The transmission of bits as in an ordinary CS service will not work in such a case. Thus voice and data services must be treated separately.

2.1.4.9 Precedence and pre-emption

Both precedence and pre-emption are based on the priority parameter. When the network is low on resources and two applications send a request for a new connection, the network will first serve the request related to the highest priority parameter. This is called precedence and means that all requests will be served according to their signalled priority. In a distributed network precedence may not be accomplished 100%, as this would require an extensive information exchange between all switches in the network. In particular, this is a problem related to connectionless packet switching as the transmission of a packet may not be preceded by any kind of request for permission to transmit.

If the network receives a high-priority request for a connection, but has no resources available, it may use pre-emption. This means that a lower priority connection, if present, will be disconnected in order to free resources for the high-priority connection to be established. The low-priority connection will be forcedly disconnected without any warning, but the parties engaged in that connection will be informed of the incident, possibly with a notice of the reason for disconnection.

Precedence and pre-emption may be used to achieve a number of different results:

- obviously, they are used to distinguish between applications of different significance by giving them different service quality according to priority;

- the network may apply these mechanisms for controlling the network traffic. In this way high-priority applications may be guaranteed a high probability of delivery and reasonable transit delay even in stressed situations where offered traffic exceeds the network capacity;
- applications depending on a number of information classes of different importance may apply different priority to the different information classes. In this way the application is assured that the most vital information is given a better chance of being served by the network (since the network is not forced to serve all requests). This helps in maintaining a minimum level of operation with a graceful degradation of application-system performance in situations when the nominal network capacity is being exceeded.

When discussing the functional architecture in Section 2.2.4 we will see that both precedence and pre-emption may be applied at different service layers.

2.1.5 *Mobility*

There are a number of different aspects of mobility, all of which are associated with different problems. Any definition of mobility is related to the fact that users are not located at fixed positions forever. Generally, mobility is the element which makes the design of radio networks really challenging.

2.1.5.1 *Global mobility*

Mobility in global terms should be understood as the need of users to be connected to a network wherever they are located on the planet Earth. Very few mobile radio system operators have a network with true global coverage, but some future satellite systems may achieve this.

For other network operators, global mobility requires co-operation with other operators so that users may gain access to a large number of networks. The transition from one operator's network to another may go unnoticed by the user or not, depending on what kind of roaming service is available.

2.1.5.2 *Local mobility*

Users may be bound by a network operator's area of coverage, but may change position within this area. If this relocation results in loss of contact with the original base station, contact with the network must be established through another base station. This is called location management. When relocation results in loss of contact with the base station during the active traffic phase, the connection must be transferred from one base station to another. For mobile telephone services this process, called handover, should go unnoticed by the user. Preferably, handover should take place before the connection to the base station is lost. This is achieved by constantly measuring the quality of transmission between the mobile unit and several base stations.

Handover often leads to challenging system design problems. In a rough terrain, with small base-station cell sizes, handover will be more frequent. Increased mobility and traffic volume (requiring smaller cells) both contribute to the same effect. This topic is further discussed in Section 2.3.

2.1.5.3 Mobile equipment

Another aspect of mobility is related to the mobile equipment itself. Obviously, to the user, mobility also means that the equipment must be small and of low weight. The requirements are looser if the user has the use of a vehicle, although a user on foot will require his communication equipment to be as small and light as possible. This leads to challenges in design to achieve a high degree of miniaturisation and low power consumption.

For distributed systems, mobility adds an element of uncertainty in both the location of the users and how to reach a destination when the connectivity changes. The system aspects of mobility are more thoroughly treated in Section 2.3.

2.1.6 Traffic modelling

The nominal capacity of a network, or its communication links, generally is not the same as its actual user information transfer capability. In order to be able to analyse the performance of a network, either prior to its realisation or in order to find the cause of some undesired behaviour, it is necessary to have some means of modelling the information flow within the network. This is constituted by user-generated information and network-generated control information. The latter is more easily modelled as the detailed behaviour of the network is given by its specifications. But some kind of description of the user-generated information is required.

The type of analogue information that can be carried by a digital channel depends on the digital channel capacity. As capacity increases, enhanced quality can be achieved for a given service. Table 2.3 gives examples of different services and the approximate channel capacity required. The scope of this Section is not to discuss channel capacities. We assume that a fixed number of digital channels are available, and then consider network dimensioning and performance measures as seen by the

Table 2.3 Approximate bandwidth required by different services

Service	Required bandwidth
Voice	3 kHz
High fidelity music	15 kHz
FM radio broadcast	200 kHz
Television	5 MHz

network users. Circuit-switched data has much in common with the services in Table 2.3, a continuous stream of information during the data transfer phase with low fixed delay, although packet-switched data has a quite different nature.

Generally, service invocations are originated by users at arbitrary instants of time, and these instants of time depend entirely on each user's own particular needs.[1] Here we use the term service invocation as the general term of addressing all kinds of services (data, voice, etc.). The service-invocation rate is described by the interarrival times, i.e. the time between two successive phone calls or data packets. Mathematically, this forms a stochastic process.

The occurrence of each service invocation corresponds to a point on the time axis at which the communication network has to take some action. Generally, a period of time will pass before the tasks are performed (due to the network's limited capacity) and the network's ability to serve requests is defined as the service rate μ. The reciprocal μ^{-1} is defined as the service time. For voice, the service time is the same as the duration of the conversation.

When the manner of arrival and service time are of interest simultaneously, as they usually are, it is more appropriate to use the word traffic. Offered traffic is the traffic a network would handle if no congestion occurred. However, there may exist time periods where all network resources are fully occupied. Within these time periods the network is not able to handle additional service requests. Lost traffic refers to the traffic that is refused service; carried traffic is the traffic that is serviced by the network.

The unit accepted internationally for describing voice traffic is the Erlang (abbreviated E) after the Danish mathematician A. K. Erlang. Traffic is said to be 1E if the average number of simultaneous calls (either carried or offered) during a period is unity. For example, a circuit that is

[1]This is a simplified model very commonly used in communication network analysis. In real life many dependencies between service invocations are found. Typically, these might be responses generated due to information received which require some action to be taken.

occupied continuously by one call during the period carries one Erlang, which is its maximum throughput.

Traffic dimensioning of the classical telephone network is done by specifying the highest acceptable call congestion probability during the busy hour. For example, when the offered traffic is 100 E the call congestion shall be less than 0.01, and this implies that the lost traffic will be less than 1E. With the introduction of land mobile network (LMN) dimensioning becomes much more complex, mainly as a result of two factors: a radio channel is used and terminal mobility is allowed. Mobility introduces different signal and interference levels, the radio channel quality becomes time variant and the user may experience altered voice quality during a conversation. If the radio channel quality drops below a certain level, the voice connection can no longer be maintained and it is aborted by the network. Aborted connections release the granted resources before the requested holding time expires and therefore the call congestion decreases. However, most users will be of the opinion that a high abort probability is more frustrating than a high call-congestion probability.

Turning to packet-switched data traffic, offered traffic is specified by a packet arrival distribution and a packet length distribution.[1] Carried traffic is specified by throughput (the amount of information which successfully passes through the network) and the service time. The service time of packet-switched data is the end-to-end network transit delay. To simplify the discussion we use the simple network model of Figure 2.14.

Networks providing packet-switched services accept bursty traffic and use buffers to queue incoming packets. In theoretical network models each user has one queue of infinite size and one server. The server takes the responsibility of sending user packets on a radio channel. Seen by the user, new packets can be served even in time periods where the packet arrival rate is higher than the server's transmission rate. In real life, however, as the offered traffic increases the queue length also increases and at some point in time buffer overflow occurs. At this point the users must stop generating new packets, or the network will have to throw away the new packets or clear some of its buffers. In both cases lost traffic is the result. This is illustrated in Figure 2.15, indicating network throughput as a function of the mean offered traffic.

Two different throughput curves are shown. The 'without loss' curve addresses the case where the users do not accept that packets are lost; throughput and offered traffic are identical. (In practice, buffer overflow can be prevented by implementing a flow-control mechanism between the

[1]Packet-switched data traffic is not a continuous stream and has to be specified in statistical terms. However, in the following only mean values of these distributions are considered.

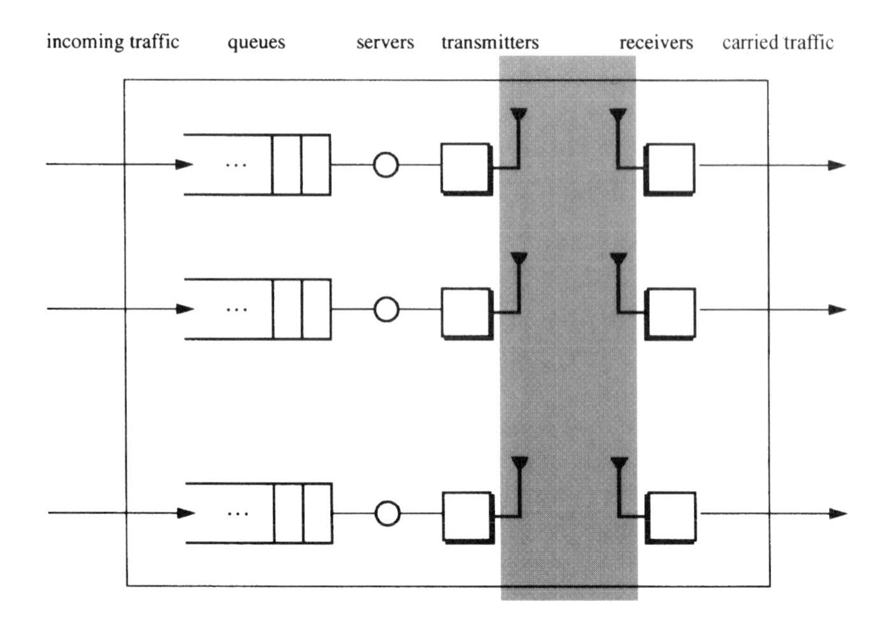

Figure 2.14 Network model for packet-switched data traffic

users and the network.) Packets sent on the radio channel are automatically retransmitted if no acknowledgement is received within a time limit from the addressed receiver. The 'with loss' curve addresses a service that does not need to guarantee delivery.

Figure 2.15 also shows the typical course of the network service time which is the sum of two components: the queuing-time delay and the server transmission-time delay. At low arrival rates packet queuing is seldom needed and the average service time is approximately identical to the

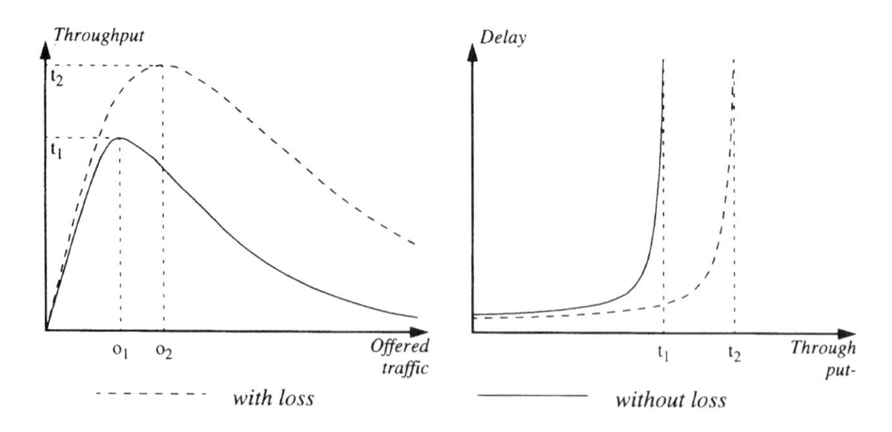

Figure 2.15 Schematic example of throughput and service time (delay) characteristics

server's transmission time. Increased offered traffic leads to increased queue length. When the offered traffic increases beyond the server's service rate, network saturation occurs with infinite queue lengths and packet delays. Figure 2.15 marks the saturation point (maximum throughput) for our two cases as $t_1$and t_2. t_2 is greater than t_1 for two reasons: 'without loss' implies that lost packets have to be retransmitted and acknowledgement packets must be returned by the receivers. In both cases extra radio-channel capacity has to be used and the network saturation point is reached at a lower traffic level. These issues will be considered in detail in Chapter 5.

2.1.7 Frequency planning

A network covering a vast geographic area with a large number of base stations does not have enough bandwidth to be able to allocate exclusive parts of the frequency band to each base station. Thus, frequency re-use is absolutely necessary. As mentioned in Section 2.1.1 the radio range is limited and a hexagonal regular structure is the most efficient one with respect to complementary but nonoverlapping radio coverage. This is shown in Figure 2.16, where the base-station cell area is a regular hexagon inscribed within the circular radio coverage area (assuming a terrain

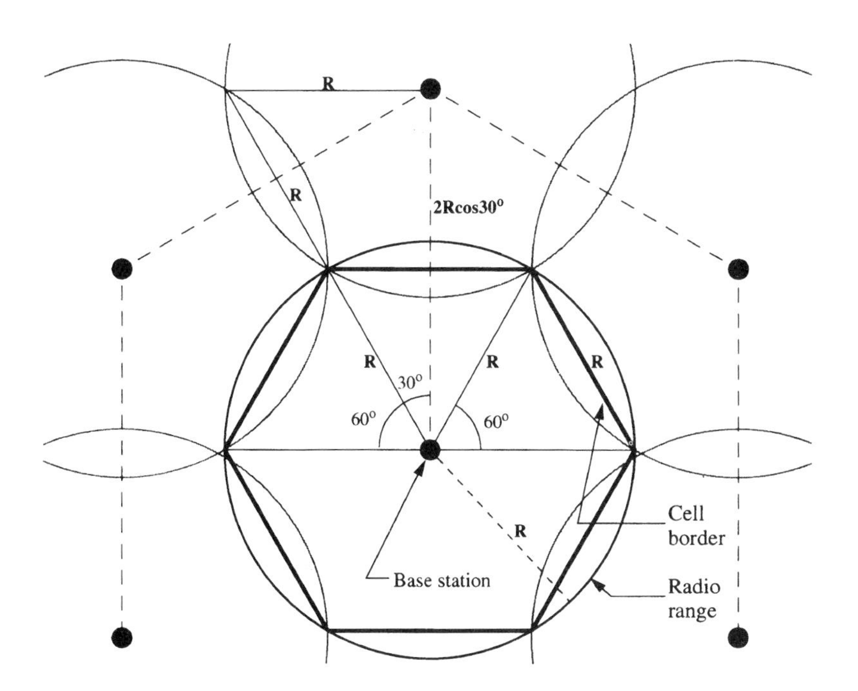

Figure 2.16 The placement of base stations for the most efficient total coverage

giving regular radio range). The side lengths of the hexagon equal the radio range R, and the base stations are placed at the angles of another regular hexagon with side lengths equal to $2 \cdot R \cdot \cos 30°$. This regular structure of base stations gives a complementary service coverage area with the minimum overlap of radio coverage areas.

The minimum number of noninterfering channels (frequencies) in this regular structure is seven. Figure 2.17 shows how to allocate the seven channels to the different base stations in order to avoid interference. Of course, this is an ideal case which is rarely found in real life. Due to terrain irregularities the cell structure (associated with the coverage area) will often be irregular, making frequency planning a more difficult task requiring sophisticated computer tools based on theoretical propagation models and information about the actual terrain (mainly topography).

The use of spread spectrum may simplify frequency planning. Due to its inherent nature SS is less susceptible to interference from other networks than a narrowband system. This makes it possible to allocate channels with less dispersed frequencies to adjacent base stations. Another option is the use of an extremely large channel bandwidth which enables the allocation of orthogonal spreading codes to the different base stations. Systems based on such an SS philosophy are being introduced in the USA [12]. The finding and use of orthogonal spreading codes are discussed in Section 4.3.4.

2.1.8 Interoperability

Interoperability is the ability to communicate between different systems or between users serviced by different systems. Interoperability can be

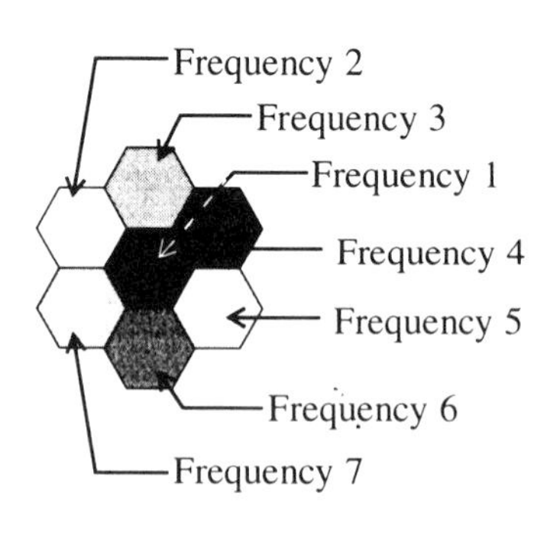

Figure 2.17 Frequency plan with efficient frequency re-use: seven cell cluster

implemented, either by making sure that all systems comply with the same set of protocols and parameters or by introducing different forms of gateways translating between the different systems. For either solution it is important to remember that interoperability between different radio networks, in the sense that a member from one network can be fully serviced by another when trying to connect to its subscribers, is much more than just sorting out the signal modulation. The whole range of network management and maintenance may be involved.

Interoperability between different public systems is commonly taken care of by a network interworking function (IWF) placed, for example, between a public switched telephone network (a fixed network) and one or more mobile services switching centres (the mobile networks), with its base stations and mobiles as shown in Figure 2.18.

The fixed public network will have to recognise a called number as a mobile number and route it to a mobile switch. The mobile switch may interrogate a location database and the call is routed via the correct base station according to the current position of the called mobile. The routing itself may take place through the fixed network. Thus a situation with the need for on-the-air interoperability does not exist, and the most important questions relating to interoperability are linked to possible inefficiencies in existing networks and their signalling systems for establishing the necessary connections.

One typical function located in the IWF between the PSTN and a mobile network is voice transcoding. Digital PSTN uses 64 kbit/s PCM coding, and digital mobile networks use a wide variety of low-rate vocoders.

For private and distributed networks such as military mobile radio systems, on-the-air interoperability seems to be a costly and never-ending theme. Never ending, because it never seems possible to agree to any standardisation which both allows for the advantages of new and upcoming technology and at the same time is not in conflict with any system just being fielded (and supposed to be in use for the next 25 years). With several generations of systems, even from different nations, the requirements for

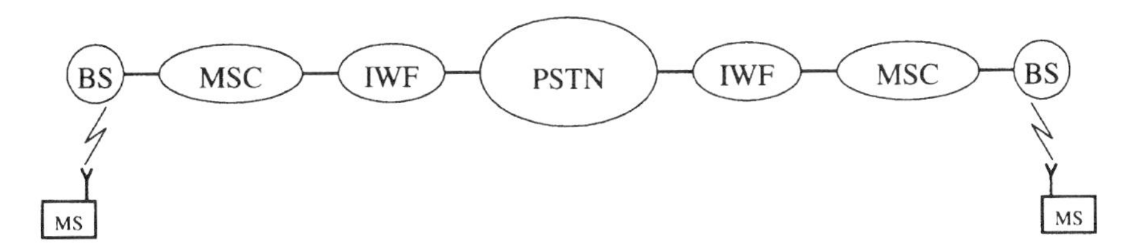

Figure 2.18 Example of public network systems interworking architecture

everybody to be able to connect to everybody directly on the air is continuously debated. The cost is particularly related to the demand for backward interoperability, meaning that a new system should be able to function along with the older systems. From time to time these demands for on the air interoperability and backward interoperability put such severe restrictions on new and more cost-effective concepts, that compromises using a specific interoperability mode have to be accepted. For new generations of spread-spectrum mobile radios, such a limited interoperability mode has for example been an FM channel which through standardised hailing techniques is able to connect to the older FM systems and to other but different spread-spectrum systems. Such a solution, called FM hailing, is shown in Figure 2.19.

2.1.9 *Compatibility*

Compatibility or the ability to co-exist becomes important as soon as the different systems are fielded. The EMC (electromagnetic compatibility) problem experienced a blooming in the mid 1990s when the new GSM cellular telephones became everyone's possession. Many stories were told about how GSM handsets affected medical electronic equipment, triggered automobile airbags and so on. How will the new spread-spectrum mobile radio system influence other systems in use and how may the systems already in use restrict the implementation of new systems?

Some of the key characteristics of spread-spectrum modulation (further discussed in Section 4.2) are simply related to how it can reject interference and the ability to operate with negative signal-to-noise ratios. Those characteristics which are related to channel path loss modelling and radio coverage in the presence of noise and interference are touched upon in this Chapter. A more comprehensive analysis of the interference-

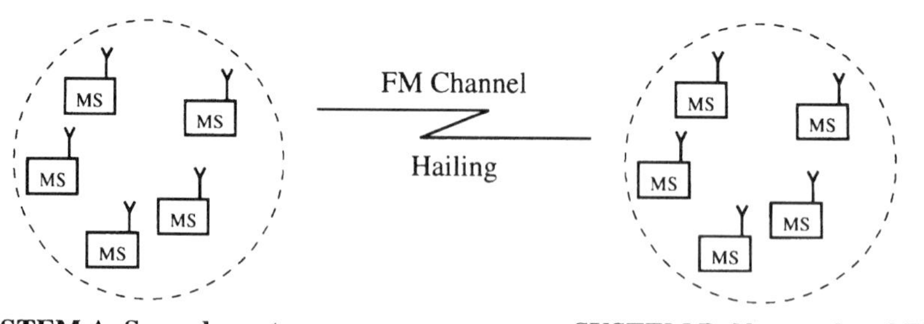

Figure 2.19 This is an example of limited on-the-air interoperability using a specific common channel

rejection characteristics of spread-spectrum techniques is given in Section 4.2.3.

However, an example of how a spread-spectrum system reacts to interference in the form of a conventional radio system is illustrated in Figure 2.20. The spread-spectrum system operates at UHF frequencies with a bandwidth of 140 MHz. The interference comes from a 2.5 MHz FM radio link signal moved around within the spread-spectrum bandwidth. The upper curves show the bit error rate as a function of S_{SS}/S_C (signal strength ratio between SS and conventional radio), for three different frequency offsets of the interfering conventional radio. The lower curves give the corresponding probabilities of detection for the three cases, also as functions of S_{SS}/S_C with $\mathrm{BER} = 10^{-3}$ and the false alarm rate below 10^{-3}.

Although it is the interference rejection capability which traditionally has been one of the advertised major characteristics of spread-spectrum

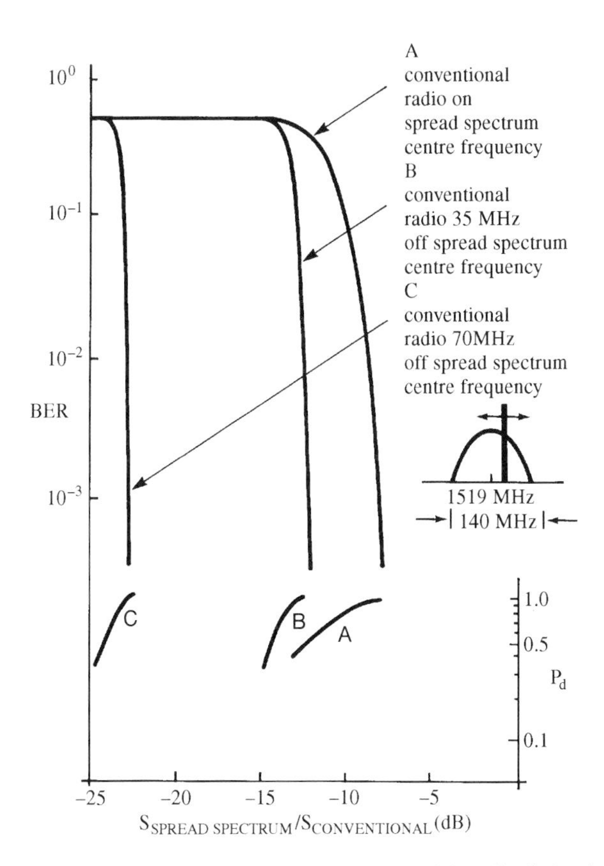

Figure 2.20 Performance of a spread-spectrum system with radio-link noise present

systems, a matter of great importance is the compatibility between spread-spectrum transmitters and existing conventional receiving systems. This matter is of importance because it is likely to indicate how accepted widescale use of spread-spectrum systems will be. It is thus necessary to develop some acceptable procedure for determining the electromagnetic compatibility (EMC) of spread-spectrum transmitters and narrowband and wideband conventional receivers. In other words, it is necessary to determine the response to spread-spectrum signals received by conventional radio systems.

The conventional radio-receiver systems will in large consist of a front end, an intermediate frequency (IF) amplifier and some baseband data processing. The front end and the IF amplifier (under the assumption of linearity) may be represented by a bandpass filter with a bandwidth equivalent to the IF amplifier bandwidth. In practice, the interfering signals are accompanied by receiver noise and the desired perturbed signal. If the perturbations are severe enough the conventional system will not be able to work alongside the spread-spectrum system. When the spread-spectrum waveform enters the conventional radio receivers one may talk about four categories of response waveform:

(i) an undistorted response which is essentially the same as the input translated to the IF frequency; this response may cause problems if the interfering signal resembles the desired signal;
(ii) an impulsive response with impulses resulting from changes in the input signal; the response is characterised by peak power and spacing of the impulse;
(iii) a noise-like response with overlapping and not distinguishable impulses, producing spiky or noisy waveforms; experiments have indicated that the characteristic of the response produced by a spread-spectrum signal may not be as close to a Rayleigh distribution as the response produced by a noise-test generator;
(iv) a CW- (continuous wave) like response with a bandwidth of less than the bandwidth of the IF amplifier and with frequency components determined by the spread-spectrum chip rate and the total number of chips in the code.

The responses are illustrated in Figure 2.21.

If the spread-spectrum transmitter system is using frequency hopping, the response of the conventional receiving system will be most sensitive to the frequency essentially on tune. The spread-spectrum system can then be treated as a single-frequency signal transmitted in bursts of length equal to the dwell time on each frequency and with a spacing given by the average time to sequence through all of the available frequencies.

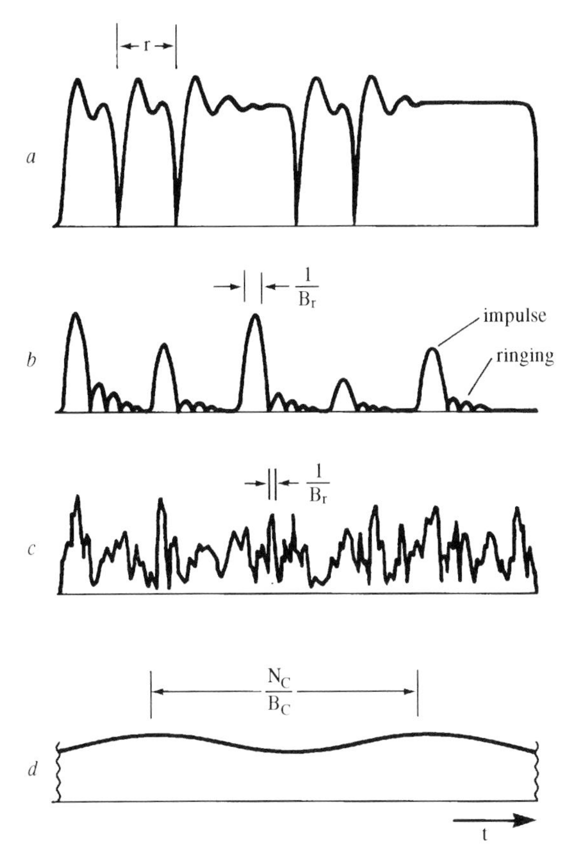

Figure 2.21 *Typical response waveforms in a conventional receiving system [16]*
a undistorted
b impulsive
c noise-like
d CW-like

Depending on the particular conventional receiver system in question, it will be important to select a meaningful measure of interference power and an appropriate interference threshold.

Let us consider another example from a real scenario. A conventional 2.5 MHz bandwidth UHF telecommunications network is operated according to ITU-R recommendations. A direct-sequence spread-spectrum system with a bandwidth of 140 MHz and a processing gain of 30 dB is to be operated in the same geographic area. The question asked, under the assumption that the response of the conventional receiver to the spread-spectrum transmitter is 'noise-like', is with what power and at what distance from the conventional system may the spread-spectrum system be operated? The acceptable amount of spread-spectrum noise depends on the conventional receiver sensitivity and the required noise-power ratio. For the particular scenario under investigation it was possible to draw a contour around the conventional receiver inside which the spread-spectrum transmitter should not operate. The result is illustrated in Figure 2.22.

2.1.10 Threat combating/electronic disturbance

The threat to a radiocommunication system depends heavily on its use. Military systems have long been subject to a wide variety of hostile threats ranging from jamming (hostile interference) and eavesdropping to intrusion. Two different terms are used to characterise the threat to our communication system from an opponent:

(i) *Electronic countermeasures* is the term used to describe deliberate actions taken in order to disturb the network. Here, the aim is mainly to disturb communications, attempting to prevent its successful operation. This may be done by inhibiting the information from flowing smoothly, but may also include the modification of network information or the introduction of false information. One of

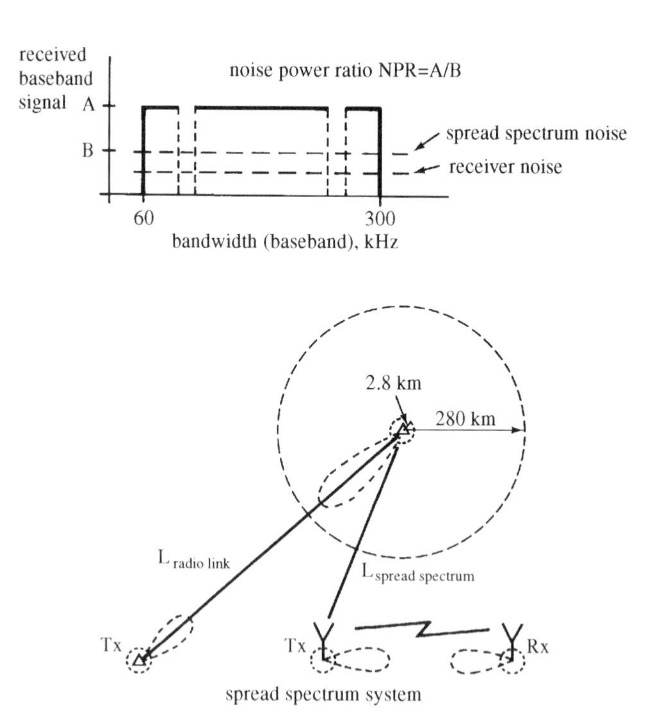

Figure 2.22 Operational requirements according to ITU-R for a conventional receiver and a spread-spectrum transmitter

the most used actions is deliberate interference, also called jamming.

(ii) *Electronic support measures* are used to characterise the activities directed at obtaining information about the network and its users. The purpose is to extract and make use of the information contained in the received victim signals and includes interception, localisation of radio stations and traffic and signal analysis. Figure 2.23 gives an example of how an electronic support system may present relevant information to the electronic-warfare operator.

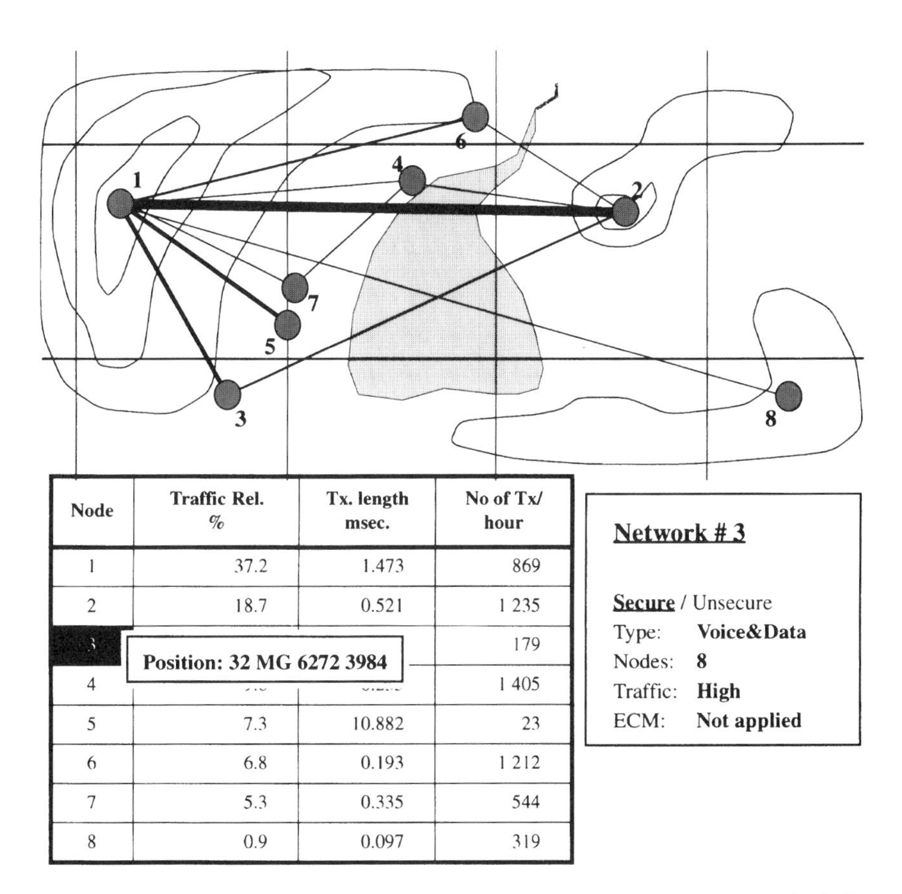

Node	Traffic Rel. %	Tx. length msec.	No of Tx/ hour
1	37.2	1.473	869
2	18.7	0.521	1 235
3	[illegible]	[illegible]	179
4	[illegible]	[illegible]	1 405
5	7.3	10.882	23
6	6.8	0.193	1 212
7	5.3	0.335	544
8	0.9	0.097	319

Figure 2.23 An example of output from an electronic support system; the circles indicate radio nodes and the width of the lines indicates amount of traffic

[1]The terms electronic counter counter measures and electronic protection measures are also used for this function. These terms cover both transmission and communication security.

In order to protect the communication system from these threats, certain actions may be taken at the design stage.[1] These actions are separated into two categories:

(i) *Transmission security*. This is the protective measures used to combat interference, network mapping (detection and localisation) and traffic analysis, and mainly consists of two methods: the ability to operate with a poor signal-to-noise ratio and the aim of hiding the transmission (reduce the probability of intercept).
(ii) *Communication security*. If the detection of the radio signals cannot be prevented, another precaution is the protection of the actual information transferred in the network. This is mainly achieved through cryptographic methods.

Civilian systems are rarely exposed to deliberate jamming, but they may be subject to unintended interference, eavesdropping, intrusion and fraud. In a communication system fraud is related to legal users (or intruders) making use of the system while not paying the cost of the use. Intrusion may be prevented through the use of authentication before allowing a user to send or receive information from the network.

Although spread spectrum inherently possesses a certain ability to hide information, e.g. through the use of long (not publicly known) direct-sequence codes, this hardly protects the information from an intelligent or resourceful opponent. The use of cryptography will prevent anyone not possessing the correct key from understanding the information content and also make identification of the transmission difficult. But it will not prevent detection of the transmission in itself. However, if spread-spectrum modulation is introduced a certain resistance to localisation, signal and traffic analysis and interference can be achieved.

In most cases, however, spread-spectrum modulation alone will not give sufficient survivability to a mobile radio network and it becomes important to develop a proper network structure taking into account the actual radio paths and the interference situation. The structure will influence the electromagnetic radiation and thus the ability to detect the signals, analyse the traffic and interfere with the network. In the case of damage to parts of the network through accidents or electronic warfare actions this should not seriously hamper the traffic. This requires a structure where calls are automatically routed and rerouted through undamaged parts of the system as indicated in Figure 2.24. This in turn requires that the network is autonomous and does not rely on a central control unit.

It is generally expected that the demand for security and survivability will increase in the future for all kinds of private as well as public mobile

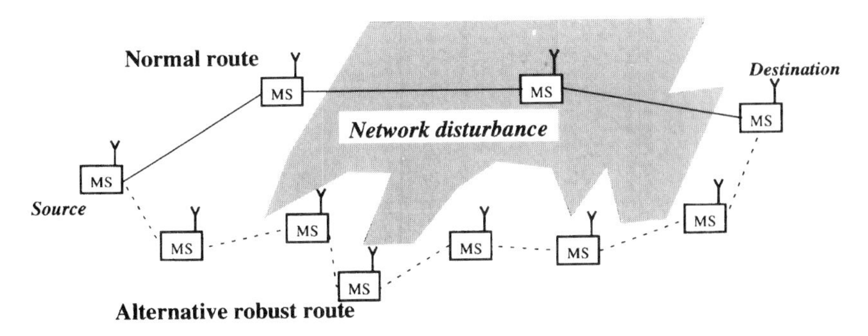

Figure 2.24 Network rerouting as a means of combating electronic countermeasures

radio networks. Security of communication systems is further discussed in Section 2.5.

2.1.11 Survivability

Survivability represents the system's ability to maintain operation when exposed to any kind of disturbance. This is not limited to electronic interference, jamming and malicious intrusion, but covers physical destruction, malfunction of network components, traffic exceeding the network's nominal capacity etc.

The survivability of a radio system is, of course, vital for military applications, but also for many civilian applications this is a significant issue. For search and rescue operations the loss of communications may result in casualties, and for commercial applications a large economic loss may be the result. Thus uninterrupted operation is an important task for all communication systems.

Increased survivability of a radio network may be achieved through a number of different efforts in order to combat different problems:

- the electronic threat (interference) may be faced with measures such as robust modulation, the use of spread spectrum, increased output power and various network functionality such as autonomy and dynamic routing;
- the network's vulnerability due to vital functions located in central network components may be reduced by simply duplicating these components;
- malicious intrusion is best combated through the application of cryptographic methods used for information hiding and/or access control with authorisation;
- in systems normally depending on a centralised base station (even with backups) survivability as seen from the user may be increased by

offering a direct mode of operation, as mentioned in Section 2.1.3. This secures a certain service availability even outside the normal service-coverage area in addition to being a backup when the base station is out of operation.

2.1.12 Integrated services system

The user desire to be able to access a number of different services in the same radio network is well accepted in these days of multimedia. But this imposes new problems for the system designer as these services may be quite different in nature, see subsection 2.1.4.1. For instance, voice and video both constitute a constant information flow, but computer applications in general produce bursts of data. Some applications require low delay but may accept loss, but others do not accept loss although they may accept longer delays. Actually, all these applications may be serviced by a single system, but the problem lies in designing efficient protocols to give a high utilisation of the limited bandwidth.

This problem must be faced as hardly any future communication system will not be presupposed to serve both voice and data applications. Flexibility is probably a keyword in this connection. In proportion to efficiency there is probably most to be gained in the design of the lower functional parts of the radio system; the physical layer with modulation etc. and the medium-access layer. This will be addressed in subsections 2.2.4.2 and 2.2.4.3.

Assume that voice is best served by circuit switching while the nature of the data applications favours a packet-switched service. The integration of these two services at the medium-access layer through frequency or time division is illustrated in Figure 2.34. The choice of an FDMA solution has one important consequence compared to the TDMA solution: the radio is unavailable for PS services as long as it is involved in a CS connection. It is possible to make the FDMA radio available for PS services in this situation, but that requires an interrupt with a short break in the CS connection. If short enough, such interrupts may cause minor deterioration of the CS service to be acceptable to the user. Anyhow, the TDMA solution probably is more elegant.

2.2 System definition and architecture

The main intention when building a system is to meet the given requirements. There is no obvious single solution to this. Many different implementations are possible. To assure the user (and others) that the system will be built according to the requirements, it is necessary to give some kind of description of the intended system before it is built. This description must serve two different purposes. First of all, the user shall be able to recognise that this system seems to be what he had in mind, and see that his requirements are met. At the same time this description shall serve as a blueprint for the system builder. Then, hopefully, the real system will live up to the user's expectations.

Such a description is usually called a system architecture. As we shall see, the architecture may have to consist of different parts, depending on how the system is viewed. The system user usually needs a totally different description than the system builder, as they consider different aspects to be important.

2.2.1 System definition

Figure 2.25 outlines a communication system with the relationship between user terminals and the subnet with mobile terminals (MT), also called nodes. User terminals (terminal equipment) are connected to the communication system through these nodes. More than one user terminal may be connected to the same node. Nodes may also exist without directly-connected terminal equipment. Such nodes may be deployed for the sole purpose of assuring network connection by relaying information. A node within the subnet does not necessarily have a connection to all other nodes in the network. But any node in the subnet may relay information intended for a user terminal connected to another network node. A subnet thus consists of a number of nodes connected in a meshed network such that any pair of nodes can exchange information, if necessary by means of relaying by other nodes.

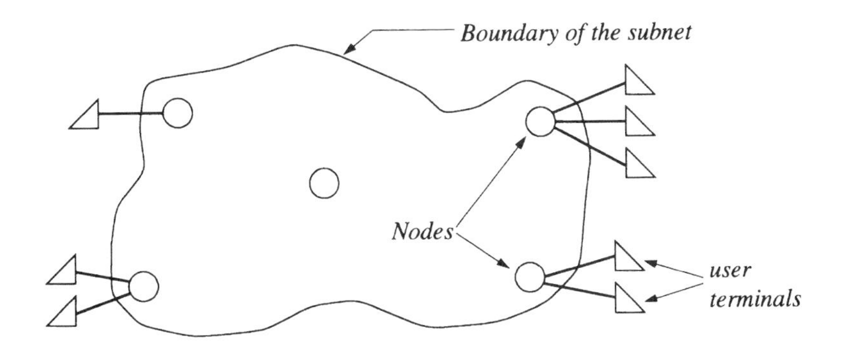

Figure 2.25 Top-level system view: relation between user terminals, nodes (or mobile terminals) and the subnet

2.2.2 Architecture

The purpose of an architecture is to give an abstract logical description to use as a reference when designing and building communication systems. The necessity of such an architecture becomes more and more important with the increasing complexity of current and future systems. An architecture can serve many different purposes in the process from designing the system, building it and using the completed system. The complexity of modern communication systems is so large that some kind of rigid structured description is required in order to avoid total chaos. An architecture must ensure that all details are controlled, and that they all fit together as anticipated. The art of communication-systems architecture has not been established for a long time. It can be dated back to the 1970s with the first results emerging a decade later. The most important result is probably the reference model for open systems interconnection (OSI). This work has been described by ISO (International Organisation for Standardisation) and ITU (International Telecommunications Union–formerly CCITT) [1].

Before going into details about the architecture, it may be useful to recognise that there appear to be three different actors on the scene: the system owner (user), the system designer and the system builder. These participants all have different needs, so it may be advantageous to make three different architectural views. In this book we will focus on the system design, thus describing the architecture from the designer's view.

It may be convenient to describe the architecture in three totally different ways:

(i) When considering the purpose of the system, a functional model is convenient. Such a functional architecture will describe the relationship with the outside world, through the services offered by the system, and the relationship to surrounding systems. It will also make a decomposition and breakdown of the required functions within a system. The simplicity of a layered model has been welcomed. Such a model is used to decompose the complicated services/functions into components with a decreasing level of abstraction. This work will be based on the ISO reference model as described in Section 2.2.3.

(ii) In addition, it may be useful to describe an architecture which identifies the main logical components of the system, called the system architecture. These are not necessarily physical components, but distributed entities or modules which must be included in the system in order for it to perform its task.

(iii) The identification of the actual physical components of the system is described in a technical architecture. This description also identifies the physical interfaces, and includes detailed technical standards to be followed. In addition, it may include information about existing systems with which the system must co-operate.

We will focus on the functional and system architectures as they are the most important ones for the system designer. The technical architecture becomes more important when approaching system implementation. Before going into detail about the different architectures for our system, let us take a look at the general communications standardisation provided by ISO and ITU.

2.2.3 OSI reference model

A reference model for interconnection of open systems has been described by ISO (International Organisation for Standardisation) and ITU (International Telecommunications Union) [1,2]. This general reference model is outlined in Figure 2.26.

This model is a functional decomposition of a communication system. The different layers perform different functions. A layer (N) offers services to the layer above ($N+1$), by enhancing the services offered to this layer (N) by the underlying layer ($N-1$).

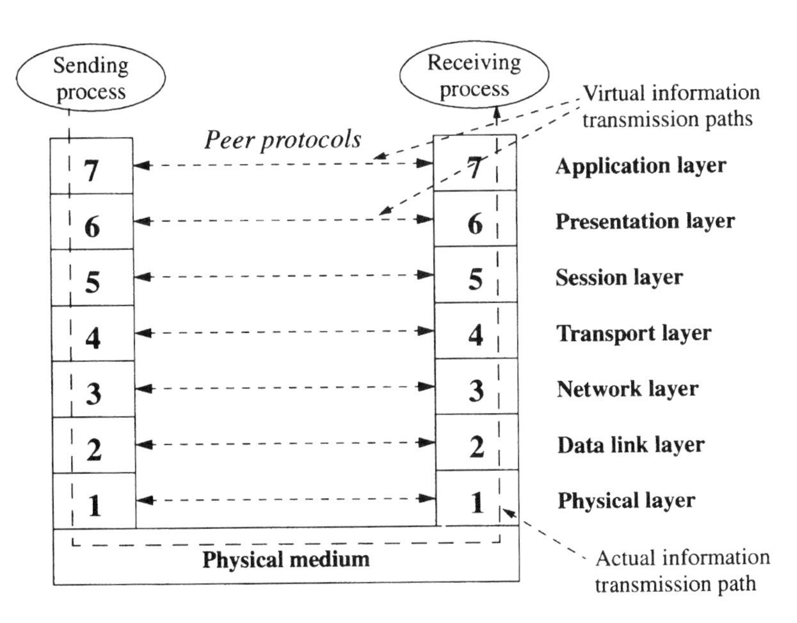

Figure 2.26 Open systems interconnection (OSI) reference model

2.2.3.1 OSI terminology

A communication system may consist of user terminals connected to a number of different communication subsystems or subnets, e.g. one terminal connected to a wire-bound subnet and another terminal connected to a mobile subnet. They are able to communicate as the two subnets are interconnected through network sublayer *3c* as described below.

Figure 2.27 illustrates the interconnection of several subnets, and introduces some terms relating to this topic. Different protocols may be used on the user access, the intranet access and the internet access. Also, different protocols may be applied to different subnets.

The active elements or procedures/functions within each of the seven layers are called entities. Entities in the same layer on different nodes are called peer entities, and the protocol describing the way in which they communicate is called a peer protocol. This communication is virtual, as the real communication is down through all the underlying layers in the sending node, via the physical medium, and up through the layers of the receiving node, as indicated in Figure 2.26.

The communication between two adjacent layers in a node (see Figure 2.28) is performed via service access points (SAP). The lowest layer, called the service provider, offers its services to the layer above, the service user. This information exchange over the interface is performed according to an agreed set of rules, and is done by exchanging an interface data unit (IDU) through the SAP. The IDU consists of a service data unit (SDU) and some interface-control information (ICI). The SDU is the information passed on to the lower entity. Together with the SDU, protocol-control information

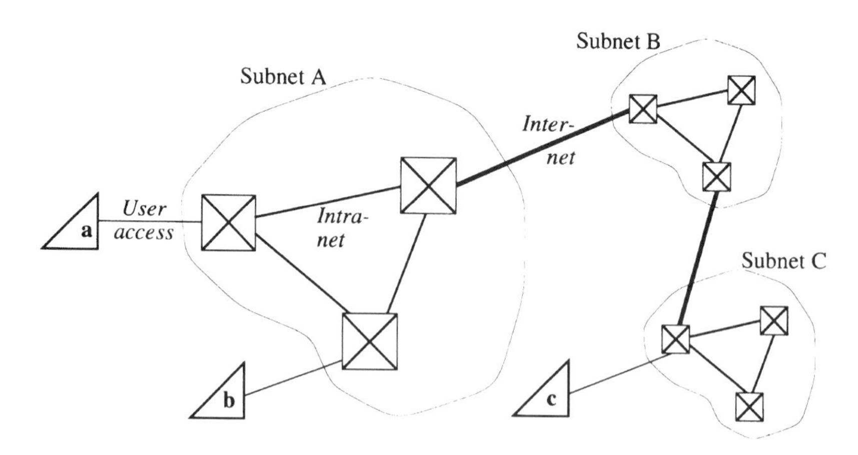

Figure 2.27 Identification of major communication system components

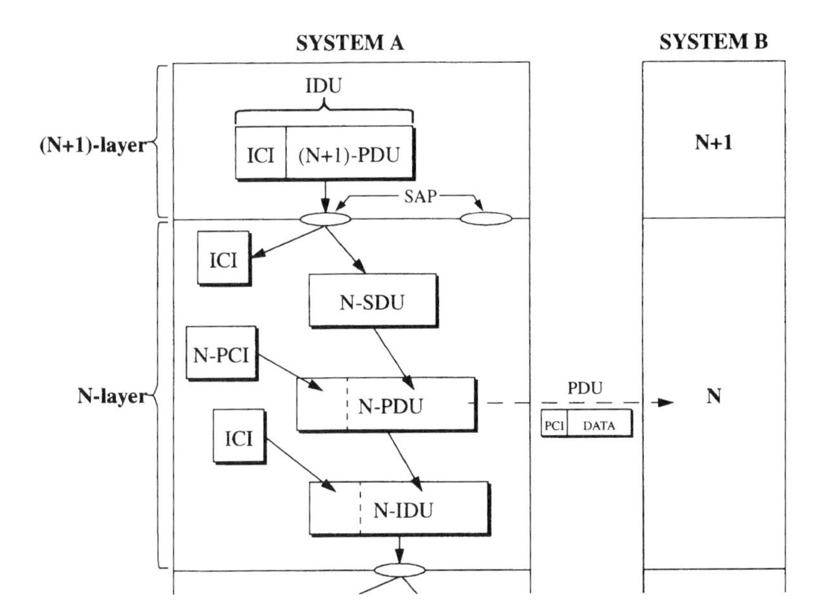

Figure 2.28 The interface between two adjacent OSI layers

(PCI) needed for the peer protocol constitutes a protocol data unit (PDU). More information on this can be found in [1], or in more understandable terms in textbooks such as [7].

2.2.3.2 Network layer

The network layer is often divided into three sublayers: *3a*, *3b* and *3c*. The *3c* layer, also called the internet sublayer, contains the functions associated with the interconnection of different subnets. *3c* may be compared to the local protocol mapping (LPM) of Figure 2.29, as it is called when interfacing the subnet protocol with the user-access protocol.

The *3b* sublayer, or subnet enhancement sublayer, is designed to harmonise subnets which offer different services. Also, this sublayer handles end-to-end control within the subnet, including flow and congestion control.

The *3a* sublayer, or subnet access sublayer, transmits and receives data and control information. It also handles the necessary relaying of data within the subnet, when the end nodes are not directly connected to each other.

The *3a* and *3b* sublayers for radio systems are treated in more detail in Section 2.2.4 and in Section 5.6. As Chapter 5 is restricted to packet switching in radio networks only, the solutions presented there are not necessarily applicable to other types of networks.

2.2.3.3 *Data link layer*

Analogous to the network layer, the data link layer is divided into two sublayers: medium-access control (MAC) and logical-link control (LLC) sublayers. As for *3a* and *3b*, both LLC and MAC sublayer functions relevant for radio networks are treated in more detail in Section 2.2.4 and Chapter 5.

The LLC sublayer is responsible for achieving reliable transmission of data over a single link (from one MS to another MS) within the subnet, with the necessary flow control, error control and retransmission.

The MAC sublayer is responsible for the channel access. The mobile subnets described in this book generally have a multiple-access channel, which means that many mobile stations may be competing for access to the same radio channel. This makes heavy demands on the access protocol, which must be efficient and fair. Much research work has been spent in the effort to study such MAC protocols for different multi-access networks in the last two to three decades.

2.2.3.4 *Physical layer*

The physical layer is responsible for the modulation and coding of the data to be transmitted on the radio channel. Also, error correction is a function which often, for practical reasons, is located in the physical layer. Radio physical-layer functions are briefly mentioned in subsection 2.2.4.2 and treated in more detail in Chapter 4.

2.2.4 *Functional architecture*

Based on the OSI reference model, an adapted reference model for a land mobile network (LMN) system is outlined in Figure 2.29. This model applies especially for packet-switched networks (or integrated-services networks), as pure circuit-switched networks might have a simpler structure. The LMN system itself offers no communication services above the network layer. Services in layers four to seven are located in the user terminal (DTE) connected to the network. As such, they are not considered to be part of the network architecture described in this book.

The communication protocol used within the LMN system usually deviates radically from the protocol used to connect the subscriber

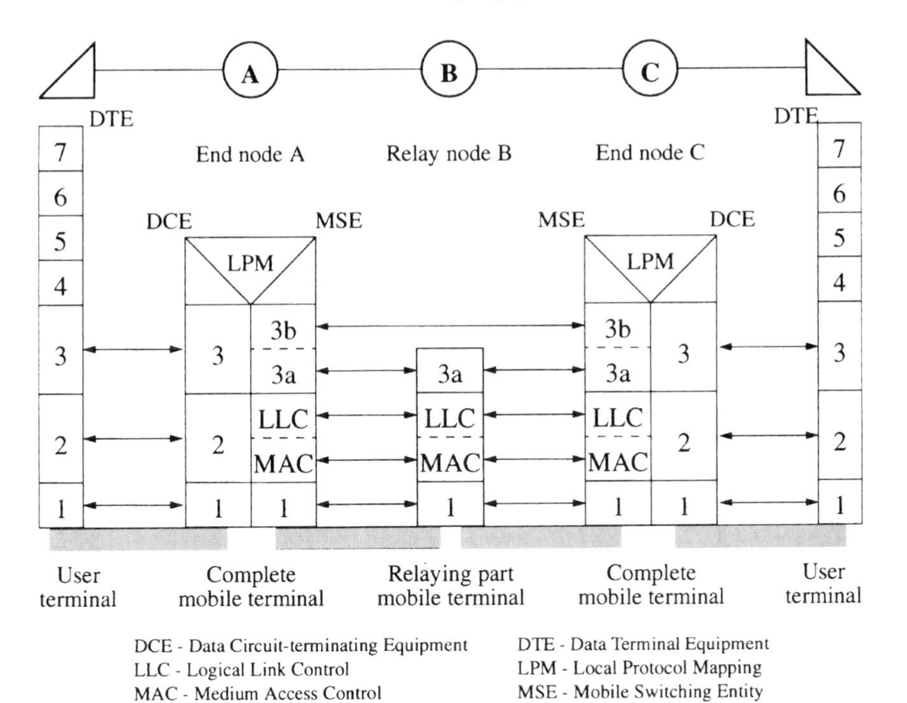

Figure 2.29 Reference model for an LMN system

equipment. The reason for this is that, compared with the subscriber connection, the LMN radio channel raises quite different challenges and thus requires different solutions. An attempt to utilise the subscriber-access protocol within the LMN system might not work at all in some cases, and would at best be a waste of the radio bandwidth resource due to its inefficiency when applied for such a purpose.

LMNs that offer both circuit- and packet-switched services often use the PS system for signalling in connection with CS services. In this case the functional model of the CS system may be simplified, if for instance no MAC layer is needed. Also, error control, retransmission and dynamic routing are often not supported by the network for CS services, leaving LLC and network layers practically empty. In such cases the required enhancement of the CS service must be (and traditionally is) performed by the end system (the subscriber terminal equipment). But some (future) LMNs may support enhanced CS services, thus requiring both MAC, LLC and network layer functions. Although the following detailed description may not be best suited to all CS systems, it is not restricted to PS systems only.

2.2.4.1 Physical medium

For a radiocommunication system the physical medium is constituted by electromagnetic waves in the radio-frequency band,[1] propagating through free space or actually most often the Earth's atmosphere. This may be a very complex matter, and signal propagation is often very difficult to predict. Generally, the signal strength decreases with the distance between the transmitter and the receiver, but no simple rules can be found to give more than approximate values. There are many factors influencing the signal propagation, and these are treated in more detail in Section 3.1.

Also, there is no static signal propagation. Two radios communicating may observe very large variations in the signal attenuation with time. This effect is strongly increased if one or both of the radios are moving. Also, there are many sources of disturbance. Due to the fact that radio waves are not contained within a limited domain, other systems operating in the same frequency band may disturb the system from a very large distance.[2] Such noise may be unintended (interference) or deliberate (jamming), see Section 3.2.

All this adds up to the fact that the radio channel may be a very unreliable transmission medium. Of course, this has a large impact on the design of the radio system which must be built to cope with this situation. This is a fact that makes radio systems differ largely from other communication systems.

2.2.4.2 Physical layer—radio

The physical layer, constituted by the radio, has to be designed to cope with the very unreliable and fluctuating transmission medium. The designer is drawn in two opposite directions. The frequency spectrum is a limited resource to be used in a best possible way. This calls for effective modulation techniques, with multilevel signalling. On the other hand, this makes the system more vulnerable to channel fluctuations and interference. These effects call for channel equalisation, forward-error control and the introduction of spread-spectrum techniques, all of which generally require increased bandwidth.

This may seem to be quite a dilemma to the radio designer. But failure to include some of these necessary bandwidth consuming techniques, in order to save bandwidth, may prove to be a disaster to the system as a

[1]No exact definition may be given. The most used frequency range today is from a few MHz (long-range HF communications) up to approximately 60 GHz (terrestrial short range or intersatellite links).

[2]This is particularly true for the lower radio frequencies.

whole. A poor quality service from the physical layer must, then, be compensated for by the higher layers. This compensation (e.g. retransmission at LLC level) may not be as bandwidth effective as if this problem had been eliminated by the radio at the physical layer.

The lesson to be learned is not to design each layer of a communication system in complete isolation. The tuning of parameters and performance of one layer will affect the layer above, relying on its services. Altogether, radio as a physical layer may provide a somewhat unreliable service, compared to other transmission media. This calls for special attention, at least when designing MAC, LLC and 3*a* protocols (in order of decreasing significance), and also to some extent layer 3*b*.

2.2.4.3 Medium-access control sublayer (MAC)

MAC protocols for packet-switched radio networks have been given much attention in recent years. This subsection is not necessarily restricted to packet switching, but the techniques mentioned are most frequently used in packet-switched radio systems. Most of them are also applicable to circuit switching (CS), although this is not given special attention. Circuit-switched radio networks are briefly handled in Section 2.4.1, and an example of a MAC reservation protocol for circuit switching is found in Appendix 2.A.

Basically, there are three different approaches to sharing a resource such as a radio bandwidth through multiple access by a number of users, as illustrated in Figure 2.30. In frequency-division multiple-access (FDMA) the channel bandwidth is separated into a number of subchannels. These subchannels all possess a nonoverlapping fraction of the total channel bandwidth, thus allowing simultaneous use of all subchannels. Another approach, time-division multiple-access (TDMA), is based on allocating all channel bandwidth to a single user, but only for a short time before handing it over to the next user. For both FDMA and TDMA there may be a need for some kind of guard band (in frequency and time, respectively) separating two adjacent subchannels, in order to prevent them from disturbing each other. Thus, it is not always possible to utilise the total bandwidth.

The third alternative, valid for spread-spectrum (SS) systems only, is called code-division multiple-access (CDMA). Here, all channel bandwidth is allocated to all subchannels for all the time. This is possible due to the fact that SS systems make use of a larger bandwidth than actually required to transfer the information. If the SS gain is sufficiently large, transmissions using different (and orthogonal) codes may take place simultaneously without (noticeable) mutual interference. The most used

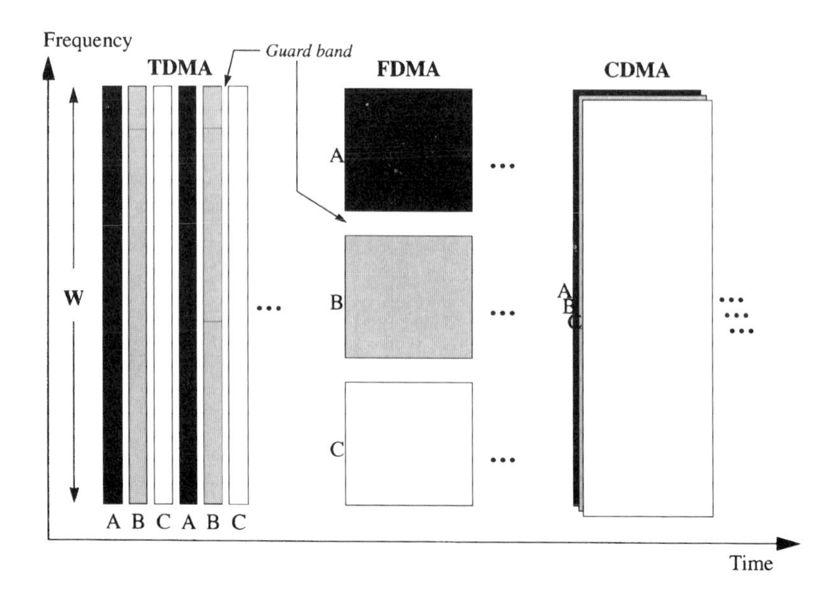

Figure 2.30 Illustrating the difference between time-, frequency- and code-division multiple-access (TDMA, FDMA and CDMA). In this example the total channel resource is equally divided between three users or subchannels (A, B and C)

technique is based on allocating a specific set of codes to each mobile station. All mobile stations know the code set of each other station in the network. Transmission to a station is performed using the code set of the desired receiver; this is called receiver-directed CDMA (RD-CDMA). Naturally, all code sets should be orthogonal. Each code set consists of at least two SS codes, or generally m codes for an m-ary system where a code is capable of representing $\log_2 m$ bits of information. Binary systems ($m = 2$) usually employ one code and its inverse to signal 1 and 0, respectively.

Also for FDMA and TDMA, SS systems require special attention. In FDMA, the use of spread spectrum may reduce the requirement to the size of the guard band. This is possible as spread-spectrum systems can work even in the presence of some interference. For TDMA systems the same fact can be exploited by designing a multiple-access protocol with an increased efficiency by allowing a larger probability of collisions than would be favourable in a narrowband system.

Centralised networks, e.g. with a base station, often employ TDMA alone or combined with FDMA, with the access controlled by the central station (e.g. base station). Mobiles apply for resources, often on a dedicated signalling channel, and the central station grants access according to demand and priority, in a totally-controlled manner. This central control may give a very effective and fair utiiisation of the radio-channel resources.

Distributed radio networks operating without a central station often use TDMA with some kind of random access. These access methods probably make the most challenging and most studied function of the radio system. The access protocol can hardly be made as efficient and as fair as in a centralised system, trying to avoid collisions which generally do not constitute a problem in a centralised system. But for many applications, in particular military but also some civilian, a distributed system which is less vulnerable and independent of the presence of dedicated resources, is often preferred. The best known random-access techniques are ALOHA and carrier-sense multiple-access (CSMA), in a number of varieties. CSMA is based on attempting to sense channel occupancy before accessing the channel, in order to avoid collisions. Both ALOHA and CSMA are described in Section 5.1. In addition, there are a number of other access techniques of this category. For more information the reader is referred to [8].

As long as all nodes are within radio range of each other, and the detection time is short, MAC protocols such as CSMA can achieve a very high efficiency combined with a good fairness. The real challenges in the design of these protocols arise in multihop networks. The most addressed challenge is that of obtaining a best possible throughput in the network. The CSMA protocol, with fairly high throughput in single-hop networks, deteriorates in multihop networks if collisions result in loss of all data (as in traditional narrowband systems). In Figure 2.31 network throughput at MAC layer (counting all hops) for a spread-spectrum system (noninterfering collisions) and a traditional narrowband system (all collisions fatal) are plotted as functions of number of hidden nodes.[1] The network of Figure 2.32 is used, creating hidden nodes by removing all but one connection from each of nodes *G*, *H*, *I*, *J*, *K* and *L* to the inner part of the network (nodes *A* to *F*). This is indicated by the dotted connections from node *H*. The network traffic is evenly distributed on all links.

Another problem of multihop networks that has not been given as much attention as throughput is fairness. In a multihop network some nodes may experience a different service quality (e.g. local throughput and delay) from the network than other nodes. In the network of Figure 2.32, the hidden node (*H*) will observe an apparently low utilised channel even at high traffic loads.[2] This will result in a higher probability of collision for

[1]Nodes that are outside the radio range of all but a few other nodes are often denoted as hidden nodes.

[2]The hidden node is unable to detect network activity from nodes other than *G*. When node *G* is not transmitting *H* assumes the channel to be free, but *G* may be busy receiving from nodes *A*–*F*.

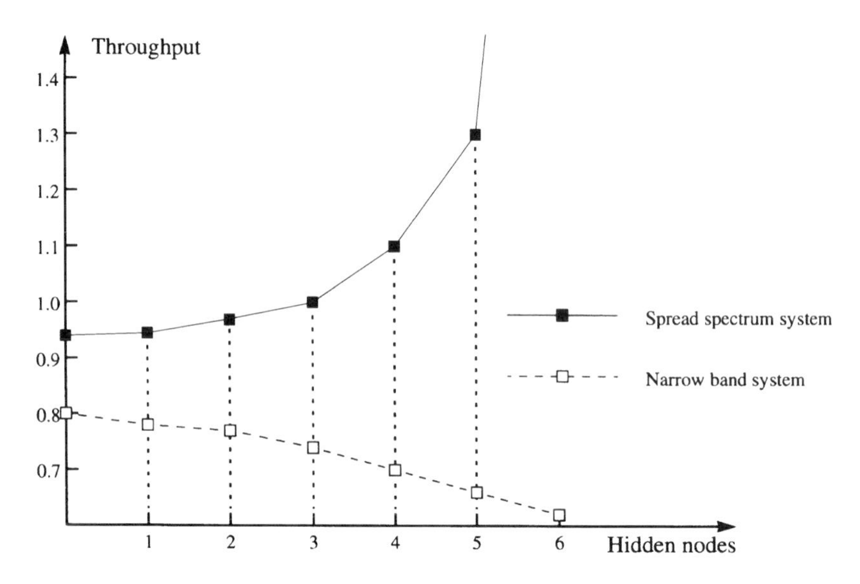

Figure 2.31 Illustrating the deterioration of the p-persistent CSMA protocol due to hidden nodes ($a = 0.01$ and $p = 0.05$)

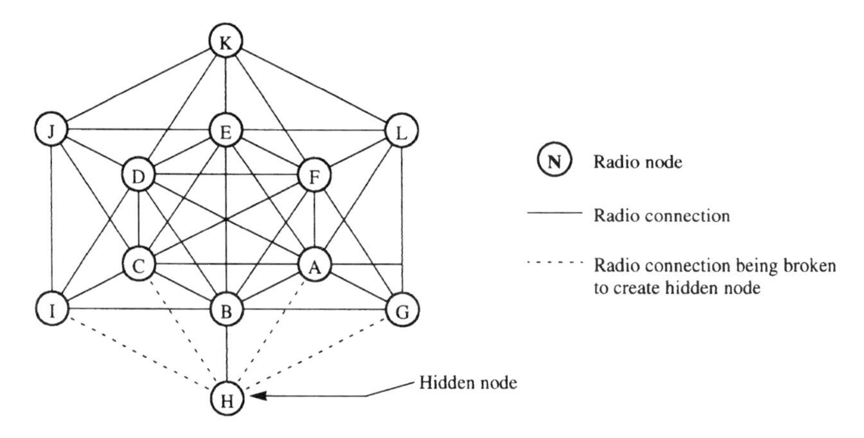

Figure 2.32 Network topology for hidden nodes simulation example. Nodes G to L are made hidden by breaking all connections except one (as indicated by the dotted connections for node H)

channel accesses from node *H*, unless the MAC protocol is designed with this in mind. An example of a MAC protocol which attempts to improve fairness is described in Section 5.1.

For direct-sequence spread-spectrum (DS SS) systems the signal may be buried in the background noise, still allowing flawless reception as explained in Section 4.2. For such systems carrier sense in the traditional way is not always possible. Activity sensing must be based on the successful correlation of the spreading codes. Traditionally, DS SS systems have initiated all transmissions with a dedicated synchronisation code (preamble). This has been particularly important for m-ary systems, as the receiver when idle would not have to search for all possible codes but only the agreed synchronisation code.

The access method based on this principle is called preamble-sense multiple-access (PSMA). Compared to CSMA this access method has one substantial drawback: detection of channel activity requires the detection of the start (preamble) of the transmission. Channel sensing (based on preamble sensing alone) at any other time during a transmission will fail. Thus, all mobiles missing the preamble will assume (wrongly) the channel to be idle. However, if the processing gain (see Section 4.2) of the SS system is strong enough the resulting increase in collision rate may not be fatal.

Previously, technology did not allow for the construction of correlators rapid enough and with sufficiently low power consumption to perform extensive code search in idle mode, at least not for mobile systems with battery power feeding. In the future, rapid and low-power technology will enable a continuous search for activity, thus allowing continuous activity detection, even for hand-held radio systems. This renders possible the use of CSMA protocols also for DS SS systems.

In wire-bound networks with one-to-one connections only, there is no need to introduce congestion control at MAC level. In a LAN or a switched radio network, on the other hand, this certainly makes sense. A MAC protocol such as p-persistent CSMA, analysed in detail by Tobagi [15], may be tuned for optimum performance as a function of offered traffic. The analytical expression for the throughput (S) of p-persistent CSMA is rather complicated, but may be simplified when the propagation delay is set to zero. This is given by eqn. 2.7, where G denotes normalised offered traffic and $0 < p \leq 1$.

$$S = \frac{Ge^{-G} \cdot (1 + pGX)}{G + e^{-G}} \tag{2.7}$$

where

$$X = \sum_{k=0}^{\infty} \frac{(1-p)^k \cdot G^k}{(1-(1-p)^{k+1})k!} \tag{2.8}$$

This is plotted in Figure 2.33 and shows that the performance of this protocol with respect to offered traffic varies with the degree of persistency. A dynamic optimum protocol may be constructed if the network traffic load can be estimated by each node. By reducing the persistency, p, when

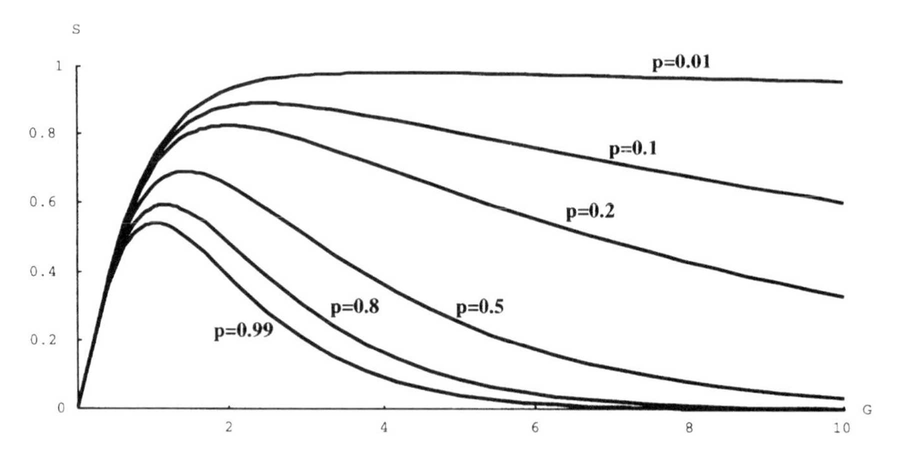

Figure 2.33 Channel throughput S for the p-persistent CSMA protocol in a narrowband system with zero normalised propagation delay (a = 0)

the traffic load increases, congestion is controlled. However, there is one problem. Nodes with diverging comprehension of the network traffic, will use different values of persistency. A node suspecting the network to be stressed (high traffic load) will be more likely to refrain from transmitting at the instant the channel becomes idle than a node with fewer neighbours sensing a lower traffic load. This way the access protocol may not be completely fair.

When a radio network is required to offer a combination of PS and CS services, this may be designed in several ways. If the radio channel has sufficient capacity to handle all services, it may be divided into time slots. Some slots are allocated to PS services, possibly using a random-access protocol with competition, while other time slots are allocated by reservation (no competition). On the other hand, if radio bandwidth is insufficient to support both services separate channels may be used for CS connections. These two schemes are illustrated in Figure 2.34. Preferably, the resource allocation should be dynamic, according to service demand at any time.

2.2.4.4 Logical-link control (LLC) sublayer

For CS systems the LLC layer is usually found to be more or less empty. But as mentioned in subsection 2.1.4.8 an error-correction function may be applicable. In PS systems, on the other hand, a more interesting list of LLC functions should be considered.

The LLC layer of a radio network generally has to operate with a much higher bit error rate (BER) than most other networks. This will have an impact on the protocols and the tuning of the protocol parameters. For

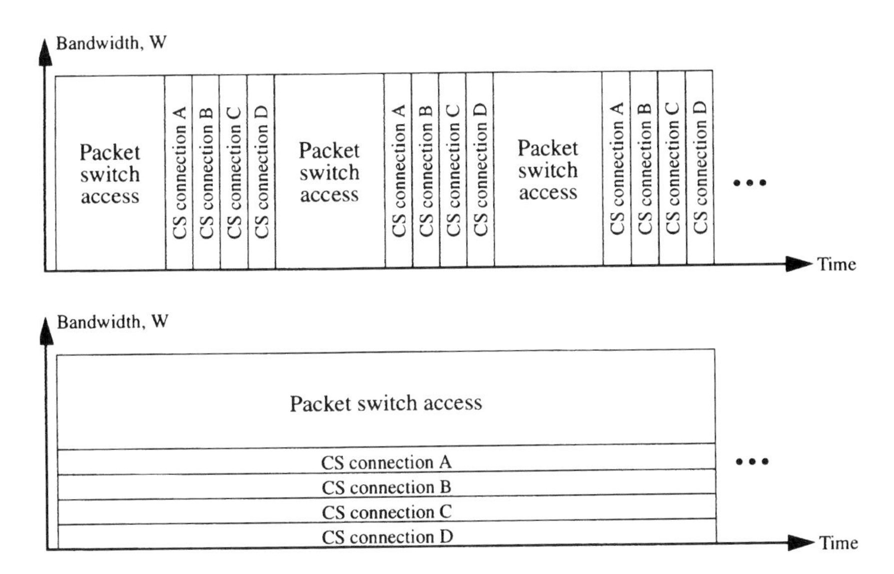

Figure 2.34 *Integrating PS and CS services into the same network, two different implementation schemes: frequency and time divided*

instance, an improved error detection mechanism may be necessary. This was mentioned when discussing the residual-error probability (REP) which is one of the quality of service (QoS) parameters described in subsection 2.1.4.8. An error-detection code length of 16 bits is standard for most wire-based communication links. As seen from Table 2.2, this code length may readily give a REP of the order of 10^{-5} for a poor radio connection. This is hardly acceptable for most applications; thus a longer code will be required for such radio networks. Also, in order to maintain a certain throughput in the case of high BER, radio systems to be used for data communication must be equipped with some kind of error-correcting codes. Block codes such as Reed–Solomon are frequently used for this purpose. This topic is discussed in Section 4.4, but let us consider an example to illustrate the benefit of forward-error correction (FEC).

Figure 2.35 shows an example comparing two different radio link strategies, error detection only (CRC) with retransmissions and forward-error correction (Reed–Solomon) in combination with CRC. The first comparison is done using the same radio bandwidth and data rate. Obviously, pure error control is advantageous at all but very high bit error rates due to the simple fact that the use of FEC results in a reduction of the available data rate. However, if the bandwidth remains the same, but the data rate is increased (to compensate for the reduction of the available data rate) in the case of FEC, we see that FEC may be favourable.

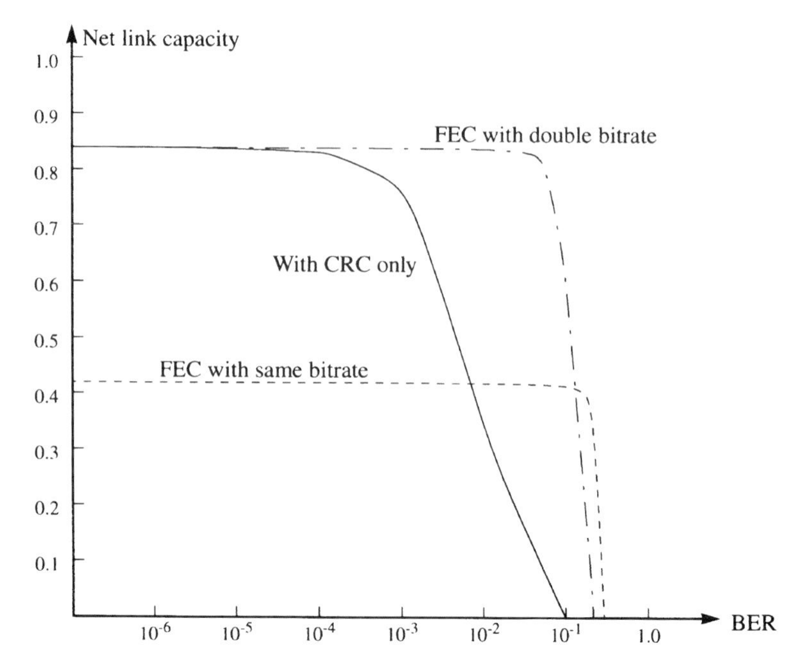

Figure 2.35 Throughput as a function of bit error rate (BER) for a link with error detection only, compared to a link with FEC; block length = 100 bits, 16 bit CRC or RS(200,100) combined with CRC; white (random) noise is assumed

Networks with a number of nodes sharing a single channel may differ substantially with respect to optimisation of the LLC protocol parameters, compared to networks with dedicated links. For instance in packet switching, the number of outstanding packets[1] on a link probably should be kept very low, even limited to one. There are three reasons for this:

(i) Due to the more unreliable channel there is a high risk of packet loss. If a go-back-N protocol is used, this may result in a reduced throughput since the loss of one packet may result in the retransmission of several packets. A large number of outstanding packets should be combined with a selective-repeat protocol.

(ii) Since many nodes share the same resource, allowing a link to have more than one outstanding packet may also influence the fairness. At least this is true if a node is allowed to send many packets in the same

[1]The term *outstanding packets* is used to indicate the maximum number of packets that a node may send without receiving an acknowledgement of reception. When this limit is reached the transmitter must wait for acknowledgement and eventually take some action to retransmit if no ack is received.

channel access. Also, in many cases, the packet sent needs relaying by the receiving node in order to reach its final destination. The attempt to send more packets would then compete with the relay node for channel access, possibly resulting in a higher risk of collision.

(iii) A restriction in the number of outstanding packets may also serve as a cheap flow control mechanism. There may be situations where a centrally-located node becomes the natural relay for a number of other nodes. This relay node is more susceptible to being overloaded with relay packets if all its neighbours are allowed to transmit several packets without asking for permission.

A number of different acknowledgement procedures have been studied. The protocol that has been found to be the best for the networks described in this book is called stop-and-wait with immediate acknowledgement (SWIA). This protocol requires that the receiver as soon as possible after successful reception of a packet issues an acknowledgement. It might then be beneficial to allocate a certain time slot in the access protocol for acknowledgements only. Of course, this protocol will not be efficient unless a rapid error correction/detection with subsequent issue of an ack(nowledgement) may be performed. Also, the ack signal must be as short as possible. The SWIA protocol is further discussed in Section 5.4.

Another function located in LLC is concatenation. This is a function which effectively may be used when a node has several (short) packets to be sent to the same link destination. By combining these packets at the LLC layer they may be treated as a single block by the MAC layer. In this way more effective channel utilisation may be achieved, provided that the allowed number of outstanding packets is sufficient to allow this operation.

Precedence at LLC layer will only be a local function. A prioritised packet ready for transmission in a node will be able to precede all other packets located in the same node, irrespective of their intended destinations. Pre-emption hardly makes any sense for packet switching as there is little to be gained from interrupting a packet being transmitted. At LLC it still may make sense as long as the present packet is still waiting for channel access at the MAC sublayer.

2.2.4.5 Sublayer 3a

One of the most important functions found at the 3*a* layer of a mobile radio PS system is routing, i.e. the process of finding the path from the source to the destination. When dynamics is introduced the quality of the different links may change frequently, affecting the choice of the best possible routes. In a distributed network the great challenge is to keep the routing tables updated in all the nodes, without consuming the total network capacity for maintenance messages only.

This is further studied in Section 5.7. Other important topics are relaying and saturation control. Relaying is further discussed in Section 5.6.

2.2.4.6 Sublayers 3b and 3c

The *3b* network sublayer is responsible for intranet end-to-end control, including retransmission strategy, duplicate filtering, priority mechanisms and end-to-end flow control. For most radio networks the layers below *3b* have added sufficient abstraction so that *3b* is more or less independent of the transmission medium. A few radio-network-specific topics related to *3b* are briefly discussed in Section 5.6. Sublayer *3c* functions relate to internetwork communications (see Figure 2.27), which is not a topic to be discussed in this book.

2.2.5 System architecture

This Section will present a general system architecture for an LMN/PMR/PAMR system. This is done through decomposition of the system into its logical main components. These logical components are not necessarily physically recognisable components, but some overlap may occur.

A top-level outline of a system architecture for an LMN system is shown in Figure 2.36. Mobile stations (providing services to mobile subscribers) are assigned to mobile subnets, mainly depending on their location and area of radio coverage. Mobile stations within the same radio subnet may communicate directly (if within radio range), via one or more MS relaying, or via the base station and the switched infrastructure.[1]

Depending on the presence of equipment (radios and infrastructure) there are different levels of autonomy. Obviously, the minimum configuration of an autonomous LMN system is the case where two mobile stations are communicating directly; additional MSs may be connected to this network. As long as all radios are within radio range to communicate (if necessary using one or more of the other radios as a relay), we say that the network is on the lowest level of autonomy.

[1]Most (mobile) cellular telephone systems do not support direct communication between mobiles; all communication must pass through the base station (and the infrastructure) even when the communicating mobiles are within radio range of each other. Generally, direct communication between mobiles is supported by PMR/PAMR systems.

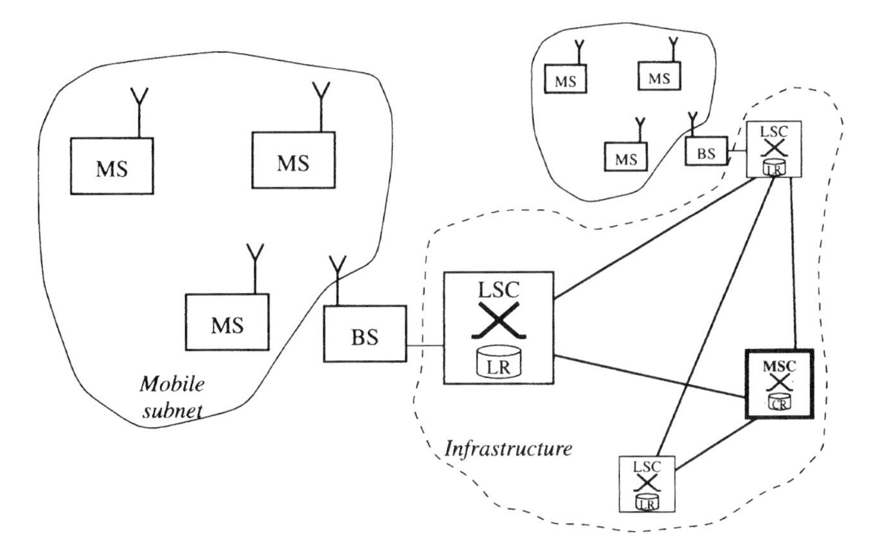

Figure 2.36 Outline of system architecture with interconnected mobile subnets

The highest level of autonomy requires the presence of an infrastructure with base stations, switches and registers and centralised control. In Figure 2.36 this is indicated by a number of interconnected switches. The trunked connections between these switches are usually not executed by the use of LMN, but by a radio link or some kind of wired connection.

Communicating radios may be connected to different switches. Each switch may also operate more than one mobile subnet. This may be ascribed to the fact that a single mobile subnet has limitations, decided or practical, as to the number of subscribers and available capacity. In order to obtain sufficient quality of service, the subscribers may have to be reallocated to a number of mobile subnets, interconnected by the infrastructure, when the traffic load within one subnet approaches the available capacity.

Figure 2.36 presents a general system architecture. This architecture can be broken down further into more detail, also indicating how the components are interconnected and showing the communication paths and interfaces. Let us first study the basic components in more detail.

2.2.5.1 The mobile subnet

The mobile subnet is composed of a number of mobile stations (MS), optionally communicating with the outside world through a base station (BS). Figure 2.37 shows a mobile network connected to an infrastructure,

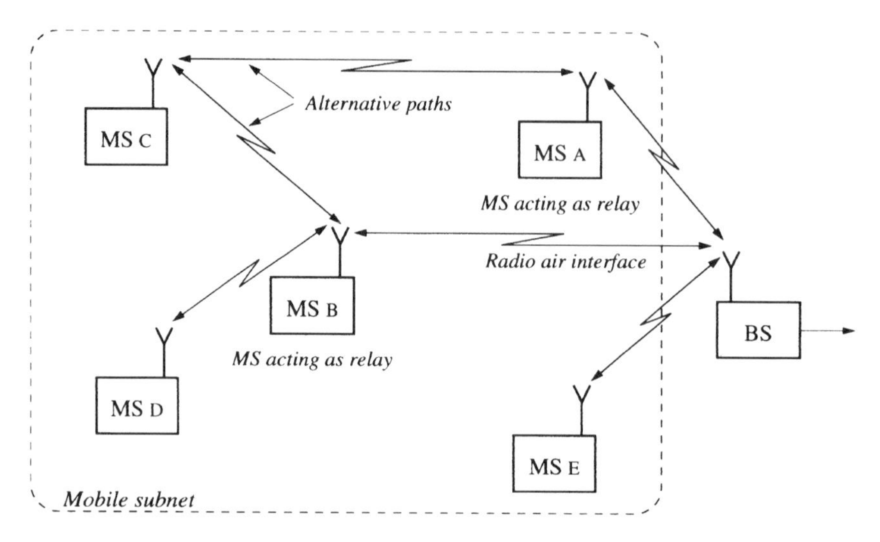

Figure 2.37 A mobile subnet connected to an infrastructure via a base station. Both nodes A and E are acting as relays due to their location

and Figure 2.38 shows an autonomous mobile subnet. The bidirectional arrows all indicate potential communication paths using the radio–air interface. In contrast to cellular telephone systems, most PMR/PAMR systems allow direct communication between mobiles. Also, relaying between two mobiles outside the radio range (of each other) is a very useful facility.

Due to the fact that control and maintenance traffic for many distributed systems may increase rapidly when the number of mobiles in the subnet increases, there is at least a practical limit (if not a system limitation) to the maximum number of mobiles in a subnet. Also, a subnet has a maximum traffic throughput capacity, and delays increase rapidly when the traffic

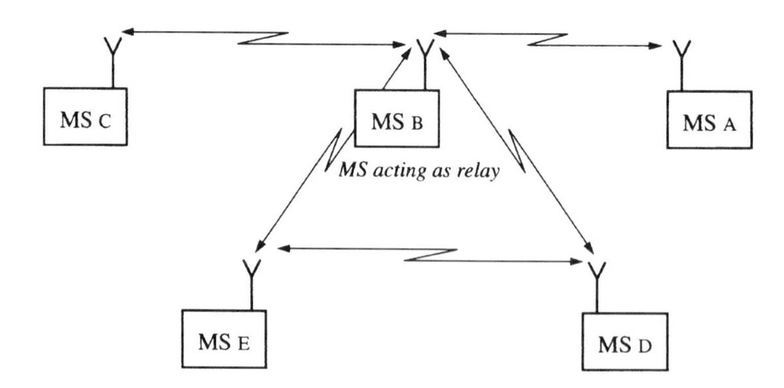

Figure 2.38 An autonomous mobile subnet operating in direct mode

load approaches the maximum capacity. These are two reasons why it may be necessary to divide users located in the same geographic area of coverage into more than one subnet. These (geographically) overlapping subnets may be connected to the same base station or to different base stations. As long as the infrastructure is completely intact (not split) this usually poses no practical problems to the subscribers. In many instances it may be convenient to split the mobile subscribers in such a way that groups of mobiles which have a large fraction of internal communication (within the group) are assigned to the same subnet. This should be done for two reasons, both to reduce traffic but also to increase survivability in case of infrastructure malfunction or destruction.

Often, the base station is not considered to be part of the mobile subnet. But for many systems the BS contains a radio part which may be very similar (at least functionally) to a mobile station. In these cases it may be convenient to consider that component of the BS to be part of the mobile subnet.

2.2.5.2 The mobile station

As previously indicated, the MS consists of a mobile termination part and, optionally, one or more items of external terminal equipment. In addition, most realisations also include internal terminal equipment, at least for the termination of voice services. A thorough composition is given in Figure 2.39.

With reference to Figure 2.39, the mobile terminal consists of the following logical components:

- a radio frequency (RF) part to handle the physical communication with the rest of the network via the radio–air interface (RAI). The RF part communicates control information with the C&M part and data with the switching element;
- a small switching element is needed in order to direct information to the correct receiver, e.g. data from the RAI may be intended either for the C&M, the internal terminal equipment or some piece of external terminal equipment;
- the control and maintenance (C&M) part is always needed in any system. It handles control towards the RAI, the switching element, the internal and external terminal equipment and the subscriber authentication (e.g. using a subscriber identity module–SIM). A database (DB) is closely linked to the C&M element, holding all necessary information. C&M may be performed locally by the subscriber, or remotely from the network;

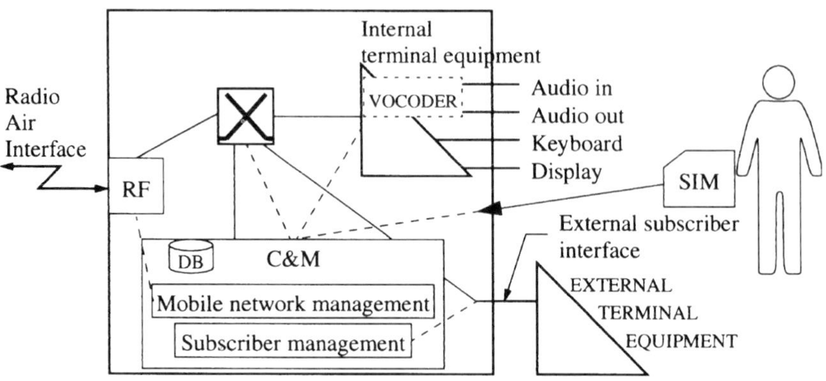

Figure 2.39 Composition of the mobile station

- the internal terminal equipment (ITE) (found in most MSs) may terminate both voice (most common) and data services. The ITE is controlled by C&M, and communicates with the switching element and the subscriber. Towards the subscriber it contains a vocoder for audio in and out via a microphone and a loudspeaker, and a keyboard for data input and a display for data output to the user;
- one or more items of external terminal equipment (ETE) may be connected to the MS through an external subscriber interface, standardised or proprietary. Both voice and data services may be terminated by the ETE, which may have the same user interfaces as the ITE (voice in/out, display and keyboard) and other optional functionality;
- subscriber identification and authentication may be performed by using a SIM card combined with a password entered through the (internal) keyboard.

2.2.5.3 *The base station*

The base station as outlined in Figure 2.40 consists of two different types of logical component, the base transceiver station (BTS) and the base station controller (BSC). A BS consists of only one BSC, but the number of BTSs may be from one to a large number (only practical limitations), depending on capacity demands.

With reference to Figure 2.40 the BSC consists of the following logical components:

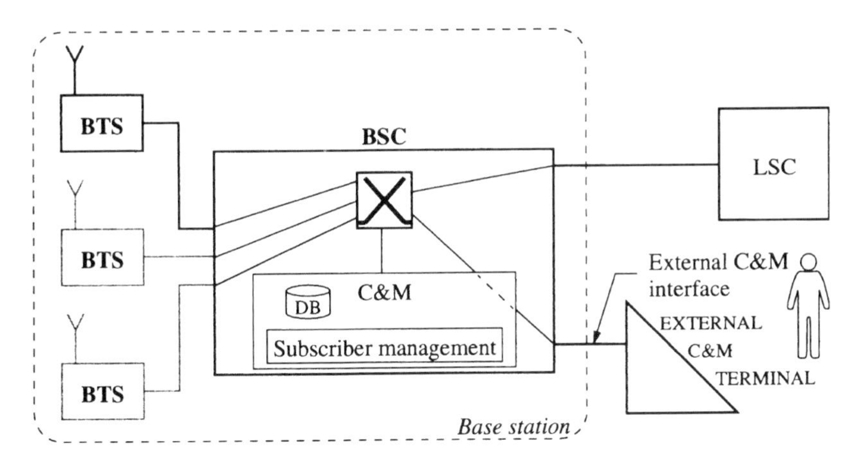

Figure 2.40 Composition of the base station

- a switching element used at least to separate user traffic from C&M data. User data switching is usually best performed locally in the BSC, but optionally this may be done in the external switching infrastructure;
- a control and maintenance system to control the mobile stations and their connected subscribers. This system may also have superior control of the BTSs. Manual control and maintenance may be performed locally through an external terminal connected directly to the BSC (may optionally be considered part of the BS), or remotely through the infrastructure from a central network control. The C&M system will also contain a database with e.g. information about all locally-connected MSs, a user-location register and control information for all BTSs;
- external connection interface to the infrastructure through a local switching centre (LSC), and interfaces to a number of BTSs.

Each BTS may be capable of handling one or more channels, and communicates with one or more MS. The BTS may be (almost) identical to an ordinary MS, or it may be realised as a totally different component. The BTS may be based on a standard MS even if it must be capable of handling several time-divided channels, but this is not possible if it must be capable of handling more than one RF channel in parallel. The BTS as outlined in Figure 2.41 is composed of the following components:

- a switching element is needed at least to be able to separate C&M traffic from user data, in the case of remote control. Switching between

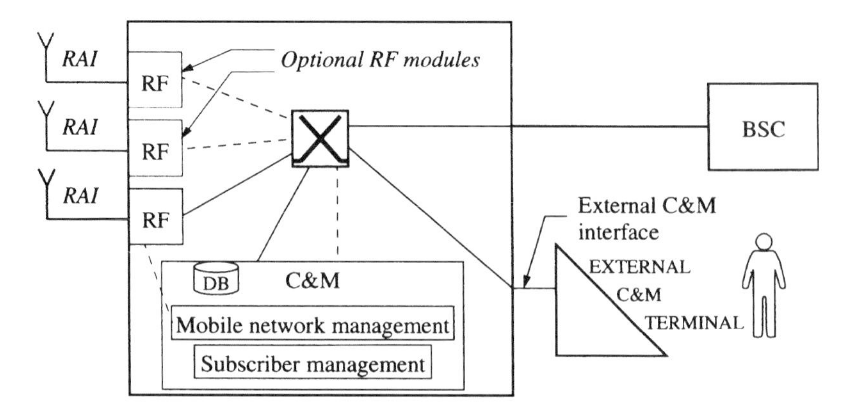

Figure 2.41 Basic configuration of the base transceiver station

two mobiles connected to the same BTS may be performed locally in the BTS or in the BSC or the infrastructure, depending on the chosen implementation. Local switching capacity increases the system survivability, but may lead to a more complex implementation;

- a control and maintenance function with databases and user-location register;
- one or more RF modules with radio interfaces to mobiles; the optional RF modules are for increased capacity;
- optionally, an interface to an external C&M terminal; C&M may also be performed remotely, but this local function may be useful at least for maintenance, trouble shooting and repair.

The architecture outlined for the base station may differ substantially from the architecture of many cellular telephone networks such as GSM (as indicated in Figure 2.42). The reason for these differences is that networks such as GSM are designed for large subscriber communities requiring a large number of network components. The reason for this distributed architecture with its standardised interfaces is that this gives network operators

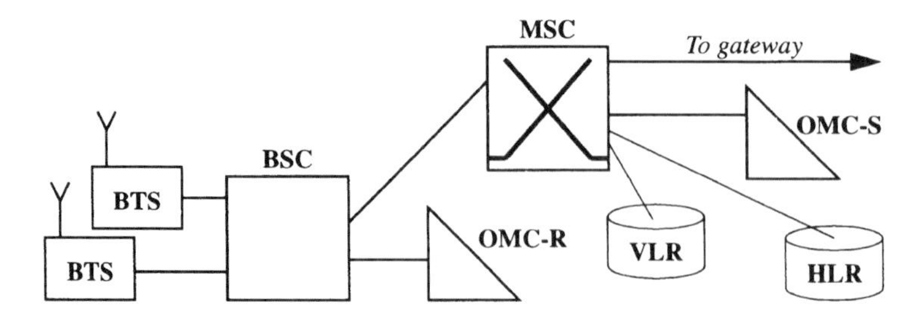

Figure 2.42 The traditional GSM system architecture with separate components

greater flexibility in choice of vendor. But also for GSM we see an evolution towards a more integrated architecture, not primarily for increased survivability but for purely economic reasons designed for smaller network operators in low-traffic-density markets [9]. A higher level of integration reduces the initial system cost when the large capacity of the original architecture is not required.

2.2.5.4 The infrastructure

Each base station normally is only able to provide service availability in a very local geographic area. A switched infrastructure is required in order to interconnect all base stations to give a regional, national or global communication system. The infrastructure may be completely integrated with the mobile communication system, or it may be a separate general communication system offering its services to the mobile system (e.g. the PSTN providing leased lines to the mobile system). The infrastructure is generally composed of a number of high-speed switches interconnected by high-capacity trunks (wire, radio link, optical fibre or satellite relay) as indicated in Figure 2.43.

In an integrated system wire-connected subscribers and mobile subscribers may communicate directly through the nearest switch (not the BS) as they are both subscribers in the same network. When the fixed and mobile subscribers are members of disjointed (but interconnected) networks, all communication must go through a dedicated gateway which connects the two networks in a service manner. This might be true even if the mobile

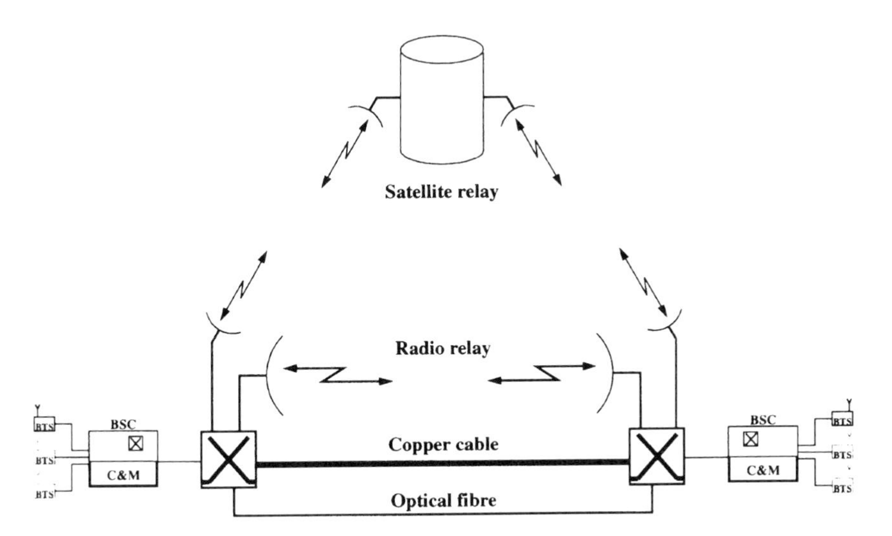

Figure 2.43 Outline of infrastructure interconnecting geographically-scattered base stations with alternative physical trunk media

network uses the fixed network to interconnect its base stations. Of course, a completely integrated system will possess an enhanced survivability as no central gateway is needed.

In the absence of the infrastructure, increased area of coverage and/or increased capacity may in many systems be achieved by (more or less) directly interconnecting two direct-mode subnets, as shown in Figure 2.44. For many systems the interconnection unit is not needed, as two mobile stations may be connected directly back to back. The two interconnected radios and the interconnection unit, which may be no more than a cable, together constitute a relay in the traditional sense.

Figure 2.45 outlines an example of a centralised mobile network. All base stations are connected to local switching centres (LSC). The LSCs are interconnected by the infrastructure. In addition, there is a main network-switching centre, the mobile services switching centre (MSC). The MSC will contain a central register (CR) holding information about the location of all mobile subscribers, whereas each LSC will hold information about its locally-connected subscribers only.

2.2.6 *Technical architecture*

The technical architecture is intended for the system builder. Its purpose is to give guidelines for system building, that is, technical rules for building the system as outlined by the functional and system architectures. The technical architecture may consist only of references to a number of technical standards where such exist. Such standards are often given for interfaces between physical system components, such as X.25 for packet switch access from a subscriber (DTE) to a network switch (DCE) or X.75

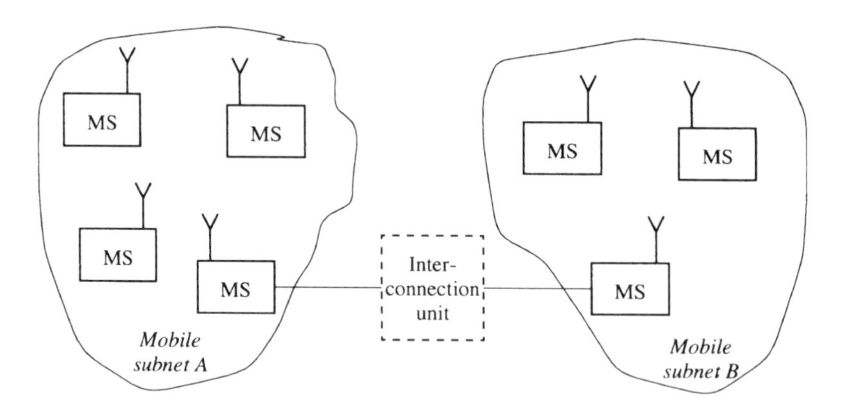

Figure 2.44 Direct interconnection of two mobile subnets. Optionally, a dedicated interconnection unit may be required

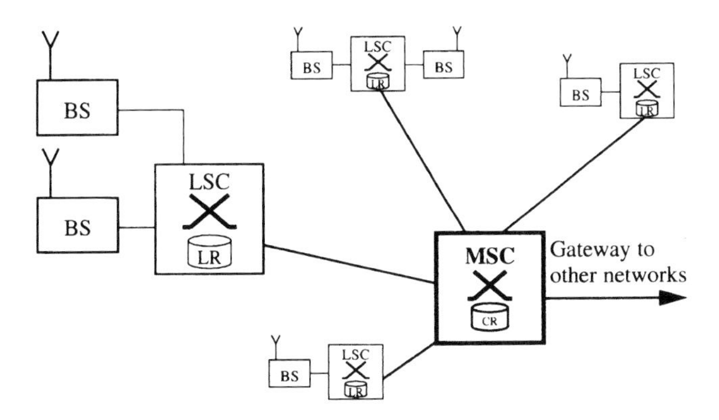

Figure 2.45 The infrastructure elements

between different network packet switches. The purpose of such standards is to enable the system operator to purchase system components from different producers and have the components successfully work together. This not only reduces prices due to competition, but also gives an improved guarantee of finding spare parts or components for expansion.

Another purpose of the technical architecture is to identify the physical components. These are not necessarily the same as the logical components. There may be technical or economic reasons to realise for instance one logical component as two physical components, or *vice versa* integrate several logical components into one physical component. The latter case may be driven by the rapid evolution of miniaturisation in production technology.

2.3 Mobility

Mobility was briefly discussed in Section 2.1.5, introducing the terms roaming and handover. Roaming will not be discussed in this book as this is mainly a topic for public mobile networks; general information may be found in books such as [13] giving an extensive introduction to GSM. In this section we will discuss different aspects of local mobility. There are two distinct situations which may arise: the subscriber may relocate between two successive communication sessions, or the subscriber may in fact be moving while communicating. The first case is related to the problem of transfer of control when a mobile loses its connection to the base station and must reconnect to another, or in distributed networks changing the network connectivity necessitating the modification of routing tables throughout the network. The second case is related to fluctuations of channel quality and handover between base stations. Finally, mobility for the user may also have an impact on the physical design of the mobile equipment. Size, weight and power consumption are keywords here.

The most important aspect of mobility is service availability at all possible user locations. This generally requires a well developed network with carefully located base stations or other means of ensuring service availability. This is mainly an issue for the service provider, but the system designer may simplify this task by including such facilities as automatic relaying by the mobile stations, increasing the effective base station area of coverage.

2.3.1 Relocation to alternative base-station

Each mobile station is connected to the network via a base station. When the mobile station moves out of the radio range of the BS, it must search for another base station. Preferably, this search should be a continuous process in order to reduce the time during which a mobile is out of contact with the network. In a cellular telephone system this is an automatic process which the user does not care about. In a PMR network there may be reasons not to let this process be completely automatic. Each user might have reasons to stay in contact with one (or a limited number of) base station, e.g. because this BS is allocated to the group to which he belongs. This is particularly important in military systems where the infrastructure is exposed to frequent break-down periods. Such a rule may ensure that a group of users who need to communicate with each other, and thus were allocated to the same BS, are still able to communicate even if the BS loses contact with the network.

When a mobile relocates to another BS this must result in the updating of location registers so that incoming calls are forwarded to the correct base station. In a centralised network correct call forwarding may be performed by the MSC knowing the location of all mobiles in the network. But for distributed systems, each local switching centre may only have knowledge about its locally-connected mobiles. Call routing must then be performed by the LSC searching for the addressee at the other LSCs. Another alternative, which may function in small networks with low mobility, is to have all LSCs constantly updating each other with information about all mobiles. Then the challenge is to obtain consistency between all the location registers.

2.3.2 Dynamic routing

In a distributed network the effort to design a routing system that can cope with a high degree of mobility is challenging. The user-location register actually constitutes a distributed database. There are two extreme ideologies; each mobile might possess information about all other mobiles in the network with their subscribers, or each mobile might know only its local subscribers.

The first case requires the frequent exchange of routing information between all mobiles. This may be done on a regular basis or it may be event driven (whenever a connectivity change occurs). This method is associated with two challenges: how to ensure consistency between the routing databases of all nodes, and how to minimise the amount of control information exchange required. This is further discussed with an example on the distribution of PROP (packet radio organisation packets) in the introduction to Section 5.7.

A key performance parameter for a packet-switched radio system is its ability to establish the path between any two mobile users. The user will therefore be interested in automatic routing and rerouting based on channel quality. The definition of quality may be a complex and integrated figure reflecting the different mechanisms which are likely to degrade performance such as propagation effects and interference. Different performance parameters that might be used to establish the best route in a packet-switched radio network are presented in Section 5.7.1.

2.3.3 Channel fluctuation

User mobility is going to make the radios experience a quickly-changing channel quality and also introduce the Doppler effect. This is related to the requirement that the radio in fact should be continuously operable during movement. This may put extreme demands on the ability of the radio to adjust to, and combat, different propagation effects. This can probably only be partly solved and only through the use of powerful signal-processing schemes supported by the latest technology. The optimal receiver for a digital signal corrupted by additive white gaussian noise is a matched filter, matched to the transmitted signal. However, if the channel introduces intersymbol interference the matched filter must be followed by channel equalisation in order to improve the receiver performance. This is, however, a suboptimal solution. Ideally, the receiver should contain a matched filter, matched to the transmitted signal convoluted with the channel impulse response. The matched filter in the receiver must then be estimated adaptively because the channel may be unknown and time-varying. This is illustrated in Figure 2.46.

Recently, smart antennas able to continuously change or adjust their radiation pattern to match the changing propagation conditions and cancelling interference, are being developed for mobile radio systems. So-

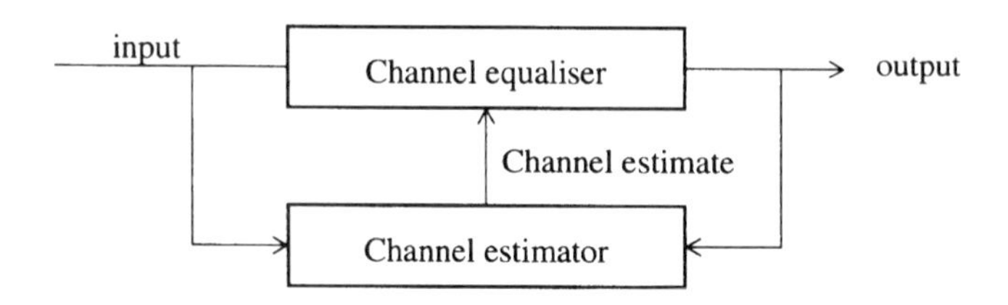

Figure 2.46 Block diagram for estimating the channel characteristics for a certain equalisation algorithm

called antenna-nulling systems placing nulls in the directions of sources of interference or jammers have been well known for many years in military systems. The combination of antenna nulling and spread spectrum has been particularly attractive because of the possibility of recognising the friendly signal as a unique spread-spectrum signal and the spread-spectrum receiver's ability to operate with signals below the noise level.

As an example a real scenario with a number of conventional VHF radio networks operated with varying loads of voice traffic and interfering with a direct-sequence spread-spectrum data link is used. The conventional VHF nets belong to military field units and the scenario is taken during a phase of preparation for attack and the attack phase. Figure 2.47 shows the power distribution for the interfering signals in the form of probabilities of different numbers of interfering signals being received above a certain signal level at the spread-spectrum receiver. These levels will for the particular spread-spectrum system correspond to a certain performance in the form of error rates and information throughput. Since the interference most of the time is made up of one conventional radio signal, an antenna-nulling system able to place one null in the direction of the interferer would be expected to greatly improve performance. Figure 2.47 shows that this indeed was the case. Antenna systems are not further discussed in this book.

Smart antenna solutions may thus offer increased gain to overcome propagation loss, as well as spatial filtering of the signals enabling reduced cochannel interference or jamming and, finally, the technique may be used to mitigate multipath effects. It will, however, be considerably easier to apply such techniques to a centralised system at the base stations, than to apply these techniques to the mobile radios in a distributed system.

2.3.4 Handover

Mobility may take the transceiver from the coverage of one base station to the coverage of another. Frequent breaks in communication may be experienced if the hold time is long compared to the time within the coverage of

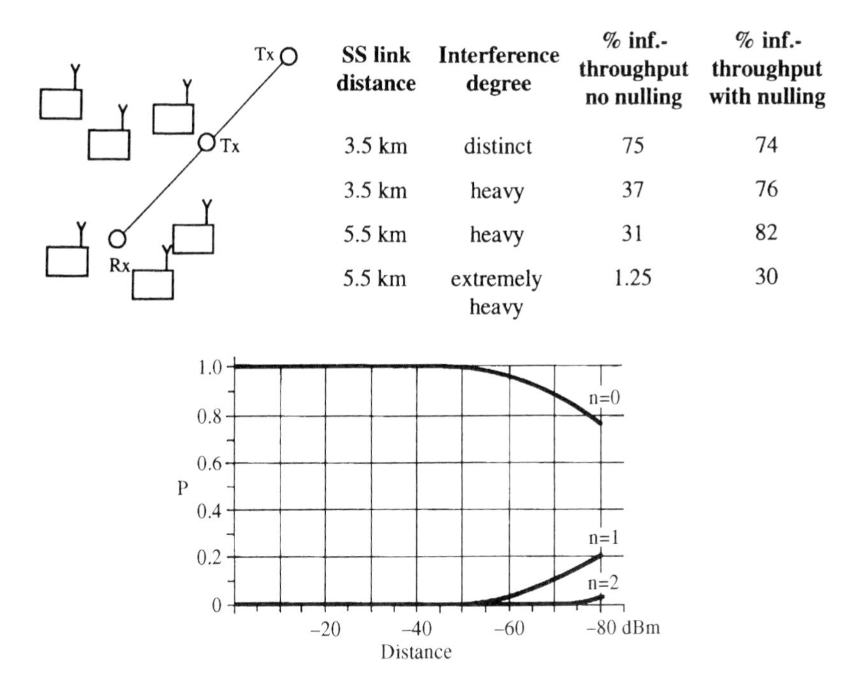

SS link distance	Interference degree	% inf.-throughput no nulling	% inf.-throughput with nulling
3.5 km	distinct	75	74
3.5 km	heavy	37	76
5.5 km	heavy	31	82
5.5 km	extremely heavy	1.25	30

Figure 2.47 Experimental scenario showing the interference-rejection properties of an antenna-nulling system integrated with a spread-spectrum radio system. The interfering conventional VHF net consists of 25 kHz voice channels and the spread-spectrum VHF data radios have a bandwidth of 5 MHz and a processing gain of 21 dB

one base station. If this is the case, the user will require some proper form of handover function from the network for uninterrupted communications. On the other hand, if the mean time spent within one cell is equal to or larger than the hold time, handover may not be required. This latter has been considered to be the case for TETRA as well as a number of PMR systems. LMN systems designed for densely-populated areas require continually smaller cell sizes in order to meet the increasing traffic requirements, thus increasing the need for efficient handover mechanisms.

The problem of handover is usually related to circuit-switched services only. For PS systems the change of base station (due to loss of contact with the connected BS) generally results in the retransmission of connectionless packets and the disconnection of any logical connections for connection-oriented services.

The amount of signalling required for handover will thus vary depending on the grade of mobility of the users. The handover procedure itself is to be hidden from the user in the best possible way and thus there is a tendency in new mobile systems to let the mobile stations themselves control most of the process, as they know the local radio environment best. At present it is common to distinguish between three phases involving both the network and the mobile stations as shown in Figure 2.48.

In first-generation mobile telephone systems such as NMT (Nordic Mobile Telephone) the handover function was exclusively controlled by the base station. In the second-generation systems such as GSM the handover function is handled through a co-operation between the BS and the mobile. Other LMN systems with even smaller cell sizes, e.g. DECT (Digital European Cordless Telephone), implement a handover procedure where the mobile itself decides which BS to connect to. This is due to the fact that small cell sizes lead to very frequent handover, thus not leaving enough time for the signalling required by a joint (BS and mobile) or solely BS-controlled handover system.

2.3.5 Mobility and user equipment

The requirement for mobility may be interpreted both as an ability of the radio to establish the necessary connections anywhere within a specified area and a need for the radio itself to be easy to carry and be of small size and weight, powered by batteries and using a small, efficient antenna. The demands on the individual mobile radios in a distributed system and the associated strong need for advanced and powerful signal processing stay in direct contrast to the wish for small size, low weight and long battery operation.

Behind the notion of low-power design lies much more than reduced supply voltages and modern miniaturised circuitry. In fact, at the logic and

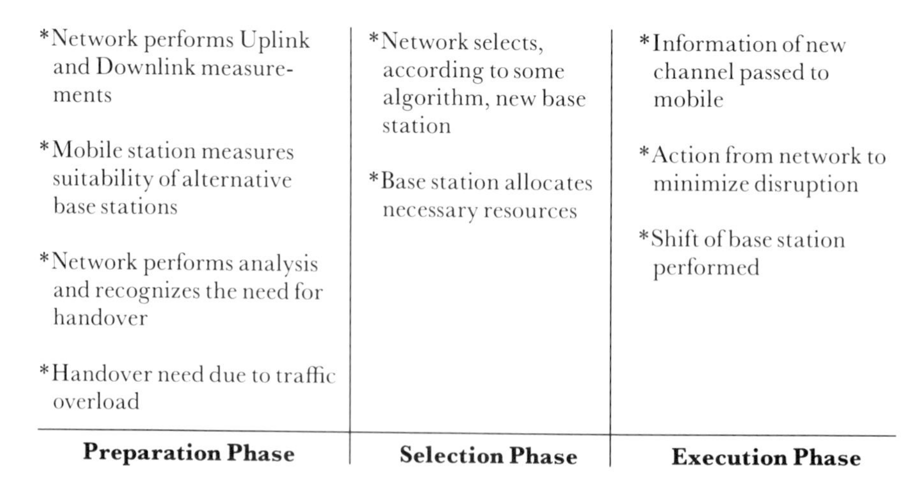

Figure 2.48 The handover process for a mobile system using base stations

device level only moderate power savings may be achieved. The work to achieve low power must therefore start right at the time of system specification and address points like functional modularisation and partitioning, algorithms and architecture.

The first point is directly connected to the operational use of the radio system in the field, deciding at what time the different functions of the system are used and grouping them in a way which may enable complete modules to be put in a state of power down. The purpose of partitioning, that is, deciding what functionality should be realised in hardware and what should be solved by software, is always to implement as much functionality as possible in software. Generally, it will be advantageous to delay partitioning as long as possible but this requires a close co-operation between hardware and software development in order to change between hardware- and software-based solutions in the mode process. The second point addresses the efficiency of the required signal processing and involves issues like complexity and concurrence. The final point covers issues like the use of parallelism in processing, pipelining and the need for redundancy.

The need to combat propagation effects like fading by output-power margins and the need for small antenna sizes with generally low efficiency also challenges the battery power budget. In fact, the antenna is often the most critical system component for achieving sufficient radio range. Some users of the radios, as for example rescue personnel and field soldiers, may not be able to optimise the use of the radio for best possible propagation conditions and they also need a short and robust antenna design. A person carrying a radio may go down on his knees or lay down flat, both positions which may be very challenging for the antenna design. The antenna design may also have large impact on the mechanical construction of the radio set itself. For instance, one antenna design to cover subbands within the complete VHF band (30–300 MHz) is a problem. It is possible, but the gain will not be the best in all parts of the band.

2.4 Switching in wireless networks

As introduced in the previous sections, switching in wireless radio networks is associated with a lot of problems not found in fixed networks. The challenges are even greater in distributed networks with no master control centre. In this section we will discuss some of the problems associated with circuit-switched systems and then terminate this topic. In addition, packet-switched radio systems will be introduced, but only briefly, as this is the topic of Chapter 5.

2.4.1 Circuit switching

Most modern cellular telephone networks provide circuit-switched (CS) services which are based on TDMA or FDMA, or a combination of these two access schemes. Future systems may, to a larger extent, also use CDMA.[1] Appendix 2.A outlines a simple TDMA protocol, called packet-reservation multiple-access (PRMA), suitable for CS networks. Common to all such systems is the fact that all challenges (after the connection is established) are related to mobility and the necessary handover process between base stations. As long as the mobile is stationary (stable signal quality) during an active session, the task of providing such services is rather simple and problem free.

The evolution of multimedia applications will require an increased range of services from wireless networks. Due to the inherent nature of the different applications, both PS and CS bearer services must be provided in order to obtain an efficient utilisation of the available bandwidth. Of course, a PS teleservice may be provided, based on a CS bearer service, but this is far from the most efficient approach. The combination of PS and CS should be performed in a flexible manner, allowing the network to adapt to fluctuating demands. Reservation mechanisms are frequently used in centralised networks. Most systems using TDMA allow the mobile to use spare time to monitor other base stations. Thus, the mobile can choose the base station with the best signal quality and initiate a handover procedure. In the process of switching from one base station to another (in the midst of an information transfer), it is important to keep bit synchronisation without bit slip. For voice and video transfer some disturbance during handover may be accepted, but for some CS data services the consequences of bit slip may be greater.

Distributed single-hop networks have no central base station to control the reservation process. This increases the conflict potential. Two nodes attempting to reserve the same channel have no way of knowing about the conflict unless they are informed by another node. This may be implemented by requiring an acknowledgement from the called node, using the (hopefully) reserved return channel. Lack of acknowledgement may be interpreted as an unsuccessful reservation. Both PRMA and dynamic TDMA are acceptable protocols, assuming that the called node uses the reserved slot in the first frame to acknowledge the reservation. Collisions must be resolved in some manner minimising the chance of a new collision at the next channel-reservation attempt.

[1]The first commercial cellular networks using CDMA commenced operation in 1995/96. CDMA-based PCS networks were also operational from 1996/97.

Most CS protocols reserve a channel for the duration of the call. For voice services we know that there are many natural breaks of short duration. In addition, the two (or more) parties involved in the conversation seldom speak simultaneously. Transmission during these voice breaks are a waste of channel capacity. But an access protocol that is based on resource reservation only when voice activity is detected leads to a requirement of very rapid (re)establishment of the CS connection at the start of a voice spurt, in order not to deteriorate the signal quality. Such protocols may be designed to work in centralised, and probably also in distributed, single-hop networks.

Another level of challenge is introduced when turning from single to multihop distributed networks in order to extend the service coverage area. In a multihop connection, channel capacity must be reserved for each radio link when establishing the connection. The greatest challenge will be to find an efficient reservation protocol that can handle the problem of hidden nodes. The protocol outlined in Appendix 2.A will not work particularly well in a multihop environment since the network nodes will have a different comprehension of which slots (channels) are reserved and which are free. Probably some kind of reservation protocol actively involving both parties of the connection must be introduced. Such a protocol may be based on the same principles[1] as found in the MACAW (multiple-access collision-avoidance for wireless LAN) protocol [14] and in the IEEE 802.11 standard described in Section 5.1.

But bear in mind that this is not sufficient to guarantee proper network operation in a dynamic network when mobiles start moving, and thereby affecting the network connectivity, before the established connection is terminated. Trying to design an efficient channel-reservation protocol for such dynamic multihop circuit-switched networks will probably give you a lifetime challenge.

2.4.2 Packet switching

Modern cellular telephone systems also offer some kinds of packet-switched service to the subscribers. There are two different ways of obtaining this service:

(i) The PS service may be applied on top of the CS connection that these networks offer. This means that the CS connection is established for the duration of the session but only used whenever there is

[1]Before transmitting, a short RTS (request to send) reservation packet is sent to the intended receiver. A response CTS (clear to send) from the receiver is required before transmitting the data packet.

a packet to send. Obviously, this is a waste of network resources as most PS applications do not have a constant stream of packets to exchange.

(ii) The separate signalling channel frequently used to establish CS connections may be used for the transmission of short self-contained messages. Due to the limited capacity of such signalling channels, the use of this service should be limited so as not to deteriorate the establishment of the CS connections.

Another way of providing PS services in a radio network is letting a number of radios share the same channel, reserving the channel for each packet separately. This topic is introduced in Section 2.4.4.

2.4.3 Message switching

The concept of message switching is not found replacing circuit or packet switching at network level in any modern communication system. There are probably no benefits of message against packet switching, at least as long as PS allows reasonably long packets. For instance, the X.25 standard for PS has an optional maximum of 4096 bytes of user data in a packet. In systems with a long round-trip delay, e.g. geostationary satellite systems, the possible benefit of longer transmissions is outweighed by allowing a larger number of outstanding frames/packets at link/network layer. Generally, message switching is implemented as a message-transfer system at the application layer, according to ISO's OSI reference model.

2.4.4 Packet switching in radio networks

In this book we understand by the term packet switching in radio networks a network where a number of radios share a common channel through a packet-based access protocol, offering packet-switched services to its users. Strictly speaking, this term covers everything from centralised to completely distributed networks. The complexity of such networks increases with the introduction of distribution, multihop and mobility/dynamics.

Sections 4.2.3, 4.2.4 and 4.4.3 will address some of the radio-design characteristics that will have great impact on the design of efficient and stable PR protocols. Chapter 5 will be entirely devoted to discussing different aspects of distributed packet-switched radio networks. These probably are the kinds of PS networks associated with the greatest design challenges.

2.5 Network security

Network security was briefly addressed in Section 2.1.10. As mentioned, security of communication systems originally was a military domain. But the widespread use of communication systems for all kinds of information (personal information, corporate information, financial transactions etc.) probably makes civilian systems just as exposed as the military systems, and then by a larger community of attackers. Our modern society has become increasingly dependent on computers and communications. This makes us very exposed and enables a criminal or terrorist with small means to cause losses of immense economic value. Thus, security should certainly also be taken into consideration for civilian public and private networks not directly involved in financial operations.

Many wire-based private communication systems may be protected through physical control of all network components. For instance, a corporate LAN restricted within controlled premises is probably only exposed to the threat from disloyal employees. A radio-based communication system is particularly open to attack, since the transmission channel in principle is accessible to anyone possessing a radio receiver or transmitter. That is why radio systems generally are given a more thorough protection using various measures. Up to now (1997) security of civilian communications systems has not been a hot topic, apart from the need to prevent fraudulent use of the network. This is probably not due to the lack of a threat, but rather an unwillingness by the system operators to pay the cost of this insurance before an incident has actually occurred. More extensive information about network security may, for instance, be found in [17].

2.5.1 *Potential threats*

Let us first categorise the potential attacks towards a radiocommunication system. This is also illustrated in Figure 2.49:

- interruption is an attack on the availability of the communications system; this may be achieved through the destruction or disturbance of network components or through the disturbance of communication links (jamming);
- interception is an attack on confidentiality; this is related to acquiring information about message contents, traffic flow, localisation through direction finding etc.;
- modification is an attack on the integrity of the information; this is related to substituting the correct information (e.g. in a message) with false information;

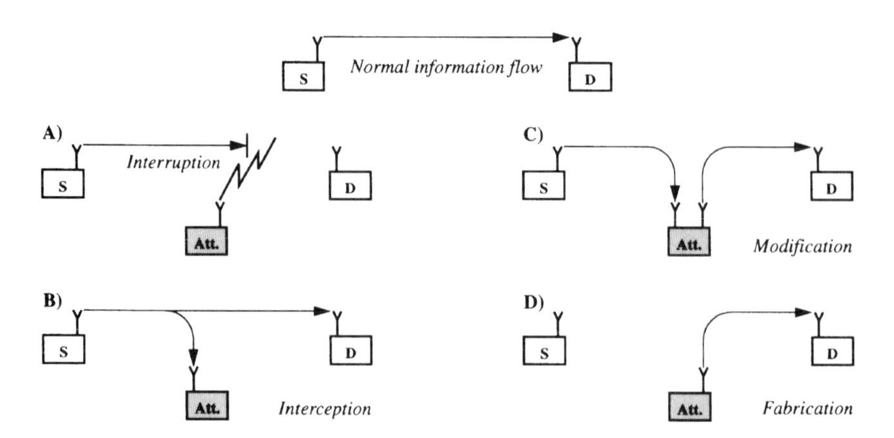

Figure 2.49 Illustrating potential attacks on radiocommunication systems

- fabrication is an attack on the authenticity of information; this is related to inserting counterfeit information.

Intercept is a passive attack which may be performed with the aim of retrieving message content or traffic analysis. Interruption, modification and fabrication are all active attacks. Generally, a trusted military radio system is required to possess countermeasures to all types of attacks, probably with interception (including both reveal of message content, localisation and traffic analysis) being given most attention, followed by interruption. For civilian systems the setting may be different, and modification and fabrication are probably just as important as interception. In this case localisation and traffic analysis are rarely of any importance.

2.5.2 Countermeasures

Let us now turn to the aspect of finding countermeasures to combat the different threat aspects previously identified.

2.5.2.1 Interruption

For a radio system where control of all network components may be secured, the actual threat associated with interruption is related to hostile interference (jamming) of the communication links. As already mentioned, spread-spectrum techniques may successfully be applied to combat this threat. In addition, there are several other efforts which may help to combat interruption. Error coding may be helpful, restoring messages destroyed by part-time interference. Strong FEC techniques such as Reed–Solomon may, for instance, be very effectively applied in combination with frequency-hopping spread spectrum to combat jamming in a military scenario. By ensuring a sufficient link budget when deploying network nodes, perhaps also using narrow-beam-directed antennas with large gain, a jammer is required to apply even more output power or locate closer to the network. All these efforts contribute to complicating the task for the jammer. Also, by applying e.g. packet-switched techniques to the network, information may automatically be routed outside disturbed links when only part of the network is affected. Finally, it should not be forgotten that interruption of radio networks generally is preceded by interception. Unless network activity is detected there is generally no reason for an adversary to engage interruption.

2.5.2.2 Interception

The special problem of radio networks is that anyone with a radio receiver in principle may intercept, and that there is no easy way of detecting this sort of attack. Spread spectrum may also give some protection against interception. It may give a certain protection against revealing message content when faced with a less sophisticated adversary, but is far from secure when faced with a technically-advanced and financially-resourceful enemy. In this case, only stringent cryptographic methods are sufficient. Obviously, the securing of voice transmissions requires digitisation before encryption can be applied.

When it comes to detection of activity, localisation and traffic-analysis spread-spectrum techniques may provide some protection. The detection of network activity by an adversary may to some extent be prevented through the use of both direct-sequence and frequency-hopping spread-spectrum, but also through effective power control[1] and directional antennas.[2] Improved protection against traffic analysis is achieved through the encryption of protocol-control information, and aiming to keep the traffic pattern and load as invariable as possible by inserting dummy information whenever needed.

Even the best cryptographic strategies will fall short if it is possible for an adversary to detect plain information caused by unintended electromagnetic transmission. It is well known that all electronic equipment emits radio-frequency energy. These signals may be intercepted within moderate

[1]The probability of detection is minimised when transmission power on a communication link is reduced to the minimum acceptable level while still maintaining an acceptable service quality.

[2]Directional antennas are effectively used on one-to-one communication links. For broadcast networks this is substantially complicated, but electronically-steerable array antennas may prove to be an interesting concept for a number of applications, especially when combined with an automatic positioning system, and the exchange of positions between network nodes.

distances. One particular aspect relating to radiocommunications systems is the possible problem (due to poor design) of low-power plain text signals infecting encrypted signals, which are then amplified and transmitted. Such unintended signals may be intercepted from a far distance.

For military applications handling highly classified information there are stringent requirements (often known as TEMPEST) for this unintended transmission. The fulfilment of such requirements is a long, cumbersome and expensive task for radio equipment manufacturers. Thus, the application of such requirements should be preceded by a thorough threat analysis, revealing the potential possibility of such an attack.

2.5.2.3 Modification

Modification in a radio network presupposes that the adversary first interrupts the original transmission (preventing it from reaching its destination), modifies the message content and finally forwards the modified message to the destination, preferably without the original sender noticing. This is not generally an easy task, and the risk of detection may be rather large, at least in a broadcast-type network. Probably the most effective way of attacking through modification will be to try to gain illicit access as an accepted network member, modifying all messages being delivered for relaying. This may be achieved either by building a dedicated radio or by modifying e.g. a stolen network radio. These methods may be both expensive and difficult, depending on the complexity of the radio.

The only waterproof means of preventing modification is through authentication. Some kind of authentication may be achieved in combination with encryption of the transmitted information, but this presupposes a total crypto-key control. Assuming that the crypto key is possessed by only one or a clearly defined and controlled group of users, the source of encrypted information is given.

2.5.2.4 Fabrication

The easiest way of fabrication is through the recording of network transmissions which are replayed back into the network at a later time. Even encrypted data may be accepted by the network as correct information this way, unless it is protected by some kind of time stamp, sequence numbering or other kind of authentication.

Both modification and fabrication are probably best prevented by using a cryptographic authentication method combining the unique key (supposed to be known by one radio only), the message content (e.g. a checksum) and a sequence number or time information. To minimise the damage caused by unnoticed loss of network radios containing valid crypto keys, some kind of user authentication will also be required.

2.5.2.5 Encryption strategies

There are two different forms of encryption in general use: the conventional, symmetric cryptographic method and the asymmetric, which is also known as public key. Generally, symmetric crypto requires less computational power, but public-key crypto has a great advantage in large communication networks, especially when combining information secrecy and authentication.

Symmetric crypto with a common (global) network key must be used in broadcast-type radio networks when concealment of protocol control information (including addresses) is required to prevent traffic analysis. Of course, this does not provide authentication. Public-key crypto may be used to protect message content, and is favourable when it comes to authentication.

Loss of a radio set containing valid keys will compromise network information, at least for the time period until the key is changed, assuming that the loss of the radio has been detected or that it is otherwise restricted from receiving the new keys. Obviously, the potential damage is greatest in a network relying on symmetric crypto with a global key. The greatest information protection is achieved through the use of dedicated keys valid for a time-limited session between two end users, combined with frequent user authentication. But such a mechanism may be both more network resource wasteful and less user friendly compared with other strategies.

References

1. Open Systems Interconnection Reference Model, ITU X.200 series
2. Special issue on open systems interconnection, *Proc. IEEE*, Dec 1983, **71**, (12)
3. SKAUG, R., and HJELMSTAD, J. F.: 'Spread spectrum in communication' (Peter Peregrinus Ltd)
4. DAVIES, D. W., and PRICE, W. L.: 'Security for computer networks' (John Wiley & Sons, 1987)
5. Radio Equipment and Systems (RES); Trans-European Trunked Radio (TETRA) ETSI standard
6. LYNCH, C. A., and BROWNRIGG, E. B.: 'Packet-radio networks. Architectures, protocols, technologies and applications' (Pergamon Press, 1987)
7. TANENBAUM, A. S.: 'Computer networks' (Prentice-Hall, 1989, 2nd edn.)
8. TOBAGI, F. A.: 'Multiaccess protocols in packet communication systems', *IEEE Trans.*, April 1980, **COM-28**, (4)

9. IROFF, M. A., and CHEN, S.: 'A distributed GSM architecture for low-traffic density markets', *Mobile Communications International*, October 1996
10. EGLI, J. J.: 'Radio propagation above 40 MC over irregular terrain', *Proc. IRE*, October 1957
11. KUO, F. F.: 'Protocols & techniques for data communication networks' (Prentice-Hall, 1981)
12. SCHILLING, D. L., PICKHOLTZ, R. L., and MILSTEIN, L. B.: 'Spread spectrum goes commercial', *IEEE Spectr.*, August 1990
13. MOULY, M., and PAUTET, M.-B.: 'The GSM system for mobile communications' (Published by authors, 1992)
14. BHARGHAVAN, V., and DEMERS, A.: 'MACAW: a media access protocol for wireless LANs'. ACM Sigcomm Computer Communication Review 8/94
15. TOBAGI, F. A.: 'Random access techniques for data transmission over packet switched radio networks'. Ph.D. thesis, Computer Science Dept., UCLA, 1974
16. NEWHOUSE, P.: 'Procedures for analyzing interference caused by spread-spectrum signals'. IIT Research Institute, ESD-TR-77-003, February 1978
17. STALLINGS, W.: 'Network and internetwork security. Principles and practice' (Prentice-Hall, 1995)

Appendix 2.A A MAC protocol for circuit-switched systems

The protocol presented here is called PRMA (packet-reservation multiple-access) in the literature.

Although most modern LMNs are capable of providing both CS and PS services, and thus use PS services for establishing CS connections, the following MAC protocol may be used for purely CS systems.

A synchronous system is required, with the radio time channel separated into repeating frames, each consisting of a number of time slots as indicated in Figure 2.50.

An example of how the access protocol works is given in Figure 2.51. Slots that are used in one frame are reserved for that connection in all consecutive frames until the connection is released and the slot becomes idle.

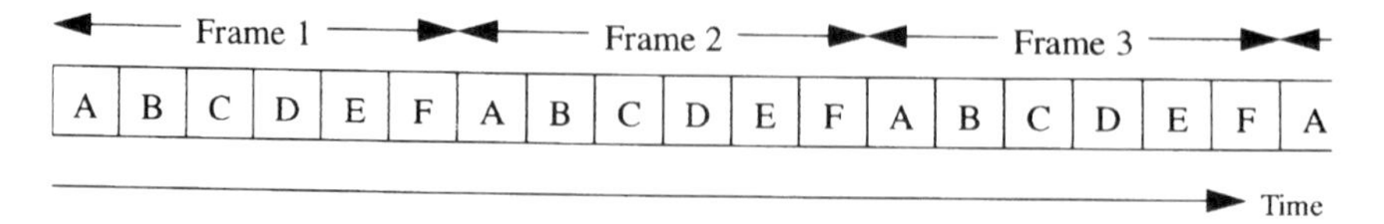

Figure 2.50 The radio channel is divided into six time slots (A to F), constituting a frame. This sequence is repeated in each consecutive time frame

	Slot A	Slot B	Slot C	Slot D	Slot E	Slot F
Frame 1	idle	Conn. 1	Conn. 2	Conn. 3	collision	idle
Frame 2	Conn. 4	Conn. 1	idle	Conn. 3	Conn. 5	idle
Frame 3	Conn. 4	Conn. 1	collision	Conn. 3	Conn. 5	Conn. 6
Frame 4	Conn. 4	Conn. 1	collision	Con. 3	Conn. 5	Conn. 6
Frame 5	Conn. 4	Conn. 1	idle	idle	Conn. 5	Conn. 6

Figure 2.51 A reservation strategy for CS connections

Slots that are idle in one frame may be accessed (for reservation) in the same slot of the next time frame. If more than one node attempts to reserve the slot, this will result in a collision, and all reservations (for this slot) fail and must be retried later. In order not to also collide in the next time slot, the colliding reservations must be spread in time according to some random mechanism. Obviously, a collision then may be followed by an idle slot if (by accident) none of the colliding parties attempts to access the next slot.

Obviously, this reservation protocol does not work unless reserved time slots are undisturbed by other nodes. This works well in a completely-connected network, but constitutes a problem in multihop networks where hidden nodes may try to access reserved time slots if they do not detect the activity, or otherwise are unaware of the reservation.

The MAC protocol used to reserve time slots may be a simple ALOHA protocol or, better, a CSMA protocol in order to reduce the rate of colliding reservations. Other access protocols may also be used.

Depending on channel capacity (slot size and frequency) and required data rate in the two directions, a separate slot may be reserved for the return channel, or the communicating parties may agree on a rule for sharing the single slot, depending on their mutual capacity requirements, as outlined in Figure 2.52.

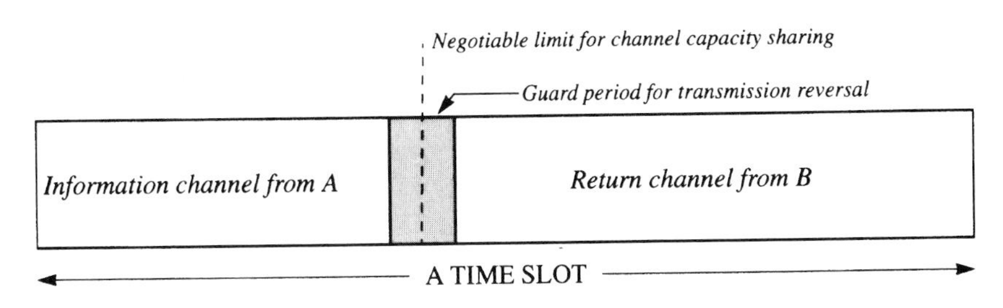

Figure 2.52 Flexible sharing of a time slot between the two directions of communication

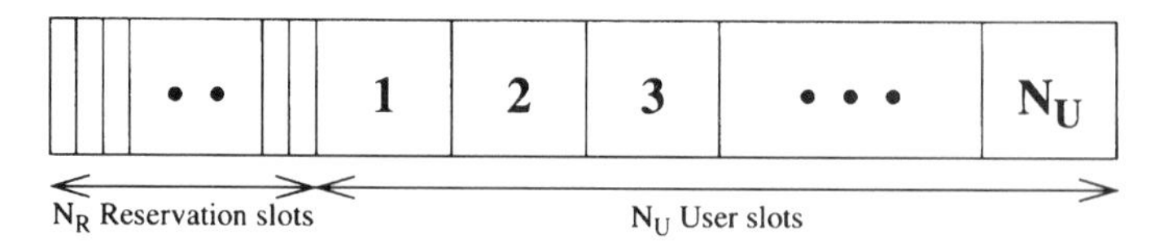

Figure 2.53 Dynamic TDMA with reservation minislots

In an attempt to increase the efficiency of the access protocol, the PRMA protocol may be replaced by a dynamic TDMA (D-TDMA) protocol. This protocol uses shorter minislots for the reservation access (see Figure 2.53). A collision thus reduces the wasted channel time. Unlike PRMA, D-TDMA is not suitable for distributed systems without a base station to handle the reservations. Most access protocols found in modern literature presuppose the existence of a central base station within range of all mobiles. The PRMA protocol may equally well be used in distributed networks, as long as this is restricted to single hop, so that all mobiles can receive all reservations. None of these protocols are suitable for multihop networks since hidden nodes may unintentionally disturb established connections in the data transfer phase.

As an alternative to this time-shared CS strategy, of course, code division may be used for SS systems. Unless a node is required to participate in more than one simultaneous connection, this may simply be done by assigning one spreading code for each node. Signalling for the establishment of a connection is performed using the SS code of the other party. In order not to be disturbed, nodes participating in a CS connection should have some means of signalling this to the other nodes in the network (which may not be listening for SS codes not used by themselves).

Chapter 3

The operational environment

A total adaptivity and match between required services and channel characteristics can only be approximated by developing realistic user ambitions based on knowledge of the radio channel and careful design. This Chapter takes a first look at some of the expected characteristics of the transmission medium, both natural and man-made.

3.1 Environmental characteristics

When radio waves propagate over the Earth, their propagation characteristics are determined by the electrical properties and the physical configuration of the surface of the Earth, including vegetation and man-made structures. At frequencies above 30 MHz which will be of interest here, it is the physical configuration that matters the most (at 300 MHz and antenna height of 3 m being horizontally polarised, the surface wave is down by 70 dB compared to the space wave), and the relevant effect on overall transmission loss is determined by the frequency concerned and the topography of the terrain. Large-scale terrain irregularities may affect the horizontal homogeneity of refractive layers in the atmosphere making radio-signal reflections possible, and small-scale irregularities on the Earth's surface give rise to reflections or scattered radio power.

The radiowaves of interest here can then be pictured as in Figure 3.1.

3.1.1 Path loss

Examples of basic teleservices as described in subsection 2.1.4.3 are point-to-point services and point-to-area services. Terrestrial fixed point-to-point

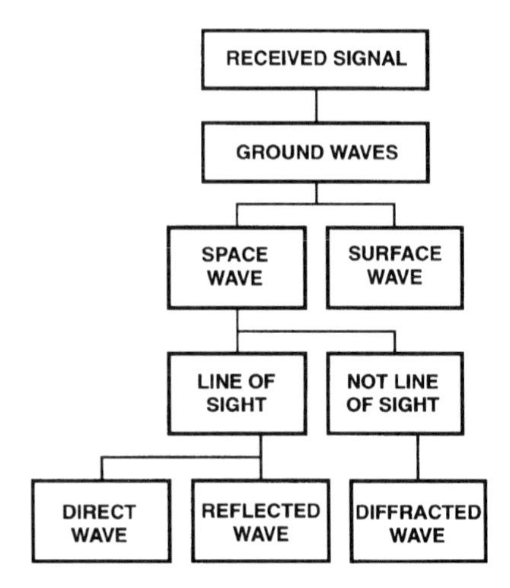

Figure 3.1 A possible decomposition of a received signal

services can be planned for with due regard to the topographical characteristic of the propagation path between the points in question. In the case of terrestrial mobile point-to-area services, propagation can be considered to take place over a multiplicity of individual paths, and the problem of determining service performance must mainly be considered on a basis of statistical coverage because of the practical impossibility of making surveys of all the individual paths in question.

Given a transmitter antenna with gain G_t which radiates with a power of P_t, the power density in the far field is given by:

$$\varnothing = G_t P_t / (4\pi d^2) \tag{3.1}$$

Placing a receiving antenna in the far field of the transmitter antenna with a gain G_r gives a received power in the receiving antenna of:

$$P_r = P_t G_t G_r (\lambda / 4\pi d)^2 \tag{3.2}$$

This is known as the Friis formula, describing the free-space path loss.

Introducing the speed of light, c, and the operating frequency, f, and assuming isotropic antennas with $G_t = G_r = 1$ gives:

$$\text{path loss} = 10 \log(P_t / P_r) = 32.4 + 20 \log f(\text{MHz}) + 20 \log d(\text{km}) \tag{3.3}$$

The Friis formula describes the path loss between a transmitting antenna and a receiving antenna in free space. For nonfree-space conditions the physical surroundings will influence the propagation of radio waves. To simplify, the transmitted radio wave is usually parted into different waves which are treated one by one and finally superimposed at the receiving site.

As was shown in Figure 3.1, the line-of-sight wave is divided into a direct wave and a reflected wave. This is illustrated in Figure 3.2.

The direct wave is the primary means of transmission for line-of-sight communication and the path loss follows that of the Friis formula. The reflected wave represents the contribution to received power due to reflections from the surroundings. Whether a reflected wave will be present or not is related to the ground topography and its interaction with the so-called Fresnel zones around the transmitting and receiving antennas of a radio path.

The reflected wave will itself consist of one reflected part and one refracted part. The reflection coefficient indicates the ratio between the incoming signal hitting the ground and the signal being reflected towards the receiving antenna. The reflection coefficient will depend on the angle of arrival, the relative dielectric constant of the Earth, the conductivity of the Earth and the frequency.

For reflections from a plane surface, it can be shown that:

$$P_r/P_t = h_1^2 h_2^2 / d^4 \tag{3.4}$$

where h_1 and h_2 are the height of the isotropic antennas at the transmitting and receiving sites, respectively, and for $d \gg h_1$ and $d \gg h_2$.

It is worth noting that the path loss of the space wave for a plane reflecting surface is independent of frequency.

Leaving the probably unrealistic plane reflecting surface assumption and introducing reflections from a rough surface gives:

$$P_r/P_t = (1-\alpha)^2 \left[\frac{\lambda}{4\pi d}\right]^2 + \alpha \frac{(h_1 h_2 + y(h_1 + h_2))^2}{d^4} \tag{3.5}$$

where α is a terrain factor which becomes 1 when the terrain allows no reflections and y is the height of the roughness which becomes 0 when $\alpha = 1$ (a plane surface), as shown in Figure 3.3.

To sum up, it would be correct to say that the path loss for the space wave will be limited on the lower side by the free-space path loss and on

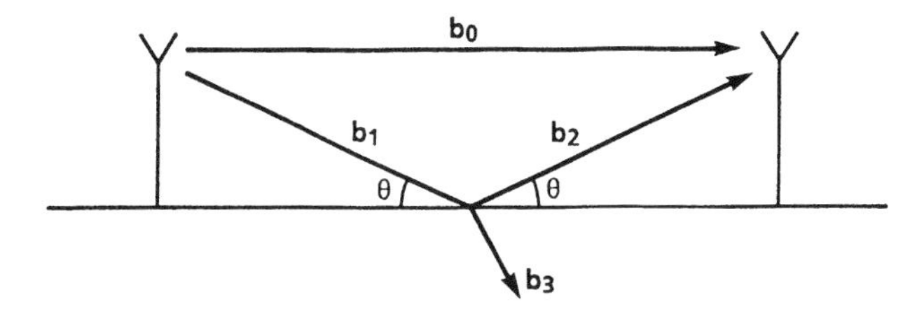

Figure 3.2 Direct and reflected wave (the wave hitting the ground being split into one reflected and one refracted wave)

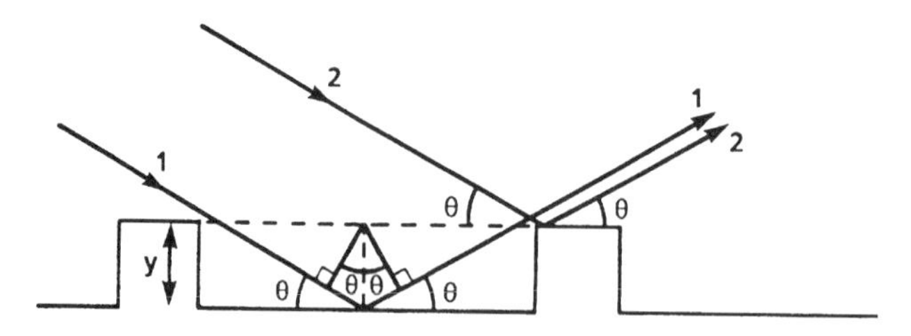

Figure 3.3 Reflection from a rough surface

the upper side by the frequency-independent loss for a plane reflecting surface. The assumption made is that the distance between the antennas is much larger than any of the two antenna heights.

We thus consider line-of-sight paths and apply the free-space path loss when there is no obstruction between the transmitting antenna and the receiving antenna. We have also considered the case when there are obstructions close to the line-of-sight path (in accordance with the definition of first Fresnel zone) causing reflections. If, on the other hand, the obstructions are of such a height that they are in line with the direct path between the antennas, or to some degree block the line of sight, we may observe diffraction. Diffraction is the phenomenon describing how waves are bent when passing sharp edges. This phenomenon is illustrated in Figure 3.4.

On one side, obstructions in the path lead to additional losses, called diffraction losses (relative to free-space loss); on the other side, as long as these losses are acceptable to the radio system, we may communicate beyond line of sight. Figure 3.5 illustrates the effect of diffraction as the transmitter–receiver path is extended beyond line of sight.

So far we have only considered reflections from the ground assuming horizontal surfaces between the transmitter and receiver antennas. Of course, if the topography is such that the terrain formations are high compared to the antenna heights, we may get reflections from vertical surfaces, as shown in Figure 3.6.

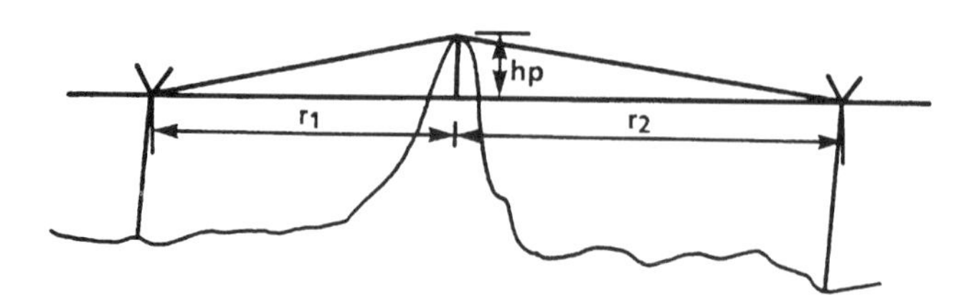

Figure 3.4 A transmission path with diffraction

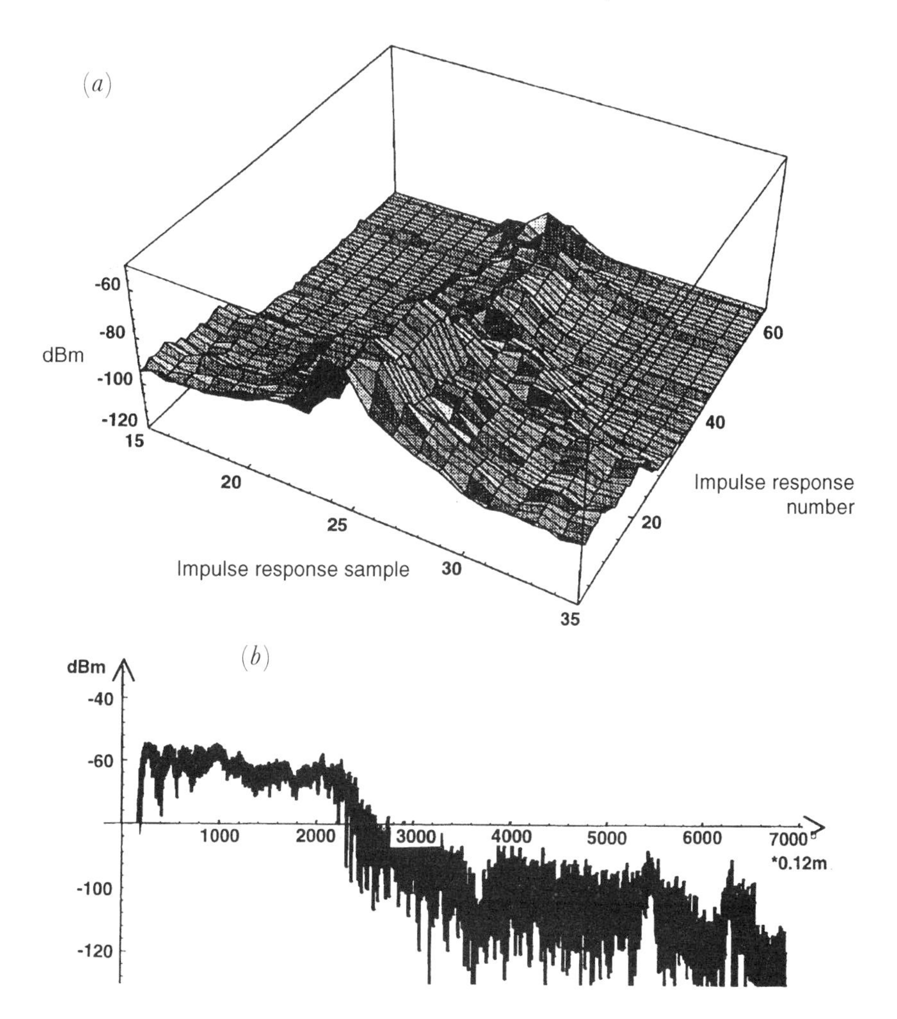

Figure 3.5 (a) median channel response as a function of distance, (b) received power as a function of distance showing the change from free-space path loss to a path loss dependent on diffraction

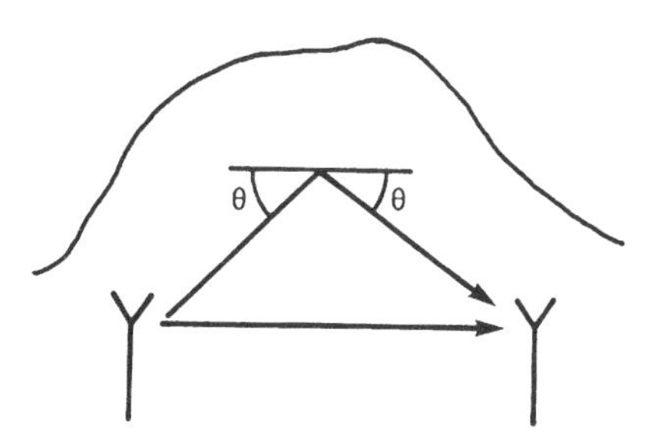

Figure 3.6 A path with vertical surfaces giving rise to possible multipath

This phenomenon is called *multipath*. These multipaths may interfere with the direct path to give a signal which can be large or small depending on the distribution of phases among the component waves, and they may be the cause of possible communication beyond line of sight (reflections around hilltops). For a real communication situation we may have multipath, multiple diffractions, or even multiple diffractions with ground reflections, making the path loss calculations more and more complex and involved. The different effects of vegetation therefore make the transmission calculations very complex and the need for empirical prediction methods and computer modelling arises. An example of the effect of vegetation is shown in Figure 3.7.

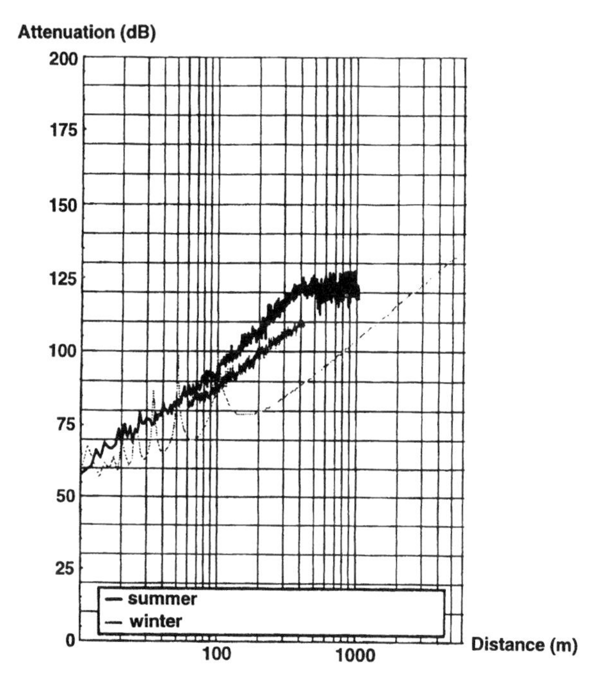

Figure 3.7 *The effect of vegetation on path loss. The reference is free-space path loss, with the upper curve showing summer values, and the lower curve showing winter values*

3.1.1.1 *Operating frequency*

The choice of operating frequency will have to take several factors into consideration. Ideally, the user would like to pick a frequency which gives full control of the path length, or service-covering area, and at the same time gives whatever bandwidth the services demand and which in a very efficient way enables emission of the necessary transmitter power using antennas sufficiently small to be convenient for any application (e.g. hand held). As we shall see in Chapter 6, real life is not like that. Increasing the frequency to provide for more bandwidth and smaller antennas at the same time provides channels which experience larger free-space path loss and require more strict line-of-sight operations. Lowering the operating frequency gives better coverage, but at the same time makes the interference due to frequency re-use more unpredictable and may introduce considerable multipath problems on systems with high data rates. Once again, compromises have to be made, and the best solutions must be individually matched to the particular system requirements. This book will, however, focus on systems generally being serviced by the VHF and UHF bands, that is from 30 MHz to 3000 MHz.

3.1.1.2 The influence of the terrain and path-loss modelling

In this subsection we will discuss the use of empirical path-loss prediction methods compared to those using some form of geographic information system (GIS) trying to calculate the diffraction losses. However, it is quite clear that only limited confidence can be placed in even the best model, and for most realistic areas with significant terrain features there will be a difference between the calculated and measured path loss. So, even using state of the art modelling based on GIS with a final resolution there is a need for incorporating some empirical correction factors depending on the characteristics of the terrain and the operating frequency. In addition, the problem with reflection from off-path vertical terrain features which generate multipath situations has been very difficult to integrate into any path-loss models and gives large variations in confidence for any particular path-point variations within one wavelength. As an example, in a Norwegian mountainous rural area the path loss varied as much as 20–30 dB just by randomising the receiver antenna position within a distance of one wavelength at VHF frequencies. Recently there have been several attempts to handle the influence of multipath. By looking at a digitised map the terrain is converted to reflections for any particular transmitter–receiver path. The reflection coefficient will indicate the orientation of the reflector as seen by the particular path. In other words, the transmitter–receiver pair is looked upon in terms of a bistatic radar.

It is a great challenge for any radio system developer to establish the technical system parameters which can meet the user requirements and in the end validate the ability of the system to meet performance requirements.

The translation from user requirements to technical characterisation is not a simple task and will require, among others, methods which provide sufficiently accurate predictions of signal strength, distortion and interference. There exist several signal strength methods which may be broadly categorised into those that are based on empirical prediction, that is some simple power-law dependence of signal strength with distance and some empirically-derived factors, for example for the influence of antenna position and the particular frequency in use, and those models which make an attempt to calculate the diffraction loss along the path between the transmitter and receiver. An example of the first category is the Okumara model [1] and for the second category a well known example is Longley–Rice [2].

For the case with no significant terrain features, empirical methods may be useful, but generally for most mobile radio areas of interest the second category is most suitable.

In the following a short description of a typical computer-based prediction system developed at FFI is explained. The system makes use of a digitised map reference and predicts the path-loss matrix, area of coverage and area of interference. The three modules are independent, making use of their respective user interfaces, as shown in Figure 3.8.

The path-loss matrix module calculates the path loss between two or more points in the terrain, and puts the path-loss values on a computer file. The area of coverage module calculates the service area for a given transmitter, and shows the results in a graphical format. The area of

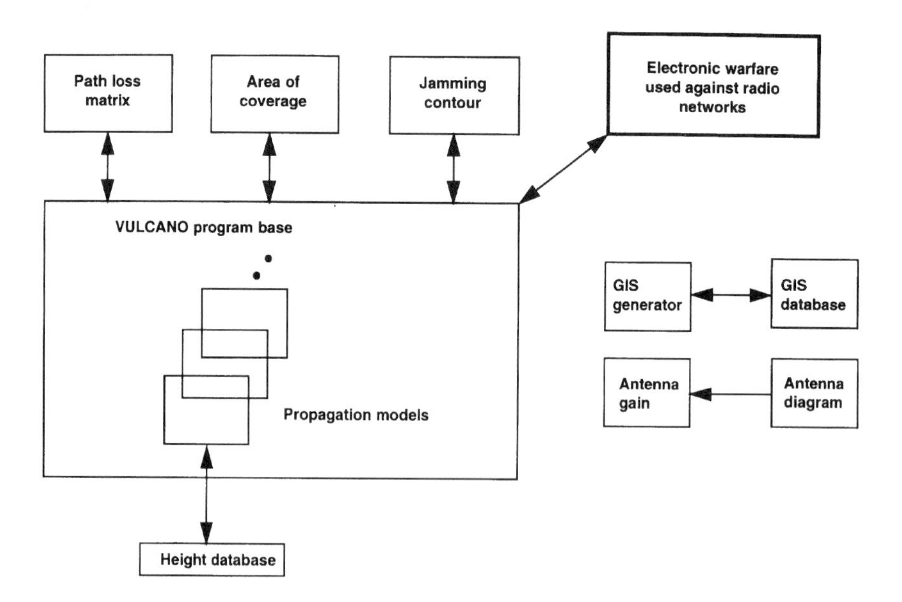

Figure 3.8 A simple overview of the FFI-developed propagation model with its different modules

interference module calculates the area where a certain precharacterised interferer must be located in order to interfere with a predefined transmitter–receiver path, and shows the results in a graphical format.

The system also has a function which calculates the antenna gain in a certain direction for a given antenna. Possible antenna diagrams are computer tabulated. The gain in a certain direction is calculated by interpolation between tabulated values.

The path-loss calculations are based on five different propagation models. The first model is an optical line-of-sight model. The second model is a line-of-sight model where line of sight is defined with respect to a defined Fresnel zone. The model checks whether the paths of interest cut the chosen Fresnel zone or not. If not, the free-space path loss, as given in eqn. 3.2, is calculated. If the path cuts the Fresnel zone, the path loss is set to infinity. The third, fourth and fifth models calculate the path losses based on the terrain profiles along the transmitter–receiver path in different ways. For the third model, calculation of diffraction from the highest point along the path takes place. The calculations are based on simple knife-edge diffraction and only take into account the point in the terrain profile that will attenuate the signal the most. In addition, the model adds the free-space path loss for the distance in question. The fourth and fifth models also take into account the losses due to reflections. The estimations of reflection losses differ slightly, however, for the two models. The last two models also estimate the signal component due to the surface wave. The fourth model allows for multiple diffraction with reflection and takes into account all terrain heights on the profile between the transmitter and receiver. The reflection loss is estimated based on a diffraction loss and the free-space path loss is added. Using a parameter called 'minimum effective antenna height' allows for the estimation of the surface-wave signal component. The antenna parameter is a function of frequency, polarisation and the characteristics of the Earth. The fifth model is equal to the fourth model with the only difference being how it calculates the reflection losses, which now are estimated based on the mirror principle.

The influence of weather, important for frequencies above 2000 MHz, is divided into five groups: clear, fog, rain, dry snow and wet snow.

The total input to the path-loss matrix model then consists of the following parameters:

0 choice of propagation model
A weather
B choice of Fresnel zone for the influence of reflection
C the resolution for the terrain-height data
D transmitter frequency and minimum antenna height

E the transmitter(s) position(s)
F the transmitter(s) antenna height(s)
G the receiver(s) position(s)
H the receiver(s) antenna height(s)
T run calculations
U run calculations with antenna diagram (only valid for one transmitter and one receiver)
V the transmitter position(s) in UTM metres
W the receiver position(s) in UTM metres

An example of the application of the model is shown in Figure 3.9.

When the area-of-coverage module is run, additional information on the receiver-required signal-to-noise ratio, total noise level (external and internal generated) and receiver noise figure (represented by bandwidth, temperature, internal receiver noise above kTB) is needed as input information. The output is then an area of coverage which can be represented as an overlay to a map, if necessary.

For the area-of-interference module, the interferers' position, frequency, power, antenna diagram and bandwidth are required. An example is shown in Figure 3.10.

The FFI model has been used to calculate expected system performance for a number of situations. A few lessons to be learned about the results of prediction compared to the results of measurements should be given.

One of the major deficiencies with most prediction programs is the lack of ability to take full account of the presence of possible multipaths. The prediction programs have problems both when it comes to taking into account the influence of off-path terrain features and with the limited resolution of the Geographic Information System (GIS) databases. It may be said that up to recently the prediction programs worked well for situations without multipath, but so did simple $1/r^4$ calculations, but for terrain with a multitude of multipaths the prediction models just as often over estimate path losses as under estimate them, and in the end perhaps do not greatly out-perform some simple $1/r^4$ model. As mentioned earlier, however, some new approaches to treating the terrain as differently-oriented reflections have been showing promising results compared to measured results. Figure 3.11 shows such an approach where the terrain is divided into reflection surfaces. In such a way the possible multipaths can be grouped according to their delay and the influence on a particular communication system can be estimated.

The situation where no account is taken of the effect of multipath may be illustrated by three examples. The first refers to test measurements in open rural areas with few marked multipath reflections. The measurements were performed for the three frequencies 34.5 MHz, 85 MHz and 390 MHz from transmitters located in northern, southern, eastern and western directions, each time for some 12 km. The results are shown in Figure 3.12 and the measured values are shown to fit well to a simple $1/r^4$ path-loss law.

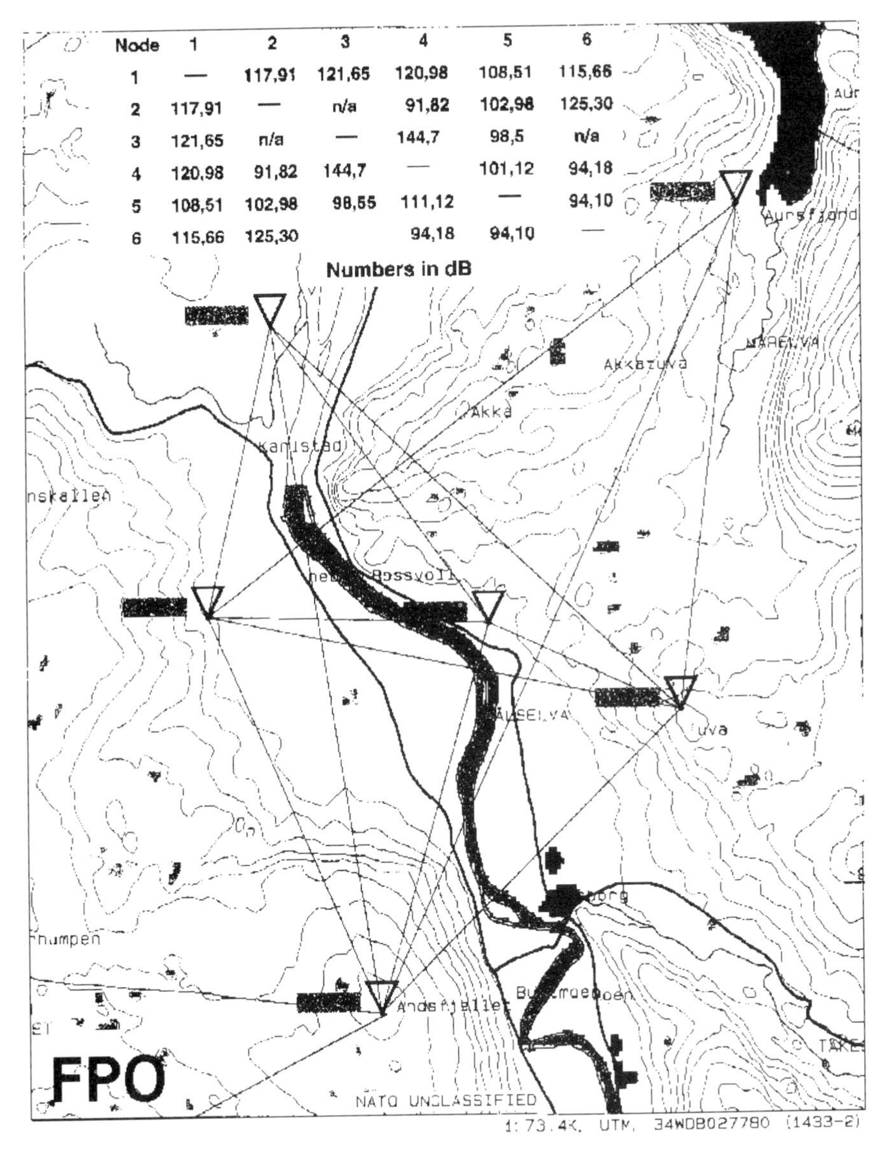

Node	1	2	3	4	5	6
1	—	117,91	121,65	120,98	108,51	115,66
2	117,91	—	n/a	91,82	102,98	125,30
3	121,65	n/a	—	144,7	98,5	n/a
4	120,98	91,82	144,7	—	101,12	94,18
5	108,51	102,98	98,55	111,12	—	94,10
6	115,66	125,30		94,18	94,10	—

Figure 3.9 Example of the path-loss matrix for a certain Norwegian scenario of transmitters/receivers at VHF frequencies (courtesy of Teleplan)

The second example is taken from a hilly and mountainous rural area, with the presence of some rivers and lakes. The presence of multipaths is frequent. This time the measurements are performed for two frequencies 34.5 MHz and 390 MHz and for different paths of approximately 10 km lengths. Figures 3.13 and 3.14 show the correspondence

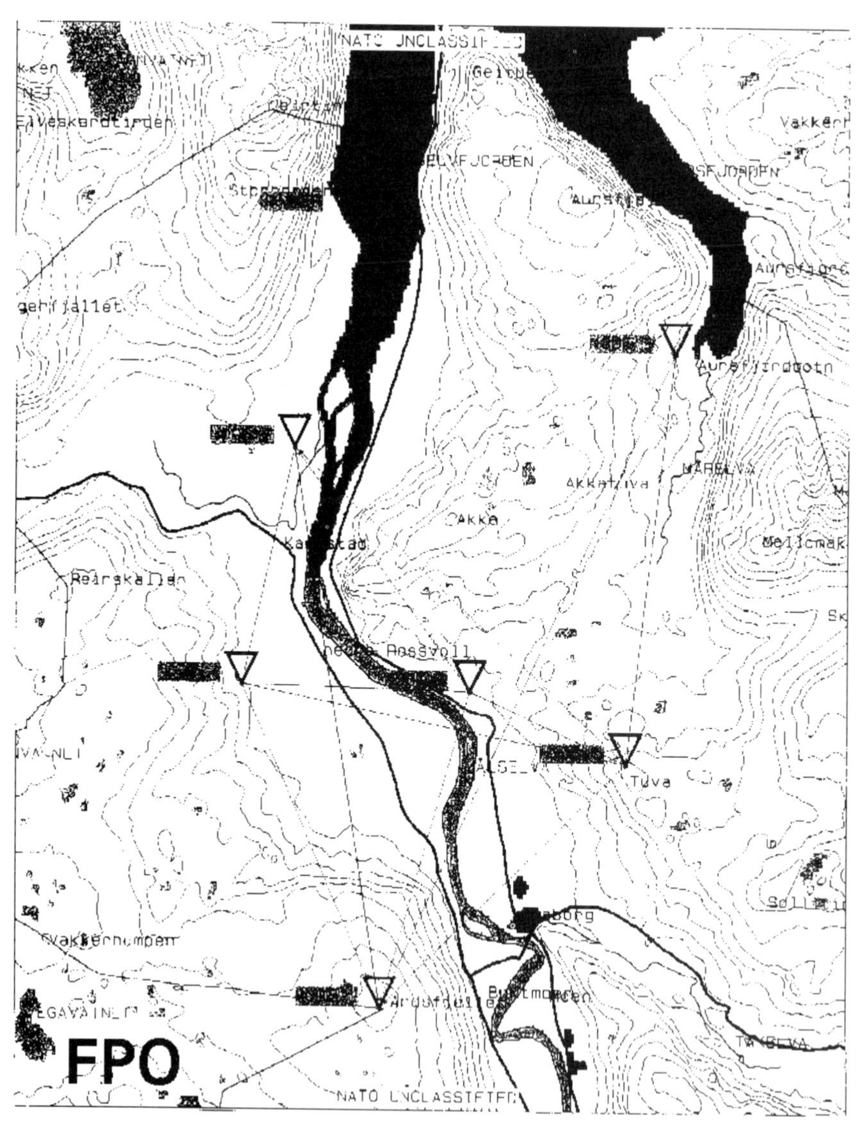

Figure 3.10 *The effect of interference (jamming) on transmitter–receiver paths from the scenario in Figure 3.9. The red lines show where the interference has degraded the information channel to an unacceptable level and the dotted lines show where only one-way communication is possible (courtesy of Teleplan)*

Figure 3.11 a The terrain as reflectors; the various shades of grey indicate the reflection coefficient

between the measured values and the values obtained with the FFI prediction model. Better correspondence between measured and predicted values is probably not possible without a more sophisticated treatment of multipaths.

The third and last example illustrates the effect of vegetation, and is shown as the difference between measurements taken during summer

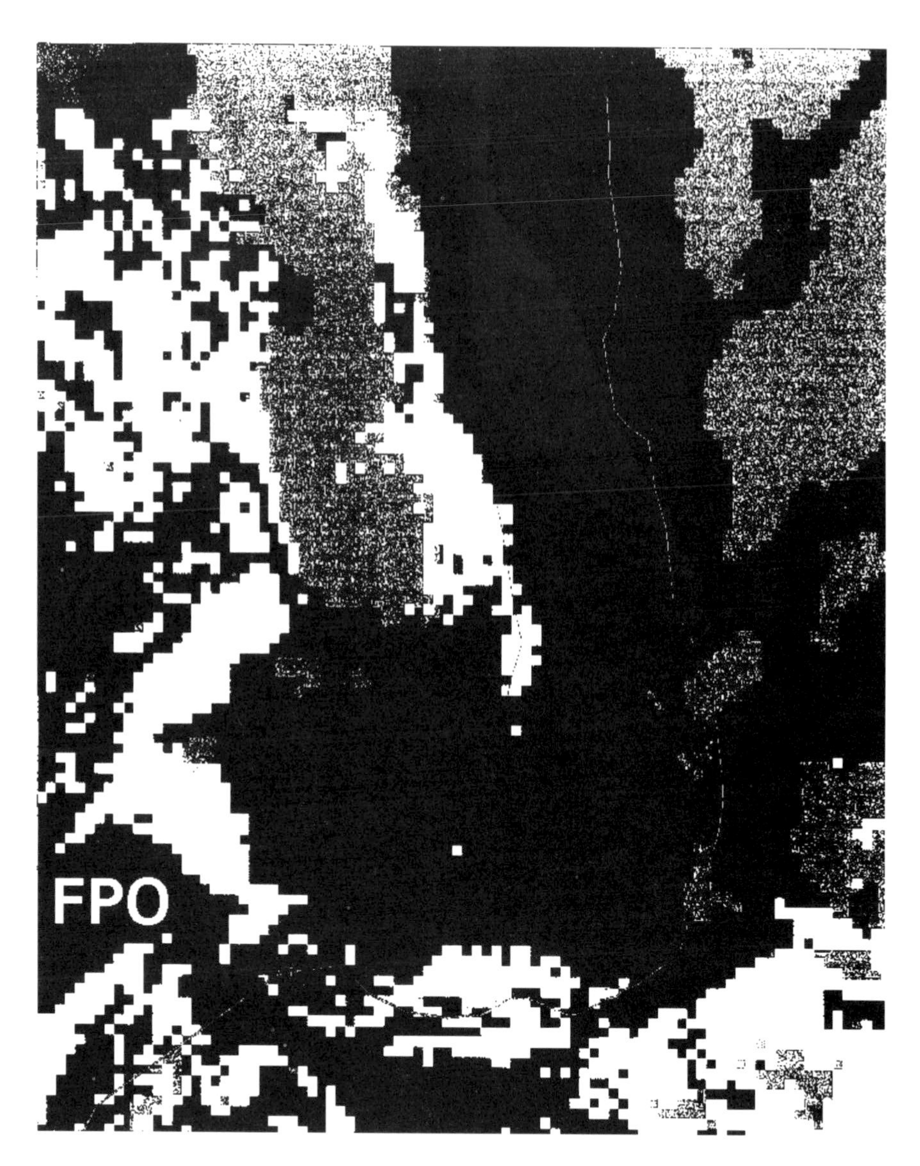

Figure 3.11 b The area where multipath to a varying degree may introduce problems

and during winter, when the trees and terrain are covered by snow. The illustrations shown in Figure 3.7 were measurements taken in a rather flat terrain with mixed forest; Figure 3.15 shows measurements taken in hilly terrain without trees. The measurements are taken at 2500 MHz and for distances up to 1 km, and show less path loss during winter. The winter measurements also better follow the theoretical free-space path loss.

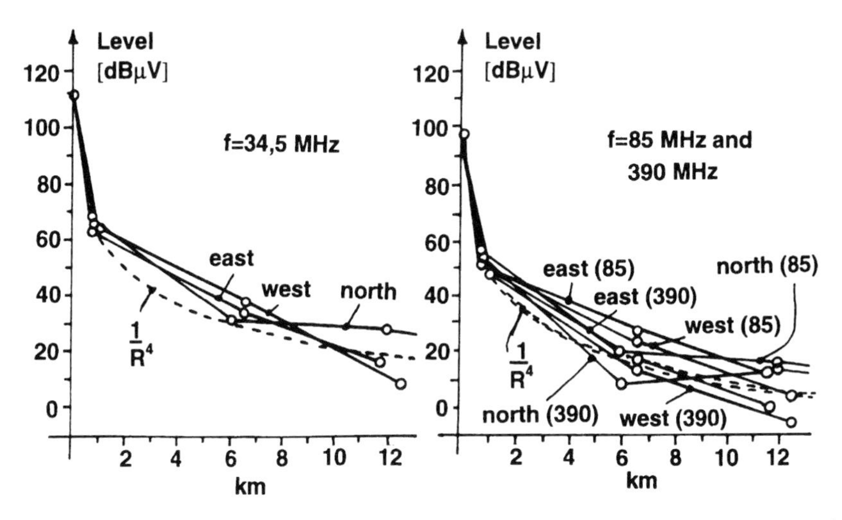

Figure 3.12 *Comparison between measured values and predictions according to a* $1/r^4$ *power law*

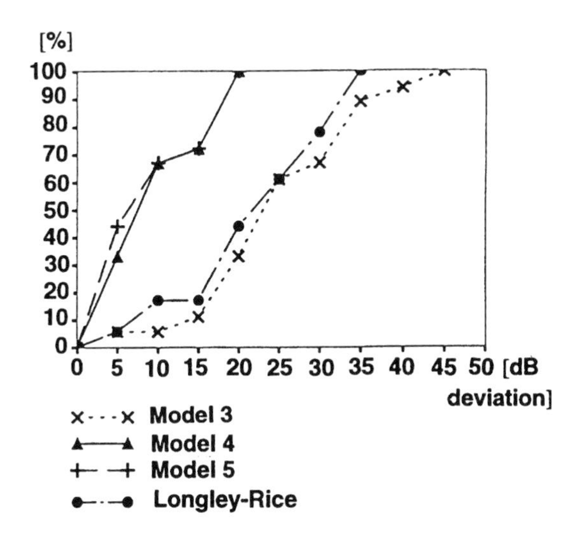

Figure 3.13 *The percentage of calculated values that differ from measured values at 34.5 MHz*

These examples have not been given to discourage anybody from the development and use of path-loss prediction models, but rather to point at the importance of supplementary measurements, and to encourage every effort in establishing and improving models which, based on some form of GIS database, will be able to treat the effects of possible multipaths. Multipaths will be more thoroughly considered in the sections that follow.

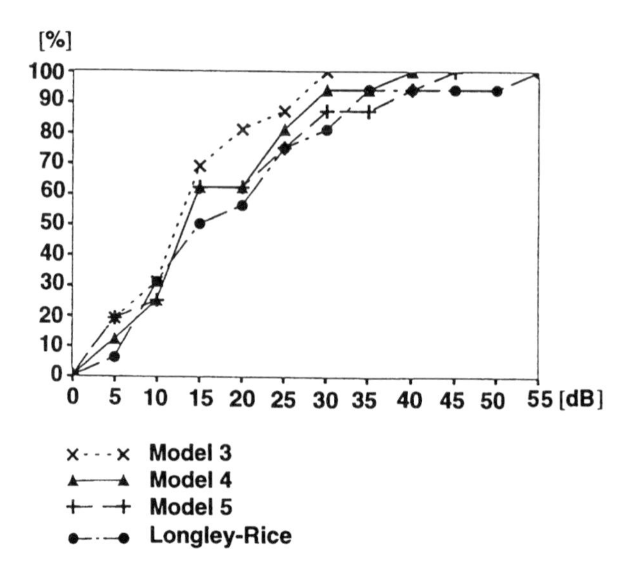

Figure 3.14 *The percentage of calculated values that differ from measured values at 390 MHz*

3.1.2 Channel imperfections

3.1.2.1 Slow fading

The path loss will experience a variability with receiver location above what can be calculated due to changes in range. Such a variation has been illustrated in Figure 3.5, where the transmission path varies from line of sight to diffraction due to shadowing by the terrain. Such gross variation in the overall path between the transmitter and receiver, giving rise to variation in path loss experienced by any receiver moving along the path, is often called slow fading. Observations show that, statistically, such variations follow a log-normal distribution. This fading is thus also called log-normal fading.

3.1.2.2 Multipath fading

At the end of Section 3.1.1 the phenomenon of multipath was introduced as the possible cause of interference with a direct path, and as the cause of possible communication beyond line of sight. Each path has different attenuations, carrier phase and delays.

The impulse response for a time invariant multipath situation can be expressed as:

$$h(\tau) = \sum_{k=1}^{K} \beta_k \delta(\tau - \tau_k) \tag{3.6}$$

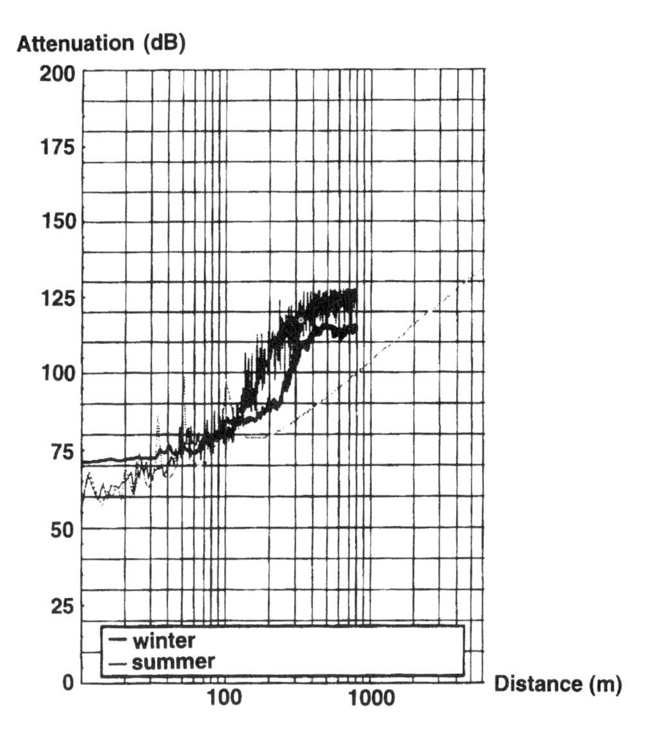

Figure 3.15 *The effect of vegetation on path loss in hilly terrain*

where each of the K paths has a complex amplitude coefficient, β_k, $k = 1, \ldots, K$, and a delay, τ_k, $k = 1, \ldots, K$, and $\tau_m = \tau_k - \tau_1$ is called the multipath spread.

The corresponding transfer function is:

$$H(f) = \sum_{k=1}^{K} \beta_k e^{-j2\pi f \tau_k} \tag{3.7}$$

Multipath delay spread thus gives rise to two effects, time dispersion and frequency-selective fading. Time dispersion stretches the signal (the signal takes different times crossing paths of different lengths before adding at the receiver antenna) so that its duration is greater than that of the transmitted signal. Frequency-selective fading shows up as a filter which attenuates certain frequencies more than others. If the signal contains closely-spaced frequency components (within the coherent bandwidth) these will experience the same attenuation (frequency-flat fading), and signals with frequencies more separated may experience different attenuations for the different frequency components (frequency-selective fading).

The effects of multipath will thus depend on the bandwidth of the signal. For narrowband signals it is possible to combat fading by transmitting more power. For a digital radio system one might define the frequency-flat fading margin as the decibel difference between the unfaded signal strength and the signal strength at the critical error rate. For wideband signals the frequency selective fading may result in amplitude and phase distortion of the transmitted signal and the degradation cannot be combated with more transmitter power.

Figure 3.16 illustrates such a case where the flat-fading margin is equal to 41 dB for a critical error rate of 10^{-3}.

For a high data rate and bandwidth the Figure illustrates how unsatisfactory operation can occur for fades of much less than 41 dB. A high data rate and large bandwidth system based on flat-fading calculations would therefore perform much below specification. If the transmitter power is increased to increase the flat-frequency fading margin such a wideband system would be somewhat insensitive to transmitter power when it comes to outage time.

Thus, although the dispersive nature of the channel is only of secondary importance when compared to the average value of the faded signal across the channel for narrowband signals, wideband signals may also be very sensitive to the dispersive effect of multipath. Frequency-selective fading is thus encountered when the bandwidth used exceeds the coherence bandwidth of the channel. This situation is even more likely to occur when spread-spectrum direct-sequence modulation is applied, since

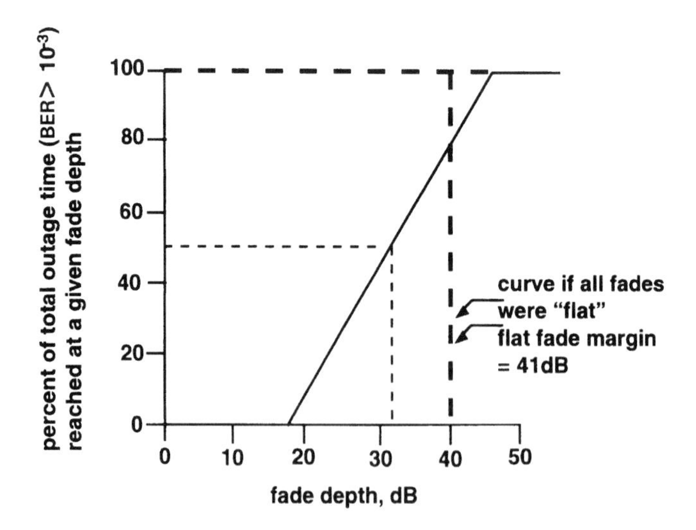

Figure 3.16 The outage distribution for a certain bit error rate (10^{-3}), and a data rate of 45 MHz being transmitted in the upper UHF band

inherent in this modulation is a large bandwidth expansion. However, when the bandwidth is large enough, the receiver may be able to make the distinction between the different multipaths in such a way that intersymbol interference occurs. Now spread-spectrum modulation can be applied as an efficient antimultipath technique to reduce the effect of intersymbol interference. The situation for narrowband and spread-spectrum signals for the case of multipath is shown in Figure 3.17.

A neccessary condition is, however, the ability of the spread-spectrum receiver to resolve the multipaths. Unresolved multipaths may give rise to fading and delay distortion.

As was mentioned in Section 1.3.3 and will be elaborated upon in Section 4.3.5, the spread-spectrum receiver is designed to operate at negative signal-to-noise ratios, the exact value being determined by the processing gain and final error rate. If concealment of transmission and transmission privacy are required properties then it is important that the fade margin can be kept low in order for the system not to be forced to operate with positive signal-to-noise ratios for long period of times. For multipath delays larger or slightly less than the inverse of the spread-spectrum system bandwidth, the spread-spectrum modulated signal will offer good frequency diversity, minimising the effect of deep-fading minima.

For time-invariant channels the multipath will thus result in no distortion of narrowband (with respect to coherent bandwidth) signals, and we are talking about frequency-flat fading. For wideband signals we may

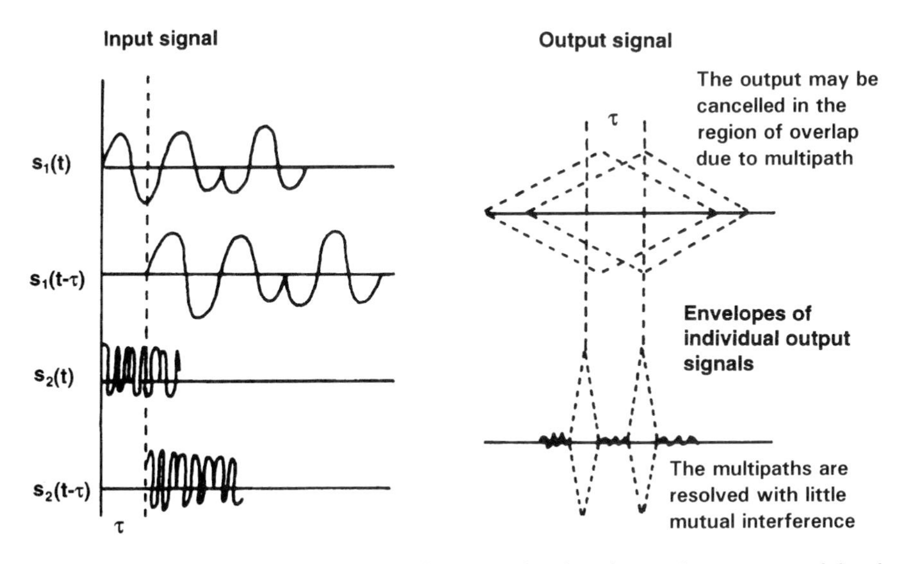

Figure 3.17 The input–output signals for conventional and spread-spectrum modulated signals during multipath

experience frequency-selective fading giving rise to distortion of the signal and at the same time we may due to the high signalling rates have dispersion giving rise to intersymbol interference.

So far, all the paths have been considered time invariant. However, multipath reception is time variant if the properties of the communication medium and the positions and aspect angles of the reflecting objects relative to the transmitter and receiver vary with time. Under such conditions we have time-selective fading and frequency dispersion.

Now β_k from eqn. 3.6 becomes time dependent so that $\beta_k = \beta_k(t)$, $k = 1, \ldots, K$ and:

$$h(\tau; t) = \int_{-\infty}^{\infty} H(f; t)e^{-j2\pi f\tau} df \tag{3.8}$$

and

$$H(f; t) = \int_{-\infty}^{\infty} h(\tau; t)e^{-j2\pi f\tau} d\tau \tag{3.9}$$

Most of the time $\beta_k(t)$ will not be known, but considering not too large time intervals it may be considered as a stationary stochastic quantity with a probability density function for the real and imaginary parts being statistically-independent gaussian variables with zero mean and the variances σ_k^2, $k = 1, \ldots, K$.

Time-selective fading can be the reason for signal distortion because the channel changes its characteristic while transmission is going on. Because the channel is time variant, frequency dispersion in the form of Doppler spreading will occur. The received signal bandwidth will differ from that of the transmitted bandwidth.

For signals of short duration no time variations of the channel will be seen, and we have a time-flat channel. When the Doppler frequency becomes larger than the Doppler resolution of the receiver, that is for signals of longer duration, a widening of the received spectrum can be observed.

The effect of multipath on transmitted signals can be summarised in the following way. If one single symbol of duration T is transmitted with a bandwidth of $B = 1/T$, then we have no selective (with respect to frequency) fading if $T \gg \tau_m$, where τ_m is the multipath spread. The symbol can be coherently processed for a period T, if $T \ll 1/B_d$, where B_d is the Doppler spread. Thus for the case:

$$\tau_m \ll T \ll 1/B_d$$

there will be no loss of coherence over the symbol time and no frequency-selective distortion of the symbol.

Both inequalities are satisfied simultaneously if the spread factor $L = \tau_m B_d \ll 1$, and an optimum symbol time $T = \sqrt{\tau_m / B_d}$.

On the other hand, if the inequalities are violated this may put restrictions on the ability to use a matched filter and the ability to use system design when trying to estimate channel characteristics.

Since we are not talking about only transmitting one symbol of duration T, but rather continuous communication, we also need to establish the conditions for which we have no time smear, that is intersymbol interference or frequency smear which may give rise to adjacent channel interference. Again, we have the same inequality as for the non-selective fading, that is $T \gg \tau_m$ and $T \ll 1/B_d$.

We have already mentioned the effect which spread-spectrum direct-sequence systems may have by using bandwidths likely to experience frequency-selective fading, and its resistance to intersymbol interference as long as the multipath delay is longer than the duration of the spreading code-chip duration.

To finalise the discussion of multipath effects, let us go through one simplified but illustrative example, considering a certain channel characteristic and a spread-spectrum direct-sequence system as shown in Figure 3.18.

At the transmitter the spreading code can be denoted $p(n) = \pm 1$, with a pulse shape $a(t - nT_c)$ where $a(t) = 1$ for $0 < t \leq T_c$, and $a(t) = 0$ otherwise.

If this kind of signal is used to modulate the carrier frequency, the spectrum may take the form of a $\sin x/x$ curve centred at the operating frequency.

Consider now the simple two-ray model, with a response $(1 + \alpha\delta\tau_m)$; the received two signals making up one received code chip may look like:

$$S(t) = \sin 2\pi f_c t + \alpha \sin 2\pi f_c (t - \tau_m) \tag{3.10}$$

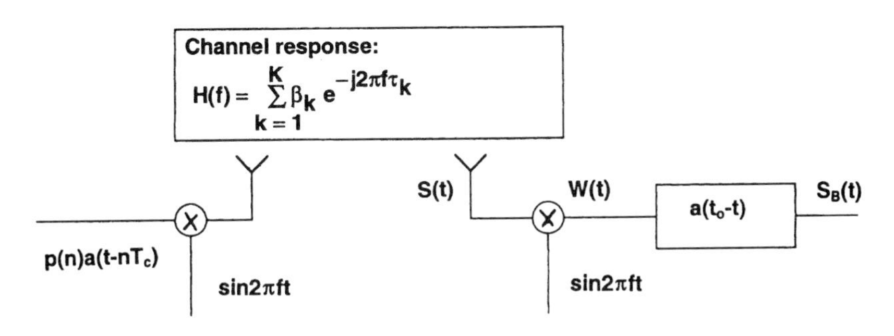

Figure 3.18 A simple spread-spectrum system operated across a multipath channel

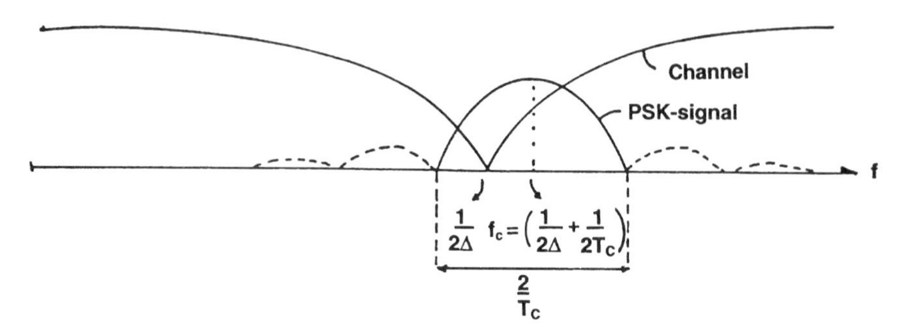

Figure 3.19 Functions for the two-path model and the spread-spectrum signal spectrum

where the first term is defined for $0 < t \leq T_c$, and the second term for $\tau_m < t < T_c$. Here, $f_c = (k/2\tau_m) + (n/T_c)$ where $k = 1, 3, 5, \ldots$ defines the particular channel notch (the frequency selectivity of the multipath channel) and n determines the position of the centre frequency relative to the channel notch. α is the relative strength of the two path signals and varies from 0 to 1. Figure 3.19 shows the channel response and the transmitted $\sin x/x$ spread-spectrum signal.

In the time interval $0 < t \leq \tau_m$ the last position of the previous code bit (chip) will interfere. Now for a pseudonoise-modulated spread-spectrum system it is acceptable to assume that there will be as many 0s in the spreading code during the message transfer as there will be 1s. Thus, on average, the signal deformation in the time interval $0 < t \leq \tau_m$ will be zero.

Removing the RF terms (multiply the received signal by the carrier frequency as shown in Figure 3.18) in the detected signal for the two-path model gives:

$$
\begin{aligned}
W(t) &= 1 && 0 < t \leq \tau_m \\
&= 1 + \alpha \cos 2\pi f_c \tau_m && \tau_m < t \leq T_c
\end{aligned}
\tag{3.11}
$$

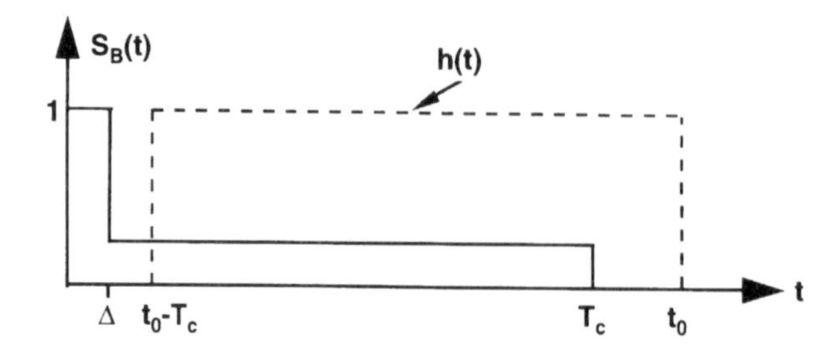

Figure 3.20 Received signal for two-path model

The signal is fed to a filter with a response $h(t) = a(t_0 - t)$ which matches the pulse shaping in the transmitter.

The result can be illustrated as shown in Figure 3.20.

After filtering the result is:

$$S_B(t) = \int_0^{\infty} h(t - \tau) W(t) d\tau \tag{3.12}$$

which for our model gives:

$$\begin{aligned} S_B(t) &= \int_0^{\tau_m} a(\tau + t_0 - t) d\tau + \int_{\tau_m}^{T_c} (1 + \alpha \cos 2\pi f_c \tau_m) a(\tau + t_0 - t) d\tau \\ &= T_c \left(1 + \alpha \left(1 - \frac{\tau_m}{T_c} \right) \cos 2\pi f_c \tau_m \right), \quad \text{for } t = t_0 \end{aligned} \tag{3.13}$$

The signalling across the multipath channel will result in a system degradation established both at the correlator input (spreading code-chip level) and at the correlator output. To simplify the calculations we will assume that the correlator is implemented as a linear matched filter for which the results should be identical. Thus we assume that we hereafter are looking at the correlator output. Then we rewrite eqn. 3.13 to read:

$$S(T) = R(0) \left(1 + \left(1 - \frac{\tau_m}{T_c} \right) \cos 2\pi f_c \tau_m \right) \tag{3.14}$$

where R is the correlation function for the spreading code.

Then for $\tau_m < T_c$ the second path may destructively interfere with the correlation peak. In addition, the peak may be smeared out and sidelobes can occur. All these effects will upset the synchronisation procedures and degrade the performance. The effect on the correlation peak for different positions of the channel notch relative to the spread-spectrum signal spectrum is shown in Figure 3.21.

3.2 The presence of noise and interference

The effect of noise and interference is important for the signal-detection process and is treated in that context in Chapter 4. However, since the previous sections have looked into the signal from a path-loss and received-power point of view, noise and interference are briefly mentioned here from the same point of view.

If the interfering signal is characterised among others by its total available power σ^2, evenly spread on the operating frequency band of the system, then we are talking about power-limited interfering signals. Such

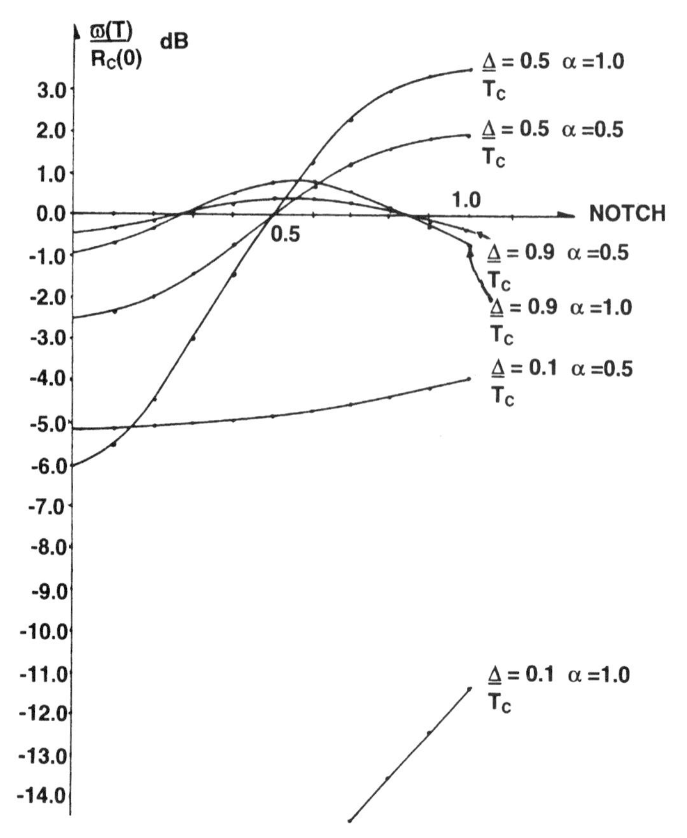

Figure 3.21 *Loss/gain in correlation peak for a spread-spectrum system experiencing frequency-selective fading*

signals are to be expected in the case of unintentional (other users) or intentional (in military applications a jamming signal) man-made interference which comes from a limited number of transmitting sources. Each of these sources is power limited because it has finite primary power and the transmitters have amplifiers with a maximum output power. The propagation of the interfering signals will be similar to that discussed for the wanted signal, but there will generally be no correlation between their path characteristics.

Frequency is a scarce resource and the mobile radio systems must try to make efficient use of the available frequencies. Channel separation should, at least, be kept at a minimum, and for many systems it may be necessary to re-use the frequency channels from one geographic area to another. In such situations it is necessary to remember that the signals are both the carriers of the wanted information and interference. Now, since the required signal level for satisfactory reception of the information is generally well above the levels at which the signal may interfere with other transmissions, the area of coverage for any particular system is much smaller than the area of interference which it may represent. This is shown in Figure 3.22.

The question of increasing the transmitter power for some transmitters to improve their coverage immediately means that some other system may reduce its coverage, or if it chooses to increase its power proportionally, be back to the starting point when it comes to area of coverage. This is illustrated in Figure 3.23.

In total the resulting interference level is increased to any other parties. However, for spread-spectrum modulated systems, the receiver may operate with negative signal-to-noise ratios and the interference effect is also significantly reduced for the case where users of the same system use different spreading codes.

Going back again to Figure 3.22 showing the area of coverage compared to the area of interference, it is obvious that a spread-spectrum wideband system, designed for signals to be received with negative signal-to-noise ratios, will to a lesser extent interfere with conventional systems, since the spread-spectrum signal will be buried in the background thermal noise at some distance from the spread-spectrum transmitter. Thus, the area of coverage may in fact be larger than the area of interference.

Looking a little more closely at the effect which a spread-spectrum signal may have on some conventional narrowband system, it is necessary

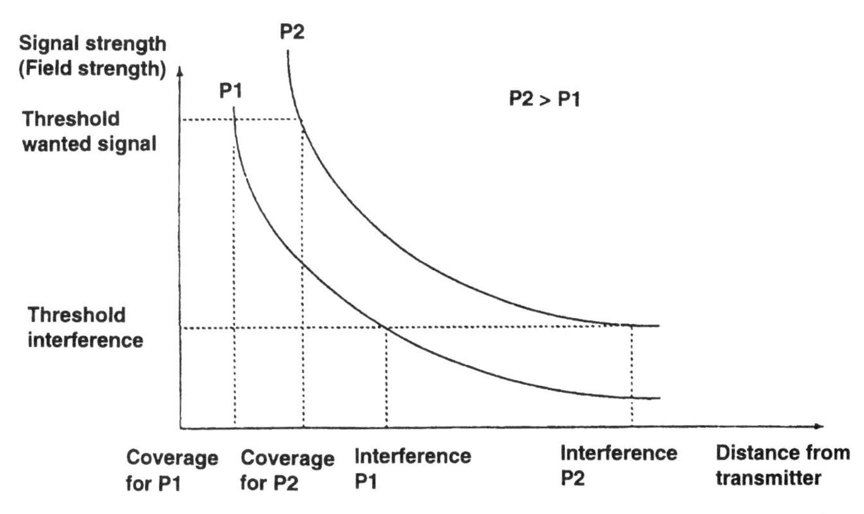

Figure 3.22 The area of coverage compared to the area of interference for two transmitters transmitting with a power of P1 and P2, respectively

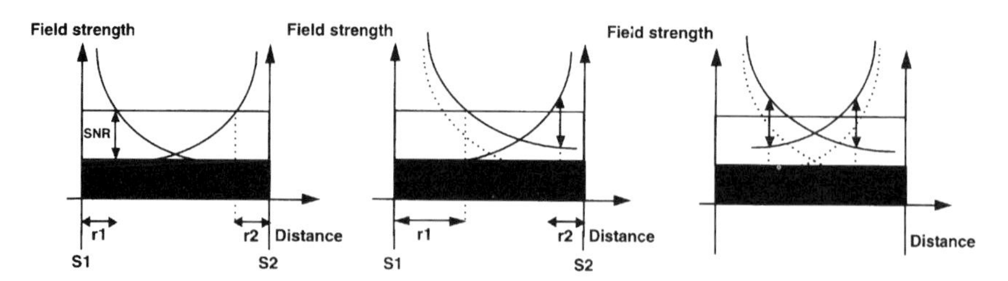

Figure 3.23 Illustration of how the systems may degrade from a situation where both receivers are limited by thermal noise to cases where one of the transmitters and then both transmitters increase output power to extend area of coverage

to develop the spectrum of a spread-spectrum signal. A general binary signal with fixed bit rate can be made equivalent to a sum of time-displaced bits. The spectrum for the binary signal may therefore be calculated by complex addition of the spectra for each individual bit.

Mathematically, the time signal for each individual bit can be expressed as:

$$s(t) = \pi\left(\frac{t - \pi}{T}\right) \tag{3.15}$$

where the π function describes a square pulse of width T and delay τ.

Transforming to the frequency plane gives:

$$S(f) = T\,\frac{\sin \pi f T}{\pi f T}\,\mathrm{e}^{-j2\pi f\tau} \tag{3.16}$$

The spectrum for a binary signal at a certain frequency may now be developed by a complex summation of the spectral values of each individual pulse. Generally, the spectrum of a binary signal will then vary around a $\sin x/x$ mean value. If the binary signal is periodic, the frequency spectrum will be a line spectrum, where the distance between the lines is given by $1/p$, where p is the period.

Now the same total signal power in the white noise as in the spread-spectrum noise gives:

$$\int_B N_o df = \int N_{DS} df = P_s \tag{3.17}$$

where the white-noise spectral density is given by $N_o = P_s/B$ and the spread-spectrum spectral density is given by $N_{DS} = (P_s/B)K(E(f_x))^2$, and K is the normalisation constant, as illustrated in Figure 3.24.

Then:

$$K \int_B (E(f))^2 df = 1 \tag{3.18}$$

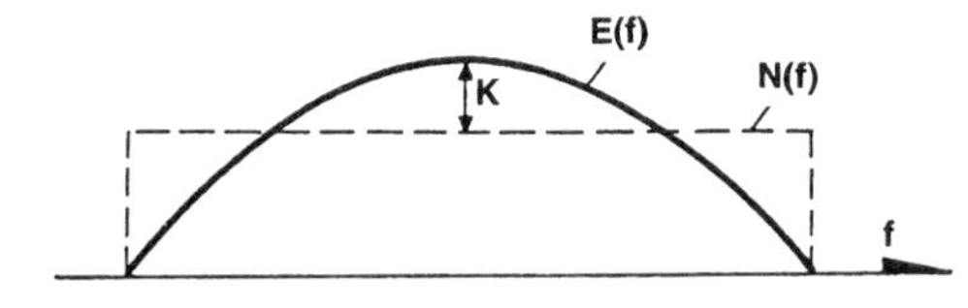

$E(f)$ is the mean value function for the spread spectrum signal spectrum

Figure 3.24 The frequency spectrum of the spread-spectrum signal and equivalent white noise

If the spread-spectrum signal has an MSK (minimum shift keying) spectrum, then $K = 3.9$ dB.

This means that the noise spectral density for the direct-sequence spread-spectrum distribution at the centre frequency is 3.9 dB above the white-noise spectral density when the total power of the spread-spectrum noise and the white noise are the same.

The noise power from the spread-spectrum signal within a bandwidth B_Δ within the spread spectrum passband is given by:

$$P_\Delta = \frac{P_{DS}}{B} \int_{B_\Delta} E^2(f)df + 3.9 \text{ dB} \tag{3.19}$$

where $E(f)$ is the mean-value function for the MSK spectrum.

Then, if the bandwidth of the conventional system is much less than the spread-spectrum bandwidth but much larger than the ratio of the spread-spectrum bandwidth to the total number of bits in the spreading code (the latter to make sure that a signal within the conventional system bandwidth for some time period is determined solely by the spreading code used during that time period), $E^2(f)$ will for a narrowband system be a wide and slowly-varying function, and the spread-spectrum noise may, to a good approximation, be looked upon as a frequency-independent noise density across the bandwidth of the conventional narrowband system with a power spectral density of:

$$N_{DS} = \frac{P_{DS}}{B_{DS}} E^2(f) + 3.9 \text{ dB} \quad \text{W/Hz} \tag{3.20}$$

This is illustrated in Figure 3.25.

If the spread-spectrum noise and the white noise are comparable in strength, then they should be added on a power basis as a total background noise.

On the other hand, a conventional narrowband interferer will be suppressed by the processing gain of the spread-spectrum system, once again reducing the area of interference.

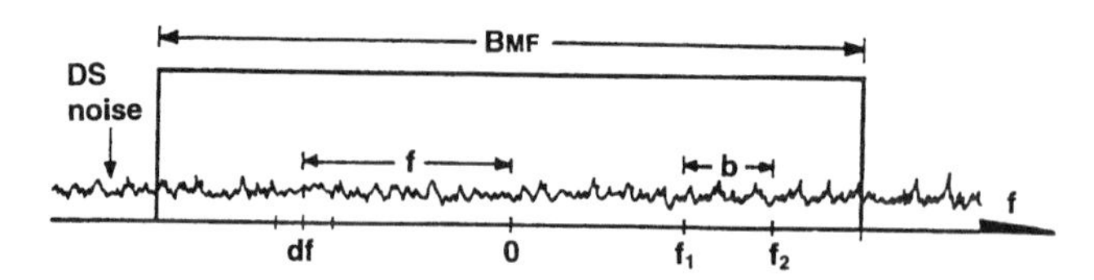

Figure 3.25 The noise from a direct-sequence spread-spectrum system within the bandwidth of a conventional narrowband system

Assuming the spread spectrum receiver makes use of a matched filter, the correlator impulse response will be a time-reversed replica of the spreading code:

$$h(t) = S(NT - t)$$

and thus

$$H(f) = S^*(f)e^{j\omega NT}$$

The level of continuous-wave interference at the detector will then be given by the amplitude function $|H(f)| = |S(f)|$, where $S(f)$ is the frequency spectrum of the spreading code. The receiver amplitude function will thus be the same as the frequency spectrum of the spreading code. It can be shown that the maximum deviation from the mean value may give considerable loss compared to white noise at unfavourable frequencies. This is illustrated in Figure 3.26.

If the conventional interfering signals are frequency modulated by a conventional channel signal which varies quickly in relation to the correlator processing time, then the interference level will be averaged around $\sqrt{N}$. A spread-spectrum signal may therefore be more robust against an FM-modulated interference than for a CW signal. If the spreading-code period covers several data bits, then the situation for CW and FM will be

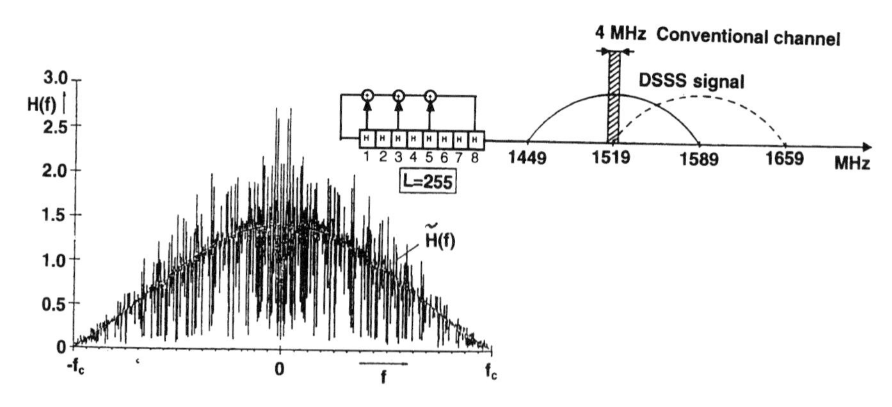

Figure 3.26 Example of how the correlator transfer function may look for a correlator of length 255. $H(f)$ is normalised to $\sqrt{255}/2$

more equal because one unfavourable frequency for one data bit may be favourable for some other data bit, and so on.

To sum up, the situation for a spread-spectrum system and a conventional narrowband system and the areas of coverage and interference may be illustrated as in Figure 3.27.

It is necessary at this point to remember that the use of spread-spectrum modulation does not extend the possible transmission distance owing to the fact that the receiver can operate with negative signal-to-noise ratios. In a wideband noise situation the gain obtained by being able to operate with negative signal-to-noise ratios is exactly balanced out by the increase in noise experienced in the receiver when going from narrowband to wideband. Spread-spectrum modulation in such a noise situation does not extend the distance of communication.

Now if the other users are other members of the same spread-spectrum system and the interference-rejection property is alternatively used as a code-division multiple-access capability, then the situation shown in Figure 3.23 will also be altered. As long as the other user interference is small compared to the thermal noise (the size of the other user interference will depend on the spread-spectrum code design), a transmitter's increase in output power to extend its distance can be combated by an increase in the other transmitter output power. However, from the point where the other user interference becomes dominant, we have a bandwidth-starved

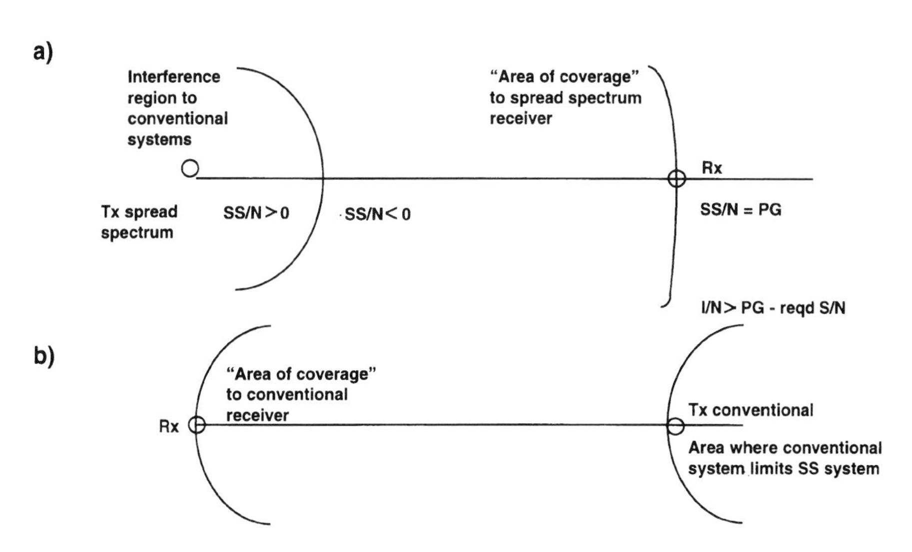

Figure 3.27 *The figure illustrates*
a how the spectrum spreading in a spread-spectrum system reduces the interference to other conventional systems
b how the spread-spectrum system operates with negative signal-to-interference (other conventional users) ratios

condition in the code-division multiple-access sense, and we are back to the situation pictured in Figure 3.23. This book will take a closer look at code-division multiple-access (CDMA) networks in Section 4.2.3.

Non power-limited interference exists in the form of thermal noise and in the form of man-made impulse noise coming from sources like motor ignition systems and different electrical equipment. Impulse noise tends to decrease as the frequency increases.

In a real situation, a combination of power-limited and nonpower-limited interfering signals will occur. If the power-limited signals have a power of σ'^2 spread uniformly across the bandwidth, B, of our channel and the nonpower-limited interference has a spectral power density of $N''_o/2$ within a bandwidth much larger than our system bandwidth, then the total spectral power density becomes:

$$\frac{N_o}{2} = \frac{\sigma'^2}{2B} + \frac{N''_o}{2} \tag{3.21}$$

For wideband spread-spectrum systems it is worth repeating that an increase in bandwidth, B, only reduces the spectral power density as long as the term $\sigma'^2/2B$ is not considerably smaller than the term $N''_o/2$.

If we consider the interfering power (and waveform) to be under the interferer's control, then we must achieve sufficiently high energy per information bit (E_b) in order to obtain the necessary bit error rates. Such a large E_b will probably make the contribution from $N_o/2$ very small and the final bit error rate will be determined under the influence of the interferer.

Thus if the energy per information bit in the wanted signal is denoted E_b, then the bit error rate will mainly depend on $(E_b/\sigma'^2/2B)$ and the influence of $(E_b/N_o/2)$ can be neglected. The bit error rate is then given by:

$$P_b = Q\sqrt{2E_b/\sigma'^2/2B} \tag{3.22}$$

which can be approximated by:

$$P_b \approx \frac{1}{\sqrt{\dfrac{4\pi E_b}{\dfrac{\sigma'^2}{2B}}}} \exp\left(\frac{-E_b}{\dfrac{\sigma'^2}{2B}}\right) \tag{3.23}$$

It is, however, at this point of value to note that it is not only the average noise power spectral density that determines the error rate. Instead of using an interferer with bandwidth equal to our system bandwidth and with power spectral density of $\sigma'^2/2B$ continuously, the interferer can be used with a duty cycle of γ but the same power spectral density. For a duty cycle of $0.5/(E_b/\sigma'^2/2B)$ the error rate can be approximated by:

$$P_b \approx \frac{1}{\sqrt{2\pi e}} \frac{1}{2E_b} / \sigma'^2/2B$$

The dramatic influence which the pulsing of the interferer has on the bit error rate is shown in Figure 3.28.

In fact, what can be observed is a manifestation of the fact that for a fixed average signal-to-noise ratio, the smaller the variance of the noise, the smaller the error probability. This fact is mentioned here, because it has a bearing on the usefulness of wideband systems, like spread spectrum on propagation channels with different forms of multipath and resulting signal fading. Wideband systems are often introduced just to combat the effects of channel imperfections.

3.3 User mobility

So far it has been shown how vertical surfaces and surrounding objects may give rise to a number of multipath signals with random amplitudes, phases, angles of arrival and time delays. In the case of user mobility this multipath geometry changes with time as the mobile systems traverse their

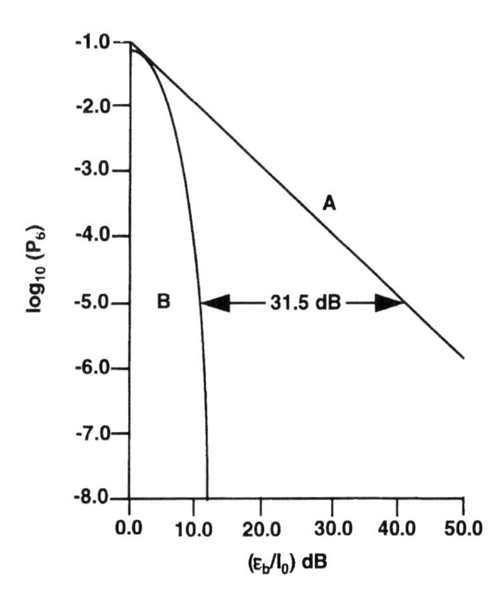

Figure 3.28 The effect of pulsing the interferer on system error probability. Curve A shows worst case pulse noise interferer and B shows continuous noise interferer

routes. The possible rapid changes show up as fast fading and Doppler spreading.

In a net each mobile radio must be able to transmit information to other mobile members of the net as introduced in Section 1.12 and further discussed in Section 2.3. To achieve such a goal the net must be given a certain structure used by the mobiles to establish connections. Hilly terrain gives the military plenty of opportunities to hide the mobiles but the price paid can be reduced connectivity. If the mobile radios are spread across a large geographic area, every radio may not be able to connect to all other radios directly. To ensure that every radio should be able to connect to any other radio, a common network with the ability to repeat or relay messages must be established. It is common to speak about two major solutions: centralised and decentralised (distributed), as shown in Figure 3.29.

Common for all public mobile systems is the idea that radio transmission is always conducted between a fixed base station and the numerous mobile users. The task for the base station is to relay information from one mobile radio to another. This limits the positions of the mobile radios since they all must have direct connections to the base station in order to operate. The radio coverage of the base station determines the maximum extension of the network. The base station is therefore usually placed high in the terrain in order to minimise the influence of the topography. Extension of the network can be achieved by connecting several base stations, for example by point-to-point radio-relay connections. Such a solution may be looked upon as vulnerable in the sense that the complete network depends on one or a few base stations.

In private radio networks, which are the theme for this book, many applications require direct mobile-to-mobile transmissions as well as all-to-all connections. A decentralised net offers more freedom when it comes to

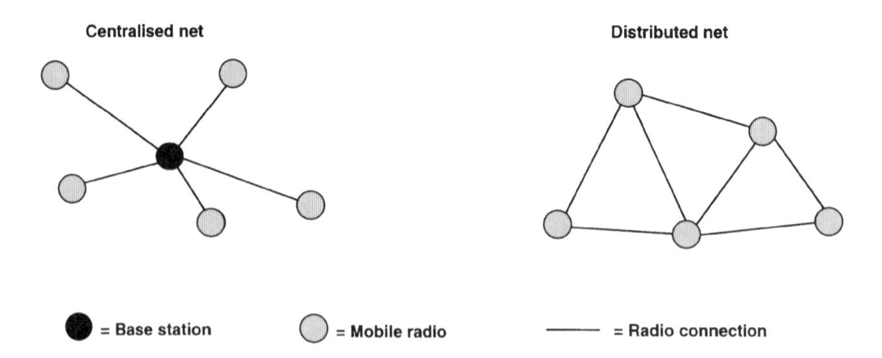

Figure 3.29 Different net structures, centralised and decentralised (distributed) nets

where the mobile radios may operate, because it is not necessary that all the radios are within radio coverage of each other. In order for mobile radio users without direct connection to be able to exchange information, every radio must be able to repeat information (relaying in multihop networks).

Let the signal that a mobile radio receives be denoted $S(t)$ and be the sum of K multipaths, that is:

$$S(t) = \sum_{k-1}^{K} S_k(t) \tag{3.24}$$

If it is now assumed that all these signal components arrive in a horizontal plane, a reference frame such as the one in Figure 3.30 can be defined.

If the transmissions are an unmodulated carrier, then the individual received signal can be given as:

$$S_k(t) = \beta_k \cos\left[2\pi f_c t - \frac{2\pi}{\lambda}(x_0 \cos\alpha_k + y_0 \sin\alpha_k) + \varphi_k\right] \tag{3.25}$$

where β_k is the amplitude of the kth path, f_c is the carrier frequency, the phase φ_k is uniformly distributed between $(0, 2\pi)$ and (x_0, y_0) is the position of the mobile radio.

If now the transmitter or receiver is moving, the signal components will be exposed to a Doppler shift. If the movement is in a positive direction along the x-axis the Doppler shift for the kth signal component will be given by:

$$f_k = \frac{vf_c}{c} \cos\alpha_k \tag{3.26}$$

This means that the signal components which arrive in front of the mobile radio will experience a positive Doppler shift, and the components arriving from the back will experience a negative shift. Intuitively the signal spectrum from the transmitter, being a delta pulse if an unmodu-

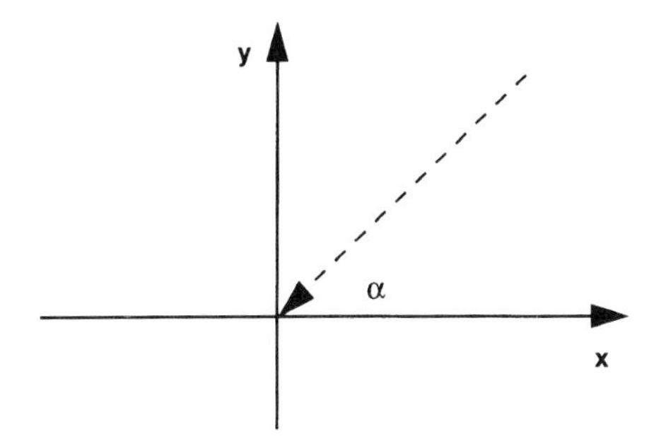

Figure 3.30 Reference frame for a multipath model

lated carrier is being transmitted, will experience a suppression or stretching in frequency. The shape of the RF spectrum of the received signal can be found by first calculating the autocorrelation function of the received signal $S(t)$ from eqn. 3.25:

$$R(\tau) = E(S(t)S(t_1\tau)) \tag{3.27}$$

The RF spectrum can then be established by calculating the Fourier transform of $R(\tau)$:

$$S(f) = F(R(\tau)) \tag{3.28}$$

An important assumption is that the probability density function of the angle of arrival is:

$$P_\alpha(\alpha) = \frac{1}{2\pi}$$

which means that one assumes that the multipath signal components are arriving from all directions α in the (x, y) plane with the same probability. Calculating the Fourier transform in eqn. 3.28 gives the classical Doppler spectrum in the form of:

$$S(f) = \frac{1}{\sqrt{1 - \left(\frac{f}{f_d}\right)^2}} \tag{3.29}$$

where f_d is the maximum Doppler shift given by $f_d = (f_c v/c)$ where f_c is the carrier frequency, v is the speed of the mobile radio and c is the speed of light.

For situations for which there exists a direct path (line of sight) in addition to the multipaths, the result is what is commonly called a Rice distribution which is the sum of the Doppler spectrum given in eqn. 3.29 and a direct-signal component. The direct-signal component arriving at an angle α_0 with respect to the direction in which the mobile radio moves, experiences a Doppler shift of $f_d \cos \alpha_0$. The resultant RF spectrum therefore contains an additional component at $f_c + f_d \cos \alpha_0$ as shown in Figure 3.31.

As the magnitude and shape of the channel impulse response will vary as the mobile radios move around it may be rather difficult to maintain communication during movement. The received signal will vary in a stochastic way and, considering the small antenna heights at the mobile radios, the signal will frequently be below the threshold for detectability in the case of obstructions between the communicating mobile radios. Under difficult situations any movements of less than one wavelength may change the impulse response. This will put heavy demands on the dynamics of any

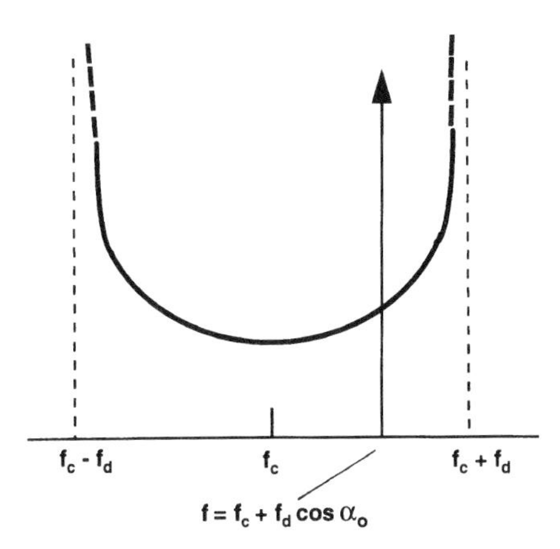

Figure 3.31 RF spectrum in the presence of a direct (line of sight) component

channel-equalisation function. In addition, the question of routing in a multihop situation will be a challenge. Channel equalisation and routing were considered in Sections 2.3.3 and 2.3.2. Characteristics of the mobile radio channel are usually described by the power-delay profiles and the Doppler spectra, as described in the next Section.

3.4 Characterisation of the channel

The communication channel may be specified by a scattering function $\sigma(r, f)$ which describes the distribution of the reflection cross section in time delay and in Doppler shift. For example, a channel can be envisaged as clusters of scatterers representing the different multipaths. Each individual path is composed of scatterers spread over a range delay and a Doppler interval. Figure 3.32 illustrates such a channel-scattering function.

The received energy is the integral of the contributions from all the scatterers affecting the channel. In addition to the scattering function, the channel can be characterised by a channel-correlation function $R(t, s, \tau, \zeta)$, where t, s form a pair of time variables and τ, ζ are delay variables. For wide-sense stationary channels the correlation functions are invariant under a translation in time, i.e. they depend on the variables t and s only through the difference $t - s$. For the uncorrelated scattering channel the scattering amplitudes of two different scatterers with different delays are uncorrelated. Hence, for a wide-sense stationary uncorrelated scatter channel (WSSUS channel) [3]:

$$R(t, s, \tau, \zeta) = Q(\tau, t - s)\delta(\tau - \zeta) \tag{3.30}$$

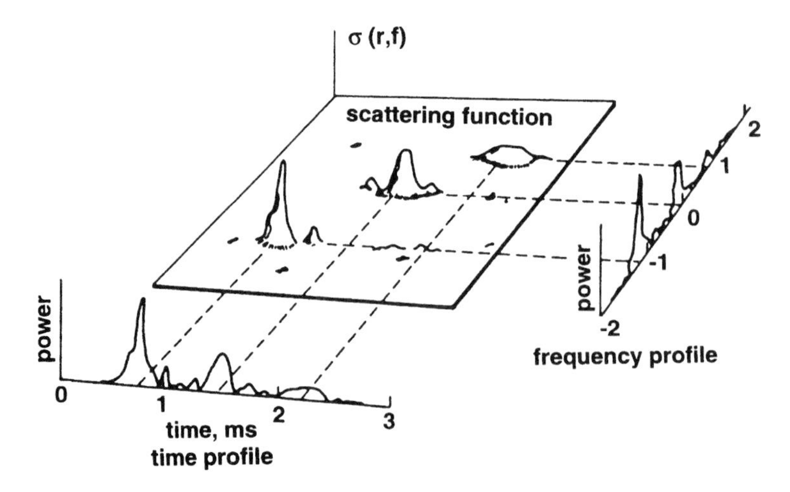

Figure 3.32 Time-frequency plane of the channel-scattering function

where Q is the delay cross-power spectral density. Taking the Fourier transform of Q gives:

$$\int_{-\infty}^{\infty} Q \exp[-j2\pi\tau(t-s)]d(t-s) = \sigma(\tau, f) \tag{3.31}$$

which is the same as the channel-scattering function with the range delay replaced by the corresponding time delay, τ.

For channels with Doppler spread only (time selective), we can define a Doppler-scattering function $S_D(t)$ and a correlation function $Q(0, t-s)\delta(\tau)$:

$$Q(t-s) = \int S_D(f) \exp[j2\pi f(t-s)]df \tag{3.32}$$

and

$$Q(0) = \sigma_x^2 = \int S_D(f)df \tag{3.33}$$

where σ_x^2 represents the mean squared value of the underlying random process of the channel. For channels with only multipath we can define a range-scatter function $S_r(\lambda)$ where λ is an increment in distance, which expresses the expected value of energy returned from an increment located at a certain distance. Total received energy can then be expressed as

$$E_{received} = E_{transmitted} \int S_r(\lambda)df \tag{3.34}$$

Taking the Fourier transform of $S_r(\lambda)$ we obtain the two-frequency correlation function:

$$R_v = \int S_r(\lambda)\exp(-j2\pi\lambda\nu)d\lambda \qquad (3.35)$$

The nondispersive channel may then finally be characterised by $\sigma(r,f) = \delta(r)\delta(f)$.

Consider a transmitted waveform $s(t)$ with duration T and bandwidth B. The waveform may be subjected to range delays of τ seconds to Doppler D Hz. If $\tau < T$ there is no apparent time spreading of the received waveform. However, the signal may still be distorted if $TB \gg 1$ and $B \gg 1/\tau$. This is because it is possible for a waveform of bandwidth B to change its value significantly in a time $\tau \gg 1/B$. This means that it is possible that returns from some multipaths will interfere. Figure 3.33 shows a segment from a PSK-modulated spread-spectrum signal, before and after transfer across a channel where $B^{-1} \ll \tau \ll T$. Thus it is possible for a spread-spectrum system to degrade even without any apparent time dispersion. If the Doppler spread B_d is much less than T^{-1}, the signal waveform reciprocal duration, there is no apparent frequency spreading of the received waveform. If $TB \gg 1$ then $B_d \ll B$ will result in no dispersion, but if $B_d T \gg 1$ the received signal will be distorted.

3.5 A designer's perspective

Fading and Doppler will be of major concern to the mobile-radio designer. In particular, the fading rate (the number of fades per unit of time) will be of interest. The fading is often described by a level crossing rate (LCR) and an average fading duration (AFD) as illustrated in Figure 3.34.

As can be seen from the Figure, both LCR and AFD will depend on the specified reference level R.

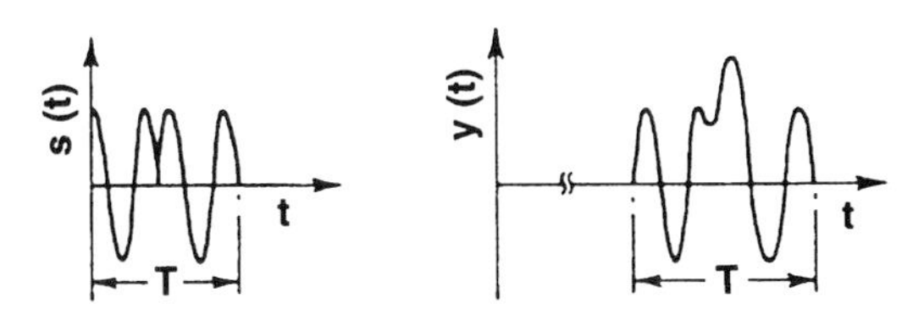

Figure 3.33 Waveform for channels for which $B^{-1} \ll \tau \ll T$. $s(t)$ is the transmitted waveform and $y(t)$ is the received waveform

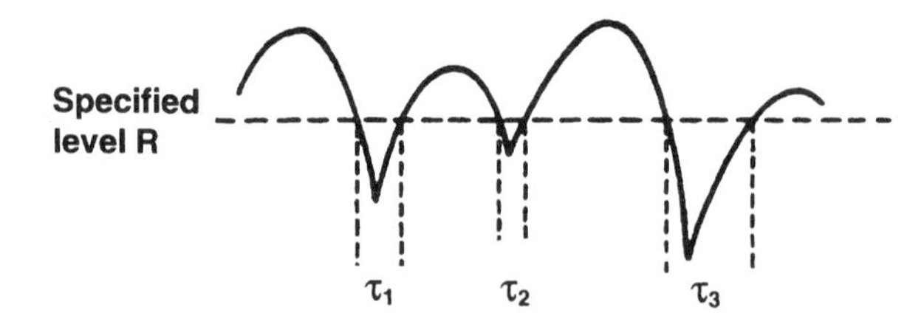

Figure 3.34 The curve indicates a signal that varies with time. The AFD parameter is the mean value of τ_1, τ_2 and τ_3

LCR will also depend on the operating frequency and the movement of the mobile radio antenna. For a vertical monopole antenna LCR can be expressed as:

$$\text{LCR} = \sqrt{2\pi}(vf/c)\rho e^{-\rho^2} \tag{3.36}$$

where f is the operating frequency, v is the speed of the mobile, ρ is the ratio between the specified reference level for R and the r.m.s. amplitude of the fading envelope. A maximum rate is reached when $\rho = 0.5$. For this value:

$$\text{LCR} = 0.976(vf/c)$$

In addition to LCR and AFD, the fading depth is used to describe how much the signal strength varies.

Two different examples can be used as illustrations:

Situation 1

Assume a stationary situation where geography, meteorology and reflecting surfaces do not change. The operating frequency is kept constant but the mobile radio is on the move. Such a situation may give a signal variation as pictured in Figure 3.35.

The rapid signal variations are due to local multipath and the much slower variations are due to movements across large distances. The latter fading will result in a slowly-varying local mean value. The mobile radio receiver will experience a continuous variation of the signal with time. The parameter LCR will increase with increased speed of the mobile and the parameter AFD will decrease for the same conditions.

If the operating frequency is set to 88 MHz and the speed of the mobile radio is changed from 50 km/h to 100 km/h then the mean time between crossing the level R for $\rho = 0.5$ changes from 251 ms to 126 ms.

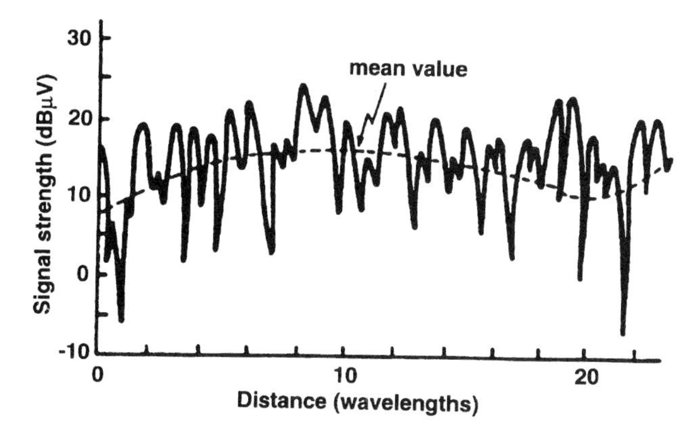

Figure 3.35 *Typical signal variations in an urban area when the frequency is kept constant and the mobile radio is moving at constant speed*

Situation 2

As for situation 1, but now the mobile is at a standstill while the operating frequency is changed. If the frequency is changed stepwise then the received signal strength varies stepwise. This is illustrated in Figure 3.36.

If the difference in frequency between changes is larger than the coherence bandwidth the signal strength is statistically uncorrelated. Whether the signal will remain constant at each individual frequency will depend on the coherence time $T_{coherence}$ which is defined as [4]:

$$T_{coherence} = 9/(16\pi f_d) \text{ (with } f_d \text{ defined as in eqn. 3.29)}$$
$$= 9 \cdot c/(16\pi f v) \text{ (where } c \text{ is the speed of light)} \qquad (3.37)$$

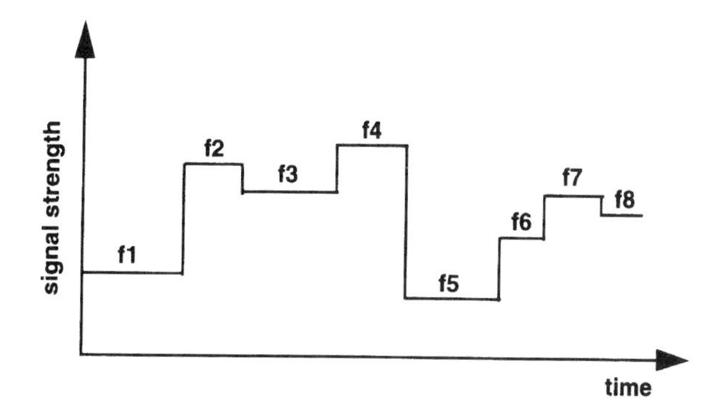

Figure 3.36 *An example of signal variations in a receiver which is at a standstill but where the operating frequency is changed*

Thus, if the mobile is kept at a standstill as assumed, the signal will always be constant at each frequency. If the mobile however is starting to move, a minimum coherence time will occur at maximum operating frequency and maximum speed. For $f = 88$ MHz and $v = 100$ km/h, $T_{coherence}$ will be 22 ms. For most frequency-hopping systems this will be sufficient for assuming a constant signal level at each hopping frequency. For $f = 2640$ MHz and $v = 330$ km/h, $T_{coherence}$ will be reduced by a factor of 100 which may be insufficient for assuming a constant signal level.

The parameter LCR will increase when the rate of frequency change increases and the parameter AFD will decrease.

Both situations will experience time-selective fading, even though the outset in situation 1 is space selective and situation 2 is frequency selective. The variations in signal amplitude may be the same for the two situations, but when and how the signals are changed will be different.

3.6 The balancing act

Rather than giving a comprehensive and thorough theoretical treatment of the mobile radio channel, which can be found in numerous excellent textbooks, this Chapter has tried to highlight the characteristics of the channel and its use most likely to influence any mobile radio system design.

Spread-spectrum modulation gives a system certain characteristics and by itself or in combination with various signal-processing schemes is very useful for matching the performance and services wished for and the channel.

The thought of providing services in the field comparable to those of a fixed network relying on nothing more than low-quality radio channels, is only worth following up because of a never ending technological cornucopia.

However, the task of designing a distributed-mobile spread-spectrum radio system is still full of challenges, compromises and a balancing act with operational desires on one side and the realities of life in the form of an imperfect channel on the other.

Even though it may be possible to combat certain channel imperfections and interference using advanced signal-processing and modulation techniques, the establishing of a radio network with reliable services still requires a lot of thought at the network level. This is the theme for the chapters to come.

References

1. OKUMARA, Y., *et al*: 'Field strength and its variability in VHF and UHF land mobile radio services', *Rev. Electr. Commun. Lab.*, September–October 1968, **16**, pp. 9–10
2. LONGLEY, A.G., and RICE P.L.: 'Prediction of tropospheric radio transmission loss over irregular terrain — a computer method'. ESSA technical report, ERL 79–ITS 67, 1968
3. BELLO, A.P.: 'Characterization of randomly time-variant linear channels', *IEEE Trans.*, 1963, **CS-11**, (4), pp. 360–394
4. STEELE, R.: 'Mobile radio communication' (Pentech Press, 1992) p. 123

Chapter 4

Radio transmission system

This Chapter deals with the various aspects of systems design that were considered to be the province of radio engineering. With the introduction of digital communications, several new themes have been introduced, such as signal coding and new modulation techniques. In addition, digital technology has caused significant changes to radio architecture due to factors such as shielding and power consumption. The explosion of commercial radiocommunications has also made possible the mass production of high-performance radio chip sets. Several books are available on specific aspects of digital radio such as continuous-phase modulation and forward-error coding, some of which are referred to later for further study. Our purpose is to give a general description of radio transmission systems, with more detail on specific areas relevant to spread spectrum communications. Direct-sequence spread spectrum systems will be given special attention.

4.1 Transceiver description

4.1.1 Definition and performance characterisation

In modern communication systems, the transceiver itself is usually only one of several system components. Even though many of the trade offs with respect to the propagation environment are defined by this unit, system optimisation is not guaranteed merely by optimising the radio parameters such as bit error rate, data rate or sensitivity. However, traditional radio performance measures are significant and play an important role together as overall system considerations.

The term transceiver is used here to refer to the basic radio element of a mobile terminal or similar elements at a base station if present. However, considerations of performance, cost and lifetime might well lead to different design solutions for mobile terminals and base stations.

The antenna is usually considered part of the radio unit, since RF solutions are highly dependent on which type of antenna is to be used. Antennas are again a compromise between performance, user convenience and to some extent cost. User convenience is usually a major factor for mobile terminals, especially for hand-held equipment. Concern about the influences of radiated power on the human body is also starting to receive more attention and may require some untraditional thinking.

Users interact with any system via terminals. In principle, there is a clear interface between the transceiver and the connected user terminal. This interface should give the user control over user data, protocol data and control information. In practice, this is not the situation. For instance, the terminal controls digital user data, although analogue sources are still dealt with separately by the radio unit (interface II in Figure 4.1). Thus, digitisation of the signal is done inside the radio unit and not in the user terminal.

Only when it comes to data terminals is there seemingly a clear interface. There are several standards covering a range of functions, from simple HDLC to X.25, ISDN and even more advanced protocols. Through increased focus on multimedia applications it also seems that voice- and video-coding standards should be implemented in the user

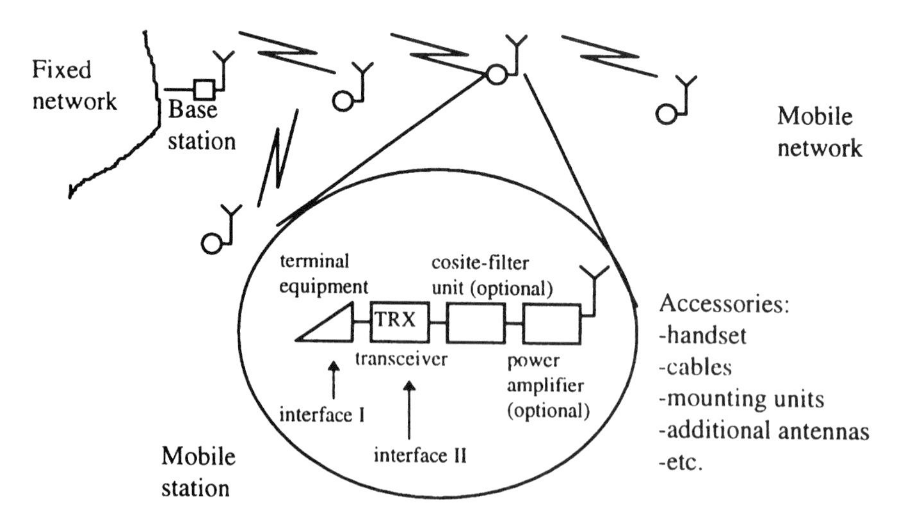

Figure 4.1 Components in a radio transmission system

terminal. Voice coding may well cease to be of prime interest to radio engineers in the future, although very low-rate coding could well be an exception to this.

Radio architecture has gone through two distinct generations and is fast moving into a third. After the pioneer days of radio, the superheterodyne radio principle has dominated analogue radio design, together with traditional AM and FM modulation techniques. The superheterodyne principle was invented by Edwin H. Armstrong in 1918, and today there are several types of superheterodyne structure. They differ mainly in the number of conversions, with the double-conversion superheterodyne being the most common.

A second-generation architecture can be found which resulted from the introduction of digital signalling. This generation is much more diverse than the previous, but the digital baseband can be seen as a characteristic of the second generation. Analogue-to-digital (A/D) and D/A conversions are usually made at baseband, but some of the latest technology performs these conversions at IF. This generation uses digital signal processing (DSP), which opens up a whole new range of capabilities and has given radiocommunications a major performance boost. Cost, power consumption and processing power have been limiting factors, but this is changing rapidly as technology progresses.

Thus, a third generation can be foreseen using a software-radio concept. A software radio should be a fully programmable radio, both with respect to functionality and signal formats. It would permit multiband RF, digital IF and multimode baseband. One of the aims of this concept is to develop an architecture which makes it possible to adapt radio performance to user needs in a dynamic network environment. We can easily think of such a unit as a managed object in an intelligent network, making it possible to remotely tailor data rate, channel coding, frequencies etc., depending on the type of service required. The software-radio concept is particularly interesting for military communication in view of its interoperability. One can imagine a standard radio hardware platform, where proprietary signal formats can be licensed, similar to the situation for computer programs.

Design criteria and performance characterisation have changed with changes in architecture and technology, and have certainly changed with the transceiver becoming a part of service-integrated networks. Various performance criteria can be divided into two main groups: those related directly to radio propagation and those (such as bit error rates, protocols etc.) concerned directly with information retrieval. In the following section some common performance criteria are mentioned with comments on their characteristics.

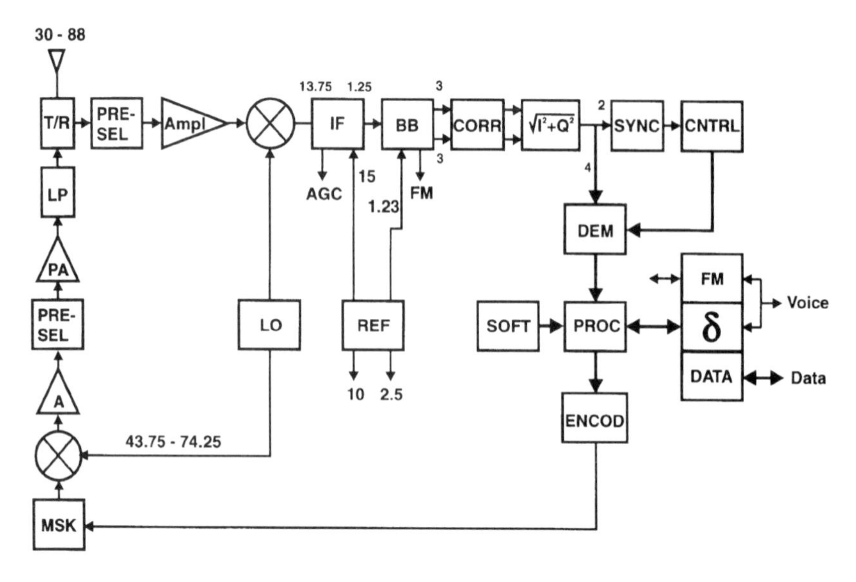

Figure 4.2 Typical second-generation radio architecture

Spurious transmissions: At present, most radio transmitters perform modulation at baseband, convert this to IF, and then reconvert this frequency up to the carrier frequency. In any radio which uses this principle rather than direct modulation of the RF carrier, some unwanted mixing products (spurious) will be generated in the process. IF and RF filtering will minimise these products, but they cannot be eliminated completely. The number of spurious frequencies with significant power varies with the carrier frequency in use, and is likely to create unwanted disturbance in a dense electronic environment. This is a particular problem with frequency-hopping radios, and results in unwanted artefacts for the GSM system.

Sensitivity: Radio sensitivity is limited in many ways. First of all, a thermal white noise background (−174 dBm/Hz) is always present, increasing proportionally with receiver bandwidth. Then there are internally-generated frequency components within the RF bandwidth. This is a particular problem when sensitive analogue components are placed in hybrid analogue/digital circuits. Finally, we have externally-generated unwanted signals which include everything from jamming to cosmic noise. This external noise level is very sensitive to the exact location of the radio. In places with a great deal of electronic equipment, these noise levels are high, and consequently the designer does not have to be too concerned with internally-generated noise nor with the noise factor of the front-end amplifiers of the radio. Apart from practical aspects such as insertion loss and noise figure, it is theoretically the signal format, including coding, which determines radio sensitivity and therefore the range of a digital radio.

Selectivity: Selectivity is defined as the ability of a radio to reduce the effect of unwanted signals, while receiving a weak wanted signal. The selectivity of a radio depends upon three main factors: IF-response, local oscillator (LO) purity and preselection filters. In a superheterodyne radio, the main filtering takes place at IF, usually in a narrowband crystal filter. This filter defines the radio response at the desired frequency compared to that at close neighbouring channels. The purity of the local oscillator also plays a major role in determining the selectivity. The reason is that after the mixer, the unwanted interference inherits the noise sidebands of the local oscillator enabling these to enter into the IF passband if the interference is strong enough and close enough to the IF centre frequency. This happens even if the interference is a perfectly-clean sinusoid. The preselection filter is also important when it comes to total selectivity. Ideally, it should tolerate watts of power and have zero insertion loss and a bandwidth comparable to the bandwidth of the signal. In addition, it should be precisely tuneable over the whole desired band. Selectivity is increasingly important as frequency bands become more crowded.

Signal-to-noise ratio (SNR) and/or bit error rate (BER) performance: Although sensitivity tells us something about the range performance in a noise-free environment, an SNR requirement tells us how well the radio will perform in an environment with external noise sources, assuming that the noise is white, which is rarely the case. In most cases, however, SNR will give a good indication of the relative performance of different radio systems. Traditionally, audio SNR was used as a figure of merit using a given r.m.s. signal (or carrier) input voltage. With digital signalling, SNR is usually linked with bit error rate (BER). Either BER is given as a function of the SNR of the input RF signal, or for a given maximum BER, the required received SNR is stated. BER is the same as probability of error and SNR is directly related to the term E_b/N_0. These relationships are covered further theoretically in Section 4.3.

Block error rate: When designing radios for mobile networks, block error rate is more useful that bit error rate. Block error rate is made up of three factors: probability of message synchronisation or message detection, probability of correct block identification and probability of correct user data. Block error rate is thus a parameter that will influence protocol design directly. This also implies that packet or block length at the radio level is not merely a matter of convenience with respect to error coding. Another aspect of this is the balancing of error probabilities between the various elements which make up a packet. A preamble that is long compared to the rest of the block will result in valuable time and capacity being wasted on unusable data. On the other hand, an insufficient preamble will mean that signals which should be readable will not be recognised, again necessitating retransmission.

This discussion is far from complete, and is only meant to indicate some of the parameters which must be taken into account when characterising transceiver performance. These and other factors will typically be a part of a system-requirement specification, but note that their relative importance will vary depending on the user scenario. Also, parameters of this type do not necessarily tell much about system performance in a complete mobile radio network.

Radio-performance parameters are treated further in other sections, where spread spectrum and other techniques are reviewed in the context of packet-switched networks.

4.1.2 Radio architecture and functional units

To introduce spread spectrum modulation in terms familiar to communication and radio engineers, a communication link divided into its functional blocks according to a second-generation radio architecture is presented in Figure 4.3. The subunits are characterised by their function, and the particular configuration of functional blocks has been more or less fixed during the development of steadily more sophisticated and advanced systems. Each improvement has added complexity to the communication system and required new theoretical methods and mathematical tools. In addition, it will be found that for various systems the relative importance of each of the functional blocks will be different.

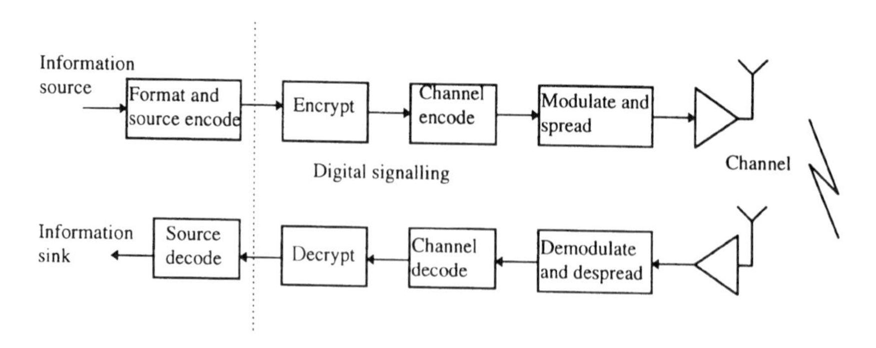

Figure 4.3 Functional blocks of a spread spectrum communication link

4.1.2.1 Formatting and source coding

In the previous generation of radio systems, the whole radio architecture was based on analogue signalling and was basically designed for audio. As the need for transferring digital data increased, various types of modem were designed to connect to the existing analogue radios using the audio channel. As the digital technologies evolved, and as the economies of large-scale integration were developed, the radio architecture changed into providing digital channels. However, audio and video sources are still analogue, and thus have to be digitised prior to coding and transmission. In order to digitise a signal, we must first discretise time (sampling) and amplitude (quantisation). This is referred to as formatting.

Source encoding is introduced in order to remove some of the redundancy in the data stream, and so to convey the same information with a small number of bits, enabling increased data rate or better protection with addition of error correction. In other words, the objective of source coding is to convert the original symbol space to a symbol space which requires the minimum channel capacity. This is obtained when the length of each symbol is made proportional to its relative probability of occurrence. An alternative formulation is to require the input data stream to have maximum entropy. Theoretically, this can be viewed as follows.

The symbol set is denoted by $X = (x_1, x_2, \ldots, x_N)$ and the relative probability of occurrence of each symbol by $p(x_i)$. It follows that:

$$\sum^{N} P(x_i) = 1 \tag{4.1}$$

It is obvious that more information is received in the detection of a symbol with a low probability of existence than in the detection of a more probable symbol. The total mean information content (entropy) $H(x)$ of all symbols is:

$$H(x) = \sum_{i=1}^{N} p(x_i) I(x_i) \tag{4.2}$$

where $I(x_i)$ is the information content of each symbol x_i. As the requirement is that the information rate should be maximum when all symbols are equiprobable:

$$P(x) = 1/N \tag{4.3}$$

then the relationship between $p(x_i)$ and $I(x_i)$ should be logarithmic and of the form:

$$I(x_i) = \log_B \frac{1}{p(x_i)} \tag{4.4}$$

where B is the base of the logarithmic function.

This gives the following for the total mean information:

$$H(x) = \sum_{i=1}^{N} p(x_i) \log_B [p(x_i)]^{-1} \tag{4.5}$$

Usually, the base chosen is $B = 2$. By convention, the corresponding unit for the information content is bits/symbol. If the decimal logarithmic function is used, the unit is decit/symbol (or hartley/symbol), and the entropy measured by the natural logarithmic function is given as nats/symbol.

These few equations show the motivation behind source coding as practised both with speech coding and video coding. Most voice coders include digitisation, source coding and error coding as an integrated module, available either as software algorithms only or integrated on a digital signal processor. Image and video coding are equally becoming standardised, even though there are more *de facto* standards in this field rather than internationally-agreed standards. As these settle, we are likely to see the sources themselves made to produce standardised digital outputs, so that the area of source coding becomes of less interest to the radio engineer. Source coding will not be covered any further, but this is a highly specialised field with a lot of published material available for further investigation [1].

4.1.2.2 Encryption

Encryption and data security are essential to many users of mobile communication systems. In particular among those seeking a tailor-made system the security aspects are vital. For a multimode system capable of handling voice, data and other services encryption might be applied to data-transmission services only or to all services, jointly or independently.

There is not much unclassified written material available on encryption. But, a general introduction to this area can be found in [2,3].

4.1.2.3 Channel coding

The purpose of channel coding is to trade information rate capacity with information error rejection and/or protection. This is achieved by adding to the transmitted symbols a set of symbols which do not carry information, but which are derived from the nonredundant set of symbols in such a way that they can detect and also correct errors in the information carrying symbols. Theoretically, this can be viewed as follows.

The symbol list is thus:

$$X = (x_1, x_2, \ldots, x_N, y_1, y_2, \ldots, y_M) \tag{4.6}$$

where y_1 to y_M are the M redundant symbols with probabilities $p(y_i)$.

The corresponding reduction in channel capacity relative to the nonerror-coded channel is given by:

$$\sum_{i=1}^{M} p(y_i) = 1 - \sum_{i=1}^{N} p(x_i) \qquad (4.7)$$

The upper bound of the improvement in error performance will be given by the relative distribution of energy in the redundant and nonredundant symbols:

$$y_E = \frac{\sum_{i=1}^{M} p(y_i)}{\sum_{i=1}^{N} p(x_i)} = \frac{\sum_{i=1}^{M} p(y_i)}{1 - \sum_{i=1}^{M} p(y_i)} = \frac{1}{\sum_{i=1}^{N} p(x_i)} - 1 \qquad (4.8)$$

For an actual error-coding scheme, the improvement factor will be lower than the optimal figure y_E. Error coding can be achieved in various ways, some of which are described in Section 4.4.

The channel coder usually translates the incoming data stream into a set of symbols. These symbols are used by the modulator and translated into an RF waveform. The modulator may take the form of a simple amplitude or phase modulator or a more complex spread spectrum modulator.

The traditional modulator is given a binary data bit stream with only two different states, 0 or 1. This is referred to as a 2-ary symbol set. Each of these states may be represented using either amplitude modulation, phase modulation (BPSK, DPSK) or frequency modulation (BFSK). However, the channel coder may also choose to expand the signal set into a higher number of symbols given as $m = 2^k$, where k is equal to the number of bits in each symbol. The m-ary coder defines the degree of parallelity used in the modulator thus directly buying circuit complexity or processing power to increase performance. This is treated further in Section 4.3.

4.1.2.4 Spread spectrum signalling

To facilitate the transfer of data through the transmission medium, the data stream has to be converted into a suitable radio-frequency waveform. The traditional approach has been to modulate the data stream into a radio-frequency carrier and stack the individual transmissions side by side in the frequency band. Frequency, amplitude or phase modulation is used, depending on the particular system design.

The importance of spread spectrum modulation techniques is that, by a substitution of the basic RF modulator by a modulator capable of generating waveforms with an information rate orders of magnitude higher than that of the data flow, a dramatic increase in transmission properties and system capability is obtained.

In a spread spectrum system, the basic idea is to replace a single data bit with a coded waveform. This data bit can then be detected optimally only when this code is known to the receiver. The more complicated and well designed is the code, the better and more secure will be the transmission. Various techniques are presented in the following section.

4.2 Spread spectrum techniques

This section intends to give an overview of the main spread spectrum techniques and their application within mobile communication today.

4.2.1 Traditional spread spectrum

Spread spectrum in radiocommunications has its origins in military systems. Desired features such as low probability of intercept (LPI) and antijamming protection were the driving forces behind a great deal of interest in spread spectrum techniques. Most definitions of spread spectrum systems have, accordingly, been focused on decreased spectral density through bandwidth expansion and processing gain.

4.2.1.1 Time/frequency-domain coding

For the purpose of describing the differences between various techniques, it might be useful to view spread spectrum as time/frequency-domain coding. The basic idea of time-domain coding is to add a certain amount of information (a code) to an input data stream, and in such a manner that the original data stream can only be decoded by knowing the original code. As an example, this can simply be accomplished by replacing data bits by code sequences.

The use of an unknown spreading code gives the communication a certain privacy. In order for a third party to spoof the communication, not only does he need a replica of the spreading code used, but also he must align his code in time with that of the legal transmitter. The reason for this is simply that the receiver will, as soon as it is synchronised to the legal transmitter, insert a time window and then only accept correlation peaks within this window. The intelligent jammer can, however, accomplish this task by sliding his timing to capture the victim receiver.

In a bursty communication system, where communication links are established and disconnected on a random and irregular basis, there would be plenty of opportunities for an unauthorised user with knowledge of the spreading codes to spoof the system. Thus, certain procedures for constructing a legal data message may be needed. Based on these procedures, legal and illegal messages could be sorted out at the data handling level. However, in many military applications it is also considered necessary to design the code generator so as to prevent an unauthorised listener deriving the code design. This can be achieved by special nonlinear code generators or by changing the spreading codes on a random and irregular basis during a message transfer. Thus, any illegal transmitter would have difficulties in fabricating the system for any significant period.

The process of imprinting codes on a data stream in a reversible manner requires that the data flow time series is mapped into a multidimensional space with a sufficient number of free variables. Any suitable transform could be applied, but as the communication and radio engineer is assumed to be attached to the theory of frequency analysis, this will be chosen as the basis for the treatment. As an introduction and recapitulation, an outline will be given of how time–frequency representation of signals describes a time series in a manner which is fairly simple yet gives all the degrees of freedom needed.

Figure 4.4 illustrates the basic idea of coding, namely to substitute a data bit with a code matrix. The Figure shows a general code pattern in which each of the time–frequency slots is independent. Furthermore, as illustrated in Figure 4.5, the ratio T/F can be chosen to suit different

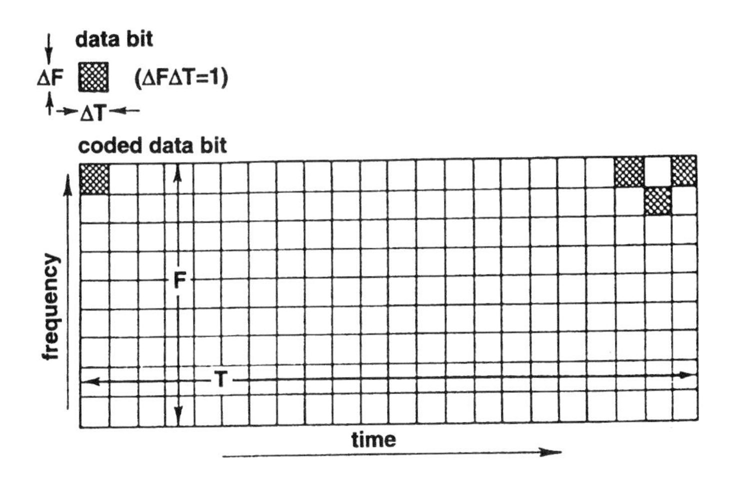

Figure 4.4 *By coding, each data bit carrying one unit of information is represented by a coded waveform carrying TF units of information*

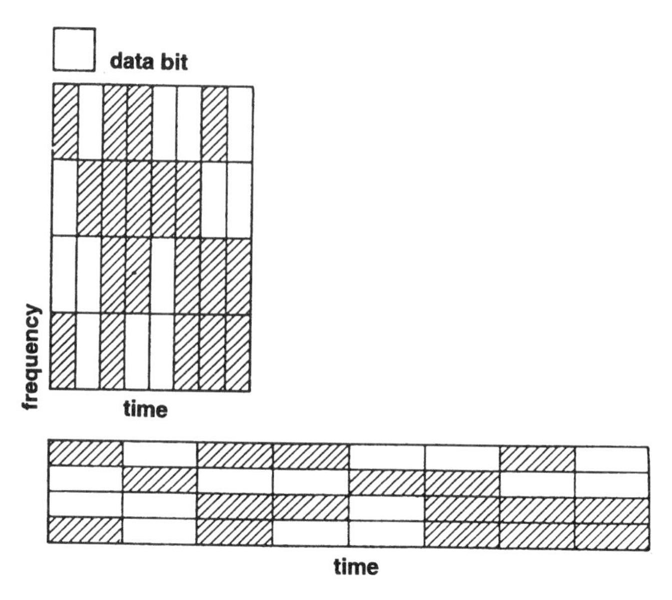

Figure 4.5 *Relative T/F ratio of any code can be adjusted to match bandwidth capability of channel or data transmission-rate requirements*

needs. The frequency F can be chosen to suit a band-allocation scheme, or the duration T can be chosen to give a required data rate.

Generalised coding scheme

A generalised coding scheme implies that no restraints are imposed on the structure of the code, and the C map will have the random structure as seen in Figure 4.6. From this it follows that there is full freedom to fight countermeasures or transmission-medium limitations (e.g. multipath).

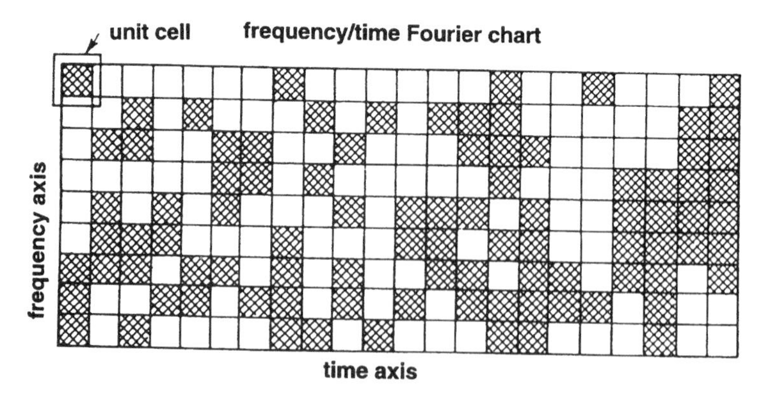

Figure 4.6 *Generalised coding scheme with independent coding along the frequency and time axes. The example shows a code with $TF = 9 \times 22 = 198$ elements*

In this case, a time–frequency map descriptor (amplitude or phase or both) is modulated by a two-dimensional code:

$$C(t,f) = a(tf) \tag{4.9}$$

where $a(\ )$ is the code function.

A generalised coding scheme is merely of theoretical interest since radio architectures so far do not provide the flexibility required. However, as architectures according to a software radio concept emerge, variety and combinations of the different techniques will be introduced.

4.2.1.2 Principles of various spreading techniques

There are three basic and fundamentally different spread spectrum techniques: time hopping (TH), frequency hopping (FH) and direct sequence (DS). In the following, these techniques will be presented as a set of different coding methods. The methods are all represented by their frequency–time maps. The choice of modulation for each frequency–time slot is not discussed here, as that would lead to more detailed analysis and is not essential for understanding the basic ideas of the coding schemes. The modulation can be a simple on/off amplitude modulation or phase-shift modulation or more complex multilevel modulation.

Time hopping

The time hopping (TH) system (Figure 4.7) keys the carrier on–off with a pseudorandom noise (PN) sequence. The bandwidth of the TH signal is a function of the keying rate, as in conventional CW transmissions. TH systems can vary considerably in the duty cycle of each transmission. This

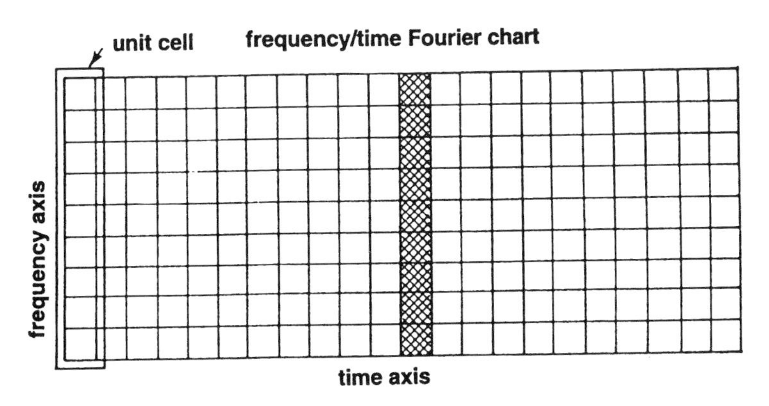

Figure 4.7 Time hopper approach. Note that the time–frequency unit cell is chosen to be rectangular rather than quadratic to conform with the standard grid; this system has processing gain, $Gp = 1$

can be as extreme as having the transmission last a fraction of a second and the keying last days between transmissions. The information is recovered by synchronising the receiver with the same timing sequence used in the transmitter. TH is not a frequency-spreading technique, however, and is mostly applied to special-purpose systems.

If only the time axis is considered, one can allow an uncertainty in the time of arrival rather than an uncertainty in the frequency. Given a time interval, T, the code function will be:

$$C(t,f) = \begin{cases} 1 \text{ for } t = a \\ 0 \text{ otherwise} \end{cases} \tag{4.10}$$

In such a system, it would be advisable to let the C function represent phase rather than amplitude to increase the total power transmitted.

Direct sequence (multitime hopping)

Figure 4.8 provides a simplified view of direct-sequence spread spectrum (DSSS). A DSSS radio works by mixing a PN sequence with the data. This mixing can be done by generating a wideband signal which, in turn, is used to modulate the RF carrier. It can also be done, which is actually more common, by modulating the carrier source with the data and then spreading the signal prior to transmission. In either case, the mixing is usually performed digitally. On the receive side, the incoming DS signal is despread by generating a local replica of the transmitter's PN code. The signal is then synchronised with this local PN sequence. The multiplication or remodulation of the incoming signal by the local PN sequence collapses the spread signal into a data-modulated carrier by removing the effects of

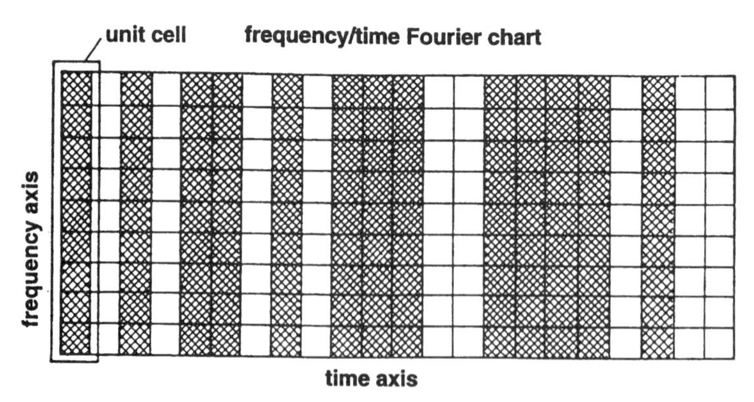

Figure 4.8 Multitime hopper or direct-sequence coding scheme. This system gives full coding along the time axis ($Gp = 22$), but no independent structuring along the frequency axis

the spreading sequence. Using correlation techniques — a measure of signal similarity — the identity of a signal that has been spread with a particular PN sequence can be discovered.

To increase the information content of the code, a set of time slots can be set in the C function. In the case where every slot is manipulated, the direct-sequence system is obtained. Here, the state of each slot is given by the code:

$$C(t,f) = a(t) \tag{4.11}$$

It should be noted that in Figures 4.8 and 4.9 the system unit cell is chosen to have the form of a rectangle rather than a square to fit the code in a C function with the same T/F ratio as the frequency-domain codes.

Frequency hopping

Frequency-hopping systems are perhaps the easiest to understand conceptually (Figure 4.9). An FH system has the freedom to jump to any particular frequency at any time. The carrier remains at a given frequency for a duration called the dwell time, and then hops to a new frequency somewhere in the spreading bandwidth. The signalling technique most commonly used in frequency hopping is m-ary frequency-shift keying, where $k = \log_2 m$ information bits are used to determine which one of m frequencies to transmit. Ideally, each frequency should be occupied with equal probability, and the probability of hopping from one channel to any other channel should also be equal. The bandwidth requirements for an FH system depend on the relative duration of the data and hopping rates.

There are actually two different categories of FH systems, which are often referred to as slow-hopping and fast-hopping spread spectrum. A

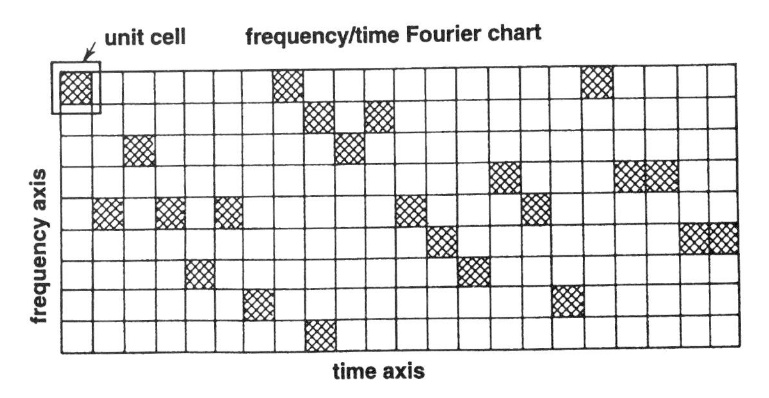

Figure 4.9 *Frequency-hopping coding scheme. Note that compared with the generalised scheme in Figure 4.6, the processing gain is reduced by a factor 9 corresponding to the number of noncoded elements along the frequency axis $Gp = 22$*

system transmitting one or more symbols on each dwell is usually considered to be a slow-hopping system. Accordingly, a system which has a hopping rate greater than its symbol rate is defined as a fast-hopping system. This means that a system transmitting 2.4 kb/s data using eight bits/symbol needs to be hopping at a rate greater than 300 hops/s to be considered to be fast hopping.

The main considerations for slow against fast hopping have been related to military threat in connection with technology and design cost. From a commercial system point of view, it is clear that there are also some fundamental differences between fast- and slow-hopping systems in how they are influenced by propagation irregularities. However, for most purposes, slow hopping is the technique which is most commonly used.

In this scheme only one frequency slot is activated at a time, and the code selects which slot is to be activated:

$$C(t,f) = \begin{Bmatrix} 1 \text{ for } f = a(t) \\ 0 \text{ otherwise} \end{Bmatrix} \tag{4.12}$$

when $a(t)$ is the code function and span $[\,a(\;)]$ is equal to the span of f.

4.2.1.3 *Frequency hopping against direct sequence*

There has been an ongoing discussion concerning which spread spectrum technique to prefer in military systems. The fact is that the answer to this question is highly dependent on a variety of factors given by the operational environment. Characterising the environment and establishing consensus about the relative importance of the various factors influencing a system is usually a difficult task, particularly since some of these factors are largely dependent on conditions like geographical and topographical placement which are not fixed for mobile users. Some of the factors to consider are:

- teleservices to support;
- compatibility and/or co-existence with other systems;
- operational area (urban environments and/or rural areas);
- Doppler frequency shifts caused by relative motions;
- interference, both narrowband and wideband;
- fading characteristics, Rayleighian or Rician;
- severe specular multipath;
- necessity for message integrity;
- response (blind acquisition required);
- etc.

With these kinds of concerns, what type of SS system would work best? One way to answer this question is to analyse each artifact in the environment, determine its effect on the system performance and, in the end, assess how the accumulation of all the errors degrades the performance of each system. Making a thorough analysis of the consequences and alternative solutions for each of these factors is, however, an ambitious task to perform.

With respect to military combat net radios, the argument has been that frequency hoppers are better off when it comes to protection against interference. Frequency-hopping systems not only have code gain against noise, but also have filter-selectivity gain for CW interferers. In other words, if the interferer coincides with a particular hop frequency, that particular hop will be destroyed and all other frequency hops will be unaffected. Thus, although a DS system may typically have 10–20 dB of protection against interferers, an FH system may have as much as 50–60 dB.

Such a comparison is not very useful, however, when considering a practical case. In an environment with several other radios, both fixed frequency and frequency hopping, a frequency hopper will encounter a large number of blocked hops due to collisions. Since in pure frequency hopping there is no protection against interference on an individual dwell, most collisions will result in data being destroyed. How much this will reduce the actual processing gain is dependent on factors like frequency planning, collocation and others that are equally hard to quantify. Another factor that can reduce the effect of frequency hopping is intelligent jamming. Dependent on hop rates and hopping bandwidth, a jammer could analyse the actual frequency band and jam any occupied frequency within microseconds. Hopping rates below a few kilohertz are therefore considered to give less protection. From this, it is obvious that there is more to a discussion on interference protection than the theoretical numbers.

When it comes to providing data communication services in both military, private and commercial radio networks, interference reduction is still of interest, but is not likely to be the most dominant feature to consider. Factors like capacity utilisation, system management (including frequency planning), access fairness etc. will be the focus for an overall system design, and which solution this leads to on the radio level is not ascertained without thorough analysis. Even though this book does not evaluate frequency-hopping versus direct-sequence systems when it comes to suitability for providing packet-switched bearer services, it is likely that in most cases the DS system would be preferred, and an FH system would be better suited to circuit-switched services.

4.2.1.4 EMI aspects of spread spectrum modulation

One desired capability in most military communication systems is to be able to operate without detection by others. Low probability of intercept (LPI) or low probability of detection (LPD) is a feature obtained by using spread spectrum. An example of this can be shown in the case of a simple energy-detecting device known as a radiometer. A radiometer detects energy received through the bandwidth, W, of its input bandpass filter. The signal or energy is then squared and taken through an integration of time, T, and the output is compared to a reference threshold to find if there is a signal present or absent. When assuming the time–bandwidth product, TW is large relative to the energy to noise power-spectral density E/N_0, which gives a gaussian p.d.f. The probability of detecting the signal is then given by [4]:

$$P_D = \phi\left\{\left[\frac{P}{N_0}\sqrt{\frac{T}{W}} - \phi^{-1}(1 - P_{fa})\right]\right\} \tag{4.13}$$

where

$$\phi(y) = \frac{1}{\sqrt{2\pi}}\int_{-\infty}^{y} \exp(-\tfrac{1}{2}\zeta^2)d\zeta \tag{4.14}$$

For a fixed false-alarm rate P_{fa}, the probability of detection can then be made smaller by reducing the transmitted power, P, or by increasing the bandwidth, W.

In analysing the effect of interference and security aspects of spread spectrum waveforms further, it is useful to consider this in terms of the degree of waveform match and degree of time synchronisation. A co-operative system will in general possess all details on the waveforms used but might not be completely synchronised. A nonco-operative system should not have knowledge of the codes used, making synchronisation aspects less important.

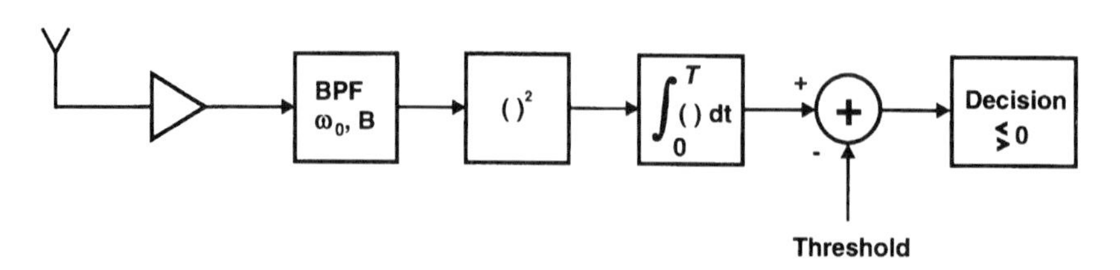

Figure 4.10 Functional energy detector or radiometer

Fundamental to the security of a spread spectrum system is the processing gain of the code, which is equal to the number of independent elements in the data C function:

$$G_p = FT \tag{4.15}$$

Secondly, there is the protection of the uncertainty of the time reference. Assuming that the uncertainty in synchronisation is δT_{sync}, this gives an added security:

$$\frac{\delta T_{sync}}{\Delta T} \tag{4.16}$$

(ΔT and ΔF are the unit elements in the time–frequency diagram.)

Thirdly, the uncertainty along the frequency axis (δF) gives the frequency uncertainty improvement:

$$\frac{\delta F}{\Delta F} \tag{4.17}$$

This gives the total efficiency factor assuming full system knowledge corresponding to the protection offered in a synchronised spread spectrum system:

$$E_{time} = \frac{\delta T_{sync}}{\Delta T} \frac{\delta F}{\Delta F} * G_p \tag{4.18}$$

Unco-operative interference (unknown codes)

For unco-operative interference, with no information as to the band allocation, the efficiency factor is $E_{time} = 1$. This corresponds to having no system knowledge. If the frequency band is known as it might be even if the spread spectrum code is unknown the protection factor will be:

$$E_{time} = \frac{\delta F}{\Delta F} \tag{4.19}$$

As the code is unknown, no further gain can be seen as time synchronisation in this case does not increase the output of the detector.

Interference aspects for co-operative systems (known codes)

If the frequency band occupied by the transmission is known, but no information as to the structure of the waveform within this corresponds to the situation, then we have interfering co-operative systems. This situation is illustrated in Figure 4.12. Those elements in the time–frequency map in which there is coincidence of the data and interfering waveforms are circled.

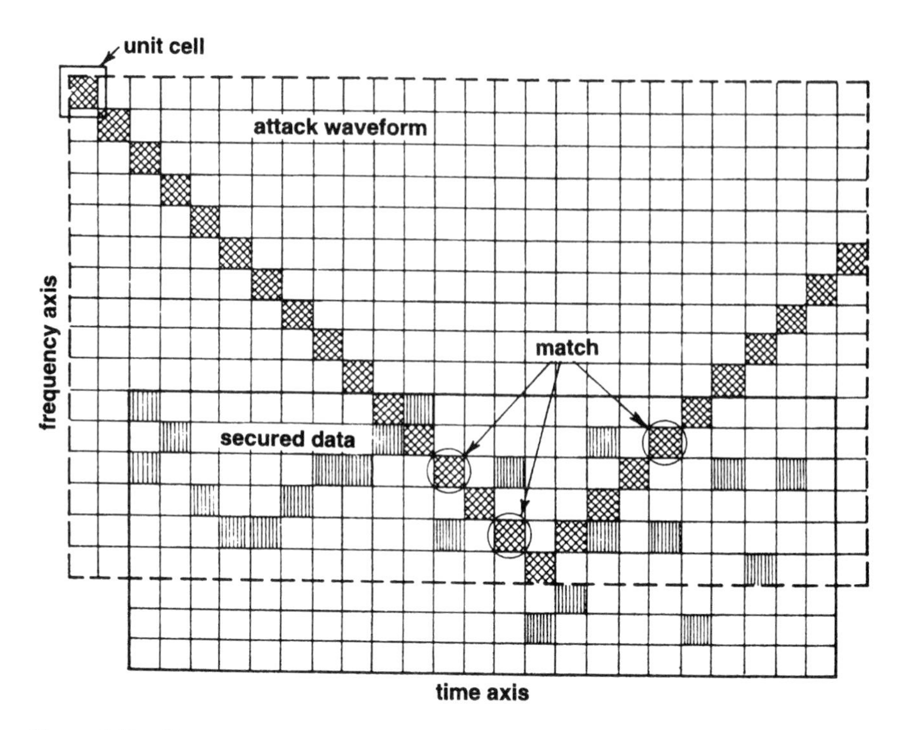

Figure 4.11 *Unco-operative interference will only to a small extent disturb a spread spectrum system, as the codes are not applied and the precise frequency and time information is not available*

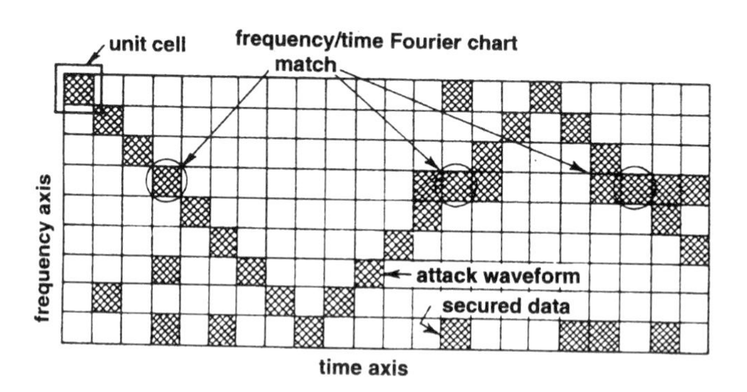

Figure 4.12 *When the main system parameters are known, a high possibility for some degree of match between codes can be achieved*

Unsynchronised systems

This situation may arise either if the legitimate receiver has failed in synchronisation, or if an illegitimate receiver does not have the information required to obtain synchronisation. Synchronisation is most often achieved by transmitting a code with sufficient complexity to ensure secure detection even if the time reference is not known in advance of the synchronisation process. The time reference can also be established by other means, such as by parallel communication links. High precision time references (atomic clocks) can also provide initial time information, which may be close to sufficient for synchronisation, and thereby reduce the time uncertainty δT_{sync}.

High data rate systems require synchronisation to better than the duration of one time slot (code chip ΔT). This might scale to distances of only a few centimetres (a 1 GHz chip rate gives a 30 cm required position accuracy). For mobile systems, synchronisation must therefore be acquired at intervals within which the mobile unit has moved only a fraction of a chip wavelength, or position or velocity has been measured. By tracking one of these parameters, corrections to the time reference can be estimated.

The code-processing gain is present because the code is now broken. This demonstrates that a spread spectrum matched filter has considerable immunity to countermeasures even if the code is accessible, because the synchronisation offers additional security. Figure 4.13 illustrates this. The attack waveform is not aligned along the time axis, and only spurious coincidences of the attack code and data code occur.

Synchronised systems

When the system is synchronised, no uncertainty exists about the location of the code on the time and frequency axes. If the code is known as well, the time-domain security has its maximum value:

$$E_{time} = Gp \frac{\delta T_{sync}}{\Delta T} \frac{\delta F}{\Delta F} \tag{4.20}$$

This situation is represented in Figure 4.14.

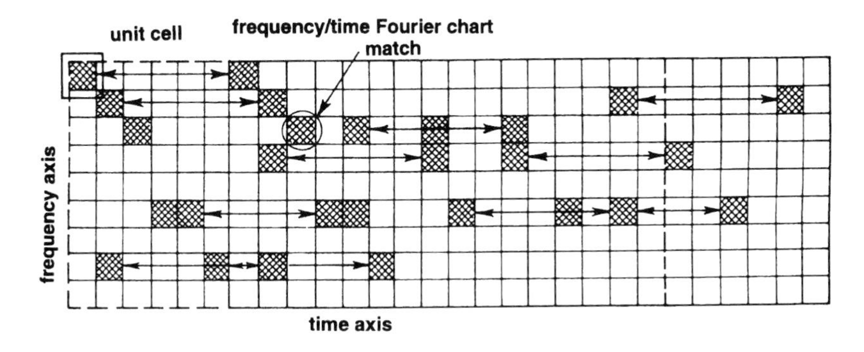

Figure 4.13 When the code is known, it is still essential to know the time axis reference points (synchronisation)

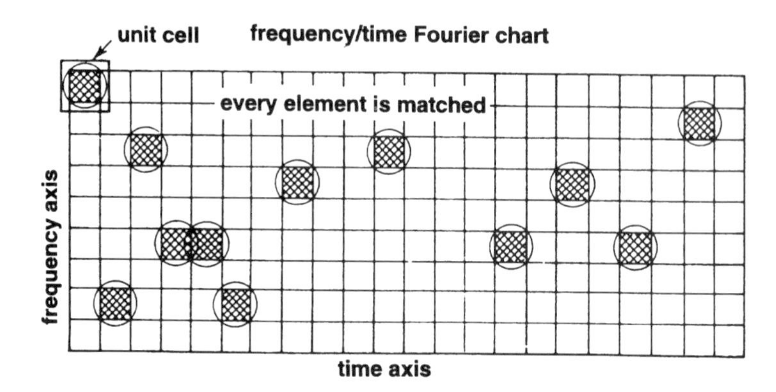

Figure 4.14 *Perfect match is obtained in a synchronised system with open codes; this implies that the data is no longer secure*

Table 4.1 *Summary of protection/immunity offered by a spread spectrum modulation scheme*

	E_{time} equals	Applicability to co-operating party	Applicability to unco-operating party
No system knowledge	1	None	The net is not accessible
Basic system knowledge	$\frac{\delta F}{\Delta F}$	None	Transmission frequencies are known
Synchronised system and basic system knowledge	$\frac{\delta F}{\Delta F}\frac{\delta T_{sync}}{\Delta T}$	Operation when code is not valid	Code is not known
Unsynchronised system with known code	$Gp\,\frac{\delta F}{\Delta F}$	Synchronisation phase	The synchronisation process is not known
Synchronised system with known code	$Gp\,\frac{\delta T_{sync}}{\Delta T}\frac{\delta F}{\Delta F}$	Normal operation	The net is broken

In summary, the data-security factor for time-domain processing for the various assumptions is presented in Table 4.1.

Consider a typical system with the following parameters:

$FT = 1000$, processing gain
$\delta T/\Delta T = 1000$, uncertainty in time equals data bit length
$\delta F/\Delta F = 10$, band allocation may be 1 out of 10

The result is that the data security is increased by a factor of 10^7. This factor may, depending on the approach chosen, lead to one of the following:

- an increase in cost of 10^7;
- an increase in processing time of 10^7;
- an increase in processing complexity of 10^7 etc.

4.2.2 Additional topics in military spread spectrum communication

4.2.2.1 Hybrid spread spectrum

Hybrid spread spectrum techniques are created by combining two or more forms of spread spectrum into a single system. There are several kinds of hybrid system. Some of the most useful are frequency hopping/direct sequence, time hopping/frequency hopping and time hopping/direct sequence.

The conceptual relationship between direct-sequence, frequency-hopping and hybrid schemes is shown in Figure 4.15 for a comparable data rate and transmission security. The direct system spreads the transmitted energy over the full frequency band by using a rapid code, and transmits one data bit at a time. The frequency hopper transmits a number of data bits while dwelling on one frequency, and then jumps to another frequency slot and continues transmission. In the hybrid system, the signal energy in each frequency slot is protected by a direct-sequence code, thereby broadening the width of the frequency band activated at any time. In Figure 4.15 the various coding schemes are given the same total processing gain, so that the conceptual characteristics of the coding methods can be easily visualised.

If the interrogating waveform's time–frequency map is $I(t,f)$, the efficiency or match between this and the code function is defined as the success factor, X_{CI}. In the general case, the I function will not have the same span in the time and frequency domains as the C function. The success factor, X_{CI}, will therefore have to be defined in the mean, averaging over every possible absolute position of the C and I functions:

$$X_{CI} = \sum_{t}\sum_{f} C(f,t)I(f,t) = C(f,t) * I(f,t) \tag{4.21}$$

4.2.2.2 Automatic channel selection (free channel search)

Automatic channel selection (ACS) is introduced in connection with military VHF combat-net radios (CNR). CNR nets belong to a large group of tactical communication systems, which are based on independent net structures without any links with other nets. Furthermore, the traffic is usually simplex and transmission is broadcast based which allows every member of the net to receive the messages. These types of net are used in

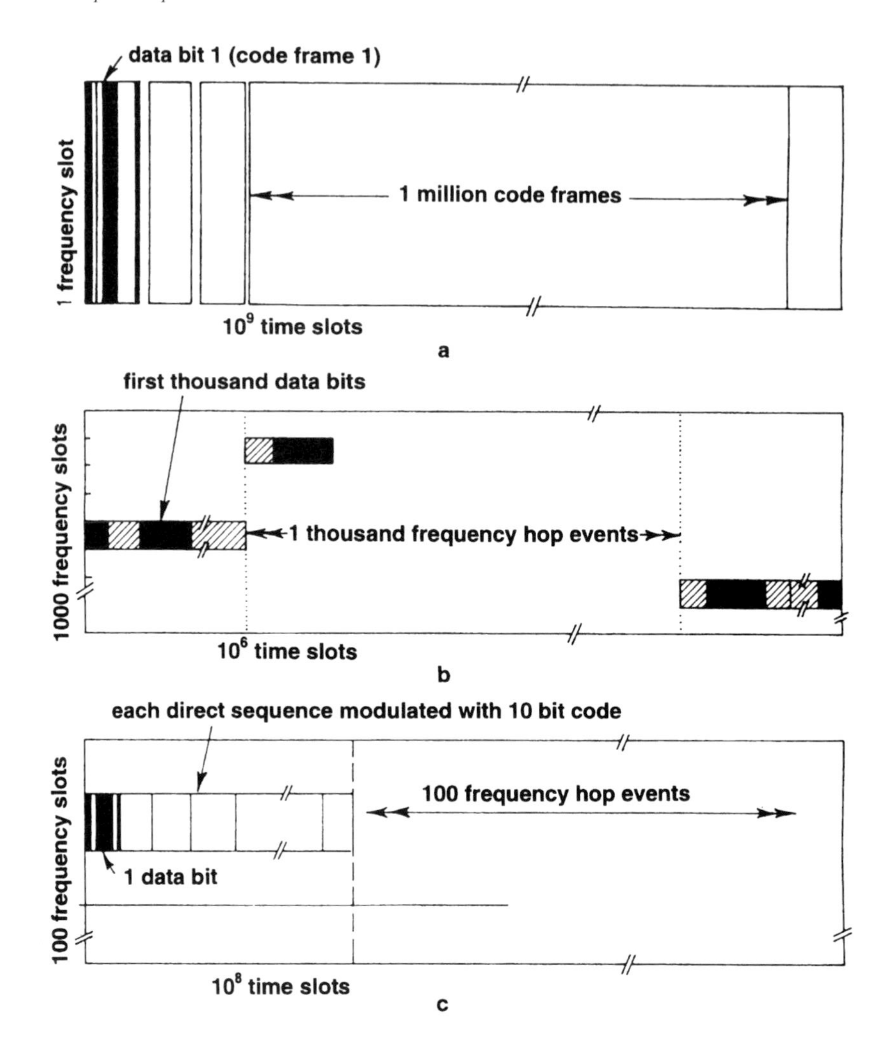

Figure 4.15 *Unified description of a direct-sequence, frequency-hopper and hybrid system, all with processing gain Gp = 1000*
a direct sequence system
b frequency hopper
c hybrid frequency hopper/direct sequence: 10 000 data bits

situations where simplicity, flexibility and an all-informed feature are important. With the introduction of digital radios, CNR nets have also been provided with options of addressing and bearer services like circuit and packet switching.

ACS is a technique intended primarily to improve utilisation and spectrum efficiency of analogue CNR nets. The large mobility of the nets requires the frequency of a certain net to be re-used over a larger distance.

This, together with relatively low traffic, results in poor spectrum efficiency.

The basic principle of ACS is to assign a bundle of frequencies instead of one frequency to a net. The same bundle of frequencies is assigned to surrounding nets. This is what improves spectrum efficiency. For a connection a transmitter first searches for a free channel (free-channel search) before starting the transmission on the chosen channel. The probability that a net does not find a free channel from the bundle depends on the number of nets which use this bundle, the amount of traffic per net and the number of frequencies in the bundle.

The number of frequencies in the bundle has to be sufficiently large in order to keep the blocking probability under a certain level. However, the time which is allowed for the free-channel search limits the number of frequencies. This acquisition time is directly proportional to the number of channels which have to be scanned. In a free-channel search (FCS), a free channel can be an unoccupied channel or the least occupied channel if all channels contain more than background noise. To measure the noise in a certain channel, originating from a jammer or another ACS-transmitter, one could for instance determine the average received noise power based on information from the automatic gain control circuitry.

ACS can also be seen as a measure against jamming, since it essentially is a spread spectrum technique (Figure 4.16). However, it is only useful for relatively-static jammed environments. Furthermore, the concept demands a correlation between the observed noise level at the transmitter and intended receiver(s), which is a reasonable assumption when dealing with some VHF systems.

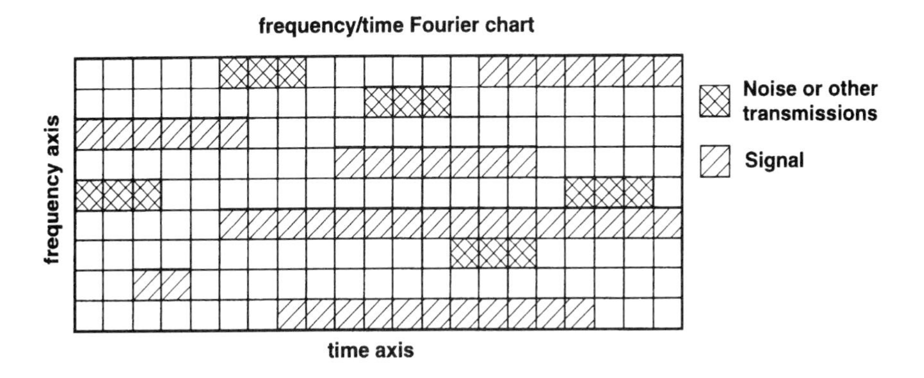

Figure 4.16 Automatic channel selection

The most interesting feature with ACS is its flexibility, adaptive frequency management and suitability for combination with both direct-sequence and frequency-hopping spread spectrum.

$$C(t,f) = \begin{cases} 1 \text{ for } f = a_{min}\,\overline{noise}\,(channel,\,t) \\ 0 \text{ otherwise} \end{cases} \tag{4.22}$$

4.2.2.3 *Narrowband spread spectrum*

Narrowband spread spectrum (NBSS), is a technique based on *m*-ary direct-sequence spread spectrum (DSSS), which is superior in performance compared to traditional DSSS. In the case of a VHF system, the use of *m*-ary orthogonal coding will make it possible to obtain large gains even with channels restricted to 50 or 25 kHz. From a spectrum efficiency point of view, this means that the total gain of the system can be maintained while reducing the bandwidth with the cost of complexity. An example of the *m*-ary direct-sequence spread spectrum principle is shown in Figure 4.17.

For binary DSSS the total gain with respect to suppression of narrow-band interference is given by the bandwidth expansion, i.e. the processing gain. By grouping two-and-two information bits into symbols, each represented by orthogonal spreading codes, we find that the chip rate of the system can be reduced by a factor of two. This means that the bandwidth is reduced accordingly and the processing gain is reduced by 3 dB. However, if the spreading sequences are kept orthogonal, *m*-ary coding itself provides a coding gain. By going from binary to 4-ary the coding gain is approximately 2.8 dB (Section 4.3). This means that when adding this gain to the processing gain, the total system gain is maintained and the bandwidth is reduced to a half. Even better, the coding gain obtained from *m*-ary orthogonal coding is applicable to the system even if the noise is white, thus enhancing the range of the system.

It should be noted that an *m*-ary DSSS radio can be made to require very short synchronisation times compared to a frequency-hopping system. Low synchronisation time is highly favourable in certain military systems such as automated weapon systems. Together with error-correcting codes, the statistical stability of a radio channel using *m*-ary DSSS provides a good means for implementing advanced data communication protocols. For practical implementations, *m*-ary signalling gives approximately 300% increased band efficiency compared to traditional spread spectrum systems.

2-ARY(BINARY):

INFO BIT	CODE
0	010110101001
1	101001010110

Data rate: 2400 bps

Chip rate: 28800 bps (BW = 25 kHz)*

Total gain: 10.8 dB

4-ARY:

INFO BITS	CODE
00	110101001010
01	010110101001
10	001010110101
11	101001010110

Data rate: 2400 bps

Chip rate: 14400 bps (BW ≈ 12.5 kHz)*

Total gain: 10.6 dB

*dependent on modulation

Figure 4.17 *Coding of m-ary direct-sequence spread spectrum. For each information bit or group of information bits (4-ary), an assigned code is transmitted instead*

NBSS provides a beneficial basis both for hybrid FH/DS modes and for automatic channel selection (ACS) modes. NATO Industrial Advisory Group (NIAG) studies conclude that frequency hopping alone will suffer from great problems both from the protection point of view (multipath, jamming) and from LPI requirements in a post-2000 operational scenario [5]. A solution with more robust characteristics could be hybrid FH/DS. However, this technique is also somewhat limited, especially with respect to bandwidth efficiency. By using NBSS in a hybrid solution, however, desired EPM capabilities and performance related to digital communications can be achieved with acceptable bandwidth usage. To provide systems with even better spectral efficiency, ACS using NBSS would be an attractive solution. ACS by itself holds the advantage of having a set of frequencies available to be used, but yields approximately a factor 2 improvement in spectrum economy compared to frequency hopping. The ultimate goal would probably end up being an adaptive system trading between hybrid FH/NBSS and ACS [6].

4.2.3 Spread spectrum in mobile radiocommunication

From providing hidden communications and protection against enemy fabrication and jamming in military systems, spread spectrum has moved into the commercial market place due to some of its less enigmatic characteristics. The main arguments have been reduced effects of multipath, interference rejection and, especially, multiple-access capabilities.

4.2.3.1 Antimultipath

One may think of the effects of multipath as follows: frequency-selective fades are caused by receiving two paths of the same signal which sum out of phase at a particular frequency. Thus, in addition to the signal's direct path, there exist one or more reflected signals impinging on the receiver. The receiver then sees more than one signal and produces a signal containing the modulation of them all. The amplitude of the signal is determined by the phase relationship between the direct signal and its reflections.

DS and FH systems deal with multipath in different ways, and are affected by it differently. The frequency hopper depends on the fact that only narrow regions of the frequency band will have deep fades. When an FH system jumps into a deep fade, the assumption is that the bit will be lost. However, the FH system is constantly hopping, and the damaged hops can usually be corrected by an appropriate forward-error-correcting code.

DS systems depend on the decorrelation of the various received signal paths. If the delay is larger than the chip rate, then all of the reflections will act as uncorrelated, independent signals. If that is the case, they can actually be decorrelated. If the multipath delay is short and well within the chip rate of the receiver, the multipath causes a flat frequency fade, which may cause the DS system to stop working correctly.

A binary DS system with a 50 kHz chip rate is susceptible to delay paths of the order of 20 μs. On the other hand, a frequency hopper providing the same information rate would typically be susceptible to multipath delays of the order of 1–3 ms. Considering a VHF system, neither would represent any major concern. However, a frequency hopper may get into trouble due to large variations in propagation delay observed at different frequencies (dwells). This is caused by a missing direct path, and reflections are dominating. For instance, this is observed with VHF propagation in hilly terrain and can cause severe synchronisation and signal-tracking problems to frequency hoppers. This is further discussed in Chapter 6, where some propagation measurements are presented.

4.2.3.2 Narrowband interference rejection

Another desired feature obtained by using spread spectrum is the ability of the system to reject narrowband interference. For the sake of this discussion, let us consider a binary direct-sequence spread spectrum radio employing a signal of the form:

$$s(t) = \sqrt{(2P)}a(t)b(t)\cos\omega_c t \tag{4.23}$$

where $b(t)$ is a binary baseband data signal and $a(t)$ is a baseband spectral-spreading signal The latter signal usually has a bandwidth that is much greater than required to transmit the data; that is the bandwidth of $a(t)$ is much higher than the bandwidth of $b(t)$. If the basic pulse shape is rectangular for both $b(t)$ and $a(t)$, then the signal defined is mathematically equivalent to binary phase-shift-key (PSK) modulation.

The data signal $b(t)$ consists of a sequence of positive and negative rectangular pulses, so it can be written as:

$$b(t) = \sum_n b_n p_T(t - nT) \tag{4.24}$$

where $p_T(t)$ is the rectangular pulse of duration T, Σn denotes the sum over all integers n which correspond to data symbols and the data symbols are from the data sequence:

$$(b_n) = \ldots,\ b_{-1}, b_0, b_1, b_2,\ \ldots \tag{4.25}$$

The binary digit b_n is either $+1$ or -1, depending on the data symbol to be sent in the nth time interval. The spectral-spreading signal consists of a sequence of positive and negative pulses, but the pulse shape (i.e. the chip) need not be rectangular. If the chip waveform is denoted by $\psi(t)$, the spectral-spreading signal can be written as:

$$a(t) = \sum_j a_j \psi(t - jT_c) \tag{4.26}$$

The chip waveform $\psi(t)$ is assumed to be a time-limited pulse of duration T_c. The sequence:

$$(a_j) = \ldots,\ a_{-1}, a_0, a_1, a_2,\ \ldots \tag{4.27}$$

is called the signature sequence or spreading code.

For convenience, we follow the usual convention whereby the signal is normalised such that:

$$T_c^{-1} \int_0^T [\psi(t)]^2 dt = 1 \tag{4.28}$$

It is also convenient to define the signal baseband $v(t)$ by

$$v(t) = a(t)b(t) \tag{4.29}$$

the product of the spectral-spreading signal and the data signal.

We consider direct-sequence spread spectrum signals for which the data pulse duration is an integer multiple of the chip duration. If $T = NT_c$ for some integer N, there are N chips per data bit, and the bandwidth of the spread spectrum signal $v(t)$ is roughly N times the data rate.

The optimum receiver for a binary direct-sequence spread spectrum system in which white gaussian noise is the only interference is the correlation receiver shown in Figure 4.18.

The received signal $y(t)$ is multiplied by $a(t)\cos(\omega_c t + \phi)$ and the product is integrated over an interval of length T.

There is no improvement in performance when we use spread spectrum on a channel in which the only interference is white gaussian noise. However, for channels with other forms of interference, spread spectrum offers many improvements over narrowband modulation. These improvements are derived from the fact that the spectral-spreading signal can be selected to give the $s(t)$ several desirable properties which can be exploited to enable the receiver to discriminate against these other forms of interference. For example, if there is an additive interference signal $I(t)$, the received signal can be written as $y(t) = s(t) + I(t) + n(t)$, where $n(t)$ is the white gaussian noise process. The effects of this interference can be reduced by choosing the spectral-spreading signal to minimise:

$$Z = \int_0^T Z(t)a(t)\cos(\omega_c t + \phi)dt \tag{4.30}$$

Now, suppose a binary direct-sequence spread spectrum system is suffering interference by a single tone having power J. This unwanted interferer or jammer places its tone directly in the centre of the receiver bandwidth. If no spread spectrum is employed, the ratio of jamming power to signal power in the data bandwidth would be I/P.

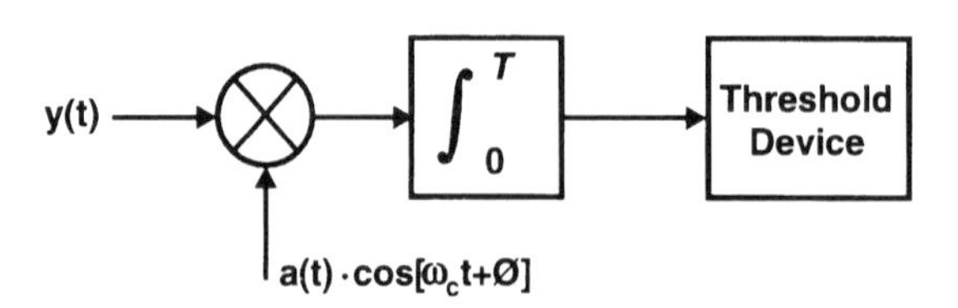

Figure 4.18 Coherent correlation receiver for binary direct-sequence spread spectrum

For our spread spectrum system the received signal is (assuming no gaussian noise):

$$r(t) = s(t) + I(t) = \sqrt{2P}a(t - T_D)b(t - T_D)\cos(\omega_0 t + \phi) + \sqrt{2I}\cos(\omega_0 t + \phi') \tag{4.31}$$

To keep it simple, assume coherent despreading, which results in:

$$y(t) = \sqrt{2P}b(t - T_D)\cos(\omega_0 t + \phi) + \sqrt{2I}a(t - T_D)\cos(\omega_c t + \phi') \tag{4.32}$$

The corresponding power spectrum of $y(t)$ is:

$$\begin{aligned} S_Y(f) &= S_S(f) + S_J(f) \\ &= \tfrac{1}{2}PT\{\text{sinc}^2[(f - f_0)T] + \text{sinc}^2[(f + f_0)T]\} + \tfrac{1}{2}IT_c\{\text{sinc}^2[(f - f_0)T_c] \\ &\quad + \text{sinc}^2[(f + f_0)T_c]\} \end{aligned} \tag{4.33}$$

where $S_J(f)$ is the power spectrum of the despread jammer. From this we can see that the data has been despread, and the jammer has been spread to the bandwidth of the spread spectrum system. The signal then goes through a filter with a bandwidth which is matched to the data bandwidth. In a superheterodyne receiver this is done by an intermediate-frequency (IF) filter. Although ideally all the signal power is passed through the IF filter, a large portion of the spread jammer power is rejected. The magnitude of the jammer power passed by the IF filter is:

$$I_0 = \int_{-\infty}^{\infty} S_J(f)|H(f)|^2 df \tag{4.34}$$

Assuming the IF filter $H(f)$ to be ideal, then:

$$\begin{aligned} I_0 &= \int_{-f_0-1/2T}^{-f_0+1/2T} S_J(f)df + \int_{f_0-1/2T}^{f_0+1/2T} S_J(f)df \\ &= \tfrac{1}{2}IT_c \int_{-f_0-1/2T}^{-f_0+1/2T} \text{sinc}^2[(f + f_0)T_c]df + \tfrac{1}{2}IT_c \int_{f_0-1/2T}^{f_0+1/2T} \text{sinc}^2[(f - f_0)T_c]df \end{aligned} \tag{4.35}$$

For large spreading ratios, $T_c \ll T$ and the sinc function is nearly constant over the range of the integration. The resulting interference power is then:

$$I_0 = I\frac{T_c}{T} \tag{4.36}$$

From this we see that the noise power of the interferer is reduced by a factor equal to the ratio between the data bandwidth and spread bandwidth, i.e. the bandwidth expansion. The processing gain is according to this defined as the inverse of the interference power reduction:

$$G_p = \frac{T}{T_c} \tag{4.37}$$

One should note that in this simplified example, there is no correlation between the interfering signal and the despreading signal, which indicates that performance is highly dependent upon the spreading sequences used.

4.2.3.3 Code-division multiple-access (CDMA)

CDMA is a fixed-assignment protocol for sharing a common channel simultaneously through application of individual spread spectrum codes. CDMA obviously increases the bandwidth efficiency significantly compared to traditional spread spectrum. CDMA also has the advantage of allowing co-existence with other systems in the same band. As long as different spreading codes are utilised, the different signals can in an ideal situation be separated by the receiver. In practice, some interference will result from the other users. Interference resulting from other users in the network is called interuser interference (IUI) or sometimes multiple-access interference (MAI).

CDMA is particularly interesting for channels which in addition have other characteristics suitable for spread spectrum, such as jamming or multipath distortion. If the number of users is kept well below maximum (eqn. 4.43), some processing gain is left for protection. If not, and traffic load is high, the network is severely degraded in the case of external interference, just like any other network not provided with spread spectrum. CDMA can be classified into two main categories: direct sequence (DS/CDMA) and frequency hopping (FH/CDMA). DS/CDMA in particular is receiving a lot of attention in connection with wireless technology. Wireless personal communication is an area of expansion. Demand for portable computers brings forward the needs of wireless voice, data and video communication. It should be noted that CDMA (or fixed allocation) is generally not preferred for packet-switched networks where random-access techniques are better [7]. This is further discussed in Chapter 3.

The conventional receiver used for direct-sequence spread spectrum can also be used for CDMA. If a traditional DSSS receiver is to perform adequately, the observed interference from other users has to be reduced as much as possible. This can be done for instance by:

- choosing codes which minimise cross correlation and autocorrelation;
- applying power control to ensure that the received power of the wanted signal is of the same order as the sum resulting from the other signals.

The near–far problem results from strong interference from other users relative to the wanted signal (due to distance, topology or fading) which creates IUI. It is not possible to resolve the near–far problem completely by appropriate spreading codes, since the output of the correlator is going to produce spurious components which are linear to the amplitude of the interfering signal. Power control can only be effective when applied to fixed-star topologies. Other solutions which can be considered include, for instance, directive antennas. Either way, it is generally difficult to predict how effective various approaches are going to be when it comes to reducing IUI, since this is dependent on factors like: number of users, topology, propagation characteristics, codes, processing gain, antennas etc.

In DS/CDMA networks all members communicate on the same frequency. Usually there is also a common data rate $R_d = 1/T_d$ and a common spreading code chip rate $R_c = 1/T_c$. Successful use of spread spectrum code-division multiple-access techniques requires the construction of spreading codes giving rise to a minimum of IUI. However, situations arise where the effects of IUI are amplified owing to operational considerations.

Consider a network operated in a master–slave configuration. Let slave station M_1 transmit the desired signal S to the master, which is d_1 km away, and let another slave station M_2 at a distance d_2 km from the master transmit a signal I which is interference to the M_1 master link. Let C (dB) be the signal/interference ratio (S/I) required at the terminals of the master station's receiving input to produce the desired output S/I. The requirement is that:

$$S - I \geq C \quad \text{(in dB)} \tag{4.38}$$

With transmitting powers of P_{M1} and P_{M2} and path losses of L_1 and L_2, respectively, this yields:

$$(P_{M1} - L_1) - (P_{M2} - L_2) \geq C \tag{4.39}$$

$$L_2 \geq C + (P_{M2} - P_{M1}) + L_1 \tag{4.40}$$

This is an explicit formula for the near–far problem and will put restrictions on where slave station M_2 can operate if the original system specification is to be met. If the transmitter powers remain constant while the value of L_2 is reduced, the inequality in eqn. 4.39 may be violated for any practical code design. The value chosen for the factor C will, however, depend on how well the codes can be designed.

The receiver will attempt to extract the individual spreading codes from a composite of many during the matched-filter processing. Both the periodic (even) and odd cross-correlation functions play key roles in calculating system performance for such situations. The wideband input signal consists of the wanted signal as well as the interfering signals, each spread by their own code. The spreading code modulating the wanted signal will match the receiver filter and the correlation peak will be sampled at a rate equal to the data rate. If the unwanted signals were totally uncorrelated with the wanted signal, then they would produce no correlation peaks at the filter output. However, the effect of cross-correlation between the local sequence and the sequences of the unwanted signals appears as cross-correlation peaks at the output of the filter. The signal/noise plus interference ratio in a CDMA system is given by:

$$[S/(N+I)]_K = \frac{E[Y^k|b_0^k]^2}{\sigma_I^2 + \sigma_n^2} \quad \text{when } K = k \tag{4.41}$$

where K is the total number of users and σ_I^2 and σ_n^2 are the variances of the other user interference and thermal noise, respectively.

From this it is clear that the actual SNR is degraded from that of a single link, where the only interference is the thermal and atmospheric noise.

An important question is, then, how much the signal power must be increased to maintain the signal/noise ratio for which the system is designed. The answer will obviously depend on the number of users, the code design and on α^r, the power ratio between the received signals. The new value for the thermal signal/noise ratio required to meet the original system specification is found to be:

$$2E/N_0 = \frac{S/(N+I)}{1 - \left[S/(N+I) \dfrac{1}{6p^3} \sum_{\substack{r=1 \\ r \neq k}}^{K} \alpha^r S^{kr} \right]} \tag{4.42}$$

E is the signal energy, defined as $\int_0^{T_d} s^2(t)dt = (A^2/2)T_d$, and $2E/N_0$ represents the thermal signal/noise ratio for a matched filter at the instant of sampling.

The increase in signal power required by each user to maintain the same performance in a multiple-access system as it would without the power increase in a single-link system is called the power cost of code multiplexing. Each user in the CDMA network is forced to increase its output power by the amount disclosed by the power cost figure.

Assuming now for a moment that all interfering users are received with the same signal power $\alpha(A^k)^2$, then $\sum_{r=1,r\neq k}^{K} \alpha^r S^{kr}$ can be replaced by $(K-1)\alpha \sum_{r=1,r\neq k}^{K} S^{kr}$.

Solving now for K, the maximum number of possible users, gives:

$$K = \frac{1}{[S/(N+I)]\alpha \dfrac{1}{6p^3} \sum\limits_{\substack{r=1 \\ r\neq k}}^{K} S^{kr}} - \frac{1}{[2E/N_0]\alpha \dfrac{1}{6p^3} \sum\limits_{\substack{r=1 \\ r\neq k}}^{K} S^{kr}} + 1 \qquad (4.43)$$

The factor $S/(N+I)$ is the threshold value, which is designed and fixed for a given performance. Thus, the number of users will for a given threshold value depend on an increased thermal SNR, α, and the code design. If the other user interference is small compared with the thermal noise, then any increase in thermal SNR will increase the possible number of users for a given $S/(N+I)$ threshold. However, when the other user interference is dominant, the second term approaches zero and the number of users is only determined by the required threshold, code design and α. The latter condition is usually called the bandwidth-starved condition. In practical situations α will, of course, not be the same for all the interfering signals.

As utilisation of digital signal processing advances, several methods are considered which take into account information about the signals from other users and do some active cancellation based on this. These types of receiver use so-called multi-user detectors. They work on the basis that they know not only the spreading codes used for the intended message, but also the codes used by others. A common way to organise coding in such an environment is to apply receiver-directed codes, which means that spreading codes are selected by the transmitter, based on which user is the intended receiver. In practice, many of the purposed receivers are too complex and suboptimum multi-user receivers have to be chosen. One type of such receivers are referred to as linear multi-user receivers. Unlike an optimum solution, they do not need to estimate power levels and their complexity is linear with respect to the numbers of users in the network [8]. In the case of multipath environments, multi-user receivers will become very complex and sharing techniques other than CDMA are likely to be better choices.

4.2.4 *The role of spread spectrum in packet-switched radio networks*

There are several properties of spread spectrum that can be exploited in packet radio networks. These properties are derived from the signal structures used in spread spectrum and from the processing that takes place in the receivers. Signal capture and multiple-access are probably the two most important ones, because of their strong influence on the throughput and delay performance. When it comes to packet-switched networks it is advantageous to trade multi-user capabilities against increased data rate (resulting in shorter packets) if effective carrier sense can be obtained. With reference to the discussion in Chapter 5, we will especially look into this possibility in the case of spread spectrum systems.

4.2.4.1 *Capture*

Capture (or capture effect) refers to the ability of the radio to demodulate at least one of a number of overlapping packets, or in other words to stay locked onto one packet in the presence of a number of overlapping transmissions. The situation envisioned here is that two or more radios transmit packets to another radio at about the same time and one packet arrives at the receiver before the reception of another is complete. If the radio cannot demodulate more than one packet at a time, which is typical for ground-based mobile systems, the goal is to provide the capability for the radio to receive one of the overlapping packets. This packet is said to have captured the receiver in question. Usually, the packet that captures the receiver corresponds to either the first-arriving signal or the strongest signal.

There are several ways of designing a spread spectrum transmitter for the application of sequences that will largely influence the capture effect. If a packet radio network is employing a receiver-oriented transmission protocol, each receiver is assigned a spread spectrum sequence and all transmissions to that receiver must take place on that sequence. The sequence timing for a given packet can be derived from the time of day or from the position in the packet, and either slotted or unslotted transmission is possible.

When two or more packets using the same phase of the same spreading sequence arrive at a given receiver at nearly the same time, there will be insufficient time offset between the two sequences to permit the receiver to distinguish between them. As a result, it is highly probable that none of the packets will be received correctly, i.e. a collision has occurred. A collision in a spread spectrum system has about the same effect as a collision in a narrowband radio. The packets involved in the collision are lost and they must be retransmitted at a later time (thereby increasing the delay).

A timing mechanism for the spreading sequences which gives better capture properties is to use the position in the packet as a reference. For such an approach, either the sequences are re-used or there must be a method whereby they are changed periodically. For a military application, periodic changing of the sequence is preferred due to transmission security. For a slotted packet radio network, the sequence changing might be accomplished by associating a randomised starting point in the original infinite length sequence for each packet slot. As a result, all packet transmissions that begin in the same slot do not use the same sequence. Even if a receiver-oriented transmission protocol is employed, two packets with transmission start times that are sufficiently far apart will arrive at the receiver as time-shifted versions of the spreading sequence. Provided that initial acquisition has been accomplished, they will not collide, and suffer random errors only. However, although the capture properties are better than for the first method, the acquisition times and acquisition probabilities are worse, which especially in the case of a frequency-hopped system will result in rather poor network performance.

Another way to provide capture in a spread spectrum packet radio network is to employ a transmitter-oriented transmission protocol. For this protocol, packets transmitted by different radios will be on different spreading sequences, no matter when the transmissions were originated. Even if the transmissions begin at exactly the same time, they will not collide. There will, of course, be random errors due to the mutual interference between the two transmissions.

This protocol has better capture properties than receiver-oriented protocols. However, the disadvantage with a transmitter-oriented protocol is that the receiver does not know which spreading sequence will be used on an incoming packet. It must search over the set of spreading sequences of all radios within range. This assumes, of course, that the receiver has the information and equipment needed to generate the spreading sequences used by these radios. Clearly, transmitter-oriented protocols place a considerable burden on the receiver.

There are some combinations of the transmission protocols which have been discussed above. Perhaps the most commonly suggested combination is based on the use of a common spreading sequence for the synchronisation preamble and address portion of the packet. The data portion of the packet can be either transmitter-oriented or receiver-oriented. Because chip synchronism has already been accomplished by the time the data portion of the packet has arrived, switching to a receiver-oriented protocol causes no extra burden on the receiver. Switching to a transmitter-oriented protocol is not much more complicated. The only additional requirement is that the receiver must be able to generate or know the sequences used by all transmitters within range.

4.2.4.2 Carrier-sense multiple-access (CSMA)

It is commonly considered that the type of channel sensing that is used to eliminate conflicts in narrowband radio networks is virtually useless in spread spectrum packet radio networks. For example, if a transmitter-oriented transmission protocol is employed, the transmitters do not know which spreading sequences to look for to see if the intended receiver is busy. By their very nature, spread spectrum transmissions are very difficult to detect by receivers which do not know the spreading sequence. If a common spreading sequence is used, a transmitter that wishes to see if the intended receiver is busy must check all packets that are being transmitted. This includes acquiring synchronisation and demodulating the address portion of the packet. Even for receiver-oriented protocols, the hidden terminal problem will limit the effectiveness of channel sensing.

Despite these problems, this section will look into the possibility of providing a spread spectrum transceiver suited for packet-switched networks. This would not only be of interest to military users, but also to user groups in general that are concerned with external interference and reliability. Combining attractive behaviours of a spread spectrum transceiver with the flexibility of a packet-switched network is therefore sought.

Carrier sense can be obtained either by some sort of energy detection or by demodulation of the signal which as mentioned earlier usually requires knowledge of all sequences and timing in the network. Furthermore, this would generally make up a receiver with significantly added complexity compared to an ordinary spread spectrum receiver.

In the case of traditional frequency hopping, neither approach is going to be very successful, at least not if the system has fairly large processing gain. The same would also to some extent apply to direct-sequence systems, but one could in this case imagine being able to create a system with fixed sequences, using the approach of demodulation. This would, however, affect transmission security, which we consider usually to be an important reason for choosing spread spectrum in the first place.

The first approach clearly represents a contradiction to the low probability of interception characteristics inherent in spread spectrum systems. However, this approach would be desirable in order to keep the monitoring circuits simple, and thus reduce power consumption and costs in mobile terminals. In order to obtain a balanced system, we need preferably moderate to low processing gains and a sufficient integration period to obtain a high probability of detection and low false-alarm rates.

As mentioned earlier, use of energy detection to provide CSMA in spread spectrum systems is in itself a contradiction. When considering actual numbers, energy detection falls into two main categories: linear methods and nonlinear methods. The latter, for example frequency-doubling detection, provide significantly-improved performance compared with traditional techniques [9]. However, there is still going to be an order of magnitude of difference between these types of detection technique compared to matched-filter detection. Now, consider in addition that CSMA based on energy detection would be very vulnerable to any kind of interference. Since all signals look the same to the detector, an unpredictable and high false-alarm rate will occur in the case of any interference either hostile or friendly. On this basis, energy detection is discarded as an alternative for CSMA detection.

When considering matched-filter detection (i.e. demodulation) of spread spectrum signals to provide CSMA, we have a classical detection problem to consider analogous to the one in radar detection theory. This would also apply when considering preamble sense or synch detection since this is analogous. In a practical case, however, code sequences will be different and will be given characteristics which influence performance. The basic detection problem is illustrated in Figure 4.19.

By observing the correlator output, we want to decide whenever a signal is present. This is done by threshold detection, similar to that shown in Figure 4.10. However, in this case the energy detection circuitry is replaced by a noncoherent correlation receiver. If the correlation value is found to exceed a given threshold, a valid signal is present and the receiver synchronises and starts decoding. This is depicted in Figure 4.20.

Threshold detection is a nonideal process, since the correlator output signal has a variance. Depending on the type of receiver, this variation can

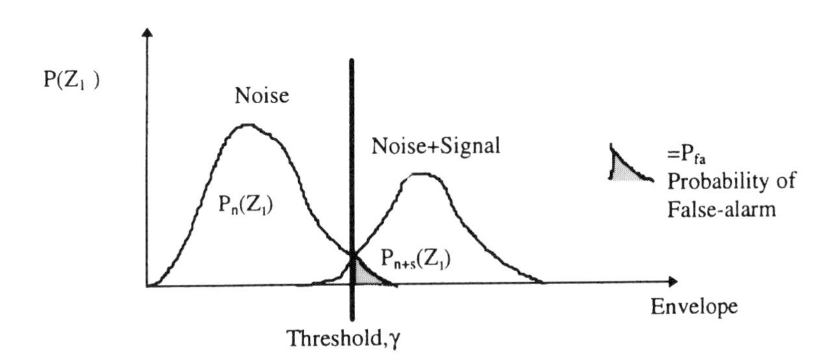

Figure 4.19 Threshold detection problem

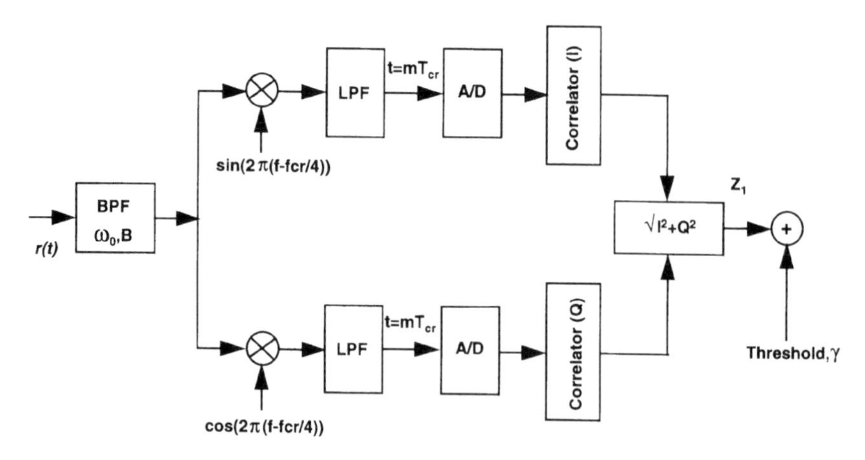

Figure 4.20 Detection by a noncoherent correlation receiver

be described by known distribution functions, which makes theoretical prediction of synchronisation and detection performance possible.

As indicated in Figure 4.19, this process is associated with a false-alarm rate (FAR), given by the area of $P_n(\mathcal{Z}_1)$ which exceeds the threshold value γ. Furthermore, the probability of detection is less than 100%, caused by the portion of $P_{n+s}(\mathcal{Z}_1)$ which lies below γ. Dependent on the signal-to-noise (SNR) ratio, this situation will either worsen or improve. This means that curves can be calculated showing detection probabilities as a function of SNR and FAR.

Let us as an example consider noncoherent detection of binary direct-sequence spread spectrum, assuming a demodulator as shown in Figure 4.20.

Let us assume that we have an AWGN channel, i.e. that $n(t)$ is a white gaussian noise process with constant spectral density $\mathcal{N}_0 = 2\sigma_0^2$. Let us further assume that the phase of any signal is such that it only appears in the I-channel whereas the Q-channel contains only noise. Each channel houses a correlator, with outputs which are identical when the signal is absent. The noise into the correlators has standard deviation σ_0. The energy in a chip is denoted by E_1 and its amplitude will be proportional to $\sqrt{E_1}$. During the correlation the noise is multiplied with the reference chips, such that the new variance per chip will be $E_1\sigma_0^2$. The noise contributions from all L chips in a code are independent stochastic variables which are summed in each correlator. If we write E_s for LE_1 and remembering that $\sigma_0^2 = \mathcal{N}_0/2$, then the total variance σ of the noise from each correlator will be $\sigma^2 = E_s\mathcal{N}_0/2$. The correlator output will now have a normal distribution with variance σ^2 and mean E_s given by:

$$P_I(x) = P_Q(x) = \frac{1}{\sigma\sqrt{2\pi}}\exp\left(-\frac{(x - E_s)^2}{2\sigma^2}\right) \tag{4.44}$$

For a noncoherent receiver we have the following decision variable:

$$Z_1 = \sqrt{I^2 + Q^2} \tag{4.45}$$

Since both I and Q have normal distributions and the same mean then $P_n(Z_1)$ has a Rayleigh distribution. If we now consider the case where a signal is present, we will still have two normal distributions, but their mean values will be different. The result is that the probability function $P_{n+s}(Z_1)$ of the decision variable Z_1, will have a Rice distribution. Knowing these conditions we can calculate detection probabilities. The values are tabulated in standard radar textbooks. An example is shown in Figure 4.21.

A decision based on a single symbol will not usually be sufficient. By summation of several symbols in a transversal filter (post detection) it is

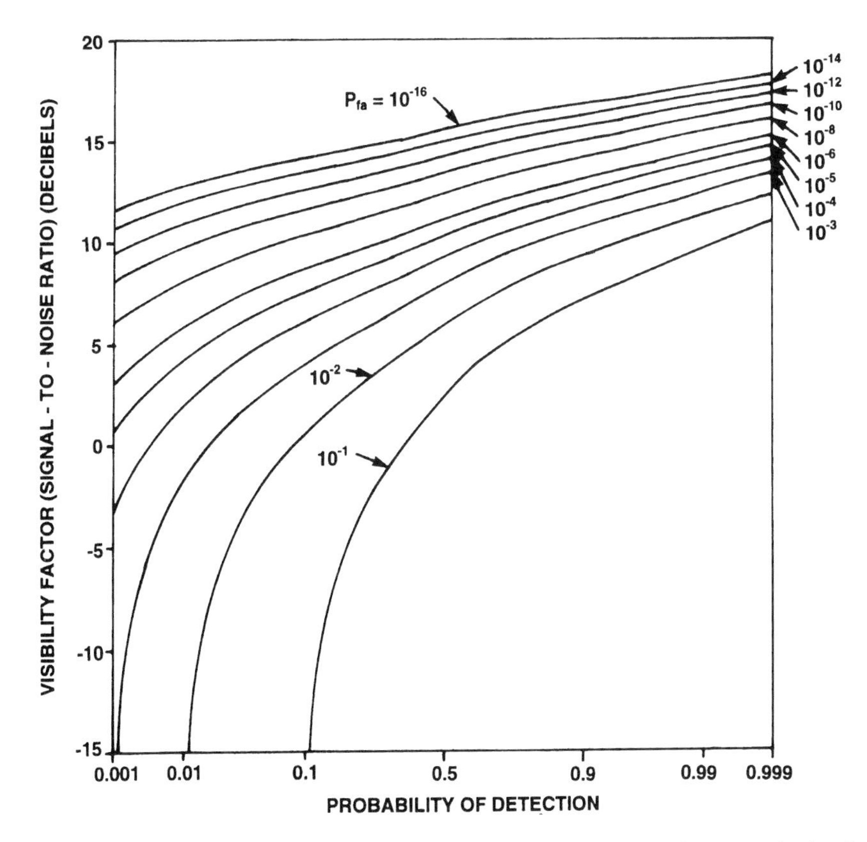

Figure 4.21 *Required signal-to-noise ratio (visibility factor) at the input terminals of a linear–rectifier detector as a function of probability of detection for a single pulse, with the false-alarm probability (P_{fa}) as a parameter; a nonfluctuating signal is assumed*

possible to improve performance at the cost of increased detection (integration) time. Summing several symbols before thresholding and detection alters the distribution of the decision variable. We now have a sum of variables:

$$Z_N = \sum_{N} Z_1(t - NT) \tag{4.46}$$

This then becomes difficult to treat analytically, and either simulations or numerical approximations will have to be used. If, however, we choose to perform a summation after each of the two correlators before the square root, then we are still in business. Results from summation of several symbols (pulses) can also be found in radar textbooks, as shown in Figure 4.22.

In the case of CSMA it is of great interest to keep the detection time as low as possible, due to overall network performance. This means that there is a trade-off to be made between detection probability (and false-alarm rate) and detection time. From this it is obvious that not all of the various spread spectrum techniques described in Section 4.2 are very well suited to effective CSMA systems. However, combining CSMA and spread spectrum is certainly of interest, particularly in the case of direct-sequence systems. We will in such a case be able to provide solutions with extremely

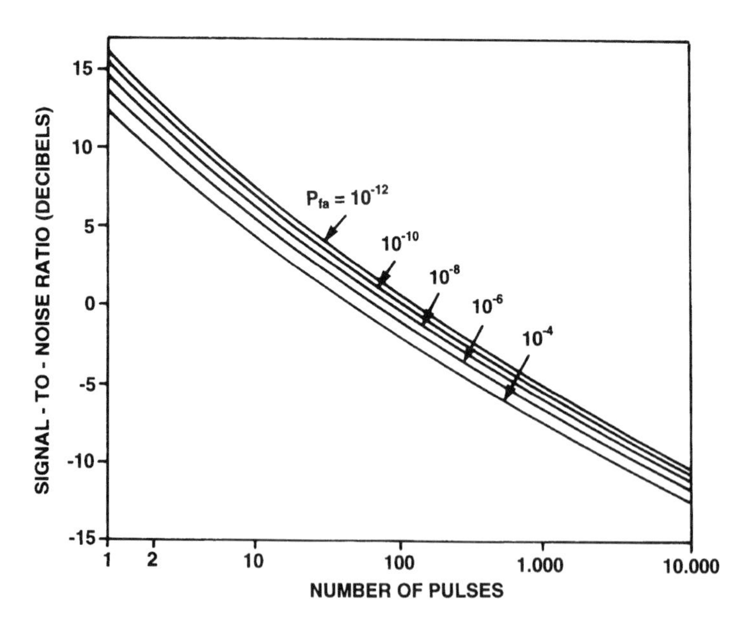

Figure 4.22 Required signal-to-noise ratio (visibility factor) as a function of the number of pulses integrated, for a linear detector; 0.95 probability of detection and nonfluctuating signal for five values of false-alarm probability (P_{fa})

reliable medium-access mechanisms, which clearly is of interest not only to the military communities.

4.3 Signal design

4.3.1 Introduction

As seen in the previous Sections, there is a wide variety of spread spectrum schemes. The variety is diverse both when it comes to the quantity of schemes and also in the range of parameters used within each scheme. These parameters are all the characteristic dimensions in time–frequency space, code structures and synchronisation approaches. This diversity seems bewildering, although in most instances the modulation type might be specified. However, the external specifications will not relate directly to a specific modulation scheme, a fact that requires a great deal of background knowledge; intuition is essential in the design process.

The experienced designer will relate specific operational requirements and system properties to each modulation and synchronisation scheme. Although the parameter space might be large, there are prevailing characteristics of the various signal formats which ultimately steer the design process into a stage where rules of thumb and simple calculations will see the design process through its initial stage. At the next stage, further changes and refinements are included directed by the requirements of the network itself and technical, regulatory and economic limitations.

This section does not intend to give a complete description of all the aspects of spread spectrum design; our aim is to present a few selected areas that are of interest to spread spectrum, and especially to direct-sequence systems. Some common digital-modulation schemes are presented. These are designed in order to obtain bandwidth efficiency in binary communication systems. However, these are also of interest to direct-sequence spread spectrum, especially in order to reduce spectral sidelobes in an efficient way. The sections on *m*-ary signalling and error control both belong to the category of channel coding, which deals with properties of pseudonoise sequences for direct-sequence spread spectrum.

4.3.2 Digital modulation

Modulation changes considerably when changing from analogue to digital signalling. In contrast to earlier solutions using a modem based on either frequency-shift keying (FSK) or phase-shift keying (PSK), a large number of schemes have been proposed especially tailored for the exchange of digital information. In the following, a description of one of the larger classes of modulation schemes, namely continuous phase modulation, is given. With respect to spread spectrum signalling, modulation on a chip level would apply especially to narrowband direct-sequence systems. As for non-spread spectrum systems, the argument for using a modulation scheme other than FSK or PSK is bandwidth efficiency. The aim is reduced spectral sidelobes which can interfere with nearby channels, and suppression of out-of-band interference. The latter is obtained because improved modulation schemes allow for sharper filtering in the receiver without distorting the signal. Also important is that these modulation schemes have a constant signal envelope, which allows for nonlinear amplifiers. This is important for mobile terminals which are powered by batteries, and for satellite systems.

4.3.2.1 Continuous phase modulation

One large class of bandwidth-efficient modulation schemes is called continuous phase modulation (CPM) [10]. CPM is designed to enable digital signalling given bandwidth- and power-starved channels. A major group among this type of scheme is characterised by a constant envelope and is defined by:

$$s(t) = (2E/T)^{1/2} \cos\left[\omega_0 t + 2\pi \sum_{i=0}^{n} \alpha_i h_i q(t - iT)\right] \text{ where } nT < t < (n+1)T \tag{4.47}$$

Information is encoded in the phase given by $\{\alpha_0, \alpha_1, \ldots, \alpha_n\}$ which are m-ary symbols with values $\pm 1, \pm 3, \ldots, \pm(m-1)$. The modulation index h_i can vary, but is in most cases fixed and equal to 1/2. The phase response function $q(t - iT)$ is given from the frequency pulse $g(t)$ as follows (normalised to $q(\infty) = 1/2$):

$$q(t) = \int_{-\infty}^{t} g(\tau) d\tau \tag{4.48}$$

The most commonly used modulation schemes are defined by the following frequency pulses:

Raised cosine, pulse length L

$$\text{LRC } g(t) = \begin{cases} \frac{1}{2LT}\left[1 - \cos\left(\frac{2\pi t}{LT}\right)\right] & 0 \leq t \leq LT \\ 0 & \text{otherwise} \end{cases} \tag{4.49}$$

Tamed-frequency modulation

$$\text{TFM } g(t) = \tfrac{1}{8}[g_0(t-T) + 2g_0(t) + g_0(t+T)] \tag{4.50}$$

$$\text{der } g_0(t) \approx \frac{1}{T}\left[\frac{\sin\left(\frac{\pi t}{T}\right)}{\frac{\pi t}{T}} - \frac{\pi^2}{24}\,\frac{2\sin\left(\frac{\pi t}{T}\right) - \frac{2\pi t}{T}\cos\left(\frac{\pi t}{T}\right) - \left(\frac{\pi t}{T}\right)^2 \sin\left(\frac{\pi t}{T}\right)}{\left(\frac{\pi t}{T}\right)^3}\right] \tag{4.51}$$

Spectral raised cosine, pulse length L

$$\text{LSRC } g(T) = \frac{1}{LT}\,\frac{\sin\left(\frac{2\pi t}{LT}\right)}{\frac{2\pi t}{LT}}\,\frac{\cos\left(\beta\,\frac{2\pi t}{LT}\right)}{1 - \left(\frac{4\beta}{LT}\,t\right)^2} \quad \text{where } 0 \leq \beta \leq 1 \tag{4.52}$$

Gaussian-shaped minimum-shift keying

$$\text{GMSK } g(t) = \frac{1}{2T}\left\{Q\left[2\pi B_b\,\frac{t - \frac{t}{2}}{(\ln 2)^{1/2}}\right] - Q\left[2\pi B_b\,\frac{t + \frac{t}{2}}{(\ln 2)^{1/2}}\right]\right\} \tag{4.53}$$

$$\text{der } Q(t) = \int_t^{\infty} \frac{1}{(2\pi)^{1/2}} \exp\left(\frac{-\tau^2}{2}\right) d\tau \tag{4.54}$$

Rectangular-frequency pulse, pulse length L

$$\text{LRC } g(t) = \begin{cases} \dfrac{1}{2LT} & 0 \leq t \leq LT \\ 0 & \text{otherwise} \end{cases} \tag{4.55}$$

L is defined as the length of the frequency pulse. If $L = 1$, the pulse is equal to the interval of the bit to be shaped; thus no intersymbol interference (ISI) is introduced. If $L > 1$, better spectral properties are achieved at the cost of reduced performance. However, this might result in a relatively better system, compared to achieving comparable spectral properties by filtering. Furthermore, coherent systems can be made to compensate for known ISI introduced in the chosen modulation scheme. In the following we will look more closely at a couple of digital-modulation schemes frequently chosen by designers.

4.3.2.2 Minimum-shift keying

The main problem with simple modulation schemes such as FSK, BPSK, QPSK and OQPSK is that all contain sharp phase transitions, which result in a lot of high-frequency components in the signal. To achieve better spectral properties, the phase of the signal has to be made continuous. The more gradual the transitions, the faster the spectral tails drop to zero. In the case of MSK (minimum-shift-keying), the phase is given a linear transition across the whole bit interval to achieve a 90° phase change [11]. This is depicted in Figure 4.23. The improvement in power spectral properties is shown in Figure 4.24, in comparison to BPSK, QPSK and OQPSK.

In terms of CPM classification, MSK is equal to 1REC. This is shown in Figure 4.25.

The signal diagram for MSK is equal to that belonging to OQPSK, and thus the performance is equal in a coherent system using a linear

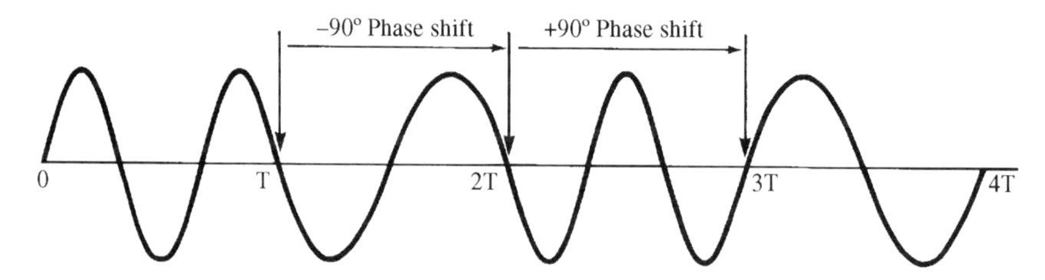

Figure 4.23 MSK modulated signal [12]

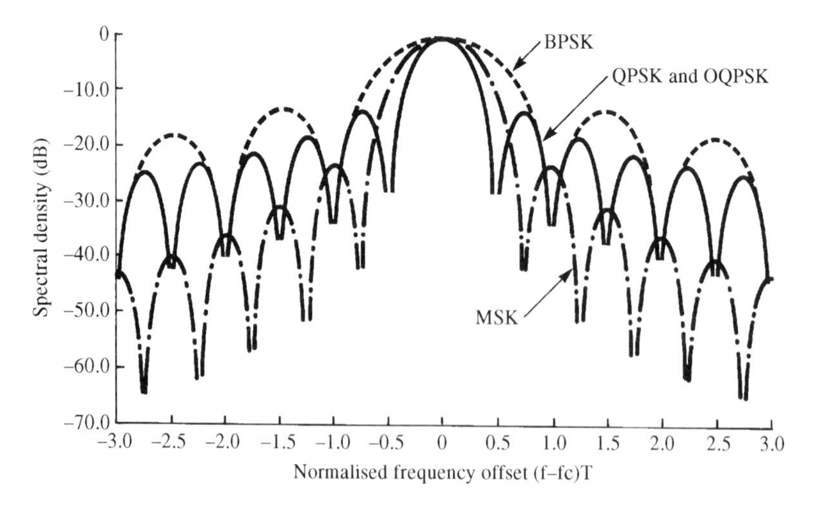

Figure 4.24 Power spectra of a few simple digital modulation schemes [13]

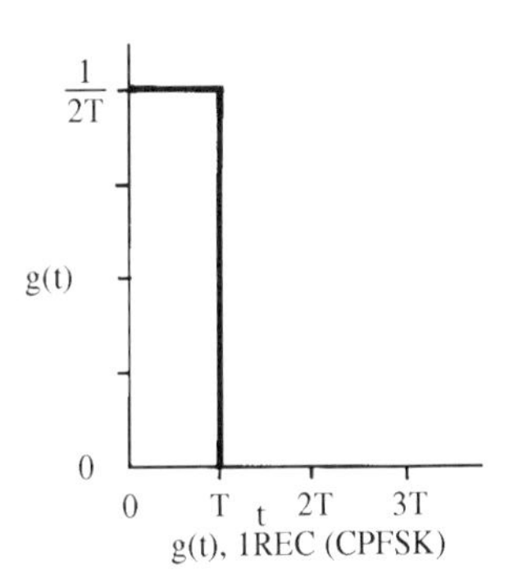

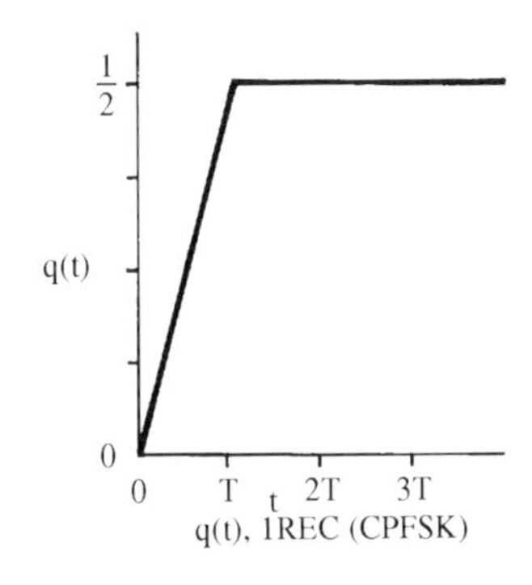

Figure 4.25 Minimum-shift keying MSK = 1REC(h = $\frac{1}{2}$), according to the CPM definition [11]

channel. In practical band-limited systems, however, MSK usually outperforms schemes like FSK, BPSK, QPSK and OQPSK.

4.3.2.3 Gaussian minimum-shift keying

In GMSK (gaussian minimum-shift keying) the phase transition is made in an interval exceeding the bit interval, resulting in even smoother transitions and further improved spectral properties [14]. The phase in a given bit interval is thus not only encoded according to the information in the given interval, but also in relation to the previous and the following bits. Accordingly, a principal GMSK modulator is shown in Figure 4.27.

In GMSK modulation the phase transitions are made continuous. This means that ISI is introduced to obtain an even more bandwidth-efficient modulation scheme, at the cost of increased bit error rate at the receiver. There are no longer distinct transitions between increasing and decreasing encoded phases, which was the case for MSK. This is shown in Figure 4.28.

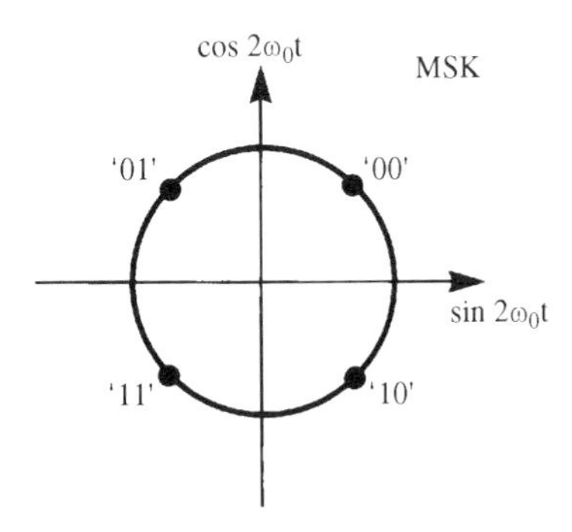

Figure 4.26 MSK signal diagram

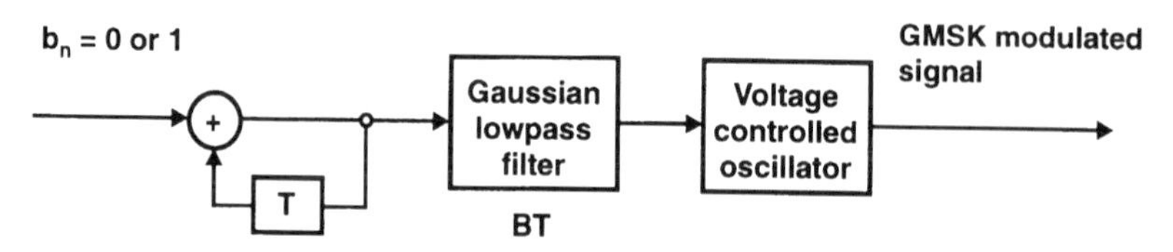

Figure 4.27 Block diagram of GMSK modulation

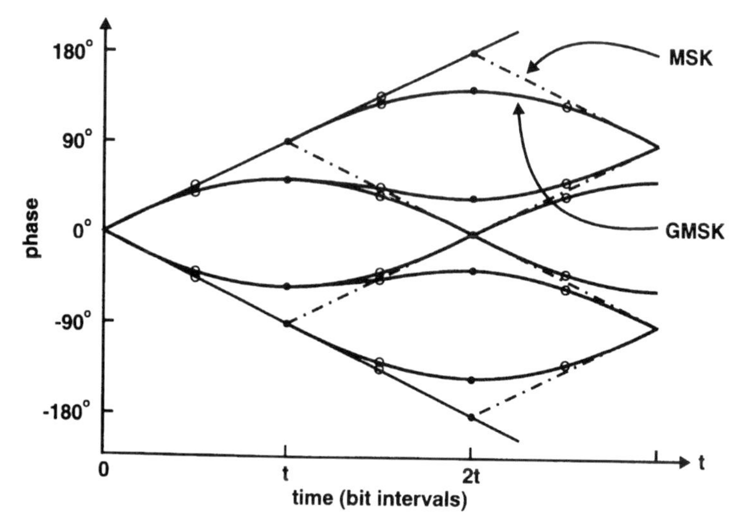

Figure 4.28 Phase transitions in MSK and GMSK modulation

A gaussian pulse is defined from minus to plus infinity and is therefore in practical implementations given a limited duration, L, in terms of bit intervals. Furthermore, it is convenient to tailor the relative pulse width to obtain the best performance in a given implementation. According to Figure 4.27, the width of the frequency pulse is a function of the filter bandwidth, B, relative to the bitrate $1/T$. This is referred to as the BT product and is usually chosen between 0.2 and 0.3 (the latter is for a GSM system). Various BT products are shown in Figure 4.29.

4.3.2.4 *Implementation*

Most CPM schemes can be implemented quite simply. One technique is to use an offset-quadrature modulator. In this case the baseband data is shaped to predetermined forms defined by values in PROM or RAM. This is shown in Figure 4.30. It should be noted that in addition to its simplicity, this modulator can be reprogrammed to produce other schemes, simply by changing the values stored in memory.

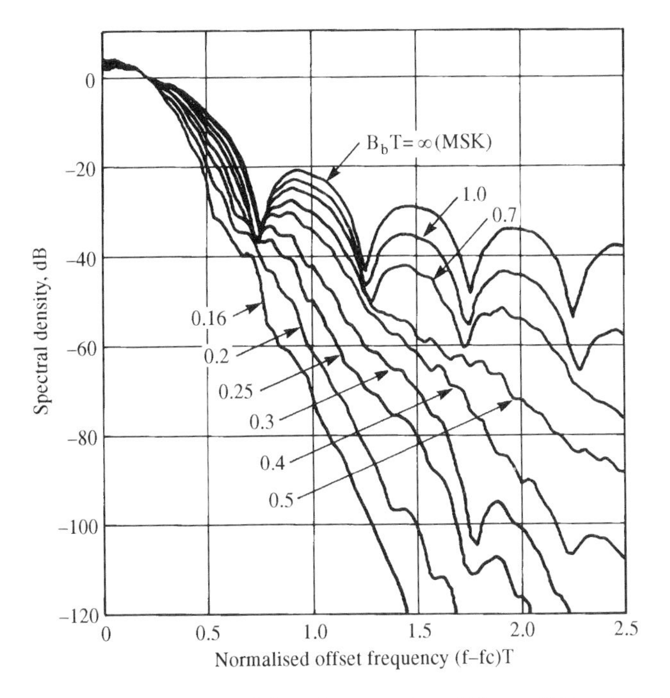

Figure 4.29 Power spectra of GMSK modulation given different BT products [14]

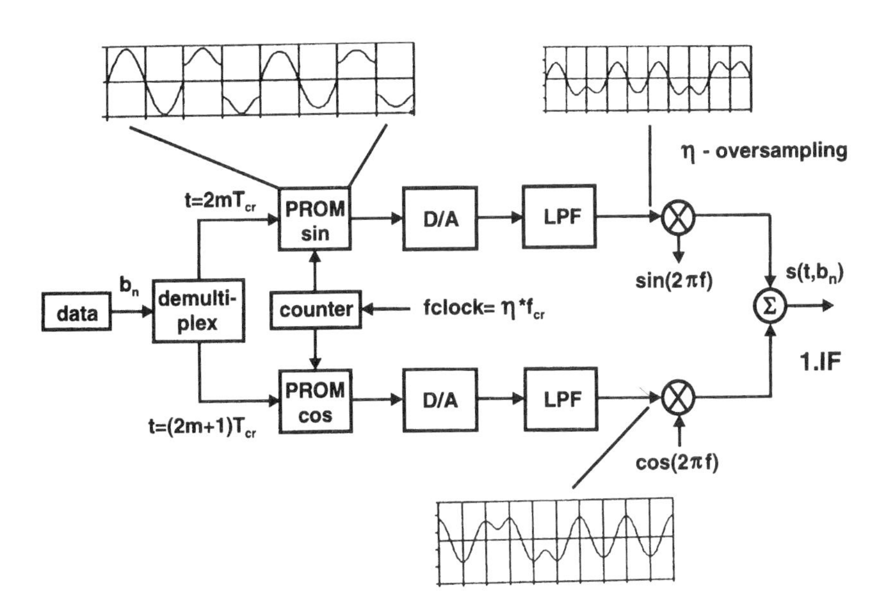

Figure 4.30 A simplified offset-quadrature modulator used to generate GMSK, $BT = 0.3, L = 3$

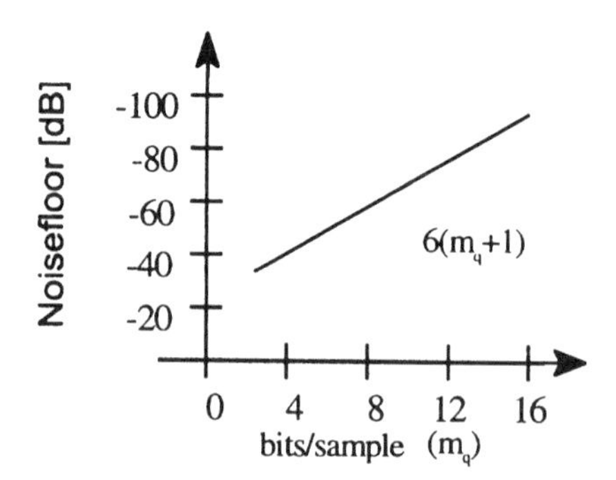

Figure 4.31 Relationship between quantisation and noise floor

By implementing a scheme using $L = 3$, the total number of possible stored shapes is equal to eight. For $BT = 0.3$, $L = 3$ is sufficient since the area outside this interval is insignificant. However, if an even narrower spectrum is sought, for instance using $BT = 0.2$, $L = 4$ is preferable. If L is chosen to be too short, it will influence the spectral tails, and what was gained by a low BT product is lost.

Figure 4.32 shows a simulation of the unfiltered spectrum from an offset-quadrature modulator. It is worth noting that the spectrum noise level produced is about -60 dB, with the eight-bit quantisation used for stored shapes.

Quantisation noise must be removed by using baseband analogue filters, irrespective of the modulation form. The sole purpose of the filter is to remove quantisation noise, and it should not affect the signal in any

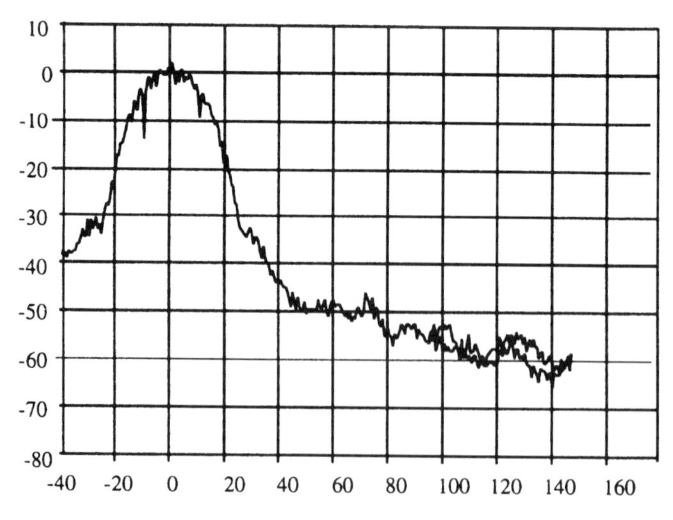

Figure 4.32 Simulated GMSK according to Figure 4.30 (chip rate = 40 kHz)

way. It is reasonable to assume that quantisation noise from the converters will behave in the same manner as white background noise. The noise background will be of the order of half a quantisation step. This implies that the background noise for a modulator with a resolution of m_q bits/sample will be:

$$N_{floor}(m_q) \approx -20\log_{10}[2^{(m_q+1)}] = -6(m_q+1)\ \text{dBc} \qquad (4.56)$$

As an example, using eight-bit quantisation, the modulator will produce a noise floor at approximately −54 dBc. Simulations and experience have confirmed that eqn. 4.56 gives a good but slightly pessimistic description of the situation for resolutions from 3–16 bits/sample.

Analogue filtering used to shape the main lobe of the transmitted signal is associated with loss. The reason for this is that practical filters will produce phase changes so that the signal is already distorted when it leaves the transmitter. Even an ideal filter, with linear phase and infinite stop-band attenuation, would alter the appearance of the chips and lead to a nonoptimal aperture. The best solution is to process all spectrum shaping of the chips using look-up tables (stored in PROM) where the trade-off between spectral properties and aperture is tailored to the communication system. Figure 4.33 shows an example of this, where analogue filtering of an MSK signal is used to obtain spectral properties compared to those of a

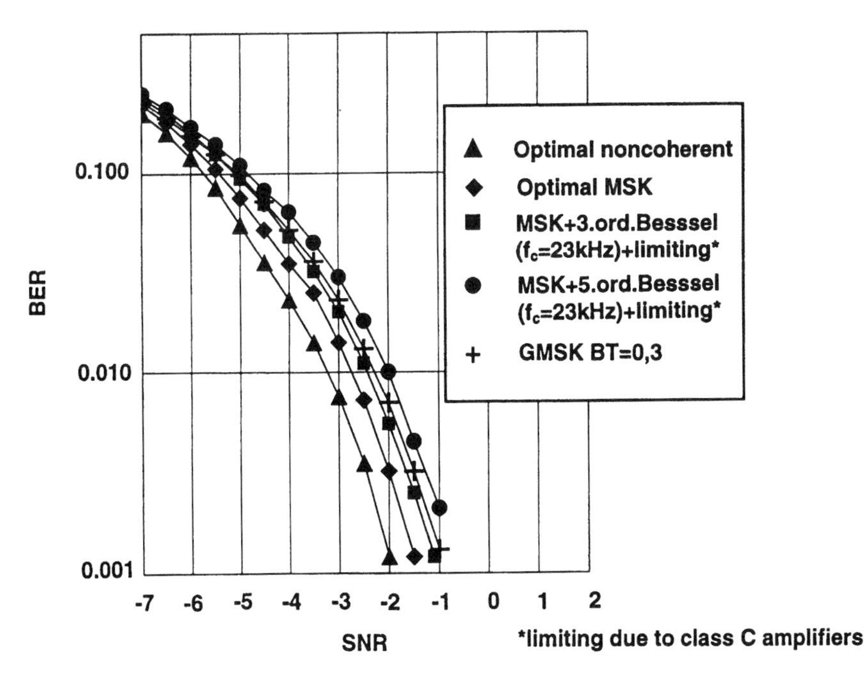

Figure 4.33 Simulated performance of filtered MSK against GMSK

GMSK signal. As shown, even moderate filtering of an MSK signal will lead to lower performance than GMSK modulation. In fact, even more filtering than shown in Figure 4.33 would be necessary in order to attain the spectral advantages of GMSK modulation.

Figure 4.33 used a simulation which was optimised for a given system with 40 kHz chip rate, corresponding to 25 kHz bandwidth. Other systems would demand different modulation schemes optimised to suit their boundary conditions.

4.3.3 *m*-ary direct-sequence spread spectrum

m-ary direct-sequence spread spectrum is a combination of two well known techniques, multilevel orthogonal signalling and direct-sequence spread spectrum. Particularly favourable combinations have retained the advantages of each technique. The combination is also referred to as narrowband spread spectrum (NBSS), Section 4.2.2. Technical progress in digital electronics now enables the use of the technique in practical systems.

m-ary signalling is a method of signal coding which falls into the category of channel coding. The main idea behind all forms of channel coding is to transform the information signals to be transmitted into an improved signal set which can better withstand any channel impairments present. *m*-ary signalling reduces the probability of error (or required (E_b/N_0) in the case of a white noise environment).

With *m*-ary signalling, k information bits are grouped to point to one of $m = 2^k$ different transmitted signals at the expense of bandwidth (binary signalling being the special case when $k = 1$). Each of these signals is a code represented by different phase and/or amplitudes or even a set of more complex signals such as, for example, spread spectrum. For the case of a spread spectrum code set, the codes have to be chosen such that they are perceived as distinctly different signals by the receiver matched filter.

Multilevel coding is based on the idea that k information bits are grouped together in order to choose (or produce) a symbol (or signal) from a companion set of orthogonal symbols. In other words, each symbol is represented by k bits, so that the total number of different symbols is:

$$m = 2^k \tag{4.57}$$

The performance gain for a communication system using *m*-ary DS SS will benefit from two separable advantages. One is an improved BER resulting from orthogonal or bi-orthogonal *m*-ary coding. The other is a processing gain resulting from the bandwidth expansion given by the direct-sequence coding in the system.

The probability for an error in a noncoherent m-ary system is given by [15]

$$P_{eb}(k) = \frac{1}{2}\frac{1}{2^k - 1}\exp\left(\frac{-E_b k}{N_0}\right)\sum_{j=2}^{j=2^k}(-1)^j\frac{2^k!}{j!(2^k-1)!}\exp\left(\frac{E_b k}{jN_0}\right) \quad (4.58)$$

Eqn. 4.58 gives the probability for a bit error as a function of E_b/N_0 and k, the number of bits at each level.

For binary communication $k = 1$, and eqn. 4.58 reduces to:

$$P_{eb}(1) = \frac{1}{2}\exp\left(\frac{E_b}{2N_0}\right) \quad (4.59)$$

This gives the probability of bit error in a binary noncoherent system with orthogonal coding, such as binary PSK.

It is also possible to use bi-orthogonal m-ary signalling in combination with direct-sequence spread spectrum. The bi-orthogonal code set is made up of an $m/2$ orthogonal code set and its antipodal pairs (i.e. antiphase). In the binary case, this would correspond to a binary antiphase signalling scheme such as DPSK. Bi-orthogonal codes theoretically lead to 3 dB performance improvement over orthogonal codes (binary case), but this decreases rapidly as k increases.

By combining eqns. 4.58 and 4.59 it is possible, for any given BER, to ascertain the improvement in E_b/N_0 over that for binary communication as k is altered. This is shown in Figure 4.35.

As an example, consider a 6-ary coding system. Grouping six bits gives a total of $2^6 = 64$ possible values or symbols. For each group of six bits entering the 6-ary encoder, one of the 64 possible symbols is selected. The total processing gain associated with 6-ary coding is 6 dB as can be seen from Figure 4.35. However, it is apparent from the curve that increasing the value of k yields a diminishing return. Thus, increasing k will rapidly lead merely to increased complexity without significant performance gain.

An increase in bandwidth is necessary when using m-ary orthogonal signalling. Use of completely orthogonal code words demands that the word length, L, is at least as great as m. In practice, this demand can be relaxed if approximately orthogonal codes are used, as discussed in Section 4.3.4.

On the other hand, it is possible to increase bandwidth by spread spectrum methods, which makes it possible to benefit from the associated processing gain. This provides an opportunity to design systems with exceptionally high performance. Figure 4.36 demonstrates two possible

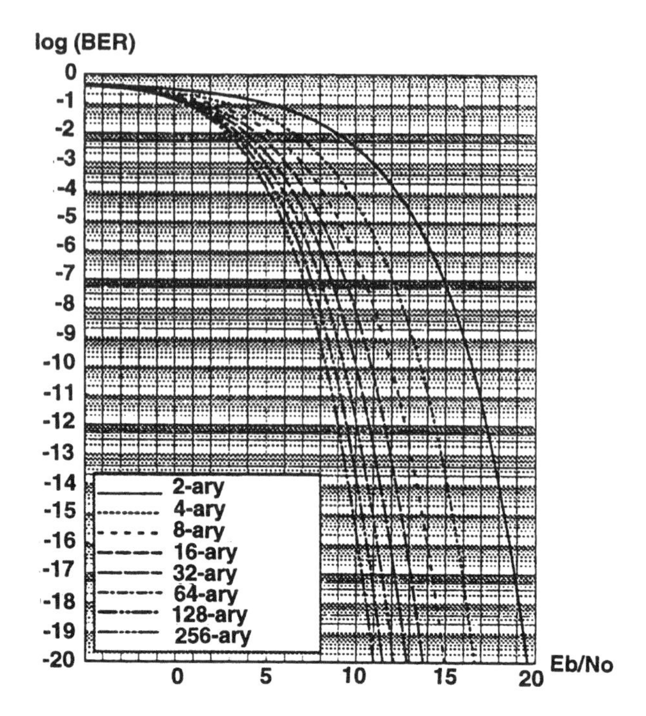

Figure 4.34 Bit error rate (BER) for m-ary orthogonal noncoherent signalling

combinations for direct-sequence systems. Using GMSK modulation, described in Section 4.3.2, it is possible to reduce the bandwidth to approximately half that of the chip stream being modulated. Figure 4.36 shows two examples where k and the orthogonal code length L are chosen to give a chip rate of 40 kHz, which leads to the desired bandwidth of 25 kHz.

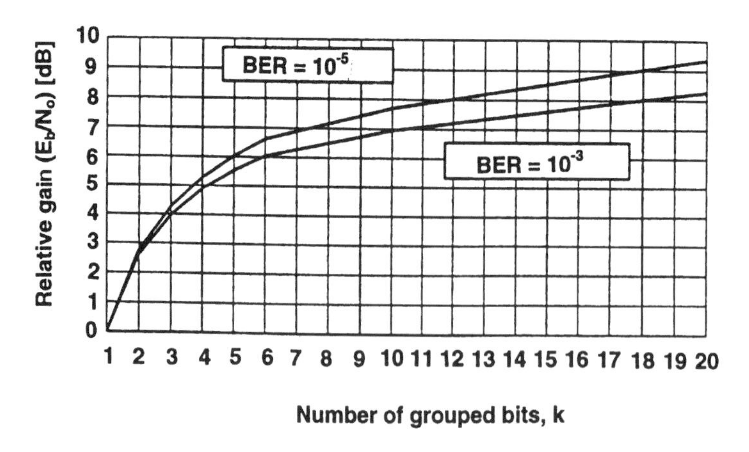

Figure 4.35 Relative gain in performance for noncoherent orthogonal m-ary signalling

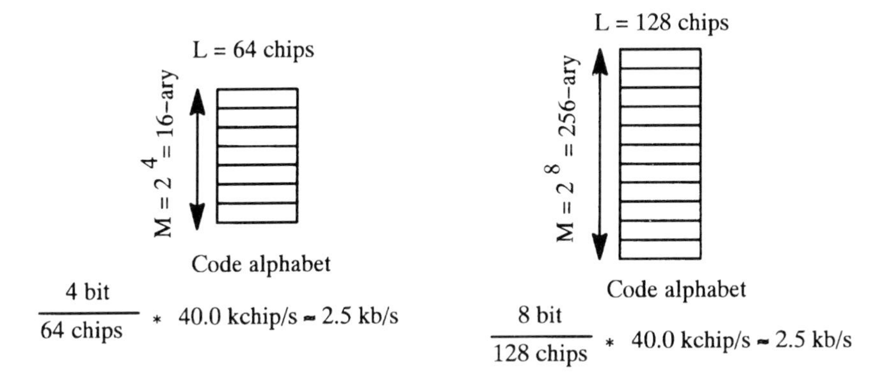

Figure 4.36 m-ary signalling being a trade-off between complexity and performance

In practice this is an optimistic estimate of the necessary information rate. In any real communication system, where complete messages are transmitted, there will be protocol overheads in the form of preamble.

It is assumed that SNR is measured across the whole utilised bandwidth. Processing gain is obtained when bandwidth is increased. In the case of DS SS it is the ratio of chip rate to data rate, since it is this increase that is exploited by the correlator.

Two examples of *m*-ary communication systems together with their respective processing gains are shown in Figure 4.37. One of these is a 40 kchips/s system giving a datarate of 2.5 kb/s using a bandwidth of 25 kHz, as shown in Figure 4.36. The other example, with 160 kchips/s, is chosen to demonstrate a corresponding system with approximately 100 kHz bandwidth.

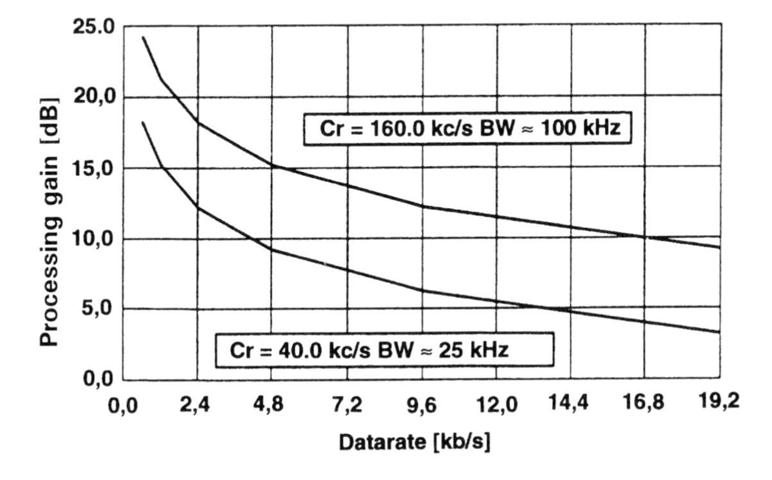

Figure 4.37 Processing gain in DSSS systems

A fundamental difference between m-ary gain and spread spectrum modulation gain is that the parallel nature of the detection process in m-ary detectors has the same effect as increasing transmitter output power. This means that all parameters that are dependent on output power or most noise conditions will be affected by the choice of the m-ary gain factor. In a channel with wideband gaussian noise such as thermal noise or competing spread spectrum systems, spread spectrum modulation does not improve the transmission quality in terms of bit error rate, but parallel processing, leading to m-ary gain, will increase the performance corresponding to the m-ary gain factor.

4.3.4 *Coding for bandwidth spreading*

Direct-sequence systems use particular codes at the transmitter to expand the spectrum, so that it occupies the whole of the available bandwidth. In code-division multiple-access the ability to differentiate between channels relies primarily on coding (Section 4.2.3). Thus the nature of the code sequences used is very important. The type of code used sets bounds on the capability of the system, bounds that can only be changed by changing the codes.

The spreading codes can be characterised by a large number of parameters, but this discussion is restricted to those which are particularly important in any design and performance analysis. Unfortunately, quantitative results are hard to compute for several key parameters, since general analytical expressions valid for a particular sequence or set of sequences probably do not exist. In these cases, evaluation of upper and lower bounds for a set of sequences can be used to calculate worst-case performance. The bounds also indicate likely gains from the effort of going through optimisation algorithms. Perhaps just as important for calculating and using bounds are the trade-offs between the values of different key parameters.

The first factor to consider is the autocorrelation of the codes. When a receiver is l code elements (chips) out of phase with the data bit edges (as will be the case initially before the data fills the matched filter) the output will be given by:

$$\begin{aligned} Y^k(l) &= \tfrac{1}{2}A^k\{b_i(a_0^k, \ldots, a_{p-l-1}^k), [n(0), \ldots, n(p-1)]\}a^k \\ &= \tfrac{1}{2}A^k b_i C^k(l) + [n(0), \ldots, n(p-1)]a^k \end{aligned} \qquad (4.60)$$

where

$$C^k(l) = \sum_{j=0}^{p-l-1} a_j^k a_{j+1}^k \quad 0<l<p \qquad (4.61)$$

is the nonperiodic autocorrelation function. For any $l \neq 0$ it is desirable that $C(l)$ is small compared with p to avoid false synchronisation.

For applications using multipath channels, knowledge of the nonperiodic autocorrelation is also important. For a two-path model the correlation process will be of the form:

$$\begin{aligned} Y^k(l) &= \tfrac{1}{2}A^k\{b_i(a_0^k, \ldots, a_{p-1}^k) + b_{i-1}(a_0^k, \ldots, a_{p-1}^k) + b_i(a_0^k, \ldots, a_{p-l-1}^k)\}a^k \\ &= \tfrac{1}{2}A^k[p + b_{i-1}C^k(p-l) + b_iC^k(l)] \end{aligned} \tag{4.62}$$

where $C^k(p-l)$ is defined in eqn. 4.61 with l replaced by $(p-l)$, so that:

$$C^k(p-l) = \sum_{j=0}^{l-1} a_j^k a_{j+p-1}^k \tag{4.63}$$

The first term on the right-hand side of eqn. 4.62 is the even or periodic autocorrelation function resulting from the primary propagation path. The next two terms are the result of the second path being delayed by l sequence chips. There are two cases to be considered. If $b_{i-1} = b_i$ then the sum of the last two terms is also an even autocorrelation function. Thus the even correlation function is related to the two odd or nonperiodic autocorrelation functions by:

$$\theta^k = C^k(l) + C^k(p-l) \tag{4.64}$$

When $b_{i-1} \neq b_i$ the odd correlation θ^k is given by:

$$\theta^k = C^k(l) - C^k(p-l) \tag{4.65}$$

The performance of a spread spectrum system in a multipath situation will thus depend on the odd and even autocorrelation properties of its spreading codes. The way in which codes should be optimised will depend on the performance criteria chosen for the system analysis.

Another parameter to consider is cross correlation. If a spread spectrum system is expanded to contain K users at the same carrier frequency in a code-division multiple-access system, the received signal can be rewritten as:

$$r(t) = n(t) + \sum_{r=1}^{K} A^r a^r(t-\tau^r) b_i^r(t-\tau^r)\cos[\omega_c t + (\alpha^r - \omega_c\tau^r)] \tag{4.66}$$

The signal may then be correlated with $a^k(t)\cos\omega t$, $k \in [K]$, which, for a digital system, gives the expression:

$$Y^k = \frac{1}{2}A^k b_i^k p + \sum_{r=1,\, r\neq k}^{K} \frac{1}{2}A^r\{b_{i-1}^r(a_0^r, \ldots, a_{p-l(r)-1}^r) + b_i^r(a_{p-l(r)}^r, \ldots, a_{p-1}^r)\}a^k \cos\beta^r + \sum_{l=0}^{p-1} n(lt_s)a_l^k \tag{4.67}$$

where $l(r) \in [0, 1, 2, \ldots, p-1]$ and β^r is the phase angle between the kth carrier and carrier r. It is assumed that $\alpha^k = 0$ and $\tau^k = 0$.

Making use of odd cross-correlation functions $C^{kr}(l)$ and $C^{kr}(p-l)$, defined similarly to the odd autocorrelations, we get:

$$Y^k = \frac{1}{2}A^k b_i^k p + \sum_{r=1,\, r\neq k}^{K} \frac{1}{2}A^r\{b_{i-1}^r C^{kr}(l) + b_i^r C^{kr}(p-l)\}\cos\beta^r + \sum_{l=0}^{p-1} n(lt_s)a_l^k \tag{4.68}$$

Again, there are two cases to be considered for each of the $K-1$ interferences influencing the kth user. If $b_{i-1}^r = b_i^r$ for some $r \in [K-1]$, where $[K-1]$ is identical to $[K]$ in the absence of the kth user, then the interference will depend on:

$$\theta^{kr} = C^{kr}(l) + C^{kr}(p-l) \tag{4.69}$$

the even cross correlation. If $b_{i-1}^r \neq b_i^r$, then the interference will depend on the odd cross-correlation function:

$$\theta^{kr} = C^{kr}(l) - C^{kr}(p-l) \tag{4.70}$$

In previous sections system performance was considered assuming perfectly-orthogonal (no cross correlation) sequences. It was also assumed that the sequences had a periodic property. As long as the receiver and transmitter were aligned and for as long as the clocks and control circuits kept the system synchronised, nonrepeated codes could be used. If, however, synchronisation is lost it will be very difficult, if not impossible, to realign the transmitter–receiver during transmission. Thus, in bursty transmissions, where the messages are fairly short, the data format usually contains a synchronisation preamble. This is a periodic sequence followed by data bits, often modulated by a long nonrepeated spreading code [16]. Then synchronisation usually occurs at the beginning of each message transfer.

In order to perform the necessary synchronisation of transmitter and receiver, the codes must have a cyclic autocorrelation function with a distinct peak. So as not to increase the possibilities of false lock and to

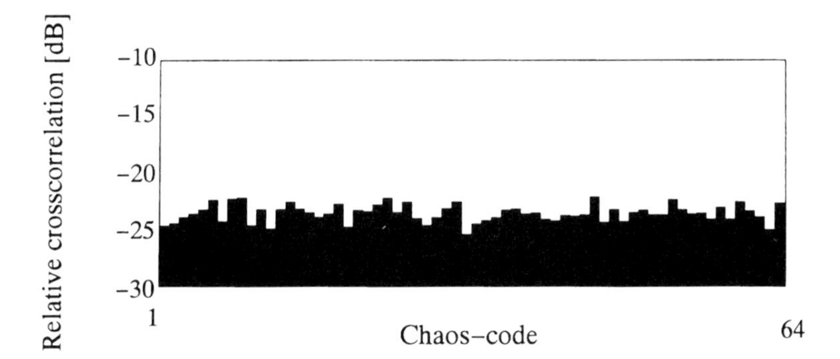

Figure 4.38 Average cross correlation with all other codes in a 64-ary chaos-code alphabet, code length 128

lessen possible multipath interference, the sidelobes of the odd autocorrelation function should be much smaller than the main peak.

In a system operated as a code-division multiple-access network, the performance is closely linked to both the even and odd cross-correlation functions. If concealment of the transmission is important, the harmonics produced by the spreading code should be fairly uniform. However, this constraint is of less importance for the higher harmonics.

In order to optimise any *m*-ary alphabet, it is advantageous to examine the average cross correlation between each individual code and all other codes. It is then possible to ascertain which codes are weakest with respect to orthogonality. The aim is to generate a code alphabet with minimal cross correlation. Figures 4.38 and 4.39 illustrate the cross-correlation properties of two code set types.

The best known direct-sequence spreading codes are Gold codes and *m* sequences. The main drawback with many codes of this type is that there is a very limited number of possible codes of a given length which meet the

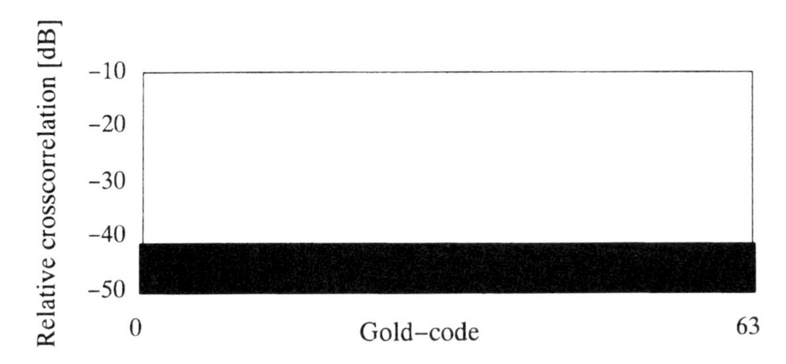

Figure 4.39 Average cross correlation with all other codes in a 64-ary Gold-code alphabet, code length 127

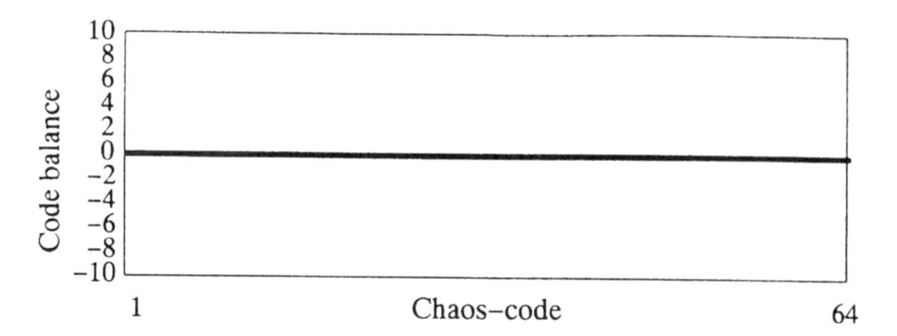

Figure 4.40 Balance in a 64-ary normalised chaos-code alphabet, code length 128

criteria for the class. A way round this problem is to generate completely random codes, know as chaos codes. These are widely used in spread spectrum communication systems. Irrespective of the class of code which is chosen, there are certain statistical properties which must be carefully controlled, since they directly influence system performance.

Balance is the measure of equality between the number of ones and the number of zeros in a code. Balance in the spreading codes will be an important parameter for most practical DSSS communication systems [17]. Imbalance (e.g. if there are more ones than zeros or *vice versa* in a code) will be associated with d.c. offset in both the receiver and transmitter. If the baseband section is a.c. coupled, as it may well be, the code will be pulled such that the signal amplitude is equally distributed around the d.c. level. The result of this is that the code fed into the decoding network will be skewed. Even worse, with any alphabet of codes with different degrees of imbalance, the baseband will be modulated since the signal always relaxes towards zero voltage. This effect will lead to a significant reduction in system performance. Figures 4.40 and 4.41 show degrees of imbalance in two given code alphabets. The chaos-code alphabet has been constructed expressly to avoid imbalance. This is not possible when using Gold codes.

Figure 4.41 Balance in a 64-ary Gold-code alphabet, code length 127

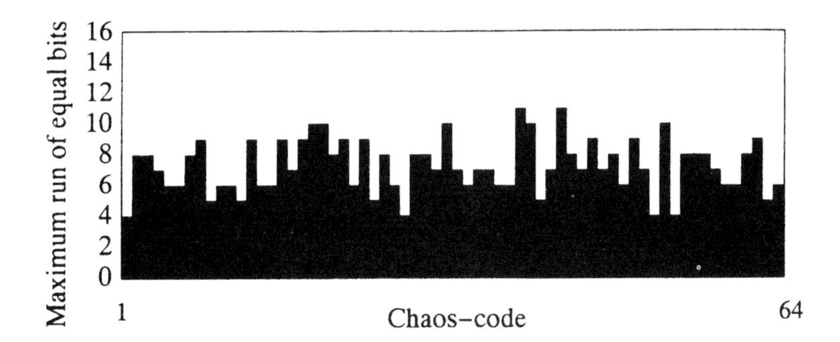

Figure 4.42 Lengths of runs in a 64-ary chaos-code alphabet, code length 128

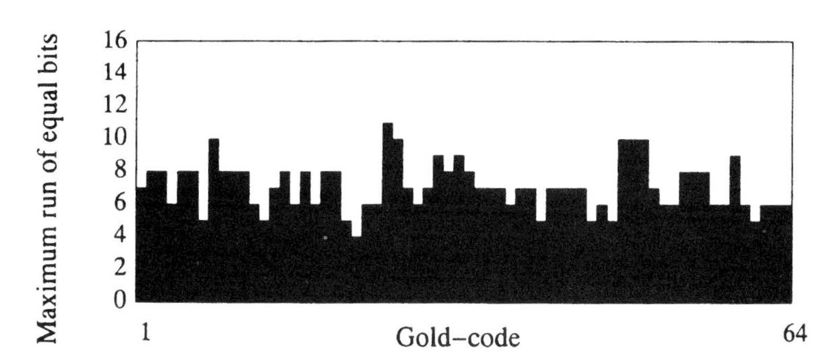

Figure 4.43 Lengths of runs in a 64-ary Gold-code alphabet, code length 127

Another characteristic of any code is the lengths of runs of ones or zeros which it contains. This is similar to imbalance, but with shorter time constant. Again, a.c. coupling of the baseband will be damaging, since the flat d.c. level of the runs will be tilted due to relaxation leading to system degradation. This can be avoided by using a d.c.-coupled demodulation network; however there are other arguments for watching the lengths of runs. For example, long runs can give codes distinct signatures. Figures 4.42 and 4.43 illustrate acceptable values for two code alphabets.

It turns out that there exist families of possible spreading codes which are easy to generate and which satisfy some of these requirements very well. However, it is necessary to apply considerable computational effort in order to obtain subfamilies satisfying all of the desired requirements.

4.3.5 Performance analysis

The point of this section is to indicate the behaviour of a matched-filter receiver. The relationship between E_b/N_0 and SNR performance is calculated for spread spectrum receivers using both white and nonwhite channel noise models.

4.3.5.1 E_b/N_0 and SNR relations

In order to obtain the relationship between E_b/N_0 and SNR we will consider a receiver model as shown in Figure 4.44. E_b/N_0 is always referred to the input of the receiving matched filter BP, and we assume that there is no intersymbol interference (ISI) in the system. This would hold for the case of MSK modulation, but not if GMSK was applied to the signals (Section 4.3.2).

A model for a receiver is shown in Figure 4.44. It is important to remember that E_b/N_0 is always referred to the input of the receiver filter BP. Furthermore:

- E_b = energy per bit = αE_c where α is the number of chips/bit and E_c is energy/chip;
- N_0 is single-sided spectral noise density in W/Hz.

It is meaningless to discuss SNR before filtering since unfiltered white noise has infinite power. We will assume that a signal-matched filter is used which is tailored to the signal in order to maximise the SNR [20].

We wish to determine the relationship between E_b/N_0 and S_m/N_m, where S_m and N_m are the powers after filtering. This can be achieved by following a chip (pulse) all the way from the transmission to reception. If a system exhibits no ISI, then the simulation for a series of pulses will be the same as that for a single pulse, since (by definition) individual pulses have no effect on each other.

It must be remembered in the following discussion that the symbols E_c, S_m etc. will refer to numerical values, and will often have incorrect dimensions.

Consider an arbitrary pulse $z(t)$. Its energy will be:

$$E_c = \int_{-\infty}^{\infty} z^2(t)dt = \int_{-\infty}^{\infty} z(f)df \qquad [\mathrm{J}] \tag{4.71}$$

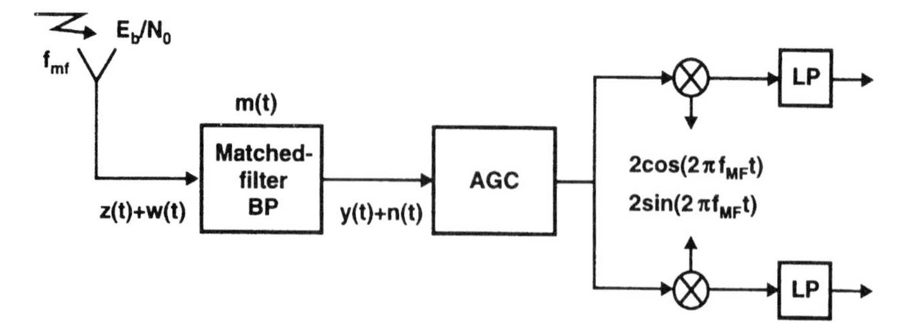

Figure 4.44 Receiver model using matched-filter channel selection

Assuming an ISI-free system and a signal-matched filter in the receiver ($m(t) = z(-t)$), the signal after filtering will be:

$$y(t) = \int_{-\infty}^{\infty} z(\lambda) \cdot m(t-\lambda)d\lambda = \int_{-\infty}^{\infty} z(\lambda) \cdot z(\lambda - t)d\lambda \qquad [\mathrm{V}] \qquad (4.72)$$

A sample at time $t = 0$ gives:

$$y(0) = \int_{-\infty}^{\infty} z(q) \cdot z(q)dq = \int_{-\infty}^{\infty} z^2(q) \cdot dq = E_c \qquad [\mathrm{V}] \qquad (4.73)$$

Signal power is the square of the sampled signal, so:

$$S_m = [y(0)]^2 = E_c^2 \qquad [\mathrm{W}] \qquad (4.74)$$

This holds for any arbitrary pulse shape which is symmetrical, and for both baseband and passband systems.

The noise power (watts) after the receiver filter will be:

$$N_m = N_0 \cdot B_N \cdot M^2(f_{MF}) = N_0 \cdot \frac{\frac{N_0}{2}\int_{-\infty}^{\infty} |M(f)|^2 df}{N_0 \cdot M^2(f_{MF})} \cdot M^2(f_{MF})$$

$$= \frac{N_0}{2}\int_{-\infty}^{\infty} |M(f)|^2 df = \frac{N_0}{2}\int_{-\infty}^{\infty} |M(t)|^2 dt$$

$$= \frac{N_0}{2}\int_{-\infty}^{\infty} |x(t)|^2 dt = \frac{N_0}{2} \cdot E_c = \frac{N_0}{2}\sqrt{S_m} \qquad (4.75)$$

The SNR can now be found by combining eqns. 4.74 and 4.75:

$$\frac{S_m}{N_m} = \frac{E_c^2}{\frac{N_0}{2}} E_c = 2\frac{E_b}{N_0}\frac{1}{\alpha} \qquad (4.76)$$

AGC will not alter SNR if it is linear, so that eqn. 4.76 will be equally valid after AGC. We can then say:

$$\left(\frac{S}{N}\right)_{MAX} = 2\frac{E_b}{N_0}\frac{1}{\alpha} \qquad (4.77)$$

For any arbitrary signal $\overline{S}/N$ is related to E_b/N_0 as follows [18]:

$$\frac{\overline{S}}{N} = \frac{E_b}{N_0} \cdot \frac{R}{B} = \frac{E_b}{N_0} \cdot \frac{1}{\beta \cdot T_b} \qquad (4.78)$$

- $\overline{S} = E_b/T_b$ is r.m.s. signal power;
- data rate is $R = 1/T_b$ where T_b is the duration of each bit in seconds;
- B = bandwidth (Hz) of the signal; since the filter is tailored to the signal this is also the noise bandwidth;
- N is the product $B \cdot N_0$ if the signal has unitary amplitude and the same bandwidth as the filter.

Eqn. 4.78 is not always too easy to use, since it is often difficult to determine the bandwidth of a signal.

In general, one can say:

$$\frac{S}{N} = \frac{E_b}{N_0} \frac{1}{\beta\alpha} \tag{4.79}$$

where β is greater that 0.5.

If a perfectly-signal-matched filter was used, only then could $\beta = 0.5$, its absolute minimum value. In other words:

$$\frac{S}{N} = \frac{1}{\beta} \cdot \frac{E_b}{N_0} \cdot r \cdot \frac{1}{s} \tag{4.80}$$

$$\frac{S}{N}[\mathrm{dB}] = \frac{E_b}{N_0}[\mathrm{dB}] + (10 \cdot \log r) - (10 \cdot \log s) - (10 \cdot \log \beta)$$

- r = code rate (≤ 1.0);
- $\beta \geq 0.5$ is a figure of merit which indicates how well the receiver filter matches the signal.

In practice, it is possible to come reasonably close to the theoretical limit $\beta = 0.5$. For example, a convolution (correlation) spread spectrum receiver has, by definition, a perfectly-tailored filter, assuming perfect synchronisation.

4.3.5.2 *Error probabilities*

Let us assume an m-ary signalling system using noncoherent detection. Performance of such a system in a white noise environment is given in Section 4.5 for various values of m, m being the number of signalling levels, which corresponds to how many orthogonal symbols are being used. The probability of bit error is thus found to be:

$$P_{be} = \frac{2^{k-1}}{2^k - 1} \sum_{n=1}^{m-1} (-1)^{n+1} \binom{m-1}{n} \frac{e^{\frac{n}{n+1} k \frac{E_b}{N_0}}}{n+1} \tag{4.81}$$

where $k = \log_2(m)$ bit/symbol. The corresponding symbol error rate (BER) is then:

$$P_{se} = \frac{2^k - 1}{2^{k-1}} P_{be} \tag{4.82}$$

This means that BER is a function of E_b/N_0, BER $= g(E_b/N_0)$. BER is then given by:

$$g\left(\frac{E_b}{N_0}\right) = \frac{2^{k-1}}{2^k - 1} \sum_{n=1}^{m-1} (-1)^{n+1} \binom{m-1}{n} \frac{e^{\frac{n}{n+1} k \frac{E_b}{N_0}}}{n+1} \tag{4.83}$$

If we then know how E_b/N_0 is varying, in other words its probability density function $f(E_b/N_0)$, then the resulting BER can be found by the following:

$$\text{BER} = \int_{-\infty}^{\infty} f(x) \cdot g(x) dx \tag{4.84}$$

where $x = E_b/N_0$. With this, a graph can be drawn showing BER as a function of mean E_b/N_0 for a given probability distribution.

Now consider the case of a Rayleigh-distributed received signal, which is a well known model in mobile communications. This is a channel model where there is no dominant direct path, resulting in large variations of E_b/N_0. This phenomenon is called Rayleigh fading due to the Rayleigh-distributed signal amplitude.

What kind of distribution is then given for E_b/N_0 when the signal level is varying according to a Rayleigh distribution?

Using a matched-filter receiver, SNR at the output of the filter is given by:

$$\frac{S_m}{N_m} = 2\frac{E_b}{N_0}\frac{1}{\alpha} = \frac{(A^2 T)^2}{\frac{1}{2} N_0 A^2 T} = \frac{A^2 T}{\frac{1}{2} N_0} \tag{4.85}$$

when assuming rectangular pulse shaping with amplitude A and duration T, and with α as the spreading factor.

If using gaussian pulse shaping instead, SNR was earlier shown to be:

$$\frac{S_m}{N_m} = 2\frac{E_b}{N_0}\frac{1}{\alpha} = \frac{\frac{A^4}{2}}{\frac{1}{2\sqrt{2}} N_0 A^2} = \frac{A^2}{\frac{1}{\sqrt{2}} N_0} \tag{4.86}$$

In both cases SNR is equal to a constant k_1 multiplied by the amplitude, A, squared:

$$\frac{S_m}{N_m} = 2\,\frac{E_b}{N_0}\,\frac{1}{\alpha} = k_1 \cdot A^2 \tag{4.87}$$

or

$$\frac{E_b}{N_0} = k_2 \cdot A^2 \tag{4.88}$$

where $k_2 = \alpha k_1/2$.

The probability density function of a Rayleigh distributed stochastic variable z is given by:

$$fz_2(z) = \frac{z}{\sigma^2} \cdot \mathrm{e}^{-\frac{z^2}{2\sigma^2}} \qquad (z \geq 0) \tag{4.89}$$

Now assuming Υ to be a function of z: $\Upsilon = g(z) = k_2 z^2$. This is also a stochastic variable with probability density function $f_\Upsilon(y)$. Using an inverse function $h: z = h(\Upsilon) = \sqrt{\Upsilon/k_2}$. $f_\Upsilon(y)$ is then found:

$$\begin{aligned} f_\Upsilon(y) &= fz(h(y)) \cdot \left| \frac{d}{dy}\, h(y) \right| \\ &= \frac{\sqrt{y/k_2}}{\sigma^2} \cdot \mathrm{e}^{\frac{y/k_2}{2\sigma^2}} \cdot \frac{1}{2} \cdot \left(\frac{y}{k_2}\right)^{-\frac{1}{2}} \cdot \frac{1}{k_2} = \frac{1}{2k_2\sigma^2} \cdot \mathrm{e}^{-\frac{y}{2k_2\sigma^2}} \end{aligned} \tag{4.90}$$

Change of variable $2k_2\sigma^2 = 1/\lambda$ gives:

$$f_\Upsilon(y) = \lambda \cdot \mathrm{e}^{-\lambda y} \tag{4.91}$$

From this it is clear that E_b/N_0 is exponentially distributed, which is also stated in [19].

The mean value or expected value and variance of an exponential distribution is:

$$E[\Upsilon] = \frac{1}{\lambda} \tag{4.92}$$

and

$$Var[\Upsilon] = \frac{1}{\lambda^2} \tag{4.93}$$

For the system given as an example above, its BER in a Rayleigh-fading environment is shown in Figure 4.45, using an exponential probability distribution for E_b/N_0 for different values of m.

As discussed earlier, wideband spread spectrum signals can give accurate estimates of the multipath structure with respect to delay and

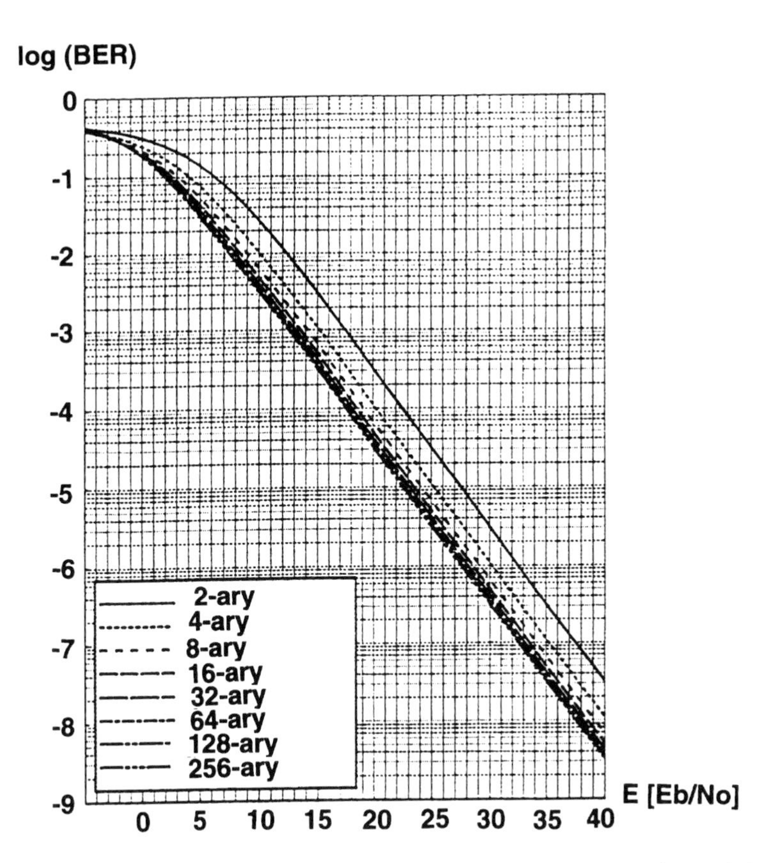

Figure 4.45 *Symbol error probability for noncoherently detected m-ary orthogonal signalling in a Rayleigh-fading environment (expected value of E_b/N_0 is given by the x-axis)*

amplitude, since the heights of the correlation peaks are proportional to peak strengths and their relative time positions in the multipath profile. This means that performance can be improved significantly. If all multipath delays $\tau < T_d$ for a spread spectrum system, then proper spreading-code choice will give negligible intersymbol interference, and a transversal filter can be used to equalise the channel.

Before data transmission the spread spectrum system can then transmit a sounding signal consisting of the unmodulated spreading code. The path strengths can be read by peak detectors and these values and their position in the delay line can then be used to set the taps and gains of the transversal filter to be used during data reception. The sounding procedure has to be repeated depending on the rate of change of the channel-transfer function. Different combinations of the multipath samples are possible and usually reflect the expected multipath profile. The output of the transversal filter is thus a weighted sum of certain tap outputs given by the sounding cycle.

There are several receiver structures that can be considered to improve channels affected by multipath diversity. A RAKE receiver is one commonly used receiver structure that is suited also to a direct-sequence spread spectrum system. However, experience shows that even simple transversal filter solutions taking into account one or possibly two additional paths are sufficient. Traditional channel equalisation is not going to be treated further in this text. Instead, examples of time–frequency analysis applied for this purpose are discussed in Section 4.5.2.

4.4 Error control

In mobile communication systems where power and bandwidth are at a premium, methods to control error rates and keep them as low as possible are of major interest. There are several methods for error control, all of which are directly or indirectly adding redundancy to the data stream. However, this does not mean that employing error control reduces overall system efficiency with respect to bandwidth utilisation. A major motive for developing various error-control or correction techniques has been the attempt to attain theoretical system performance. According to Shannon, coding schemes can exist which make it possible to communicate with zero error for E_b/N_0 greater than $-1.6\,\mathrm{dB}$ within a memoryless channel with white gaussian noise. In any practical system it is not possible to reach the Shannon limit, because bandwidth requirements and implementation complexity become excessive. A lot of research has been invested particularly in forward-error correction coding schemes, and their performance applied to digital communication systems has been thoroughly analysed theoretically [20]. This Section will present the subject of error control in general. It will then describe in some detail certain aspects of error control with relevance to spread spectrum communications and packet-radio networks.

In energy-limited systems, such as deep-space communication systems, concatenated coding schemes have been implemented which can operate in the region of $E_b/N_0 = 0$ to $1\,\mathrm{dB}$ for bit error probabilities down to 10^{-6}. As an example, the Galileo spacecraft achieved a BER of 10^{-6} at $E_b/N_0 = 0.91\,\mathrm{dB}$, using a combination of convolutional and Reed–Solomon codes [21]. This is an extreme case, and would not be cost effective for most mobile communication systems. Most of these operate at present with E_b/N_0 of several dBs less. Theory states that uncoded coherent BPSK signalling requires approximately $E_b/N_0 = 9.6\,\mathrm{dB}$ to attain bit error rates of 10^{-5}. In theory, this E_b/N_0 can be improved by 11.2 dB when optimal coding is used.

Coding gain is defined as the difference between the value of E_b/N_0 required to obtain a given error rate without coding and the value when using coding. Coding gain is usually determined by plotting the probability of error versus E_b/N_0 for both coded and uncoded operation.

In contrast to processing gain as defined in spread spectrum systems, coding gain will improve system performance (range or bit error rate) assuming white gaussian noise. Most coding schemes also help combat channel imperfections other than white noise. It can thus be viewed as complementary to spread spectrum signalling. Apart from allowing transmitted power to be reduced, error coding has no effect on power spectral density and so does not improve the chances of keeping transmission hidden.

4.4.1 Methods for reducing bit error rate

One can think of three basic ways to reduce bit error rates. The first of these is to add some redundancy to the data stream in such a way that it is less vulnerable to certain channel deficiencies. Repetition/redundancy coding with interleaving is an example of this technique. At the receiver this would imply de-interleaving together with some form of recombination such as majority vote. This technique was quite popular in early military digital radios. The advantage was its simplicity and the significant performance improvement obtained in the burst-noise dominated channels commonly encountered with frequency-hopping spread spectrum radios. Alternative solutions were purposed using combinations of convolutional coding such as [22].

A second technique is to include parity bits both for error detection and correction. This allows error correction at the receiver without any retransmission of the signal. This is known as forward-error correction (FEC). FEC can be utilised both to improve performance in white noise (i.e. range improvement) and/or to prevent burst errors. There are many different forms of FEC, all based on fairly deep mathematical theory and usually requiring the implementation of complex coding and decoding algorithms. An exact prediction of the performance for any particular code is usually difficult to calculate, and is usually obtained by simulation. However, bounds on average performance can often be used to obtain an approximate prediction.

There is a major division among forward error correction codes between convolutional codes and block codes. In convolution coding, redundant bits (or symbols), determined by preceding information bits,

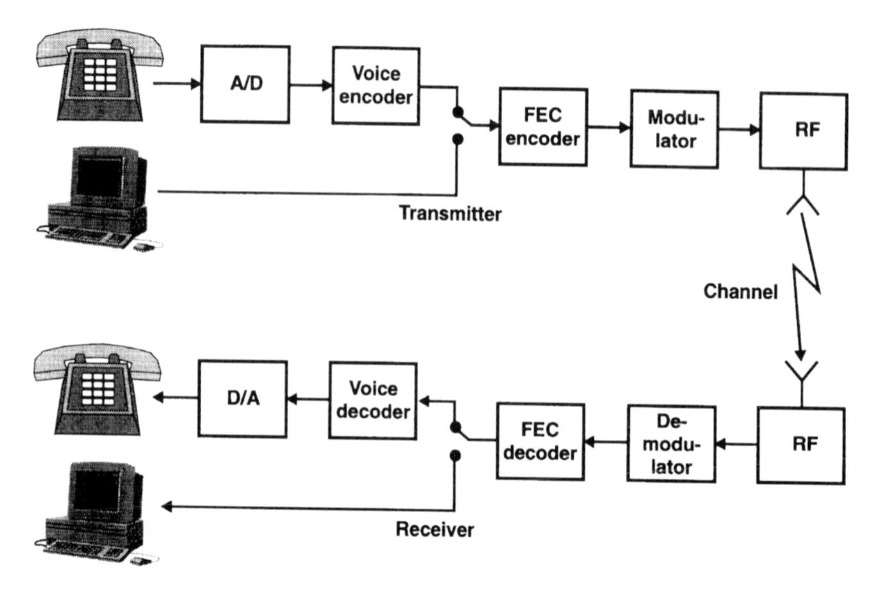

Figure 4.46 Example of forward-error correction in a digital communication system

are added continuously to the bit stream. An example of a convolutional code is shown in Figure 4.47, where the datastreams g_1 and g_2 are generated to be interwoven, modulated and then transmitted. Coded bits are generated by modulo-2 addition applied to five information bits at a time, and produce an output of twice the rate of the input stream. Each input bit fed into the register produces two output bits, so the code is thus said to be a half-rate ($R = 1/2$) code. It also has a constraint length $v = 5$ which is equal to the length of the shift register (memory of the code). As the number of stages in the shift register of Figure 4.47 is increased, the number of gating combinations giving streams g_1 and g_2 will increase dramatically. One can also alter the rate of the code by generating additional strings to include and/or making several shifts into the register at once.

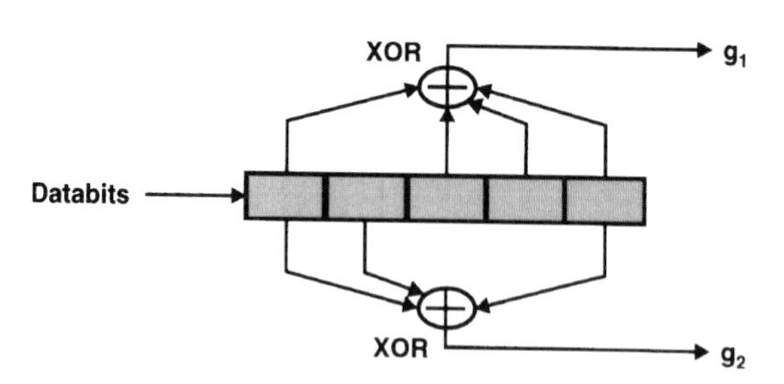

Figure 4.47 Convolutional code with constraint length $v = 5$ and rate $= 1/2$

There is no established way of determining optimum combinations, and convolutional codes are therefore usually found by computer search.

Convolutional codes are a subset of so-called tree codes. Decoding of a convolution code is analogous to determining which path through a trellis best matches a possible originating sequence. The Viterbi algorithm, which represents maximum-likelihood decoding of the trellis, is usually used for this purpose. Some examples of performance are shown in Figure 4.48. These show that as constraint length increases, so do the correcting capabilities of the code. However, decoding complexity increases drastically with constraint length and values greater than $v = 10$ are not practical. It is worth noting that as the code rate is decreased (from $R = 3/4$ to $R = 1/3$) performance increases with the performance curves becoming less steep. However, there is a limit to this improvement and at some rate below $R = 1/3$ performance will start to worsen as the code rate is even further decreased.

Convolutional codes are very effective in white noise environments, with moderate to low requirements for data integrity corresponding to bit error rates in the range of 10^{-3} to 10^{-7}. They are particularly useful when reliable soft decision information is utilised. However, bursts of incorrectly received bits will occur due to poor reception or to unreliable soft decision information in any system. As mentioned, complexity restricts constraint length in any practical implementation. Since the decision window of the decoder is equal to the constraint length, bursts longer than this will cause a complete collapse of the decoding algorithm. This means that in practice the performance of convolutional codes will be severely degraded in channels subject to severe interference and/or fading. Interleaving the transmitted data is commonly used as a way to overcome this problem, but this is a nonoptimal approach and, for channels with predominantly nonwhite noise, other coding schemes are likely to give better results.

Block codes are the other subcategory of forward-error correction codes. The best known of these are Reed–Solomon codes, which combine an effective decoding algorithm with near optimal correction capabilities when errors result from burst noise in combination with white noise. For this reason, Reed–Solomon coding is widely used in military tactical communications.

Reed–Solomon coding is usually applied to symbols rather than individual bits, and the information symbols to be transmitted are divided into blocks of k symbols. An algorithm is then used to generate redundancy symbols from each block. These symbols are added to the block to produce a coded word of $n = k + r$ symbols (see Figure 4.49). This coded word is then modulated to form the transmitted signal.

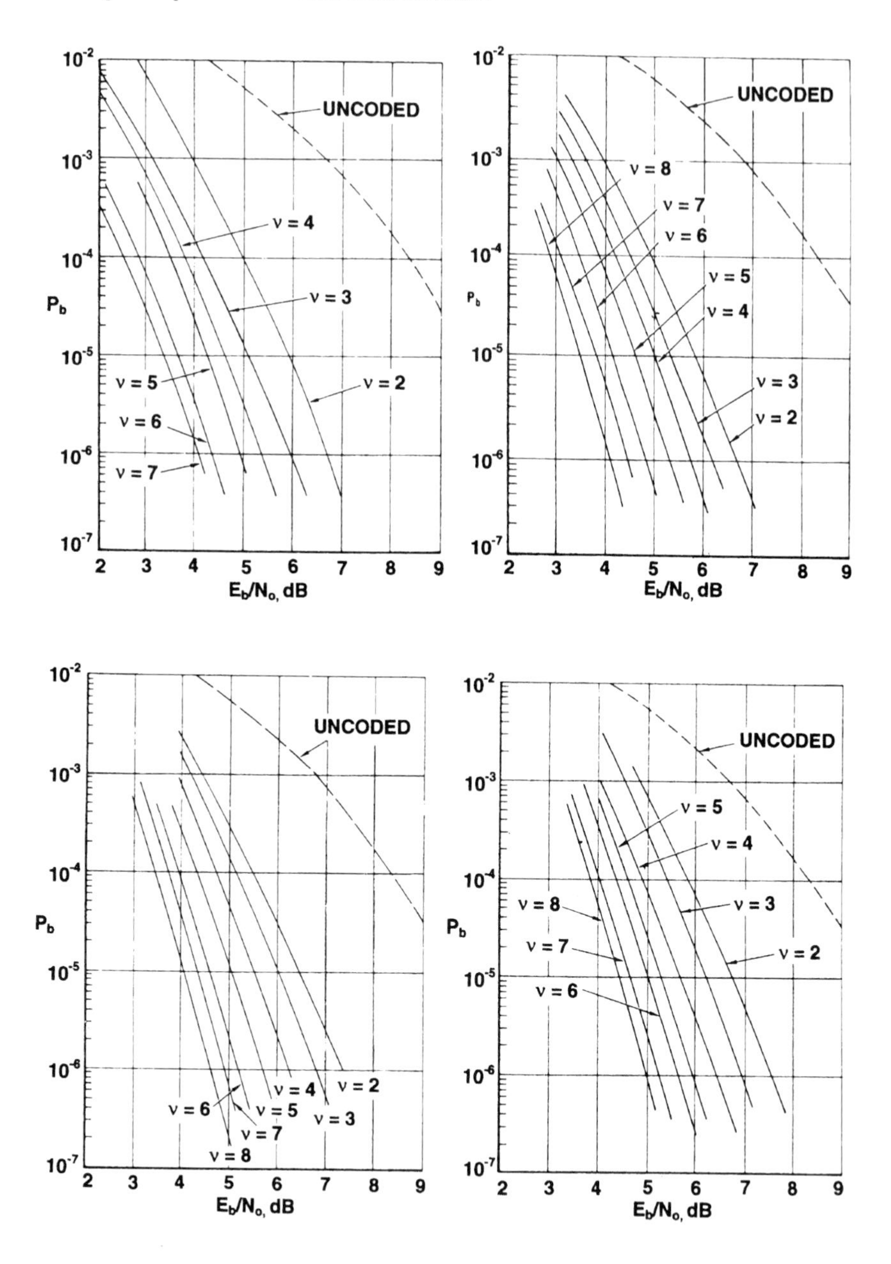

Figure 4.48 Convolutional code performance with PSK modulation over an AWGN channel with soft decision Viterbi decoding [20]
a $R = 1/3$
b $R = 1/2$
c $R = 2/3$
d $R = 3/4$

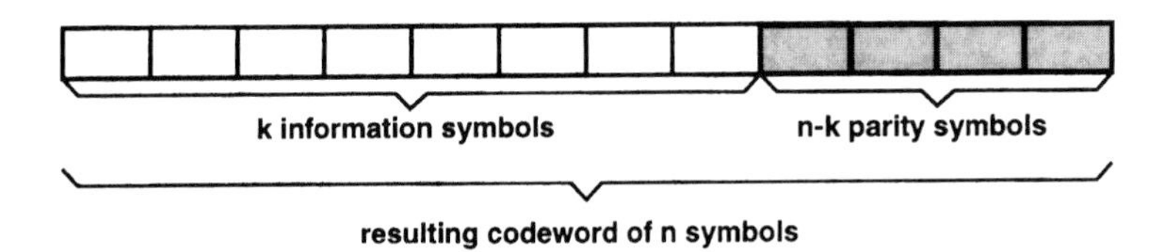

Figure 4.49 *Block code with rate $R = 8/12 = 2/3$, i.e. 33% redundancy, blocklength $n = 12$*

The code rate of a block code is given by $R = k/n$ (a low rate indicates more redundancy). It is possible to correct a maximum of $(n - k)/2$ errors. A code which achieves this is referred to as a maximum-distance code. All Reed–Solomon codes are maximum-distance codes. The reason that only $(n - k)/2$ rather than $(n - k)$ errors can be corrected is that one correction symbol is required to localise the error, and a second symbol to perform the correction. If information about the locations of the errors (erasure information) can be obtained from elsewhere in the receiver, the full $n - k$ numbers of errors can be corrected.

Reed–Solomon codes are actually a subset of a larger class of error-correcting codes, namely BCH codes. BCH codes are defined by the same characteristics and performance qualities as given above for RS codes. The reason why RS codes have become so popular is the fact that for this subset efficient coding algorithms can be found and easily implemented. Some examples of performance are given in Figure 4.50. It is worth noting that it has been shown that a relatively broad maximum of coding gain against code rate for fixed n occurs roughly between $1/3 \leq R \leq 3/4$ for BCH types of codes [23]. However, in order to optimise performance for bursty channels, tailoring of rate and block length has to be done for each individual application. A communication system that provides several types of bearer service will therefore usually want to exhibit a programmable implementation.

The performance of both convolutional and block codes can be improved significantly by utilising additional information such as signal strength, phase irregularities etc. There are basically two types of information that can be made available. First, the decoder can provide either soft- or hard-decision information to the decoding algorithm. Hard decision means that some specific permitted value is assigned to each demodulated bit or symbol. For example, a single bit will be demodulated as either 1 or 0. With a soft decision, a demodulated bit would be given a probability. For example, the bit could be replaced by a three-bit value, giving the decoder one of eight possible values depending on the received signal quality. Convolutional codes are able to improve their signal-to-noise

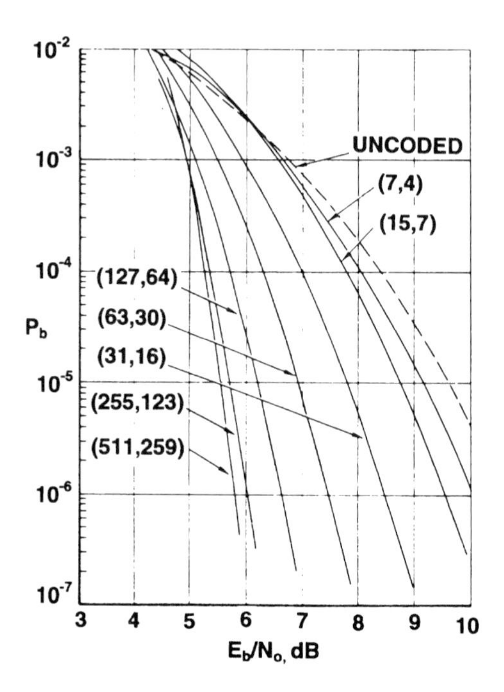

Figure 4.50 Performance of BCH codes with coherent BPSK modulation [20]

threshold by more than 1 dB for high error rates when utilising soft information. This approach is used for example in Trellis-coded modulation and can be very profitable.

A decoding algorithm can also be given so-called erasure information. Erasure information arises when the demodulator or inner code has determined that a symbol at a given position is likely to be wrong. This information can be gathered from a combination of signal-to-noise estimates, the level of the received signal (i.e. from AGC information), by signal decoding information such as correlation values etc. Erasure information is of great value to block codes, since it means that the locations of errors are predetermined, so that the error-correcting capabilities can be fully utilised. It is not possible to elicit reliable erasure information in a white noise environment unless an inner code is used in a concatenated system.

A third method of error control is to add parity bits to the data stream in order to detect errors, requesting retransmission if necessary. This method in fact consists of two separate mechanisms, namely an error detection and a repeat request, and so is referred to as automatic repeat request (ARQ).

Some well known types of ARQ are go-back-N and selective repeat. In a go-back-N system, the receiver requests retransmission of the last N

blocks of data, when a block is received in error. An example of such a system is the EUROCOM D/1 tactical area system trunk-group protocol which uses go-back-4. The advantage of such a protocol is that blocks do not have to be individually labelled, and thus give simpler algorithms than selective-repeat protocols. Disadvantages are that more data than necessary must be retransmitted, so that the maximum round-trip delay on a link has to be limited to allow the erroneous block to be repeated. This is why go-back-N algorithms are not suitable for systems such as satellite links.

In selective-repeat systems, ARQ is usually negative and the receiver asks the transmitter for retransmission of one specific block. This ensures that only blocks with errors are retransmitted, but requires more protocol overhead on each block. Which method to choose is therefore a trade-off between minimising the loss of throughput due to protocol overhead (long blocks) or due to repeats (short blocks).

ARQ and FEC can be used at different levels of error control, since most FEC algorithms also can detect whether all the errors can be corrected or not. Packet-switched networks are good examples of systems that might use hybrid ARQ and FEC at the physical layer or, alternatively, FEC at the physical layer and ARQ at the link-layer protocol. This is further discussed in Section 4.4.3.

Mobile links frequently suffer from debilitating Doppler and multipath fading effects. This would suggest that a block-coding scheme would be attractive as a way to combat the bursty nature of the channel. Another approach would be to consider convolutional coding combined with interleaving, keeping in mind the fact that link delays greatly influence the performance of packet-switched networks. If a coding scheme is implemented in such a way that it is necessary to decode and re-encode at each intermediate node in a multihop network, the delay associated with interleaved convolution coding will probably be unacceptable. In such situations the solution may be to avoid interleaving by using a more suitable block code tailored to the noise statistics. This is also one of the reasons why an m-ary superchannel is considered in the following.

4.4.2 Concatenated coding techniques

The basic concept behind concatenated coding is to add more levels of coding to a system. A common technique is to use a concatenation of codes with two or more layers of coding. A simple system using two levels of coding is shown in Figure 4.51.

The system consists of an inner code, which in combination with the channel can be viewed as a superchannel to the outer code. The inner code

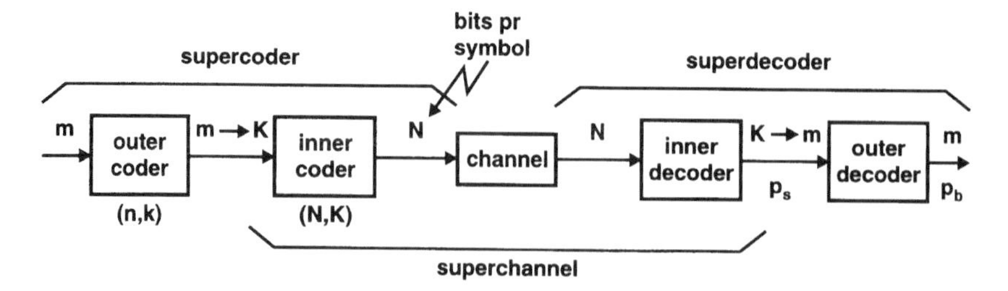

Figure 4.51 A conceptual concatenated coding system

can be utilised not only to correct errors, but also to provide the outer code with erasure information. The inner code can be either binary or m-ary. Furthermore, the inner code should ideally be designed to alter the statistics of any remaining error in such a way that they can effectively be corrected by the outer code. Most concatenated coding schemes aim at combining the best properties of different coding techniques, and are merely ways of simplifying the treatment of the overall coding scheme. By separating the problem into more manageable elements, it becomes easier to optimise design performance. In most cases, such a treatment gives the added advantage of reduced complexity, and thus reduces the cost of implementation.

This section will be limited to a discussion on using Reed–Solomon codes as the outer code, and orthogonal m-ary codes described in Section 4.3.3 as the inner code, since these would be the typical case of an NBSS, aimed at providing a basis for spread spectrum packet-switched mobile networks.

Let us first consider the super channel. The probability of symbol error P_{es} (SER) for noncoherently detected m-ary orthogonal signalling is given by Sklar:

$$P_c = \frac{1}{m}\exp\left(\frac{-E_s}{N_0}\right)\sum_{j=2}^{m}(-1)^j\binom{m}{j}\exp\left(\frac{E_s}{jN_0}\right) \tag{4.94}$$

where m is the size of the symbol alphabet, $E_s = (\log_2 m)E_b$ is the energy per symbol if E_b is the energy per bit, and

$$\binom{m}{j} = \frac{m!}{j!(m-j)!} \tag{4.95}$$

Dependent on the value of m, there are several outer codes to consider, Reed–Solomon coding being the most popular. Knowing the probability of symbol error p_c (c for coded) of the symbols that are presented to the decoder, we can calculate the probability of symbol error after decoding, $P_{es}(RS)$.

A statistical expression for $P_{es}(RS)$ is:

$$P_{es}(RS) = \sum_{i=0}^{n} (P_i)(P_{es}(RS)|i) \qquad (4.96)$$

where i is the number of symbol errors in a block of n symbols, P_i is the probability of exactly i errors occurring in a block, $P_{es}(RS)|i$ is the probability of symbol error after decoding, given there are i errors in the block before decoding and P_i is given by:

$$P_i = \binom{n}{i} p_c^i (1 - p_c)^{n-i} \qquad (4.97)$$

because: the probability that i symbols are in error is p_c^i. Then $(n - i)$ symbols must be correct. The probability of this occurring is $(1 - p_c)^{n-i}$ since the probability of a correct symbol is $(1 - p_c)$. There exist t different combinations of i errors out of n as given by eqn. 4.97.

$P_{es}(RS)|i$ varies with i, the number of errors in a block. If i is less than or equal to t errors in a block (t is the RS decoding limit) the decoder will correct all errors. $P_{es}(RS)|(i \leq t)$ is then zero. If more than t errors occur, however, one of two things can happen:

(i) The decoder fails to decode and returns the corrupted sequence (block) of symbols as it is. $P_{es}(RS)|(i > t)$ is then simply given by i/n. Decoding failure is flagged.

(ii) The decoder returns a valid sequence other than the correct one, but flags correct decoding. This is called a decoding error. The exact value for $P_{es}(RS)|(i > t)$ for this situation can be calculated, because the relationship between the various sequences is known. But let us just assume that it equals some fraction α.

For i greater than t, let the probability of decoding error be denoted P_E. Then the probability of decoding failure is $(1 - P_E)$.

Summing up, then, $P_{es}(RS)|i$ is given by the following:

$$P_{es}(RS) = 0 \qquad \text{where } i \leq t$$

$$P_{es}(RS) = aP_E + \frac{i}{n}(1 - P_E) \qquad \text{where } i > t \qquad (4.98)$$

The probability of decoding error, P_E, varies with p_c, the code rate and block length. When p_c is close to 1, P_E takes on its maximum value $\sim 10^{-7}$ for the highest rate (worst case) code. As p_c decreases, P_E decreases more rapidly, e.g.: for $p_c = 10^{-3}$, $P_E = 10^{-18}$.

Conclusion: P_E can be ignored as it will have no influence of importance on the calculation in eqn. 4.98 for a practical case. $P_{es}(RS)|(i > t)$ hence reduces to i/n.

Substituting eqns. 4.97 and 4.98 with P_E = zero into eqn. 4.96 gives the probability of symbol error after decoding for an RS(n, k) code:

$$P_{es}(RS) = \sum_{i=t+1}^{n} \left(\binom{n}{i} p_c^i (1 - p_c)^{n-i} \right) \frac{i}{n} \tag{4.99}$$

where p_c is the probability of symbol error before decoding.

Now we are in the position to calculate $P_{es}(RS)$ for a given E_b/N_0 (our operating point). What value do we use for p_c, the symbol error rate at the input of the decoder? For a valid assessment of coding gain, the result $P_{es}(RS)$ must be directly comparable to P_{es}, the symbol error rate without coding for the same E_b/N_0.

Coding gain is the change in decibels from the level required for a certain BER before coding, to the level that gives the same BER after coding.

This implies that the data rates must be the same before and after coding. Coding introduces redundancy: for every data symbol that is transmitted, another k/n parity symbol requires channel capacity, hence reducing the energy for each symbol on the channel by k/n. The symbol error rate p_c at the input of the RS decoder is hence found at a performance curve that is shifted 10 log n/k to the right with respect to the systems performance curve.

Without shifting curves we can obtain p_c directly from the systems performance curve: subtract 10 log n/k from the operating point marked with an asterix, and read off the SER at this new E_b/N_0 on the systems

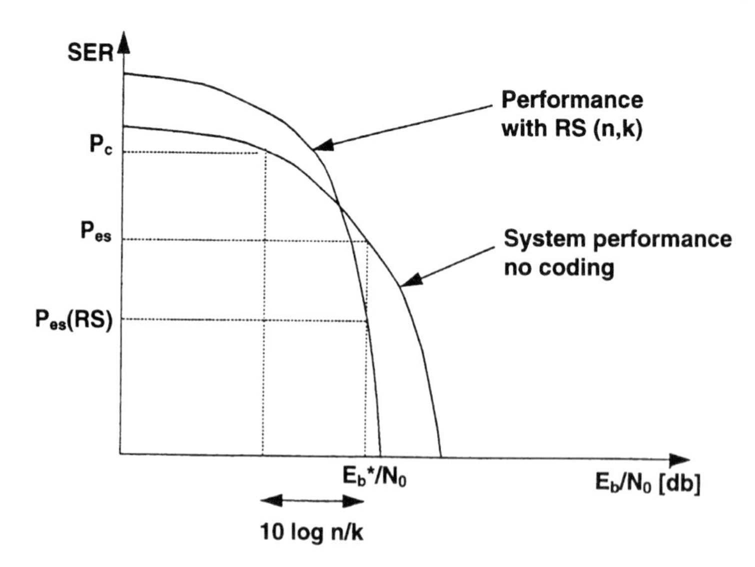

Figure 4.52 Illustration of RS performance calculation at E_b^*/N_0 (eqn. 4.99)

performance curve. This SER is the p_c that gives the correct $P_{es}(RS)$ at the given operating point. This is illustrated in Figure 4.52.

Let us consider a noncoherent system consisting of a 256-ary innercode, with various options for the RS outer code. Bit error probabilities for the superchannel are shown in Figure 4.53, according to eqn. 4.94.

Figure 4.54 shows various RS performance curves in BER against E_b/N_0 superimposed upon the systems performance curve of Figure 4.53. Consider the RS(60,30) code. For E_b/N_0 smaller than 4 dB we get a negative coding gain. Why is this? A low-rate code adds a lot of

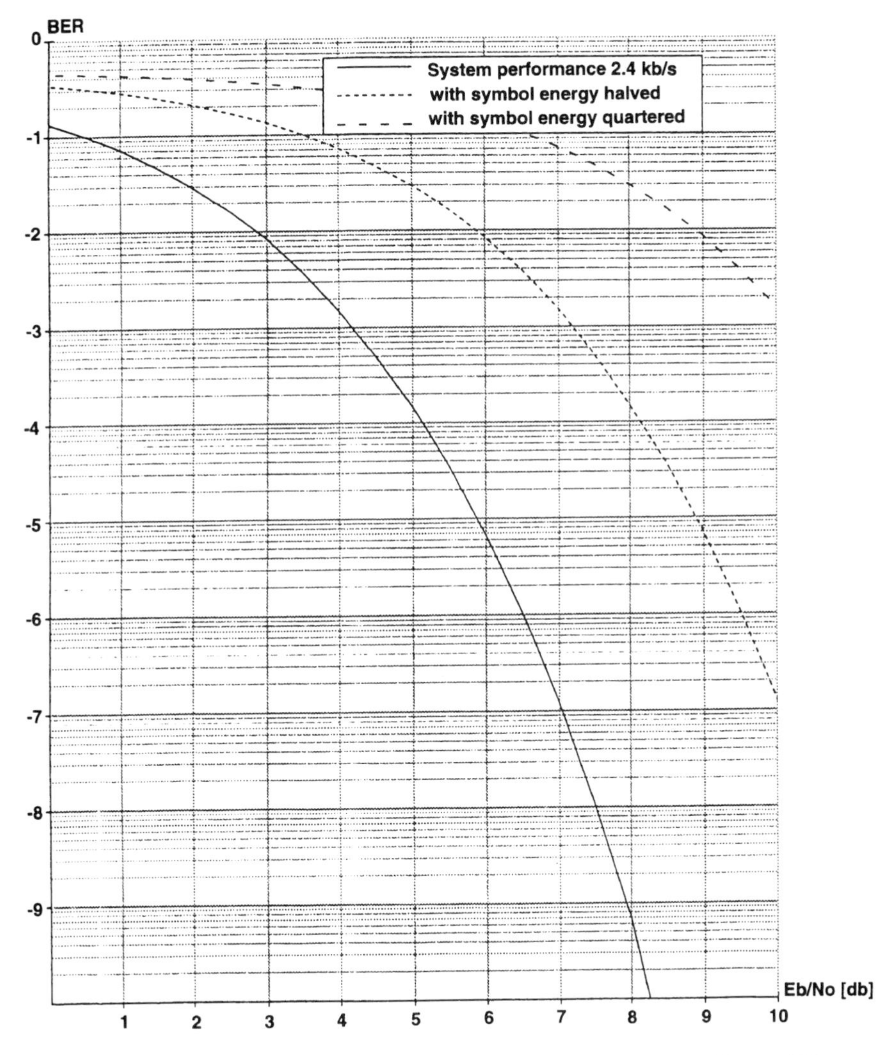

Figure 4.53 Performance of noncoherent orthogonal m-ary signalling with m = 256

redundancy, and the energy per channel symbol is reduced accordingly. For a half-rate code such as this we get a halving of the channel symbol energy. This will, of course, generate more errors than t in each block, the RS decoder will fail to decode and the overall BER will be degraded with negative coding gain as a result. At the cross-over point of zero coding gain, the RS code corrects exactly the extra number of errors that it itself introduces. For every RS code there is, hence, an E_b/N_0 limit below which the coding gain is negative. We need to increase transmitter efficiency above this limit until the average number

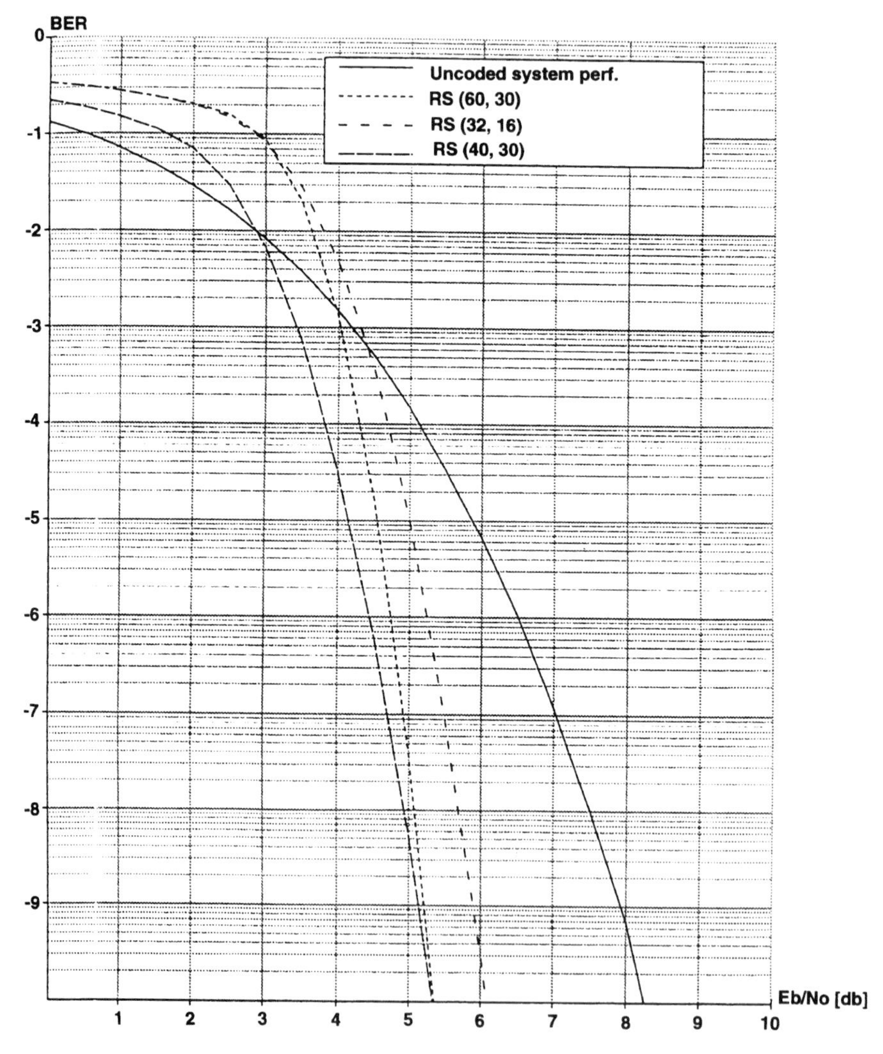

Figure 4.54 Performance of various concatenated coding schemes in AWGN, RS outer and m-ary inner code

of errors per block drops below t; only then will the error-correcting capability of the RS code be effective. This limit is higher for low-rate codes, but these codes make up for it by yielding massive coding gains at low BERs, due to the very steep slope of the waterfall curve.

It should be clear, then, considering the above and Figure 4.54, that for every BER there exists an optimum code with respect to coding gain for our system in AWGN. If coding gain at low BERs is the priority, a high-rate code should be used, e.g. RS(40,30) with a coding gain of ~ 0.75 dB at BER $= 10^{-3}$. But if BERs down to 10^{-9} and better are required, without a giveaway of dBs in efficiency, the expense of a strong low-rate code should be worthwhile, e.g. RS(60,30) with a coding gain of a little less than 3 dB at BER $= 10^{-10}$.

4.4.3 Error coding in packet-switched radio networks

As mentioned in previous sections, the design of error-control schemes in a radio intended for use in a packet-switched radio network has to be seen in connection with network protocols and network performance. Optimisation of bit error rates at the radio level itself does not guarantee an optimum system performance. So what are the factors and interdependencies to consider?

First of all, any network will obviously gain from a reduction of interference and noise encountered on its links. Burst noise can be reduced by applying coding such as Reed–Solomon. How severe this problem is will give some indication of the rate of the code which is to be applied. In addition, we know that the stronger the code, the more redundancy there is introduced (up to a certain level), which reduces the effective throughput on the channel. However, the alternative is that without coding most packets will be in error and will have to be retransmitted one or several times. In this case, the effective throughput is thus even more severely reduced. There is also a break-even and optimum point to be found according to the network performance in a given error environment. Most of the error correction has to be done by the forward-error-correction code, but not all (typically 90–95% of the packets should be error free). A few packets must be allowed to be in error and to be handled by higher level protocols.

The block and packet size are also parameters linked with the above discussion. First of all, packet sizes have to be addressed with respect to the foreseen application. For applications with short messages and critical time requirements, packet sizes and block lengths might result from this. If not, there might be some alternatives available to play with at the lower levels. In general, the relationships are such that the longer the blocks, the stronger the code performance, but also the longer the delay introduced. Furthermore, the longer the blocks, the more capacity is lost in the case of retransmissions, which means that even more responsibility should be carried by the forward-error-correction mechanisms at the radio level.

Another aspect of network performance is the effect of improving error performance in white gaussian noise, which is also a strong feature of most forward-error control coding schemes. Let us consider an example, where we have implemented a coding scheme which improves performance by a moderate gain of 1 dB at an accepted bit error rate. What can be gained from this? Any one of the following can be argued to be the case:

- in a VHF system, this would correspond approximately to a six per cent range extension and 12% increased coverage. Depending on the scenario, a 12% increased coverage might be significant. This could for instance result in a decreased number of multihop connections within the network, which would increase throughput as a result of fewer transmissions, or network connectivity would be better and thus result in fewer collisions. We know that all medium-access protocols suffer severely from hidden nodes;
- a 1 dB increase of noise margin is introduced, which corresponds to 1 dB increased capture range for the packet-switched radio system;
- data rate can be increased by 26%;
- transmitted power can be reduced by 26%.

Gain can only be exploited as one, or as a reduced combination of the above, and those which are of interest will vary in each case. Furthermore, in this simple example we have not taken into account the amount of redundancy that had to be introduced in the coding to obtain 1 dB gain. If this is of the order of 26%, it is quite obvious that we have basically introduced complexity and cost without obtaining any substantial improvement in the case of white noise. However, as we have seen in earlier sections, strong coding techniques are capable of introducing very large gains, with relatively low amounts of redundancy, so that in a real case the numbers would be quite different from those in this example. With respect to packet-radio networks, the aspects relating to connectivity and capture range are of most interest, but are also the most difficult to evaluate with respect to network performance, which is further discussed in Chapter 5.

In the case of codes like Reed–Solomon, the code can and should be implemented to identify blocks or packets that cannot be corrected (decoding failure), so that this information can be utilised by an ARQ scheme on the link level. If not, the protocols have to introduce additional redundancy (CRC) to enable this task. In short packets this might be of importance and might, for instance, be used to give the protocol data unit (PCI) a separate error check. As errors are detected almost instantly by forward-error correction, the time wasted on an erroneous packet is less, and the radio can thereby return to its channel sensing mode faster. This will result in a more effective medium-access performance, especially in situations with high error rates. In military systems, this would be of importance.

Another desirable feature of several error-correcting codes, is that they are capable of knowing how many errors they have encountered. This might be exploited as a quality measure for the link. It cannot be expected to replace other link measurements, but it can certainly be treated as additional information which can improve the basis on which routing in the network is done.

It has earlier been shown that block codes improve in performance as block length is increased, which is the case with both white noise and burst noise affected channels. We also know that coding delay is increased accordingly, which will reduce the efficiency of the network. Another feature that should be mentioned with respect to error coding in packet networks is that long block lengths also become uneconomic in systems with large variations in message lengths submitted by the applications. For short messages, the blocks thus have to be stuffed or padded to fit the larger size used by the error-correcting code. In most systems there is therefore a need to implement circuitry which can operate with various block lengths, and one should at least have one short code to deal with acknowledgement packets.

It is not possible to give any analytical relationship between the various factors described in the general case. Simulation of network performance is the only way to evaluate resulting performance. A fundamental understanding of the various factors involved and their relative importance will make it possible to estimate some reasonable values to consider, and will thus make it possible to reach a balanced system design in a reasonable amount of time.

4.5 Technology forecast and new trends

There has been a growing interest in digital signal processing, which has led to the disclosure of several techniques also aimed at mobile communication. Of major interest are applications related to interference cancellation, channel equalisation, interuser-interference cancellation and in our case methods to improve carrier-sense capabilities. Two major subjects within signal processing are time–frequency analysis and higher order statistics. The novelty of time–frequency analysis is that the time and frequency aspects are handled simultaneously with a single function. A great deal of international research activity in this field has resulted in better understanding of both the theory and application. Section 4.5.1 briefly approaches the theory, and Section 4.5.2 indicates how time–frequency analysis can be applied practically in order to suppress narrow-band interference in a radio receiver. Section 4.5.3 introduces the use of high(er) order statistics and indicates how particular facets of the theory can be used to improve detection in a spread spectrum receiver.

4.5.1 Time–frequency analysis applied to radiocommunications

4.5.1.1 Fourier transform

The classical technique for processing a radio signal is to use a Fourier transform (FT) or more correctly the discrete Fourier transform (DFT), which is the digital approximation used in a practical system. An FT provides maximal information about the spectrum of the signal. An analysis of this spectrum indicates which measures may be suitable for the removal of undesired components. A standard procedure has been to calculate the parameters of a filter through which the signal is passed. The filter can sometimes be made adaptive, so that it can at least to some extent, follow undesirable components with variable frequency. This approach exhibits a number of weaknesses. First, the FT gives no information about the chronology of the signal. For example, a short noise spike will be spread over the whole spectrum, without any possibility of determining the time at which it occurred. In general, all alterations to the characteristic properties of the signal within the analysis window will be spread throughout the spectrum and so will be lost for possible processing. If the noise component occupies a significant proportion of the signal bandwidth, it will not be possible to design an effective filter. It sounds like a good idea to analyse and then return to the time domain using the inverse FT (IFT). Unfortunately, this is not a practical solution for short-duration signals. This is because the DFT used in any real system gives a poor representation of the spectrum for a short sequence. An inversion with the inverse DFT will not reconstruct the original signal. Also, the problem of nonstationary signals has not been solved.

The concept of performing correlation in the frequency domain then appeared, in the form of the question: for a set of coding sequences (which will be referred to as a code bank) with given correlation properties, what are the properties of the Fourier-transformed code bank? The answer is as follows:

'Let C be a real code bank with maximum relative cross correlation ρ. *An FT of this code bank results in another code bank having the same maximum relative cross correlation'*

This statement is fairly easy to prove, and shows that a code bank keeps its correlation properties in the Fourier domain. The consequence of this is that one can take the FT of an incoming sequence, process this in the frequency domain and then correlate the result with the FT of a code bank. This technique avoids the errors introduced when using an IFT.

This is an ingenious concept. Signal processing can be simple, for example by using a thresholding algorithm and setting the values of frequency components above a certain frequency to zero. One avoids the IFT, with its attendant problems. Nonetheless, this technique has its weaknesses. First, it depends on taking the FT of a code sequence. For code sequences longer than about 200 chips this can work. However, a typical frequency hopper often exchanges less than 200 chips per frequency dwell. In this case, an FT will give a poor estimate of the frequency spectrum. A second problem is that the FT gives a result which is complex (i.e. the FT of a real code bank is a complex code bank), so one is forced to use a time-consuming complex correlation. Finally, the technique does not solve the problem of nonstationary signals as mentioned earlier. For example, a short noise pulse is spread over the whole spectrum and cannot be removed without distortion of the signal. Even with these drawbacks, the idea may be worth noting. It is possible that there are applications where the use of a Fourier-transformed code bank would be advantageous. Conventional Fourier analysis can give valuable information concerning the characteristic properties of a signal which is stationary during the analysis period.

Let us establish the notation which we shall use here by repeating the relationships for the FT and IFT for a signal $x(t)$. Then:

$$\mathrm{FT}_x(\omega) = X(\omega) = \int_{-\infty}^{\infty} x(t)e^{-j\omega t}\,dt \tag{4.100}$$

The index x indicates that the object function is $x(t)$. Similarly, the IFT is written as:

$$\mathrm{IFT}_x(t) = x(t) = \frac{1}{2\pi}\int_{-\infty}^{\infty} X(\omega)e^{j\omega t}\,d\omega \tag{4.101}$$

A nonstationary signal alters its characteristic properties over time. The short-time FT (STFT) examines the signal during time intervals which are so short that this variation is negligible, so that a signal can be regarded as stationary during any interval. This is achieved by multiplying the signal $x(t)$ by a window function $w(t)$ in the time domain. The STFT is then defined as:

$$\mathrm{STFT}_x^w(\omega, t) = \int_{-\infty}^{\infty} x(\tau) w^*(\tau - t) e^{-j\omega t} d\tau \tag{4.102}$$

where the superscript * denotes the complex conjugate.

The window function $w(t)$ then extends over an effective time interval, τ, which is delayed according to the elapsed time, t. The result is that the STFT becomes a two-dimensional function of both frequency and time. This is an alternative way of saying that the transformed spectrum is also a function of time.

Eqn. 4.102 can be interpreted in two different ways. The first of these is as already described, where a sampling window moves along the time axis. The alternative interpretation appears if eqn. 4.102 is rewritten as:

$$\mathrm{STFT}_x^w(\omega, t) = e^{-j\omega t} \int_{-\infty}^{\infty} x(\tau) h^*(\tau - t) d\tau \tag{4.103}$$

where

$$h(t) = w(t) e^{j\omega t} \tag{4.104}$$

The constant term outside the integral sign in eqn. 4.103 represents the delay to the window. The frequency spectrum appears as the inner product of the signal $x(t)$ and a function $h(t)$ centred about $\tau = t$. Eqn. 4.103 may now be written in the form:

$$\mathrm{STFT}_x^w(\omega, t) = e^{-jwt} \int_{-\infty}^{\infty} \chi(\tau) h^*(t - \tau) d\tau = X(\omega) H(\omega) e^{-j\omega t} \tag{4.105}$$

since a folding in the time domain is equivalent to a multiplication in the frequency domain. This shows that the amplitude spectrum exhibits the response of a system with an impulse response (characteristic):

$$h(t) = w(t) e^{j\omega t} \tag{4.106}$$

Let us now assume that $w(t)$ is the impulse response of a bandpass filter. The filter's frequency response will be given by:

$$\mathrm{FT}_w(\omega) = W(\omega) = \int_{-\infty}^{\infty} w(t) e^{-j\omega t} dt \tag{4.107}$$

Now $h(t)$ is the response for a bandpass filter with centre frequency $fo = \omega_o/2\pi$. The FT of $h(t)$ will be:

$$\mathrm{FT}_h(\omega) = \int_{-\infty}^{\infty} w(t)e^{j\omega_0 t}dt = W(\omega - \omega_0) \qquad (4.108)$$

We can now choose values for ω_o in eqn. 4.108 in order to obtain a filter bank which can be used to find the amplitude of the signal in a time window, centred at time t, on the time axis. The bandwidth of the bank is automatically defined by the duration of the time window. This situation is depicted in Figure 4.55. Let us denote the effective duration of the time window, $w(t)$, to be T seconds with a bandwidth B Hz. Then the product BT, which is the shaded area denoting the window in Figure 4.55, will be constant for any given window. Choosing a short window will give good resolution in time, but the bandwidth of the filtered signal will be large, with consequently poor frequency resolution. Conversely, choosing a long window will permit good resolution of frequency but at the price of uncertainty in time. The choice of the window parameters is thus a compromise to suit the design requirements.

Inside the windows there is no information on either time or frequency. It is only possible to discriminate between what is inside and what is outside the window. It is then desirable to have a window with the smallest BT product area. The area is fixed for any particular window, and will depend on the exact window shape, i.e. the form of the function $w(t)$.

If the function $w(t)$ has gaussian form then the product $BT = \frac{1}{4}\pi$, which is the minimum possible value. In general one can say $BT \geq \frac{1}{4}\pi$. This is usually called Heisenberg's uncertainty relationship for signals

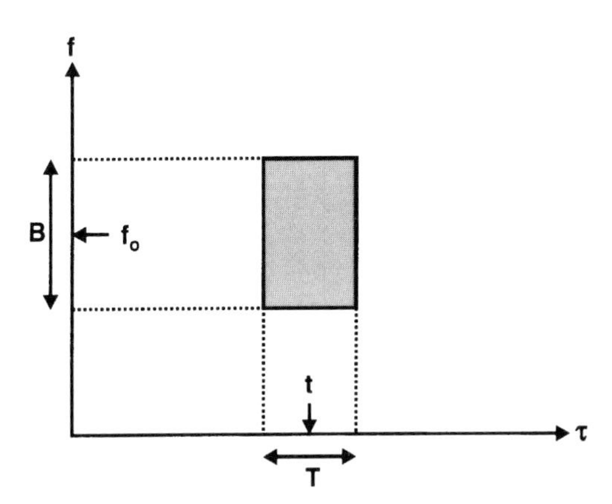

Figure 4.55 Time–frequency representation

after the close analogy to the Heisenberg uncertainty principle in quantum mechanics.

It is interesting to see the consequences of two extreme terms for the window $w(t)$. If $w(t)$ is a Dirac function (i.e. an infinitesimally short spike with infinite amplitude), then:

$$\text{STFT}_x^{\delta}(\omega, t) = \int_{-\infty}^{\infty} \chi(\tau)\delta(\tau - t)e^{-j\omega t}d\tau = \chi(t)e^{-j\omega t} \tag{4.109}$$

This gives perfect definition of time, but no information about frequency, since the complex exponential function has a value of unity for all values of ω.

At the other extreme, if we let the window function have a value of unity for all values of t, then:

$$\text{STFT}_x^{1}(\omega, t) = \int_{-\infty}^{\infty} \chi(\tau)e^{-j\omega t}d\tau = X(\omega) \tag{4.110}$$

This now yields all possible information about frequency and none concerning time. Incidentally, this is actually the situation for the conventional FT analysis.

As an introduction to wavelets we will briefly look into a popular extension of the traditional STFT, namely the Gabor transform. The Gabor transform is actually an STFT with a fixed gaussian window.

4.5.1.2 The Gabor transform

Let us consider a gaussian function of the form:

$$g_\alpha(t) = \frac{1}{\sqrt{2\pi\alpha}}\, e^{-\frac{1}{2}\frac{t^2}{\alpha}} \tag{4.111}$$

The integral of this function has a value of unity. Using the conventional half width for the exponential, α represents the resolution with which the signal can be characterised in time as:

$$(\Delta t)^2 = \frac{\dfrac{1}{2\pi\alpha}\displaystyle\int_{-\infty}^{\infty} t^{2-\frac{t^2}{\alpha}dt}}{\dfrac{1}{2\pi\alpha}\displaystyle\int_{-\infty}^{\infty} e^{-\frac{t^2}{\alpha}}dt} = \alpha \tag{4.112}$$

which in turn yields an effective width of:

$$T_s = \sqrt{\alpha} \tag{4.113}$$

The GT is then defined as:

$$\mathrm{GT}_x^g(\omega; t, \alpha) = \int_{-\infty}^{\infty} x(\tau) g_0(\tau - t) e^{-j\omega\tau} d\tau \qquad (4.114)$$

This GT transform is now a function of ω, i.e. a family of spectra with parameters t and α. If we keep the values of t and α constant and then integrate GT$(\omega; t, \alpha)$ with respect to t, then this shows that:

$$\int_{-\infty}^{\infty} \mathrm{GT}_x^g(\omega; t, \alpha) dt = \int_{-\infty}^{\infty} \left[\frac{1}{\sqrt{2\pi\alpha}} \int_{-\infty}^{\infty} x(\tau) e^{-\frac{1}{2}\frac{(\tau, t)^2}{\alpha}} e^{-j\omega\tau} d\tau \right] dt$$

$$= \int_{-\infty}^{\infty} x(\tau) e^{-j\omega\tau} \left[\frac{1}{\sqrt{2\pi\alpha}} \int_{-\infty}^{\infty} e^{-\frac{1}{2}\frac{(\tau, t)^2}{\alpha}} e^{-j\omega\tau} d\tau \right]$$

$$= X(\omega) \qquad (4.115)$$

The set:

$$\left\{ \mathrm{GT}_x^g(\omega; t, \alpha) : t \in R \right\} \qquad (4.116)$$

of the Gabor transform gives an exact decomposition of the Fourier transform $X(\omega)$, and also provides instantaneous spectral information.

In order to approach wavelets, there is an alternative way of expressing the Gabor transform. If we were to choose another window defined by $\mathrm{G}(\tau; \omega, t, \alpha) = g_\alpha(\tau - t)e^{-j\omega t}$, we can write eqn. 4.114 in the form:

$$\mathrm{GT}_x^g(\omega; t, \alpha) = (x, G(\tau; \omega, t, \alpha)) = \int_{-\infty}^{\infty} x(\tau) G^*(\tau; \omega, t, \alpha) d\tau \qquad (4.117)$$

This expresses the Gabor transform as the inner product of the signal $x(\tau)$ and a window function in the form of a pulse with a gaussian envelope. This product is superimposed on a circular function. It is now more natural to regard the GT as a weighting of the signal $x(\tau)$ with a window function $\mathrm{G}(\tau; \omega, t, \alpha)$. This interpretation is easy to visualise and leads naturally to the continuous-wavelets treatment discussed in the next section.

The operation of the GT can be demonstrated by a fractional example.

Example 4.1: Consider a sequence $x[n]$ of 1000 samples. The first 500 of these have the form $x_1[n] = \sin(2\pi n \times 0.1)$ and for the remaining 500 $x_2[n] = \sin(2\pi n \times 0.3)$.
The entire signal is then:

$$x[n] = x_1[n]\{u[n] - u[n - 499]\} + x_2[n]\{u[n - 500] - u[n - 999]\}$$

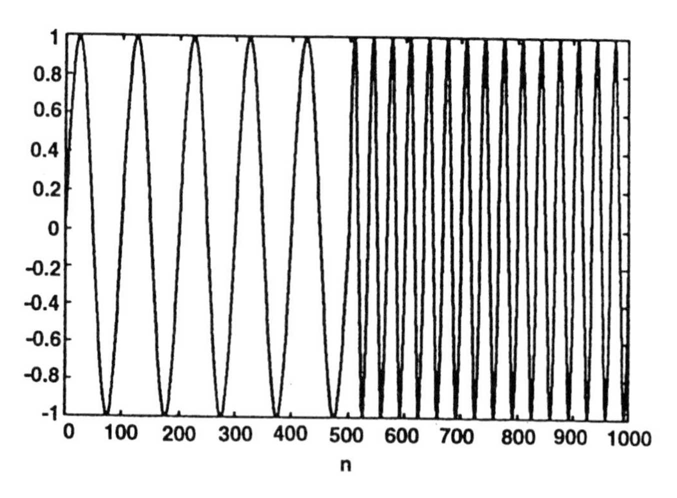

Figure 4.56 The data sequence x[n]

and is depicted in Figure 4.56. The FT of $x[n]$ is shown in Figure 4.57. This FT clearly shows the presence of the two frequency components during the sampling period, but gives no information about the times at which they were present.

The GT of the sequence $x[n]$ is shown in Figure 4.58. The window used has the form $g_\alpha[n] = (1/\sqrt{2\pi\alpha})e^{-(1n^2/2\alpha)}$, using $\alpha = 100$. The sequence of 1000 samples is divided into 15 identical intervals. At each interval a new FT is performed. The GT shows clearly that the low-frequency component is present only during the first half of the signal sequence, and the higher frequency component only in the second half.

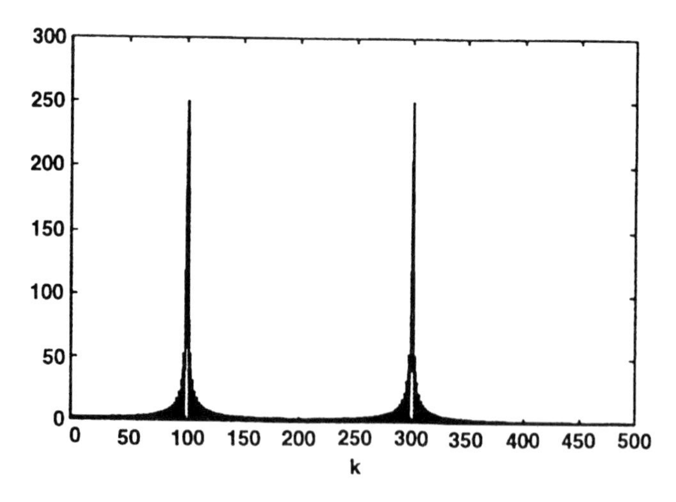

Figure 4.57 FT of the data sequence x[n]

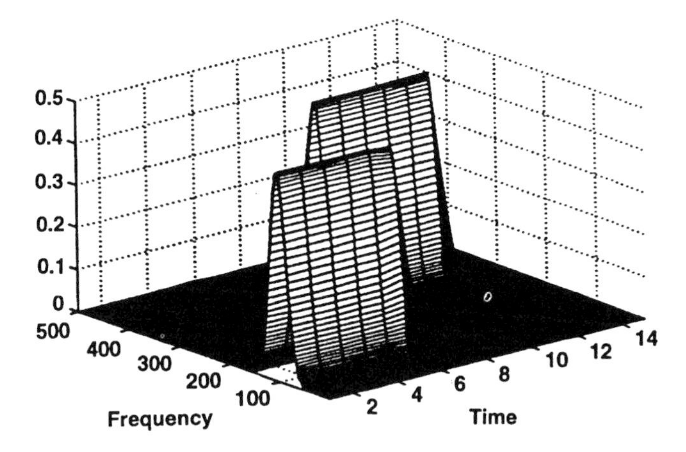

Figure 4.58 GT of the data sequence x[n]

4.5.1.3 Wavelets

In the previous subsection we considered the FFT as a transformation with a fixed rectangular window. Further, the GT was considered to be a similar transformation, but with a gaussian window. In other words, these transformations were characterised by specific base properties. From this standpoint wavelet analysis can be regarded as a similar time–frequency transformation, but with different base properties or window functions. The flexibility provided by the choice of these properties makes wavelet analysis an interesting tool for several signal-processing tasks.

The Gabor transform was shown to enable the identification of frequency components in a signal, and also shows how these vary over time. The mathematical representation in the previous subsection was based on the choice of one single window (gaussian) for any analysis. This gives resolutions in time and frequency which are constant throughout any analysis.

It is sometimes considered more desirable to have frequency-dependent resolutions. Typically, one needs accurate time information for high-frequency components and, conversely, good frequency resolution at low frequencies. This is because in general, one wishes to detect rapid changes in the signal. Such rapid changes, by their nature, must have high-frequency components. Detecting then quickly calls for good time resolution, i.e. a short time window with corresponding large bandwidth. On the other hand, the low-frequency components cannot change so rapidly and the time window can be lengthened with impunity, allowing better frequency resolution. A window near the ideal will then have a width which decreases with frequency.

Keeping this in mind, a general theory has been developed which establishes the mathematics behind this general transformation technique. The techniques described here for using wavelets for signal analysis are inspired largely by [24].

Let us start with the set $L^2(R)$ of all integrable quadratic functions:

$$L^2(R) = \left\{ f : \int_{\infty}^{\infty} |f(x)|^2 dx \langle \infty \right\} \tag{4.118}$$

Consider now the function $\psi \in L^2(R)$ which can be translated along the time axis and choose $\psi(x-k)$ $k \in z$, where z is the set of all real numbers in the set $L^2(R)$. We want ψ to give different frequency resolutions at different frequencies. This can be done by choosing several different values for ψ with decreasing width as frequency is increased. This alteration in the width of the function ψ is referred to as a dilation. So the set:

$$\psi(2^j x - k) = \psi(2^j x - k/2^j) \qquad j, k \in z \tag{4.119}$$

is a function which gives a binary dilation 2^j, and a dyadic translation $k/2^j$. All functions in the set $\{\psi(2^j x - k)\}$ appear as a binary dilation coupled with a dyadic translation of one single function, ψ. Any set $\{\psi(2^j x - k)\}$ is called a wavelet, and the generic function ψ is called the mother wavelet.

Using conventional notation for the inner product, let $f, g \in L^2(R)$. Then the inner product of f and g is given by:

$$\langle f, g \rangle = \int_{-\infty}^{\infty} f(x) \cdot g^*(x) \cdot dx \tag{4.120}$$

with the norm of f defined as:

$$\| f \| = \langle f, g \rangle^{\frac{1}{2}} \tag{4.121}$$

Note that for all $j, k \in z$:

$$\| f(2^j)(-k) \|_2 = \left\{ \int_{-\infty}^{\infty} |f(2^j x - k)|^2 dx \right\}^{\frac{1}{2}} = 2^{-\frac{j}{2}} \| f \|_2 \tag{4.122}$$

If all $\psi \in L^2(R)$ have unit length, then all functions $\psi_{j,k}$ defined as:

$$\psi_{j,k} = 2^{j/2} \psi(2^j x - k) \qquad j, k \in z \tag{4.123}$$

will also have unit length. This leads to the result that:

$$\| \psi_{j,k} \|_2 = \| \psi \|_2 = 1 \qquad j, k \in z$$

Note that the word length is used as a convenient way of saying extension along the x-axis, and will, of course, have the same dimensions as x in this discussion, $[T]$ rather than $[L]$.

Now consider the following definition.

A function $\psi \in L^2(R)$ is defined as an orthonormal wavelet if the set $\psi_{j,k}$ defined in eqn. 4.123 is an orthonormal basis set for $L^2(R)$ such that:

$$\left\langle \psi_{j,k}, \psi_{l,m} \right\rangle = \partial_{j,l} \cdot \partial_{k,m} \qquad j, k, l, m \in z \tag{4.124}$$

Let us choose a function $f \in L^2(R)$. Since $\psi_{j,k}$ is an orthonormal basis for $L^2(R)$, it is useful to consider the projection of f onto these basis vectors, $\psi_{j,k}$. This projection has the value of the inner product $c_{j,k} = \langle f, \psi_{j,k} \rangle$.

The function f can now be written as a sum of scalable basis functions, with coefficients $c_{j,k}$ which are scaling factors. This gives:

$$f(x) = \sum_{j,k=-\infty}^{\infty} c_{j,k}(x) \cdot \psi_{j,k} \qquad j, k \in z \tag{4.125}$$

Eqn. 4.125 is referred to as a wavelet series for f. Based on these relationships we are now able to define the general integral wavelet transform (IWT) for f as:

$$W_\psi f(b, a) = \frac{1}{\sqrt{|\alpha|}} \int_{-\infty}^{\infty} f(x) \cdot \psi\left(\frac{x-b}{\alpha}\right) dx \tag{4.126}$$

This definition makes it possible to deduce that the wavelet coefficients are special cases of the general IWT:

$$c_{j,k} = W_\psi f\left(\frac{k}{2^j}, \frac{1}{2^j}\right) \tag{4.127}$$

This allows the (j, k)th wavelet coefficient to be determined as the IWT of f, evaluated with a dyadic position $b = k/2^j$ and a binary $a = 2^{-j}$.

In order to perform a transform, one must be able to reconstruct the function f from $W_\psi f(b, a)$. In certain situations, this can be possible, in which case f as a function of $f(x)$ can be written as:

$$f(x) = \frac{1}{C_\psi} \int_{-\infty}^{\infty} \int_{-\infty}^{\infty} W_\psi f(b, a) \cdot \psi_{a,b} \frac{da db}{a^2} \tag{4.128}$$

The condition for eqn. 4.128 to hold is that:

$$C_\psi = \int_{-\infty}^{\infty} \frac{|\psi(\omega)|^2}{\omega} d\omega \langle \infty \tag{4.129}$$

where $\Psi(\omega)$ is the FT of $\psi(x)$.

Eqn. 4.129 further implies the requirement that:

$$\int_{-\infty}^{\infty} \psi(x)dx = 0 \tag{4.130}$$

Since $\psi \in L^2(R)$, then the integral of $\psi(x)$ must be a finite quantity. This implies that $\psi(x) \rightarrow 0$ as $x \rightarrow \pm\infty$. Since the mean value of ψ is also zero, then $\psi(x)$ has the form shown in Figure 4.59 for a typical function.

Eqn. 4.126 can be rewritten, using inner-product notation, as:

$$W_\psi f(b, a) = \langle f, \psi_{a,b} \rangle \tag{4.131}$$

Eqn. 4.131 indicates that $W_\psi f(b, a)$ is in fact a measure of the similarity between f and $\psi_{a,b}$. Since eqn. 4.130 implies that $\psi(0) = 0$, then $\psi(x)$ is a bandpass envelope analogous to the window in the Gabor transform. However, $W_\psi f(b, a)$ behaves as a variable window. Decreasing a will shorten the duration of $\psi_{a,b}$ and extend its acceptance of higher frequencies. In other words, in order to examine high frequencies one would use a small value for a, giving a short time window but large bandwidth. Conversely, for low frequencies a large value for a would be appropriate, giving a long narrow bandwidth window with poor time resolution but excellent frequency resolution. By letting $a = 2^{-j}$ the spectrum is analysed with a bank of frequency filters one octave wide and with time resolutions which improve with frequency. This is exactly what was described as desirable at the start of this section.

As a is changed, the passband of ψ will change, so a can be regarded as a variable denoting frequency. Variations in b move ψ along the time axis, so that b has the property of a variable denoting time. The concept of frequency denoted by a is not the same as that from the FT; it is merely a

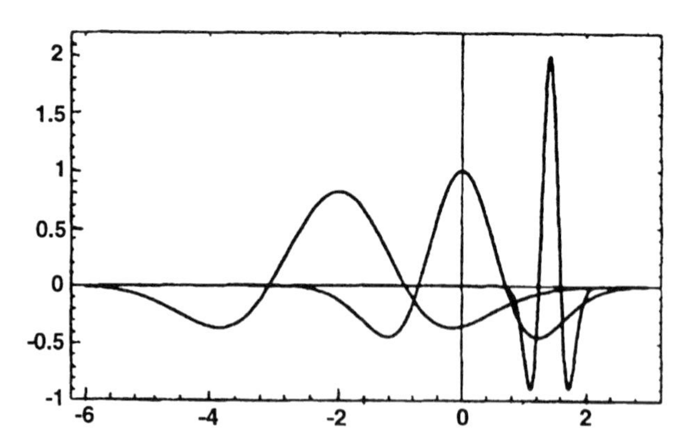

Figure 4.59 *A few wavelets obtained from the mother wavelet $\psi(x) = (1 - 2x^2)e^{-x^2} = \psi_{1,0}(x)$, which is the second derivative of a gaussian. Displayed are (from left to right): $\psi_{(3/2),-2}(x)$, $\psi_{1,0}(x)$, and $\psi_{(1/4),\sqrt{2}}(x)$ [25]*

scaling factor. Therefore, the concept of scale is introduced. A graphical representation of $|W_\psi f(a, b)|$ will be a three-dimensional figure known as a scalogram. It shows the numerical value of the IWT as a function of time delay b and scale a.

4.5.1.4 Wavelet analysis

An important factor in wavelet analysis is multiresolution. A signal can be analysed with good time resolution at high frequencies and good frequency resolution at low frequencies. We will now look into how to enable the practical application of wavelets.

When wavelets are used in signal processing it is first necessary to design two filters $h_0(n)$ and $h_1(n)$. Ingrid Daubechies has developed an algorithm for this [26]. Response curves for low-pass and high-pass digital filters with length $N = 32$ are shown in Figure 4.61.

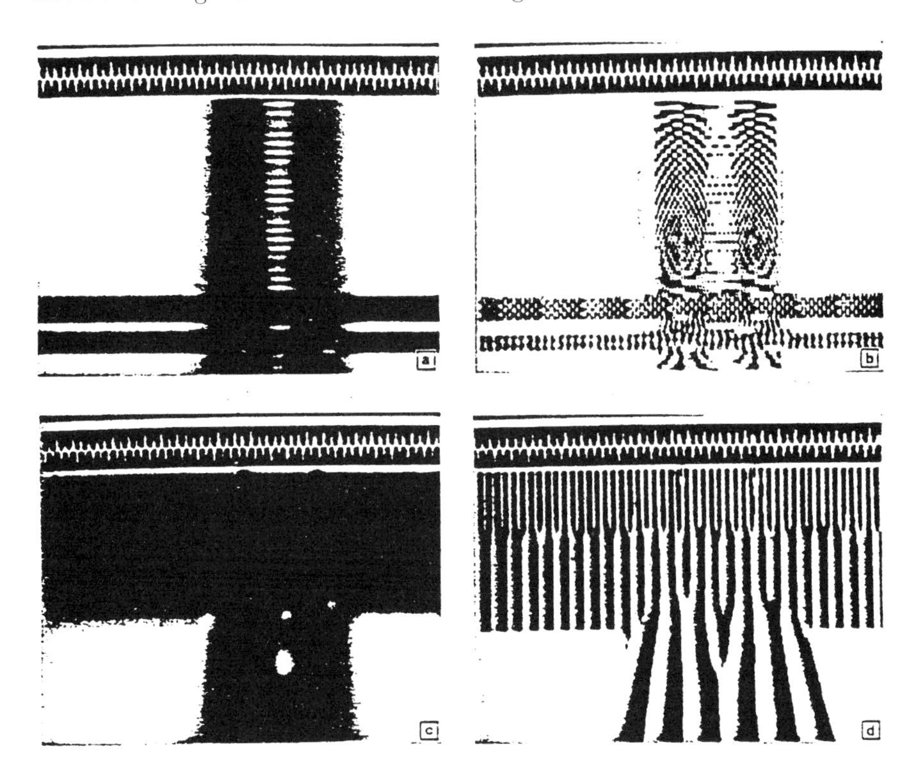

Figure 4.60 *Spectrogram and scalogram for the STFT and CWT analysis of two Dirac pulses and two sinusoids [31]*
a magnitude of the STFT
b phase of the STFT
c amplitude of the WT
d phase of the WT

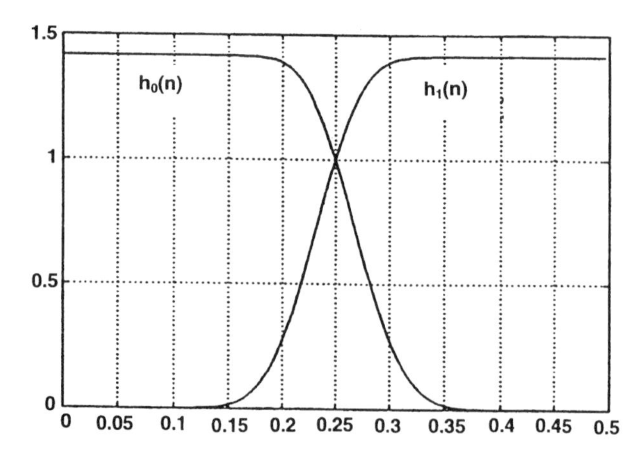

Figure 4.61 Finite impulse response (FIR) filters suitable for wavelet analysis

Assume a sequence $x(n)$ of length N. Let this sequence pass through a low-pass filter $h_0(n)$, and let it be decimated by a factor of 2. The filtered sequence will be $x_{1l}(n)$ with a length $N/2$, containing the low-frequency component of $x(n)$. Similarly, pass $x(n)$ through the high-pass filter $h_1(n)$ and reduce it by a factor of 2, yielding a sequence $x_{1h}(n)$ (also length $N/2$). This high-pass component $x_{1h}(n)$ is retained, and the low-pass component $x_{1l}(n)$ is refiltered in the same manner as $x(n)$. This gives two new sequences $x_{2l}(n)$ and $x_{2h}(n)$, both $N/4$ long. Again, $x_{2h}(n)$ is retained and $x_{2l}(n)$ is refiltered.

Thus, each successive low-passed sequence will have half the bandwidth of the previous sequence, but with only half the number of samples. This means that frequency resolution is doubled at each iteration, but time resolution is halved, since there are only half as many samples. This is shown in Figure 4.62.

Referring to Figure 4.63, the sequence $x(n)$ can be reconstructed as follows. In sequence $x_{4l}(n)$ a zero is inserted between each sample and the expanded sequence is then passed through a reconstruction low-pass filter $g_0(n)$. Similarly, the high-pass component $x_{4h}(n)$ is padded with zeros and passed through a high-pass filter $g_1(n)$. The outputs from $g_0(n)$ and $g_1(n)$ are added together, the sum being the sequence $x_{3l}(n)$. Again, the process is reiterated through $x_{3l}(n)$ and $x_{3h}(n)$, giving $x_{2l}(n)$ and $x_{2h}(n)$. These are then used to form $x_{1l}(n)$. In this manner, the outputs $x_{1l}(n)$ and $x_{1h}(n)$ are finally reconstructed and then added to form $x(n)$, the original sequence.

There is a trade-off between the length N of the data sequence $x(n)$, the length of the filter and the number of iterations. The lengths of the filtered sequences are halved at each iteration. When these lengths become comparable with the filter length N, it is ambiguous to perform the

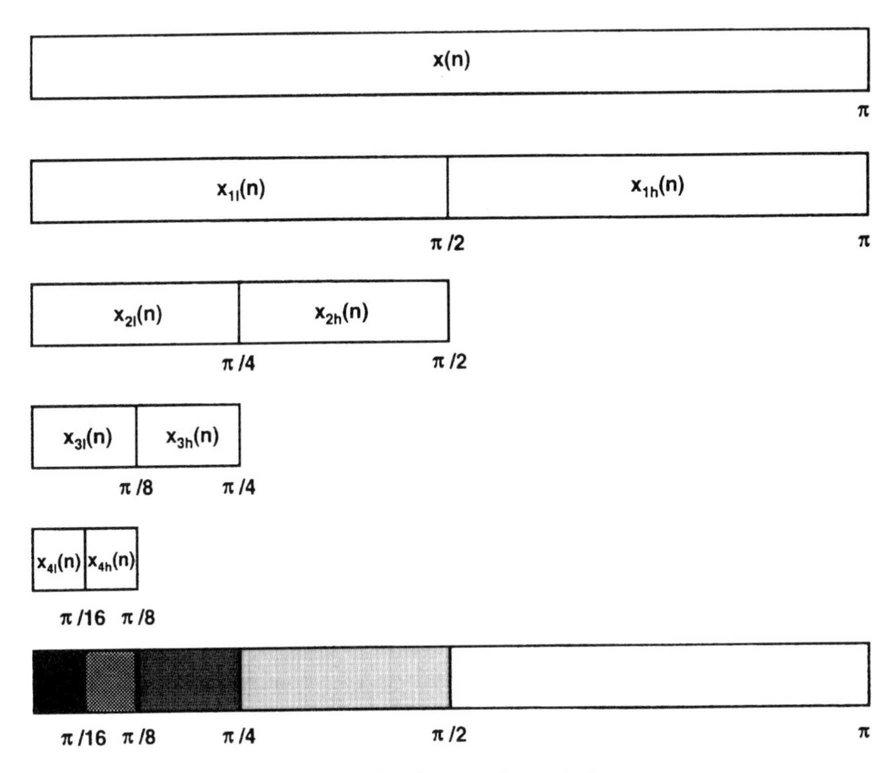

Figure 4.62 Multiresolution analysis using wavelet analysis

filtering. Even when the sequence is periodical there is still a limit to the practical number of iterations. Thus, only a relatively small number of filtering iterations are possible in any practical system with relatively short sequences and/or long filters. This limited number of iterations then sets a limit on the maximum frequency resolution that can be obtained.

4.5.1.5 Wavelets applied in radiocommunications

So far we have treated two different methods for obtaining simultaneous time and frequency information about a signal, namely the Gabor transform (GT) and the general wavelet transform (WT). The GT gives constant resolution for time and frequency across the whole signal spectrum. The WT is a multiresolution tool, with variable time and frequency resolution. It is good frequency resolution at low frequencies that is often desirable in many baseband signal-processing applications, such as voice coding.

In the case of analysis of a relatively narrow radio channel, it might not be an advantage to have better time or frequency resolution at any particular frequency in the channel. For this purpose there is a variant of

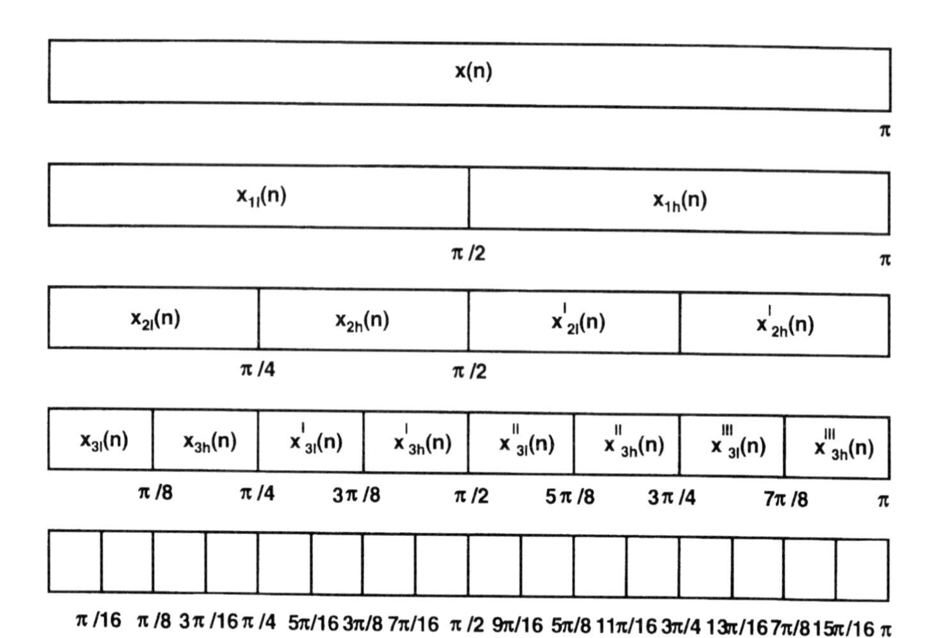

Figure 4.63 Frequency resolution using wavelet packet

wavelet analysis which gives the desired frequency-independent resolutions in time and frequency. This is called wavelet packets. The earlier mentioned Gabor transform can be viewed as a particular case of this, having a gaussian window function.

The treatment is rather like the WT described in the previous section. Again, one has the two filters $h_0(n)$ and $h_1(n)$. However, now both the low- and high-pass filtered components are refiltered instead of only the low-passed components. This is depicted in Figure 4.63. The result is that the frequency range is now divided into equal increments, so that time and frequency resolutions are independent of frequency.

Wavelet-packet analysis is a complicated technique to implement in practical systems. A way around this is to apply the theory for general filter banks in a way that matches wavelet analysis. A general filter bank is shown schematically in Figure 4.64. For our purposes it will consist of two sections:

(i) An analysis section consisting of M filters, each with $2\pi/M$ Hz bandwidth. The output length is halved at each successive filtering.
(ii) A resynthesis section consisting of M reconstruction filters. Prior to reconstruction the signal is padded with M zeros between each sample. In this way the reconstruction filters interpolate between the sample values.

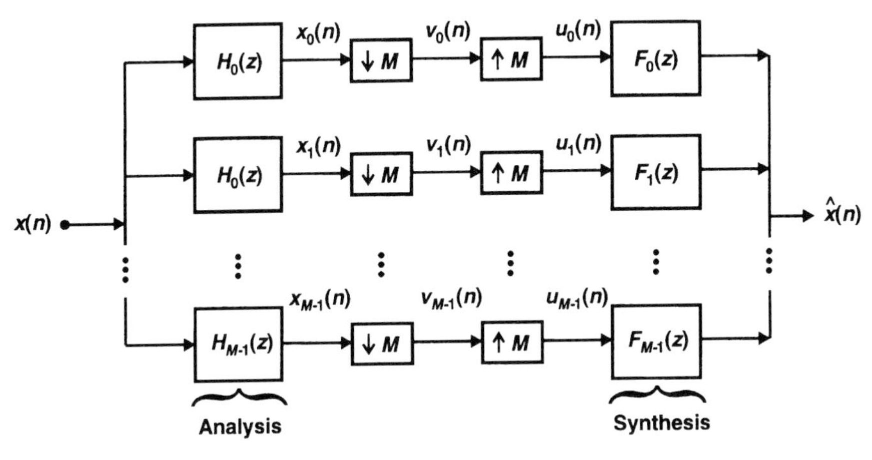

Figure 4.64 *General filter bank consisting of sections for analysis and reconstruction (synthesis)*

The signals, $v_i(n)$, between the analysis and reconstruction filter banks may now be processed as desired. It is necessary that these filter banks give (near-)perfect reconstruction. The criterion for this is that $\hat{x}(n) = c \cdot x(n - n_0)$, where c and n_0 are constants. This is achieved if the filters are constructed as para-unitary filter banks [27]. There are various ways of constructing a para-unitary filter bank, but detailed descriptions are beyond the scope of this book. However, an example of the use of para-unitary filter banks is given in the next section.

4.5.2 Interference suppression in spread-spectrum systems

As discussed in Section 4.2, direct-sequence spread spectrum systems have inherent capabilities for rejecting narrowband interference. But, even in these systems, such interference can be very disturbing. Different techniques have been proposed to improve performance under such conditions. Most of these benefit from different lengths of correlation between signal and interference, i.e. signal is broadband and interference is narrowband. Filtering (notch) is then used to reject the unwanted part. To comply with a nonstationary environment, algorithms are developed to adjust the filtering in an adaptive system. This section intends to show how the previously-covered theory relating to filter banks can be used to detect and suppress unwanted narrowband interference.

The major differences between the observed frequency spectra for most wanted signals and encountered noise should indicate that an effective signal-processing method would have to exploit information available in the frequency domain. In traditional Fourier analysis, however, all time-domain information is lost within the considered interval for the sake of frequency information retrieval. As already mentioned, a solution is then to implement filter banks which provide both time and frequency information about the signal. This process can be seen as a decomposition or mapping of a signal into the time–frequency domain as described in Section 4.5.1. From this a dynamic allocation and suppression of unwanted signal elements is possible. A major benefit with reference to spread spectrum systems is the ability to locate and track narrowband interference with respect to frequency while keeping full resolution in time.

A time–frequency transformation can be implemented as a number of analysis filters, followed by down sampling. Reconstruction of the input signal is accomplished by a corresponding set of up samplers and synthesis filters. This is depicted in Figure 4.65.

As already mentioned, perfect reconstruction can be acheived using para-unitary filter banks. To preserve time resolution, decimation is omitted and reconstruction is obtained by simple addition. This results in a simplified filter structure (Figure 4.66) where all filters are mutually orthogonal (resulting from para-unitarity). A small amplitude distortion results which is not significant in most cases.

When a received signal is an additive combination of a broadband spread spectrum signal, white noise and narrowband interference it is possible to improve performance significantly by using a simple threshold algorithm in conjunction with the filter bank pictured in Figure 4.66 [28].

The threshold algorithm will first need to detect if there is any narrowband interference present. This is done by tracking the average signal amplitude in all M subbands, and then comparing the minimum and maximum values found. If these two values differ by more than T_1,

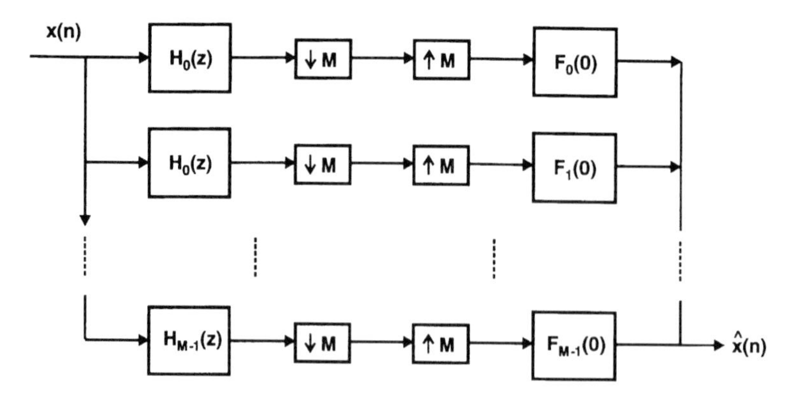

Figure 4.65 M-channel maximally-decimated filter bank [28]

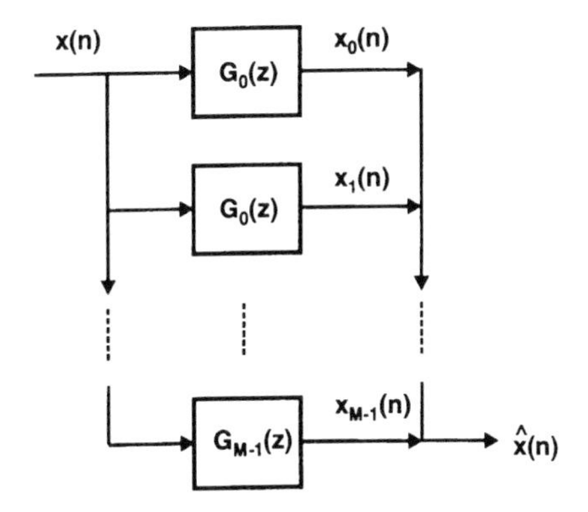

Figure 4.66 Simplified filter bank with M subbands [28]

narrowband interference is likely present. Notably, T_1 is made adaptive with respect to the average (white) noise background.

If a narrowband component is assumed to be present, each individual subband is analysed further using a sliding rectangular window, w, and average subband signal levels are compared to a subband threshold, T_2, which is set on the basis of information from the predetection threshold, T_1. If the average signal amplitude inside the subband window, w, is above the given threshold, all samples within the window are set to zero. This leaves us with three parameters which have to be tuned in each individual case, namely the subband window length, w, and the two levels T_1 and T_2. It is of importance that T_1 is adjusted so that further subband processing is done only when interference is actually present.

In order to illustrate this method, let us again consider an m-ary DS SS system with GMSK- ($BT = 0.3$) modulated chips. A principal receiver as simulated using Cossap® is shown in Figure 4.67.

In this example, white gaussian noise and a continuous sinusoid with random phase is added by the channel. The signal is decomposed using a serial demodulator followed by an eight-channel filter bank for the in phase and quadrature components. Furthermore, the following parameters are given as:

- code length, $L = 128$ chips/symbol;
- bandwidth, $B = 40$ kHz;
- 128-ary signalling, i.e. $k = 7$ bit/symbol;
- window size in the threshold algorithm, $w = 128$ chip;
- interference-to-signal ratio, $(J/S) = 0, 3.5, 6.0, 14$ dB.

Interference or jammer frequency is chosen according to eqn. 4.132, giving $f_j = 0.175$, 0.125 and 0.0. These are selected so that $f_j = 0.175$ is in

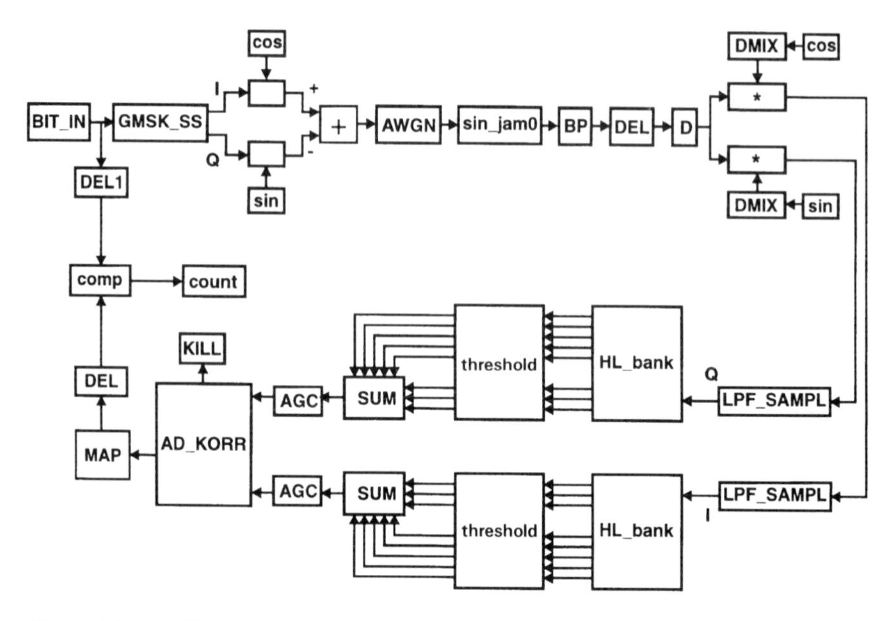

Figure 4.67 *DSSS communication receiver with a filter bank and threshold algorithm, COSSAP®*

the middle of the 3d subband; the other two are located between subbands 3 and 4, and subbands 4 and 5, respectively.

$$f_j = \left| \frac{f_{mf} - f_{interference}}{B} \right| \tag{4.132}$$

Results from simulations are shown in Figure 4.68 for various frequencies and (J/S) ratios. For comparison, the case where no antijam measures are applied (i.e. no filter bank) is also shown.

If the power introduced by the jammer is moderate ($J/S \leq 0$ dB), there is not much gain associated with the threshold algorithm. In such a case the system is not much influenced anyway due to sufficient processing gain. However, as J/S increases above 0 dB, the importance of intelligent signal processing rises accordingly. It is worth noting that performance is nearly independent of interfering power above a certain J/S ratio. In a practical system, however, a limit is encountered when gain control and limited quantisation results in loss of all useful signal information. With respect to the algorithm, it is actually better the stronger the interference. As J/S decreases, performance is dependent upon the adjustment of T_1 and T_2.

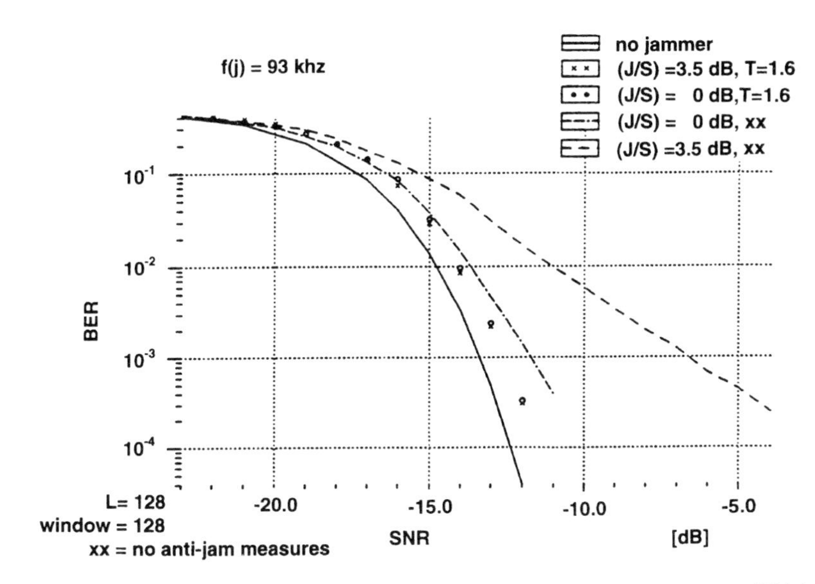

Figure 4.68 Performance for single-frequency jammer suppression in an m-ary CPM communication system

If a practical implementation is to be considered, it would probably require a more advanced and robust algorithm than the one presented here. However, this example shows that intelligent signal processing can be given a rather simple implementation through the use of filter banks.

4.5.3 Higher-order spectral analysis

This section gives a brief description of higher-order statistics (HOS), and takes a closer look at how this technique can be applied to binary DS SS systems. Our intent is to use signal processing so that all hidden information in a received signal can be made useful. One approach that we have covered is to apply some sort of either simple or complicated form of time–frequency analysis. This was an improvement from traditional techniques such as the Fourier transform.

Information contained in the power spectrum is obtained by second-order statistics (autocorrelation). With higher-order statistics (cumulants and moments) of a signal, we further process signals in order to extract additional information about signal characteristics, while retaining available phase information. This makes it possible, among other things, to identify nonlinear signals and systems [29]. This is something which also can be used in order to perform channel equalisation. Another important characteristic is that all cumulants (and spectra) of order higher than two are identical to zero for any gaussian process, something which forms the basis for the example given later in this Section.

In order to show one example of how HOS can be applied to communication systems, we will have to establish a short reference on deterministic and stochastic signals.

The kth-order correlation of a deterministic signal $\{y(i)\}_{i=0}^{N}$ is given by:

$$y_k(i_1, \ldots, i_{k-1}) = \sum_{j=-N}^{N} y(i)\,y(i+i_1), \ldots, y(i+i_{k-1}) \tag{4.133}$$

A k-dimensional Fourier transform of this gives a kth-order signal spectrum.

Now consider a finite impulse response filter as depicted in Figure 4.69, with the following relationship between the input and output signals:

$$y_k(i_1, \ldots, i_{k-1}) = \sum_{j=-N}^{N} h_k(j_1, \ldots, j_{k-1})\,w_k(i_1 - j_1, \ldots, i_{k-1} - j_{k-1}) \tag{4.134}$$

From the above we deduce that for instance the third-order correlation of the output signal can be found by folding the third-order correlation of the input signal with the third-order correlation of the filter impulse response. When $y(i)$ is stationary and its mean value is equal to zero, the following relations apply:

first-order cumulant (= *mean value*):

$$c_{1y} = E\{y(i)\} \tag{4.135}$$

second-order cumulant (= *autocorrelation function*):

$$c_{2y}(i_1) = E\{y(i)\,y(i+i_1)\} \tag{4.136}$$

third-order cumulant:

$$c_{3y}(i_1, i_2) = E\{y(i)\,y(i+i_1)\,y(i+i_2)\} \tag{4.137}$$

fourth-order cumulant

$$\begin{aligned} c_{4y}(i_1, i_2, i_3) = {} & E\{y(i)\,y(i+i_1)\,y(i+i_2)\,y(i+i_3)\} \\ & - c_{2y}(i_1)c_{2y}(i_2 - i_3) - c_{2y}(i_2)c_{2y}(i_3 - i_1) \\ & - c_{2y}(i_3)c_{2y}(i_1 - i_2) \end{aligned} \tag{4.138}$$

These cumulants can be estimated consistently using T samples like:

$$c_{3y}(i_1, i_2) = \frac{1}{T} y_3(i_1, i_2) \tag{4.139}$$

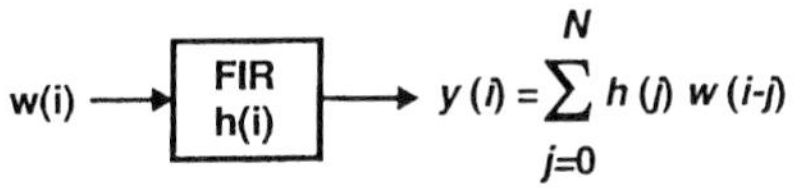

w(i) → FIR h(i) → $y(i) = \sum_{j=0}^{N} h(j)\, w(i-j)$

Figure 4.69 Relationship between input and output signals in an FIR filter

and so on. A k-dimensional Fourier transform of a kth-order cumulant will accordingly result in a signal spectrum of order k. Furthermore, it can be shown that higher order cumulants are not affected by additive gaussian noise.

Now, let us consider an example. We would like to examine the possibility of using HOS to detect if a deterministic signal is present or not. Our approach will be based upon the following [30]:

> *The kth-order correlation of the autocorrelation sequence for any signal will always produce its maximum value at the origin.*

We know that the output of a matched filter is identical to the autocorrelation function of the applied signal in the absence of noise. Now the kth order correlation at the origin is found according to eqn. 4.133, which can be seen as the energy of the signal with respect to its kth correlation:

$$y_k(0, \ldots, 0) = \sum_{i=0}^{2N} y^k(i) = E_{ks} \rangle 0 \tag{4.140}$$

We would like to decide upon the following hypothesis:

$$\begin{aligned} &H_1; \quad x(i) = s(i) + v(i) \qquad i = 0, 1, \ldots, T-1 \\ &H_0; \quad x(i) = v(i) \qquad i = 0, 1, \ldots, T-1 \end{aligned} \tag{4.141}$$

where $s(i)$ is wanted and $v(i)$ is gaussian noise.

Let us consider a matched filter receiver as shown in Figure 4.70, where by estimation of the value at the origin of the third-order correlation a decision is to be taken.

If H_0 is true (i.e. only noise is present), then the matched filter output is given by:

$$y_1(i) = \sum_{j=0}^{T-1} v(j)h(i-j) \qquad i = 0, 1, \ldots, N+T-1 \tag{4.142}$$

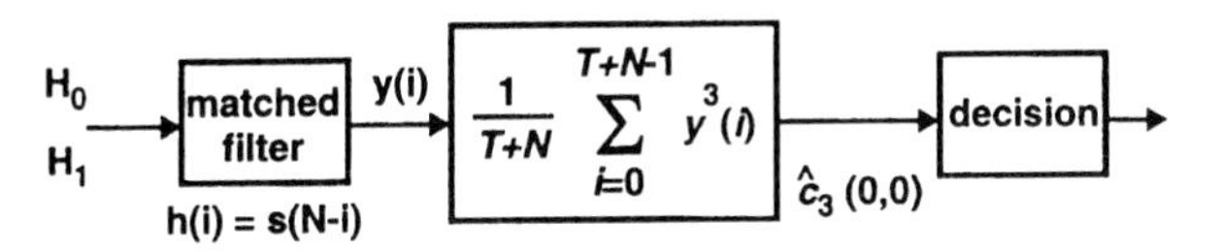

Figure 4.70 Detection based upon third-order correlation

resulting in the following statistics:

$$\frac{1}{T+N} y_3(0,0) = \frac{1}{T+N} \sum_{i=0}^{T+N-1} y_v^3(i)$$
$$= \frac{1}{T+N} \sum_{j_1 j_2}^{N} v_3(j_1, j_2) s_3(j_1, j_2) \tag{4.143}$$

As time of observation increases, this will approach the third-order cumulant $c_{3y}(j_1, j_2)$, which is equal to zero since $v(i)$ is additive gaussian noise. This means that:

$$H_0: \lim_{T \to \infty} \frac{1}{T+N} y_3(0,0) \to 0 \tag{4.144}$$

If H_1 is true, then the matched filter output consists of a signal and a noise component:

$$y(i) = y_s(i) + y_v(i) \tag{4.145}$$

where $y_s(i) = s_2(i)$ is the autocorrelation of the signal. We then get:

$$\frac{1}{T+N} y_3(0,0) = \frac{1}{T+N} \sum_{i=0}^{T+N-1} y^3(i)$$
$$= \frac{1}{T+N} \left[\sum_{i=0}^{T+N-1} y_s^3(i) + \sum_{i=0}^{T+N-1} y_v^3(i) \right. \tag{4.146}$$
$$\left. + 3 \sum_{i=0}^{T+N-1} y_s^2(i) y_v(i) + 3 \sum_{i=0}^{T+N-1} y_s(i) y_v^2(i) \right]$$

From the previously-mentioned characteristics of the third-order statistics, this turns out to be:

$$H_1: \quad \lim_{T \to \infty} \frac{-1}{T+N} y_3(0,0) \to \frac{1}{T+N} \left[\sum_{i=0}^{T+N-1} y_s^3(i) \right] + 3E\{y_v^2(i)\} S_2(0) \tag{4.147}$$

If the signal does not contain a d.c. component, i.e. $S_2(0) = 0$, we are left with the energy of the third-order correlation (E_{3s}) as the detection statistic.

Now, let us take this a step further and consider classification between two known signals, i.e. a binary system as shown in Figure 4.71. The two signals between which we want to discriminate must have equal energy, which in our direct-sequence spread spectrum system is achieved by using two antiphase spreading sequences.

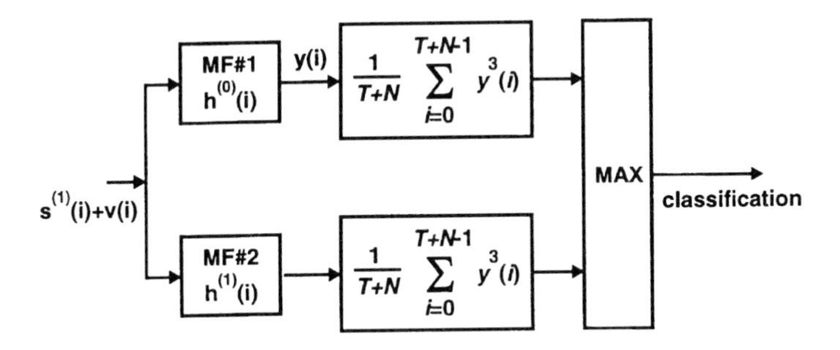

Figure 4.71 Receiver based upon matched filter and third-order correlation

The detector shown above will function on the basis of the third-order correlation as follows [30]:

> *Consider a bank of matched filters followed by kth-order correlation algorithms. The maximum value (at the origin) will always appear at the output of the filter which best matches the received signal.*

Simulated performance is shown in Figure 4.72, compared to ordinary matched-filter detection. In the case of coherent detection the latter is given by:

$$P_e = Q\left(\sqrt{\frac{2E_b}{N_0}}\right) \tag{4.148}$$

As shown, our third-order correlation detector does not perform as well as the ordinary matched-filter detector in pure white gaussian noise. This

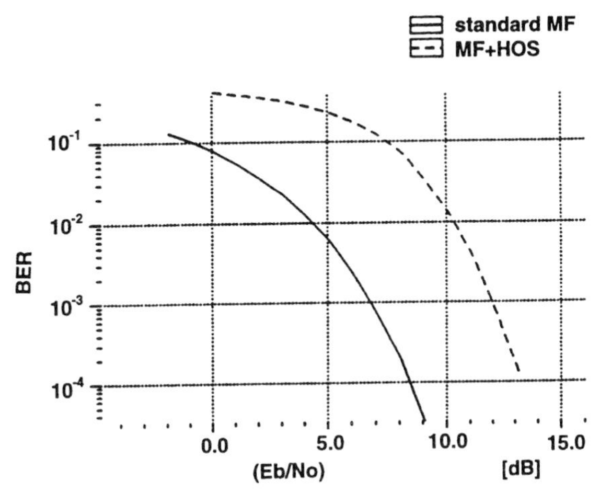

Figure 4.72 Simulated performance of third-order detection compared to traditional matched-filter detection

is not surprising since we know the matched-filter detector as a maximum-likelihood detector. However, we would expect this to be different in a coloured-noise scenario.

In order to improve performance of the third-order detector we should seek to produce a $y_s(i)$ autocorrelation function with a smoother characteristic. In a direct-sequence spread spectrum system, this can easily be achieved by replacing the matched filter part with a cumulative correlation function. This function is referred to as *cum_sum*, and is defined as:

$$K(n) = \sum_{i=0}^{n} x^{(l)}(i)h(i)$$
$$x^{(l)}(i) = s^{(l)}(i) + v(i) \tag{4.149}$$

where $h(i)$ is the stored data sequence.

This means that the correlation output will appear as an incline (or decline) and not only as a single value. An improved performance resulting from this is shown in Figure 4.73.

If we examine the results given in this example we are left with a solution that is almost comparable to maximum-likelihood detection in the case of white gaussian noise. However, the solution based on HOS is theoretically not dependent on the noise statistics, and can be expected to outperform a matched filter solution in most coloured-noise environments [30]. We know that any signal-processing algorithm optimised for gaussian models will become severely degraded in cases were this assumption is no longer valid. In spread spectrum communications, the type of signal we

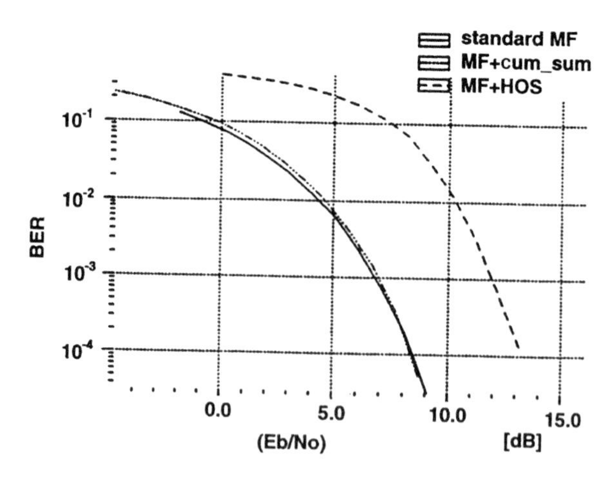

Figure 4.73 Simulated performance of third-order detection preceded by either matched filter or cum_sum

want to detect is usually known (deterministic), and the noise environment is assumed to be additive white gaussian. Obviously, this is an assumption that in many cases does not hold, which makes techniques like higher-order statistical analysis interesting in mobile communications. However, the complexity (cost) associated with this technique makes a thorough system analysis essential before implementation is considered.

To summarise, we know that any nonlinear or nongaussian system will exhibit higher-order statistical properties which, in turn, can be exploited by using suitable signal analysis. However, the performance of any practical system will be critically related to the accuracy with which these higher-order properties can be estimated. These estimates will usually exhibit excessive variances due to the finite number of samples used and the extraneous noise power. This means that full theoretical performance cannot be achieved in any practical system.

References

1. JAVANT, N.S.: 'Digital coding of waveforms' (Prentice-Hall, 1984)
2. WELSH, D.: 'Codes and cryptography' (Oxford University Press, New York, 1990)
3. STINSON, D.R.: 'Cryptography: theory and practice' (CRC Press, 1995)
4. DILLARD, R.A.; and DILLARD, G.M.: 'Detectability of spread spectrum signals' (Artech House Inc., 1989)
5. NATO Document AC/302(PG/6)D/25 NIAG-D(90) 1, annex 20
6. BERG, O.: 'Multi-role radio in the VHF-band'. Proceedings of the 6th ARFA symposium on *Use of the radio spectrum in NATO beyond 2000*, NATO HQ, Brussels, May 1993
7. TOBAGI, F.A.: 'Multiaccess protocols in packet communication systems'. *IEEE Trans.*, April 1980, **Com-28**, (4)
8. MOSHAVI, S.: 'Multi-user detection for DS-CDMA communications'. *IEEE Commun. Mag.*, Oct. 1996
9. URKOWITZ, H.: 'Energy detection of unknown deterministic signals'. *Proc. IEEE*, 1986, **55**, (4)
10. ANDERSON, J., SUNDBERG, T. and AULIN, C-E.: 'Digital phase modulation' (Plenum Press, 1986)
11. SUNDBERG, T.: 'Continuous phase modulation', *IEEE Commun. Mag.*, April 1986, **24**, (4)
12. GRONEMAYER, S.A. and McBRIDE, A.L.: 'MSK and offset QPSK modulation', *IEEE Trans.*, 1976, **COM-24**, (8)
13. AMOROSO: 'The bandwidth of digital data signals', *IEEE Commun. Mag.*, Nov. 1980
14. MUROTA, K. and HIRADE, K.: 'GMSK modulation for digital mobile radio telephony', *IEEE Trans.*, 1981, **COM-29, (7)**
15. SKLAR, B.: 'Digital communications' (Prentice Hall, 1988)

16. HJELMSTAD, J. and SKAUG, R.: 'Fast synchronisation modem for spread spectrum communication system using burst-format message signalling', *IEE Proc. F*, 1981, **128**, (6)
17. SIMON, M.K., OMURA, J.K., SCHOLTZ, R.A. and LEVITT, B.K.: 'Spread spectrum communications vol I' (Computer Science Press Inc., 1985)
18. SIMON, M.K., OMURA, J.K., SCHOLTZ, R.A. and LEVITT, B.K.: 'Spread spectrum communications vol II' (Computer Science Press Inc., 1985)
19. SEYMOUR, S.: 'Fading channel issues in system engineering', *IEEE J. Sel. Areas Commun.*, February 1987
20. CLARK, G.C. and CAIN, J.B.: 'Error-correction coding for digital communications' (Plenum Press, 1982, 2nd edn.)
21. YUEN, J.H. *et al.*: 'Modulation and coding for satellite and space communications', *Proc. IEEE*, July 1990, **78**, (7)
22. MATTISSON, P.: 'Codec for tactical radio TR-8000'. Proceedings of NRS 86, Nordic radio symposium, ISBN 91-7056-072-2
23. WOZENCRAFT, J.M. and JACOBS, I.M.: 'Principles of communication engineering' (Wiley, 1965)
24. CHUI, C.K.: 'An introduction to WAVELETS' (Academic Press Inc., 1992)
25. CHOEN, A. and KOVACEVIC, J.: 'Wavelets, the mathematical background', *Proc. IEEE*, 1996, **84**, (4)
26. DAUBECHIES, I.: 'Ortonormal bases of compactly supported wavelets', *in* Communication on pure and applied mathematics, vol. XLI 909-996 (J. Wiley and Sons, Inc., 1988)
27. VAIDYANATHAN, P.P.: 'Multirate systems and filter banks' (Prentice Hall Signal Processing Series, 1993)
28. RANHEIM, A.: 'Narrow band interference rejection in direct-sequence spread spectrum system using time-frequency decomposition', *IEE Proc., Commun.*, December 1995, **142**, (6)
29. NIKIAS, C. and RAGHUVEER, M.R.: 'Bispectrum estimation: a digital signal processing framework', *Proc. IEEE*, July 1987, **75**, (7)
30. GIANNAKIS, G.B. and TSATSANIS, M.K.: 'Signal detection and classification using matched filtering and higher order statistics', *IEEE Trans.*, 1990, **ASSP 38**, (7)
31. RIOUL, O. and VETTERLI, M.: 'Wavelets and signal processing', *IEEE Signal Process. Mag.*, October 1991

Chapter 5

Packet switching in radio networks

As discussed in Section 2.2.1, the traditional method for providing communication services to mobile users is to equip the user terminal with a radio and locate the switching function in centralised fixed equipment. An example of such a system is GSM. Here, the routing and relaying functions are located in the base stations and in the infrastructure that interconnects the base stations. A network where the switching function is distributed and located in the equipment carried by the network users provides an alternative to the traditional telecommunication systems. The first challenge met in such mobile radio systems is a highly dynamic network topology. The network operates under the direct influence of the users which set the operating conditions by deciding the operating terrain, spatial distribution and traffic distribution. The spatial distribution is a function describing where the users are located within the network coverage area. They can be located at fixed positions, at random locations specified by a random distribution, or at positions which alter over time. The traffic distribution characterises the user's packet stream to be served. This stream is specified by three components: the packet interarrival distribution, the packet length distribution and the traffic pattern distribution (a matrix showing the probability of addressing each destination within the addressing domain).

The sequence of presentation is bottom-up, according to the reference model presented in Figure 2.26, which means that we start by specifying a radio-channel model and gradually move upward in the protocol hierarchy. The exception from this is the opening Section, 5.1, which gives a tentative overview of random-access protocols facilitating asynchronous access to a shared radio channel and emphasises the significance of the network topology. A major issue here is how efficiently the different protocols restrict the multiuser interference under diverse conditions.

The spatial distribution sets the distance between the radios and, together with the operating terrain, it determines the path loss finally experienced across the radio links, see Chapter 3. The radio-channel quality is determined by the radio block error rate figure which again is the function of the signal-to-noise ratio (SNR) as described in Chapter 4. The network protocols can directly affect the SNR by regulating the number of concurrent radio transmissions. If only one network node transmits at a time, the SNR is solely determined by the background noise. The term background noise is in this Chapter used as a common term for all interfering sources not under the control of the network protocols. Interference caused by the network radios is referred to as multiuser interference. It is important to make a clear distinction between the two interference types when discussing protocol issues since only one of them can be controlled. Section 5.2 discusses the significance of the spatial distribution, multiuser interference and the radio-channel quality.

Section 5.3 considers MAC-layer design by reviewing some random-access protocols facilitating a set of nodes to share one radio channel. A problem common to all random-access protocols is the network instability which may occur without careful design and this is a major issue for the Section. One particular random access protocol is analysed in all its detail.

The MAC protocol facilitates packet exchange over a single radio channel but the probability of packet loss is too high for practical use. The LLC layer must enhance the quality of service by implementing an automatic-repeat request (ARQ) protocol. Section 5.4 outlines different LLC protocol alternatives and analyses performance for fully-connected networks (i.e., all the network nodes are within the radio coverage area of each other). The design domain is expanded considerably as the protocols must handle fragmented networks and additional functions have to be introduced at the LLC level. Section 5.5 considers the new effects coming to light when the networks topology no longer constitutes a complete graph.

A wireless radio network which only needs to provide information transfer between users having direct radio contact is said to be a single-hop network. Otherwise it is called a multihop network. The network layer has the overall responsibility for handling the traffic stream between any pair of users, at any time within the whole service coverage area. This is a demanding task and Section 5.6 presents the functionality needed. Among other tasks, store and forwarding of packets must be performed. An even more complex task is to locate the users as they move around and to find useful paths between them. This is the mission of routing. Although routing is a function within the network layer, it has been assigned its own Section 5.7 due to its complexity.

5.1 Random-access techniques and network topology

To satisfy the needs of wireless data networking, different methods have been developed for asynchronously accessing a common radio channel [24]. Random-access techniques are robust because they do not relay on any special control mechanisms (e.g. tokens) and are easily implemented in distributed systems. Their main disadvantage is the difficulty of finding a distributed control mechanism that regulates the user activity such that all links remain above the SNR threshold needed for successful packet transmissions. The efficiency of these techniques varies with user traffic and radio connectivity.

The intention of this Section is to prepare the reader for the forthcoming sections by first presenting a short overview of different access techniques and the problems encountered. Superficial reflections of their efficiency are also stated but more advanced analyses are given in later sections.

The simplest random-access method is the pure ALOHA [25]. Here, a mobile terminal is allowed to transmit unless it is already transmitting, regardless of any other network activity. This means that the demodulation of an incoming packet is aborted as soon as it gets a user data packet to transmit. An obvious improvement to this scheme is to introduce the restricting rule that a node is not allowed to start transmitting when locked onto any packet. Under pure ALOHA the vulnerable period is two times the packet length because to succeed, any transmission must not hit the previously-transmitted packet nor must any new transmissions hit the current packet. The idea of slotted ALOHA [25] is to increase the throughput by reducing this vulnerable period. A universal time axis is introduced and divided into equally-sized slots. All the network nodes must synchronise the transmission start point to coincide with the beginning of a slot. Normally it is assumed that the packet length and the slot size are identically sized and then all interfering transmissions will overlap and be of the same length. Therefore, slotted ALOHA decreases the vulnerable period by a factor of two compared to pure ALOHA and achieves higher throughput. The slot size requirement is readily fulfilled in systems using block-oriented FEC and providing fixed user packet lengths. To support transmission of other user packet lengths, segmenting of longer packets has to be performed by a higher-layer protocol. Reassembly is done at the receiver side prior to the delivery of the packet to the user.

An intuitive improvement of the ALOHA protocol is to implement activity detection, that is as long as the radio detects ongoing transmissions a carrier sense signal is kept active. This scheme is called carrier-sense multiple-access (CSMA) [25]. Implementation of CSMA requires additional radio hardware for reliable detection of a carrier. Under CSMA a node is prohibited from transmitting in the presence of any transmissions by its neighbours. Consider the event that a transmission has just ended, the radio channel changes state from busy to idle and all the nodes having a packet to send initiate a transmission. A collision will occur if two or more nodes have a packet ready for service. It is unlikely that two nodes have a packet to send simultaneously in low traffic states; however, as the offered traffic increases, the probability of collision increases and the network throughput may start to decrease for two reasons. First, overlapping transmission of the first packet increases the interference level and directly affects the probability of correct reception. Secondly, the probability of addressing another transmitting node increases. When the offered traffic has reached the rate where each node receives at least one new user data packet during a packet transmission time, every node will have a packet to send at the moment the radio channel becomes idle. The network throughput therefore decreases to zero. The following algorithm can stabilise the network:

- upon a user packet arrival, a random delay D is drawn;
- stay idle for a period of time D while sensing the channel;
- if a radio-channel state change to busy is detected during D, abort the current packet scheduling;
- when the time delay D has elapsed, send the packet.

Many different CSMA protocols have been studied in the literature but only two of them are mentioned—nonpersistent CSMA and p-persistent CSMA [25]. The idea with nonpersistent CSMA is to limit the interference between the packets by rescheduling a packet which arrives while the channel is busy. If the channel is idle upon arrival, the packet is transmitted immediately.

Under the p-persistent CSMA scheme, a busy node sensing an idle channel transmits with the probability p. The probability of nontransmission is therefore $1-p$, and the deferred packets are delayed one time slot (here one slot is identical to the propagation delay). At this new point in time, the channel state is sampled again. If the channel is idle, a new random p-number is drawn and the process is repeated. Otherwise, the channel is busy and the packet is scheduled for retransmission according to the retransmission delay distribution in use.

An example of a distributed radio network implementing CSMA is the civilian wireless local area network (WLAN) standard HIPERLAN (high performance radio LAN) by ETSI (European Telecommunications Standards Institute) [32]. Detection of the channel state is based on measurements of the received power. Its channel-access protocol, the elimination-yield nonpre-emptive priority multiple access (EY-NPMA), restricts the number of overlapping transmissions by applying a combination of active signalling and passive listening as shown in Figure 5.1. On detecting a free channel, at time instance t_1 in the Figure, each station holding a packet shall listen for n slot times where n is determined by the priority of the packet under service. The nodes with the lowest priority delay are the nodes with the highest priority and each of them emits an access burst at the end of the waiting time, given that no other access burst is detected. At t_3 the lower priority nodes know that high-priority traffic shall be given precedence, they suspend serving their packets and the system has discriminated the traffic according to the priority level. To further restrict the number of contending nodes an elimination burst of random length is sent by each node passing through a prioritisation phase. The Figure demonstrates the case where two nodes enter this state. A node completing the elimination burst senses the channel and the winning node(s) is the node which has drawn the longest elimination burst. The number of nodes remaining at t_5 depends on the elimination burst-length distribution. In order to reduce the number of nodes even more, a yield-listening phase is entered at t_5. The nodes which have not been filtered out up to now stay idle for a random listening period. The node(s) with the shortest period transmits at t_6 and the contention phase comes to an end. The transmission of the data packet completes at t_7 and the destination

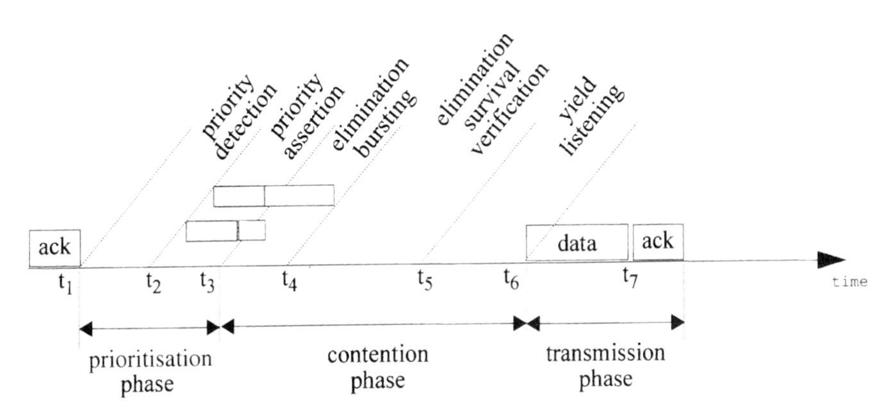

Figure 5.1 HIPERLAN channel-access technique

node returns an acknowledgement immediately upon receiving an error-free data packet.

Another WLAN standard implementing CSMA is IEEE 802.11 [31]. This standard also uses CSMA as the primary channel-access mechanism but more sophisticated techniques are included. Such techniques will be discussed in the following.

RF systems must cope with man-made noise as well as atmospheric and galactic noise. However, in practice, other RF sources may have even greater impact. Frequencies are often re-used thereby leading to cochannel interference. For instance, in the United States, wireless LANs can operate in the licence-free ISM (industrial, scientific and medical) bands and spread-spectrum techniques are used to isolate already existing narrow-band services from the new communication services [29]. A side effect of the CSMA method is that any device radiating electromagnetic energy into the operating frequency band may easily block the network. The only obstruction between this interference and the system is the propagation path loss. Consequently it could be very easy to block the traffic in CSMA networks such as the HIPERLAN.

Greater robustness is achieved by spread-spectrum signalling [1,30], where each transmission is prefixed by a preamble assigned a particular code known only by the network radios. This code can even be shifted periodically and a signal that does not contain this code will not cause a 'carrier on'. Apart from requiring a more intelligent jammer, this method gives enhanced protection due to a processing gain, as explained in Chapter 4. The first system practising this principle is the packet-radio system described in [17,18] and implemented by the Defence Advanced Research Project Agency around 1975.[1] Another military system described in [19] uses a preamble as a protection but does not employ spread-spectrum signalling.

A packet-radio system developed by the Norwegian Defence makes extensive use of spread-spectrum signalling for protection [3,4]. A preamble is added to each transmission and a receiver is only able to detect a busy channel upon receiving this preamble. Otherwise, it concludes that the channel is free. We call this method preamble-sense multiple-access (PSMA). Figure 5.2 illustrates the PSMA operation. The channel becomes idle at t_0 and all the busy nodes (i.e., the nodes holding packets to be sent) draw their random access delay D. The node with the shortest delay starts to transmit at t_1. The preamble ends at t_2 and the other nodes detect a busy

[1]Packet-switching techniques for distributed radio networks were initially developed by the military and later found applicable for civilian radio networks. Within the military domain the original name was 'packet-radio network' although the civilian domain uses the terms 'wireless LAN' or 'radio LAN'.

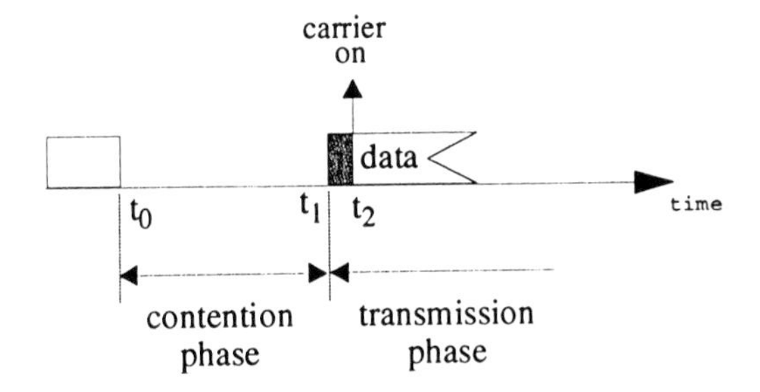

Figure 5.2 PSMA channel-access technique. An ongoing transmission can only be detected by receiving a particular code placed in the front of each packet

channel. A packet-length indicator must be inserted by the originator and is needed by the receivers for detecting the end of packet. Although PSMA improves the robustness it has two major imperfections: if bit errors occur in the length-indicator field, the receiver misjudges the packet length, and a receiver sampling the channel after the preamble has passed concludes that the channel is idle. Thus it may interfere with an ongoing transmission.

The design of radio networks becomes more complicated in the presence of hidden nodes, exemplified by node *A* in Figure 5.3. A pair of stations is referred to as being hidden if they are not within radio range of each other. Consider the event when node *A* transmits a packet to node *B*. If one or more of *B*'s neighbours initiates a transmission before *A* has completed its transmission, the outcome depends on the resulting interference level. The link $A \rightarrow B$ will experience more retransmissions as the load increases, but the reverse direction always succeeds in a noiseless system. The 'nearly complete' (NC) topology has excellent connectivity and the CSMA scheme controls the number of simultaneous transmissions efficiently. This is not the case in the star network where many overlapping

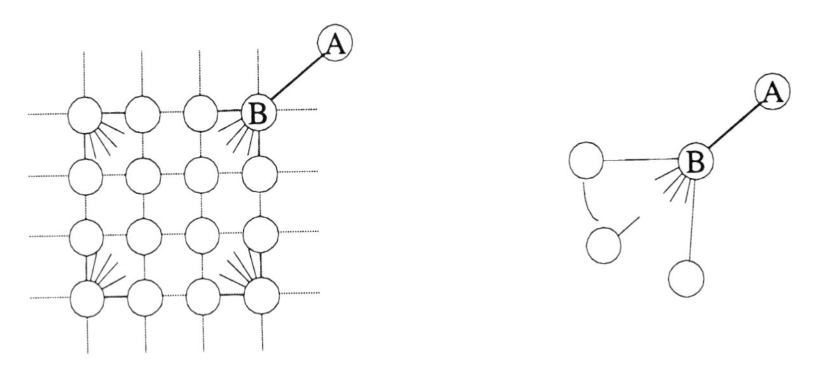

Figure 5.3 A nearly complete and a star topology

transmissions give a high interference level at node B and the link $A \rightarrow B$ becomes saturated at a very low traffic level.

A customary method for preventing network collapse is to implement an adaptive retransmission timer that introduces a random back-off delay. The effect of the adaptive timer on the star network where the edge nodes experience the same probability of packet loss, is increased back-off time and thus reduced packet loss. This is, however, not the case for the NC network. The well connected region experiences a high probability of success and uses a minimum back-off value, whereas the link $A \rightarrow B$ experiences a high loss rate and node A increases its back-off. But this is of no help as long as the other nodes continue to use a low back-off. From A's point of view, the adaptive back-off leads to decreased performance as its only effect is an increasing delay between new transmission attempts without giving the benefit of a better success rate as in the star netwok. The best A can do to maximise throughput, is to retransmit as fast as possible hoping that the hidden nodes will stay off the channel, or that the capture effect will rescue the packet.

To generalise the discussion, focus is put on the link $A \rightarrow B$ in Figure 5.4 and node A's hidden node set $H_{A \rightarrow B}$. The vulnerable time period for the forwarding operation is the entire data packet length because if one or more nodes in $H_{A \rightarrow B}$ initiates a transmission during this time, they will hit the packet. The same problem occurs when B returns an acknowledgement packet, but this time it is the node set $H_{B \rightarrow A}$ which may cause interference. This shows that the probability of success becomes a function of the packet length and the packet-arrival rate in the different regions.

Busy-tone multiple-access (BTMA) is a technique preventing hidden nodes from transmitting by having a dedicated signalling channel where the receiver can emit a busy tone during packet reception. In the classical BTMA scheme [26], a special central station issues the busy tone when it detects a busy channel. In [26] one supposes that the radio coverage area of this central station includes all the network nodes. The application of this principle to a distributed system requires that each receiver provides

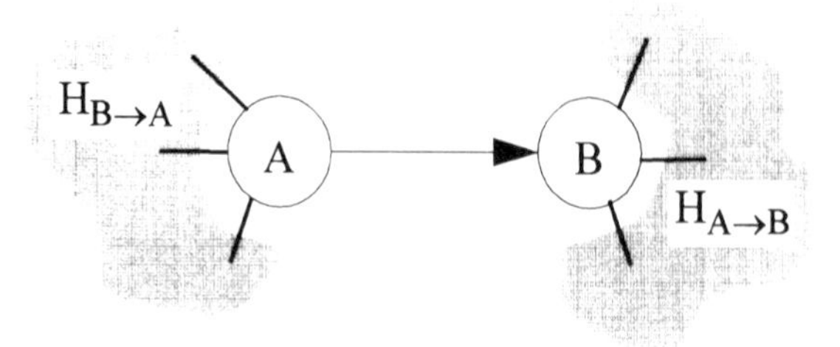

Figure 5.4 A radio link and the hidden-node sets

this functionality. Consider the implementation of this method in the network in Figure 5.5.

Simultaneous activation of the links $A \rightarrow C$ and $B \rightarrow C$ is now effectively prevented. However, concurrent transmissions on the links $A \rightarrow C$ and $G \rightarrow H$ are also prevented even though they do not interfere. The link $G \rightarrow H$ is blocked because F, which is covered by A, emits a busy tone. It is not possible to eliminate this problem. However, if the nodes analyse the address information placed at the beginning of the packet and then force the nodes not addressed to turn off their busy tone, the blocking duration is at least reduced.

The link $D \rightarrow E$ is blocked but if spread-spectrum signalling is used, the resistance to noise may be high enough for node C to tolerate the increased interference level. A further improvement might be to let node C delay the busy tone until the SNR drops to a certain threshold. This then really becomes some form of collision detection. CSMA with collision detection (CSMA/CD) is used by some wired LANs where the transmitter listens to its own transmission. At the moment when the nodes detect a bit pattern other than that transmitted, a collision has occurred and all transmissions are aborted. This method works here because the SNR levels at both the transmitter and receiver sides are practically identical. This is not the case for radio networks and therefore performance degradation occurs in a radio network under CSMA as the nodal degree decreases (which leads to an increasing number of hidden nodes).

In the absense of a dedicated signalling channel, the network designer is forced to use the ordinary data channel to curb the hidden nodes. The most important conclusions of the reflections above are that the contention is at the receiver and that a mechanism which can prevent the hidden nodes from transmitting has a significant potential for increasing the network performance. The final result depends on the channel capacity

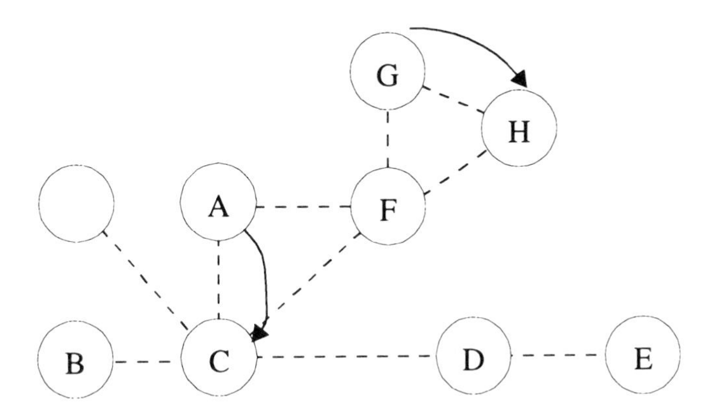

Figure 5.5 Example topology

taken by such a function. Node B in Figure 5.4 is the only node which is connected to the 'dangerous' nodes and the source node A and, consequently, is the only node that can take the responsibility of being a controller. By implementing a reservation protocol, the originator A requests access to node B by sending a request to send (RTS) packet. Node A is granted access to the channel upon receiving a special control packet called the clear to send (CTS), see Figure 5.6. This packet is broadcast by node B and also orders B's neighbours to be silent for a period of time. This method is applied in the WLAN standard IEEE 802.11 [31].

A reservation scheme should be symmetrical in such a way that it turns off all the hidden nodes $H_{A \to B}$ when A forwards the data packet, and turns off all the nodes in $H_{B \to A}$ to increase the probability of having a successful acknowledgement (ACK) packet. At time t_0 node A invokes the reservation procedure by issuing an RTS packet to B and succeeds at t_1 after a number of attempts. Having received the RTS, node B responds positively by returning a CTS packet. All nodes overhearing the RTS and/or the CTS packets shall stay off the channel for a period of time. When A receives the RTS, A has been granted reserved access and shall immediately send the data packet on the air. Node B returns the ACK without delay at t_4.

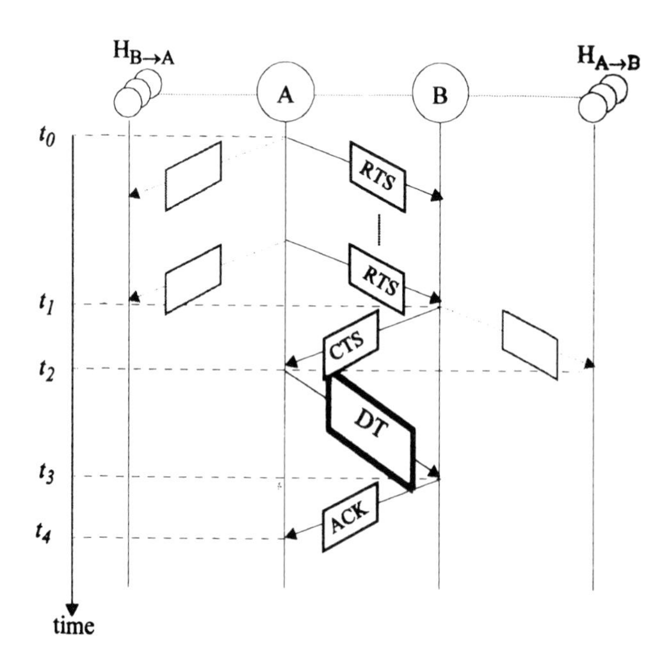

Figure 5.6 Channel-reservation technique. A random-access method is applied for transmitting the RTS packet

It is critical that all the nodes $H_{B \to A}$ stay idle during the transmission of the CTS and the ACK. Transmissions within the time period $[t_2, t_3]$ from any node in $H_{B \to A}$ cause no harm but the only practical way is to include this period in the forced idle period. The length of the data packet under service is included as control information in both the RTS and the CTS packets. Nodes overhearing an RTS and/or a CTS shall stay idle for the duration $t_{CTS} + t_{DT} + t_{ACK}$ and $t_{DT} + t_{ACK}$, respectively.

To conclude, CSMA is efficient in networks where the state of the receiving node (idle/demodulating) can be judged at the sending side. This condition is fulfilled in complete network topologies and Sections 5.3 and 5.4 consider these cases in detail. For topologies with hidden nodes a better solution might be to let the receiver control the channel access by using the BTMA or the RTS/CTS. This is studied in Section 5.5.

5.2 The spatial distribution and capture models

The signal-to-noise ratio (SNR) is a function of many variables at the network level. This section focuses on how the radio-channel quality varies as the user-traffic and the user's geographic locations change. Both have great impact on the average SNR which again determines the likelihood of receiving an error-free packet across the radio channel. A capture model is a probability model expressing this probability.

A scenario commonly used by the radio designers during performance analysis of their design is a transmitter sending to a receiver under the influence of noise. With an ambition to express the performance of each link in a network, say for a 25-node net, $25 \times 24 = 600$ links have to be considered. This is because radio links are generally asymmetric and thus have different performances. The network designer must therefore resort to approximations.

The choice of modulation and coding is usually matched to the system requirements but here no particular preference is made and for simplicity, differential PSK (DPSK) is used. The bit-error probability P_b for DPSK under white gaussian noise, is given by [34, page 274]:

$$P_b(E_b, N_0) = \tfrac{1}{2} e^{-E_b/N_0} \tag{5.1}$$

where E_b/N_0 is the SNR per bit. An important issue is how the use of spread-spectrum modulation affects the network-level design. Although spread-spectrum signalling gives no gain for gaussian channels, it may improve the performance when the channel suffers from multiuser interference. The latter is expected to be the major problem in a random-access system and the effect of gaussian noise can be neglected. With DPSK, we can set $E_b/N_0 = snr$ where snr is the SNR due to multiuser interference, and the effect of spreading on the BER can be modelled as:

$$P_b(snr, G_p) = \tfrac{1}{2}e^{-snr\cdot G_p} \tag{5.2}$$

where the processing gain $G_p = r_{BW}/r_{data}$. If the probability of error is independent from bit to bit in a packet and the SNR level is the same for all bits, then the probability of receiving an error-free packet of size m [bits] can be written as:[1]

$$P_C(m) = [1 - P_b]^m \tag{5.3}$$

Eqn. 5.3 is plotted in Figure 5.7 and it can be observed how the P_C curve has a fast transition from failure to success as the SNR level changes.

Consider the network in Figure 5.8 and receiver B in particular. All nodes are identical so the only contribution to different radio-link budgets is the path loss. A is the first node to transmit and B demodulates the signal in the presence of $k-1$ interfering sources. Using the propagation path loss model $1/r^\beta$, the SNR at node B given that B is locked onto a transmission from A, can be calculated by:

$$SNR_{S_{Tx},A\to B} = -10\cdot\log\left(r^\beta_{A,B}\sum_{\forall j\in S_{Tx}} 1/r^\beta_{j,B}\right)\quad[\text{dB}] \tag{5.4}$$

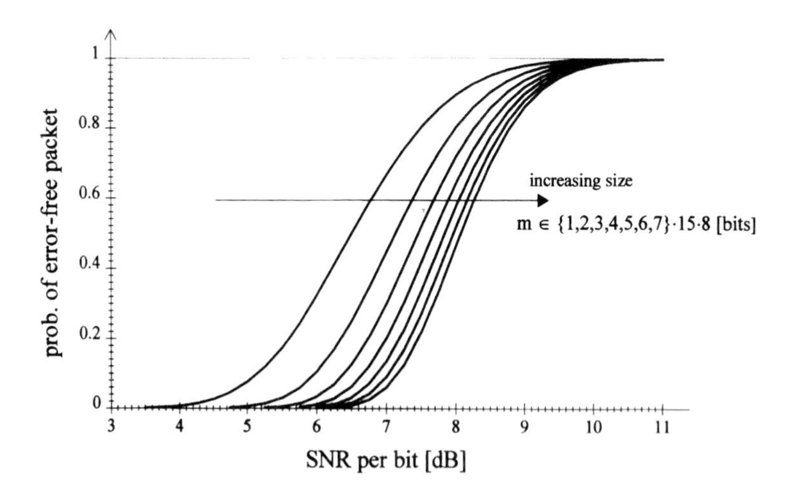

Figure 5.7 Probability of successful packet demodulation for DPSK modulation and $G_p = 1$

[1] A more precise model for P_C is given in [35] but here it is assumed that all the transmitters reached the receiver with the same power level. Therefore this model does not give us the possibility of considering different spatial distributions.

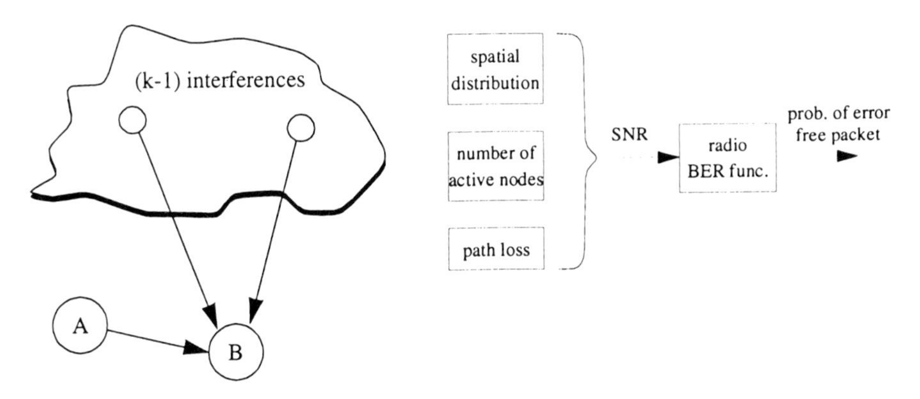

Figure 5.8 Receiving node B demodulates a packet from node A

where S_{Tx} is the set of transmitting nodes and $r_{i,j}$ is the distance between node i and node j.

S_{Tx} should actually be the set of any interfering sources, but in a random-access system it is the multiuser interference that normally contributes to reduced SNR. Therefore jamming or background noise is neglected in this section. A high multiuser interference level results if the network protocols fail to control the channel access in an orderly manner. The input traffic to a network is normally stochastic, and the SNR also becomes stochastic because S_{Tx} is related to the user traffic. In a system where only one user transmits at any time instance, S_{Tx} will be an empty set.

The capture effect in a spread spectrum system is the ability of a receiver to synchronise to the first arriving signal and reject later arriving signals. A receiver's capability to stay locked onto the first signal when the interference level increases depends on the processing gain. The capture probability C_k is the probability for a spread-spectrum receiver to successfully demodulate the transmission on the radio link from A to B given that k nodes transmit. By the total probability formulae we can write (for a system with multiuser interference only):

$$C_k = \begin{cases} 1 & \text{for } k = 1 \\ 0 & \text{for } k = n \\ \sum\limits_{(\forall i \in S_{net})} \sum\limits_{(\forall j \in S_{net} - \{i\})} C^*_{k-1}(|i \to j) \cdot P_\lambda(i \to j) & \text{otherwise} \end{cases} \tag{5.5}$$

where S_{net} is the set of all nodes in the network and n is the number of nodes. Eqn. 5.5 gives the two trivial cases $C_1 = 1$ and $C_n = 0$. In the first case only one node transmits and the node will always succeed in a noiseless system. The second case addresses the event that all nodes transmit and none of the nodes are in a state to demodulate the incoming signals. The inner sum calculates the average success for one particular node to all its possible destinations and the outer sum takes the average over all the network nodes. The term $C^*_{k-1}(|i \rightarrow j)$ is the probability of successfully demodulating the transmission from node i to node j, in the presence of $k-1$ interfering transmissions and with node i being the first node to transmit. The term $P_\lambda(i \rightarrow j)$ expresses the probability of having a transmission on the link $i \rightarrow j$. Every link must be weighted to get a correct network average because the user traffic and/or the network protocols determine which links are activated. For example, an asymmetric traffic pattern leads to more frequent transmissions among some sets of nodes and therefore affects the probability of success observed at the network level.

Let $C^*_{k^*}(|A \rightarrow B)$ be the probability of successful demodulation of the transmission $A \rightarrow B$ when k^* additional nodes transmit, given that node A was the first node to transmit (i.e., we have $k^* + 1$ transmissions simultaneously). This conditional probability can be calculated for each link in the network according to:

$$C^*_{k^*}(|A \rightarrow B) = \sum_{\forall j \in S_{k^*}} P_C(m, SNR_{j,A \rightarrow B}) \cdot P_{\lambda 3}(j|A \rightarrow B) \tag{5.6}$$

Here S_{k*} is all the subsets of the set $S_{net} - \{A, B\}$ of size k^*. The path-loss property for the different links is taken care of by the term $SNR_{j,A \rightarrow B}$, see eqn. 5.4. The effect of the modulation and coding schemes on the data block of size m is incorporated through P_C, see eqn. 5.3. $P_{\lambda 3}$ expresses the relative occurrence of the elements in S_{k*}. The user traffic and/or network protocols may give precedence to certain elements. If not, it is possible to set $P_{\lambda 3}(j|A \rightarrow B) = 1/|S_{k*}|$ where $|\ |$ is the cardinality function, and obtain the simplified expression:

$$C_k*(|A \rightarrow B) = \frac{1}{|S_{k^*}|} \sum_{\forall j \in S_{k^*}} P_C(m, SNR_{j,A \rightarrow B}) \tag{5.7}$$

Eqn. 5.6 only considers the $(k-1)$ interfering sources and not their individual geographic locations. For this reason $C^*_{k^*}(|i \rightarrow j)$ is an approximation which expresses an average over the entire network.

A relationship between the radio-channel quality (eqn. 5.3), the SNR level for a network structure (eqn. 5.4) and the capture probability (eqn. 5.5) has now been established. At the network design level the latter is of major interest. Example 1 below illustrates this relationship by means of a simple spatial distribution.

Example 1

Consider the simple radio network in Figure 5.9. All three transmitting mobiles are identically configured to continuously transmit at constant power on the same radio channel. As mobile A starts to move, its link length $r_{A,B}$ becomes a function of time so the SNR also becomes a function of time. The other mobiles are kept at fixed locations. Consider the channel quality as seen by A while it continuously moves and transmits to B. The two other concurrent transmissions are the interfering sources. From eqn. 5.4 we have:

$$\begin{aligned} SNR_{\{I_1,I_2\},A\to B} &= -10\log(r_{A,B}^{\beta}(r_{I_1,B}^{-\beta}+r_{I_2,B}^{-\beta})) \\ &= -10\log[2(r_{A,B}/r)^{\beta}] \quad \text{[dB]} \end{aligned} \tag{5.8}$$

The start position is a distance $\sqrt{2}\cdot r$ from B. Assume that A's speed distribution is constant, v, and regard the situation before passing B. Then:

$$\begin{aligned} SNR_{\{I_1,I_2\},A\to B}(t) &= -10\log[2(\sqrt{2}-vt/r)^{\beta}] \\ &= -10\log[2\cdot(\sqrt{2}-t')^{\beta}] \end{aligned} \tag{5.9}$$

where r/v is used as the time reference such that $t=(r/v)t'$. Similarly, the SNR expression after passing B is given by $SNR(t)=-10\log[2(t'-\sqrt{2})^{\beta}]$. Eqn. 5.5 gives the capture probability $C_3=C_2^{*}=P_C(m,SNR(t))$. Figure 5.10 plots $SNR(t)$ and the resulting capture probability as a function of time for $\beta=4$ and the block size $m=120$ bits.

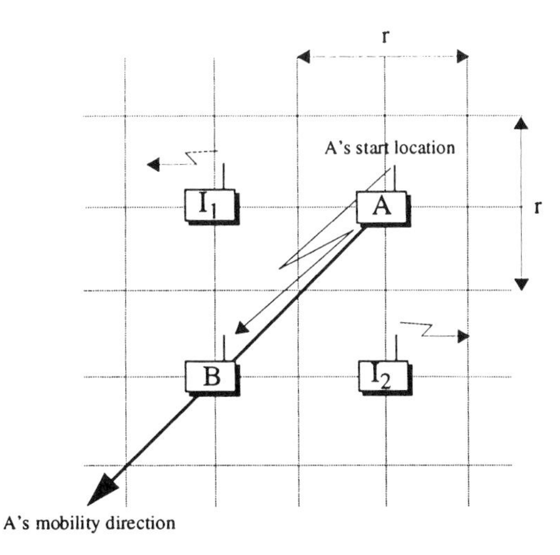

Figure 5.9 *Example of a configuration that gives a time-variant channel*

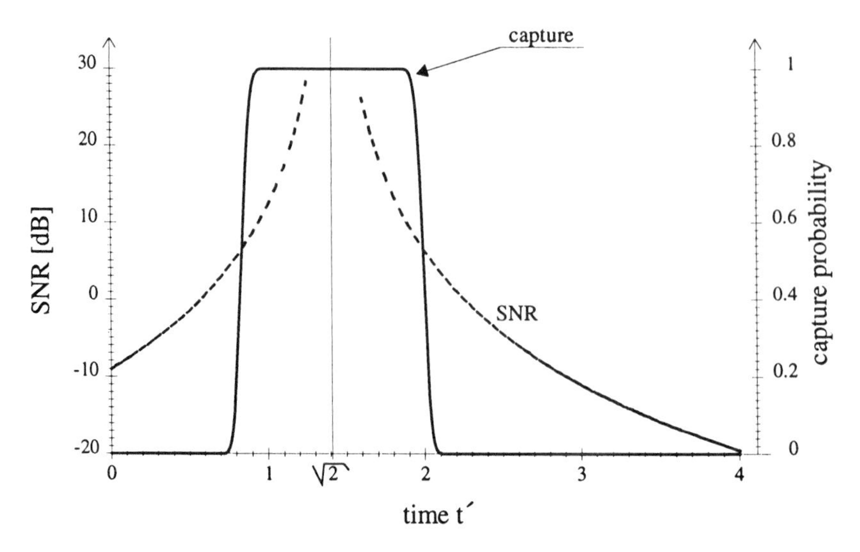

Figure 5.10 *SNR and capture probability for the radio channel from A to B*

This example illustrates the fast transition between success and failure which leads to quickly changing channel quality as mobility is introduced. Thus LMN providing seamless handover has short time limits to re-establish an ongoing conversation when the quality starts to drop. In a multihop packet-radio network the greatest challenge is probably related to the routing algorithm which has to find efficient routes for the user traffic.

Leaving the example of a time-variant channel and constant traffic, the network in Figure 5.11 with a fixed regular structure is considered next. A network of size n with a uniformly distributed traffic pattern is assumed. If the nodes are granted access to the radio channel on an equal basis, the relative offered traffic at each link is the same and $P_\lambda(i \rightarrow j) = 1/(n(n-1))$. By these simplifications eqn. 5.5 reduces to:

$$C_k = \frac{1}{n-1}\frac{1}{n}\sum_{(\forall i \in S_{net})}\ \sum_{(\forall j \in S_{net}-\{i\})}\frac{1}{|S_{k*}|}\sum_{\forall j \in S_{k*}} P_C(m, SNR_{j,A\rightarrow B}) \qquad (5.10)$$

The significance of this expression is discussed further in the following examples.

Example 2

With $\beta = 4$ we consider the radio-channel quality under DPSK modulation as seen by the stations in the nine-node network in Figure 5.12. If we partition the nodes according to the radio quality observed, we have the three sets {1, 3, 7, 9}, {2, 4, 6, 8} and {5}. Conditioned on two simultaneous

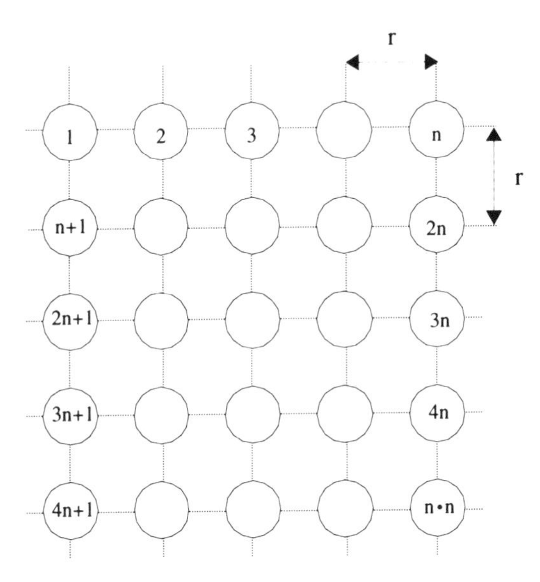

Figure 5.11 *Fixed user spatial distribution. The number of nodes is $n \cdot n$ and the radio ranges are assumed to be larger than $\sqrt{2} \cdot (n-1)r$*

transmissions, the best (e.g. at $SNR_{2\rightarrow1}$) and the worst (e.g. at $SNR_{1\rightarrow9}$) SNR values are 10.9 dB and −9.8 dB, respectively. Eqn. 5.3 gives the probability of successful delivery of a block of size $m = 480$ bits, $P_{C,2\rightarrow1} = 0.716$ and $P_{C,1\rightarrow9} = 0$ for the two levels. This shows the dramatic link quality changes across the network and it might be necessary to consider throughput on a link-by-link basis. However, generally networks contain too many links to open for this detailed level.

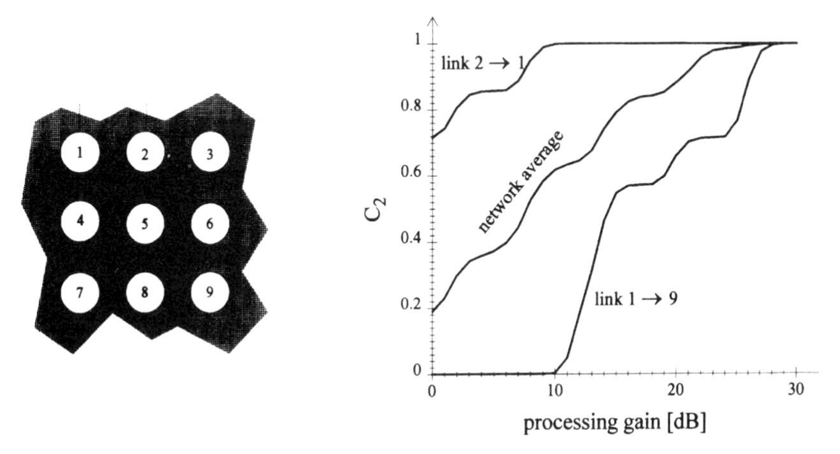

Figure 5.12 *A nine-node network and its capture probability under two simultaneous transmissions*

One advantage of spread-spectrum signalling is the decreased BER for increased processing gain (G_p). In our nine-node network, we can increase the G_p to a level where the link $1 \rightarrow 9$ can sustain one overlapping transmission. Figure 5.12 shows that $P_{C,1\rightarrow 9}$ starts to increase when G_p gets greater than 10 dB and at $G_p = 20$ dB the probability of success for a 480-bits packet is nearly 0.6. Also note that $P_{C,2\rightarrow 1}$ has reached one at $G_p = 10$ dB. The cost of spread-spectrum signalling on the other hand is the reduction of the data transmission rate which leads us to the interesting question of whether it is better to let the network protocols reduce the multiuser interference rather than to apply spread-spectrum signalling. This is a question to be considered in a subsequent section.

Example 3

The network users inject packets into the network at random times. Thus the nodes become busy with a packet to send, and the network-level protocol must regulate the access to the radio channel. With perfect regulation only one node transmits at a time regardless of the intensity of the user-input traffic. However, as will become obvious in later sections, this is not achievable in a real system. We must therefore reduce our ambition and open for the occurrence of two or more simultaneous transmissions. The average SNR level must be above a certain point if most of the packets are to succeed. Table 5.1 gives numerical SNR values for the spatial distribution in Figure 5.11 as a function of number of simultaneous transmissions.

Even for two simultaneous transmissions the average network SNR level is zero and many of the radio links are disconnected with DPSK modulation and no processing gain (here a relay function is needed). Of more interest for the network throughput is the resulting capture probability C_k. Table 5.2 shows the C_k values at different processing gain values.[1] If the

Table 5.1 Network average SNR in dB as a function of the multiuser interference represented by the number of concurrent transmissions

	number of concurrent transmissions, k		
network size	2	3	4
25	0.0	−5.6	−8.7
16	0.0	−5.2	−8.1
9	0.0	−4.7	−7.3

[1]Eqn. 5.5 is very demanding to compute for large n and k. We had to use Monte-Carlo simulations for (n, k) larger than $(25, 5)$.

Table 5.2 The capture probability C_k for different network sizes. β is set to 4, the packet size to 480 bits and the BER characteristic (eqn. 5.2) is used

network size	G_p [dB]	C_2	C_3	C_4
25	0	0.242	0.121	0.075
	10	0.587	0.377	0.276
	20	0.843	0.702	
16	0	0.230	0.105	0.061
	10	0.594	0.387	0.287
	20	0.853	0.719	0.608
9	0	0.190	0.076	0.03
	10	0.617	0.409	0.305
	20	0.882	0.765	0.671

network level is unable to keep the average number of simultaneous transmissions below two, we see that the packet-loss probability is nearly 0.8 for a system without any processing gain. As the processing gain increases, higher multiuser interference can be handled.

An assumption often used in the literature is $C_{k+1}/C_k = \text{constant} = q$ which gives the simple capture model:

$$C_k = q^{k-1} \tag{5.11}$$

In our case we should use $q = C_2$. By using this approximation in the spatial distribution we introduce an error and Table 5.2 shows that C_{k+1}/C_k is not constant. Despite the imperfection of this capture model we find it to have sufficient accuracy to be used in the later sections. The effect of background noise or jamming might be modelled as:

$$C_k = p_{noise} q^{k-1} \tag{5.12}$$

where p_{noise} is the probability that a single transmission succeeds when noise is present.

Example 4

A network must usually accept user packets longer than the maximum length that can be handled by the radio interface. When this occurs, the network protocol must divide the packet into two or more smaller packets at the originating site, transmit each across the radio link and re-assemble the set of smaller packets into the original long packet at the receiver side. This gives the network designer the option of selecting the packet size across the radio link. The probability of packet success decreases with increasing packet length as shown in Figure 5.7. Here the success is given

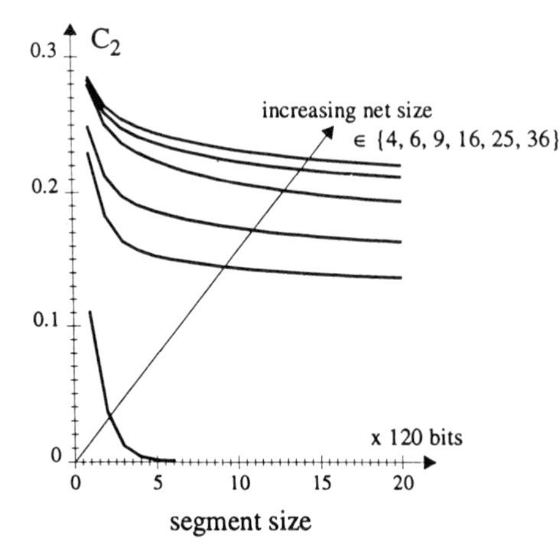

segment size [bits]	C_k		
	2	3	4
60	0.287	0.112	0.048
120	0.249	0.09	0.037
480	0.190	0.076	0.03
960	0.178	0.075	0.026

Figure 5.13 *Capture probability lag 2 (C_2) as a function of the packet length and the network size (left). C_k as a function of segment size for the nine-node net (right)*

as a function of the SNR level. The SNR level changes with the spatial distribution and the number of simultaneous transmissions, and Figure 5.13 shows the capture probability lag 2 (C_2) as a function of the packet length for some different spatial distributions given by Figure 5.11.

The cost of sending a packet over the radio channel is the bandwidth taken by the protocol-control information in order to perform packet switching (addressing, error-detecting codes etc.). As the number of segments increases the overhead also increases but the probability of success increases. Therefore, for a given user packet length there exists a number of segments minimising the radio channel capacity used for the delivery of a message. Consider the event where the first transmission is hit by exactly $(k-1)$ overlapping transmissions. The success probability of the first transmission becomes $C_k(m/n_s + o_b)$. The segment size $m/n_s + o_b$ is determined by the original message of size m which is segmented into n_s segments and o_b is the protocol-control information added to each segment. Figure 5.13 also gives a few numerical examples of the relation between the segment size and the C_k for the nine-node net in Figure 5.12.

The number of transmissions needed to deliver all the segments is n_s/C_k and the total number of bits sent (Λ) for delivering the message of size m becomes:

$$\Lambda = (m + o_b \cdot n_s)/C_k \cdot \qquad (5.13)$$

A large n_s gives a higher C_k, but introduces the larger overhead $o_b \cdot n_s$. A smaller n_s gives a lower C_k and a higher segment loss rate.

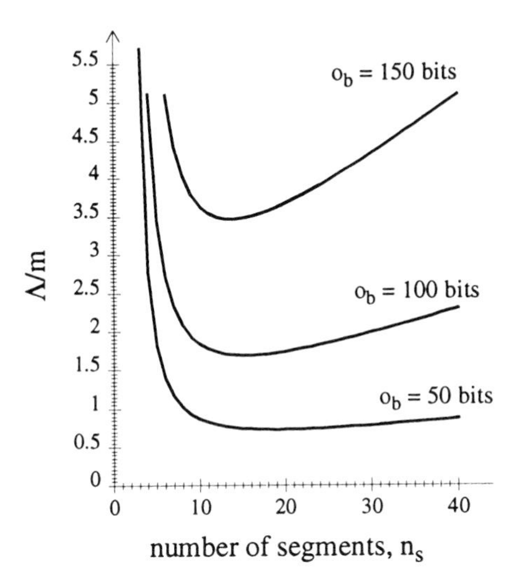

Figure 5.14 *Λ/m by eqn. 5.13 for link 6→5 and $m = 1000$ bits as a function of the number of segments*

Example 2 showed that many of the radio links in the nine-node net suffer from bad link quality and that by increasing the processing gain better connectivity could be achieved. We have just seen another alternative which is to send more but shorter packets. The total number of bits sent over a link divided by the message size Λ/m, represents the overhead to be minimised. Figure 5.14 uses eqn. 5.13 to express Λ/m for the link 6→5 as a function of the number of segments and the protocol control information size o_b. When o_b is large, much control information is added to each segment and the optimum n_s gets smaller. At the extreme is $o_b = 0$, there is nothing to pay for a transmission and the optimum n_s becomes identical to the number of bits in m, i.e. single-bit segments shall be used. This example gives only a simplified discussion leaving out many important issues at the network level. A more complete picture will be given after treating medium-access control in the next section.

Example 5

The spatial distributions looked at so far have assumed geographically-fixed users. A spatial distribution where the users pop up at random locations is shown in Figure 5.15. Here the users are normally distributed around the centre of a circle. The majority of the users shall be within the shaded area and by using $\sigma = r/2.87$, the probability of drawing an x/y-displacement larger than r is less than 0.004. Obtaining a closed-form expression for this scenario is intricate [15]. It is much easier to use Monte-

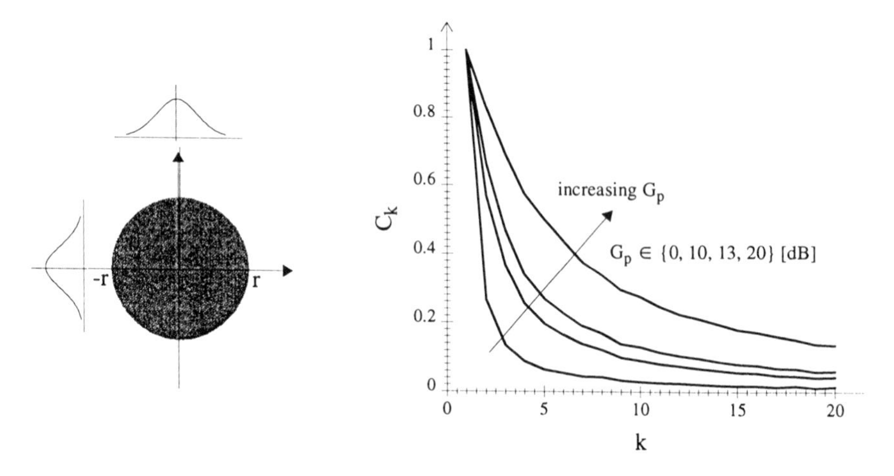

Figure 5.15 *User location distribution (left) and the corresponding capture probability C_k. The user location distribution is based on the normal $(0, \sigma^2)$ distribution. C_k is plotted against the number of simultaneous transmissions k and the processing gain G_p. The packet size is 480 bits.*

Carlo simulation and this is carried out in order to find the capture probabilities shown to the right in the Figure.

Even in this example we find that C_{k+1}/C_k is not a constant and Figure 5.16 gives an impression of the magnitude of the error by drawing simulated C_k values together with the simplified model.

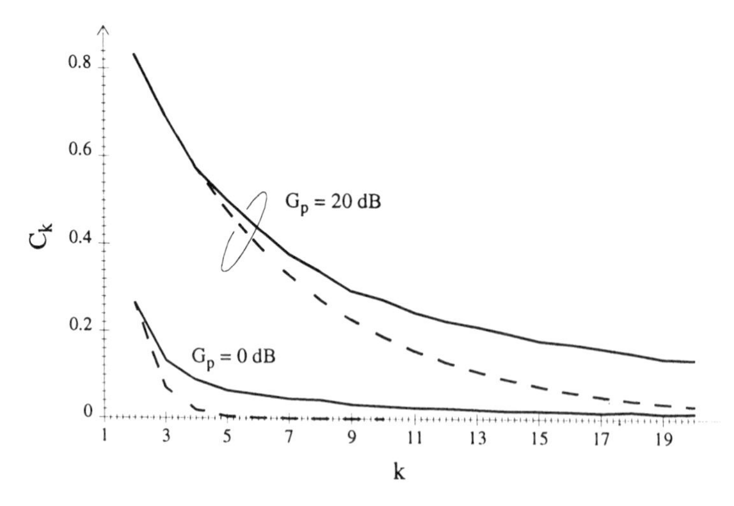

Figure 5.16 *C_k plot for normally-distributed spatial distribution by simulation (solid line) and by the simplified model q^{k-1} (dashed line)*

In summary, this section has shown that the radio-link quality generally varies over the network, even for simple spatial distributions. Consequently, the links should ideally be considered individually. Because of possible huge numbers of links, this is impractical and for complex topologies, impossible. Some simplification must be introduced and the capture model C_k (eqn. 5.5) models the packet success probability as an average over all the network links. C_k expresses the sum of many effects as shown in Figure 5.17.

Yet another useful approximation is to introduce $C_k \approx p_{noise} \cdot q^{k-1}$. Here q is an abstraction of the spatial distribution, packet length, path loss and the radio characteristics (processing gain and modulation). A procedure for taking these parameters as inputs has been devised and their impact has been discussed.

A system where the first transmission sustains any number of overlapping transmissions is said to have perfect capture ($q = 1$) as opposed to the zero capture case ($q = 0$), where any overlapping transmission leads to demodulation failure. Radios providing nonperfect capture fulfil the relationship $0 < q < 1$.

The major network design challenge is to control the multiuser interference represented by k. This becomes evident by regarding the term q^{k-1}. A real sysem has $q < 1$ (see example 3 for numerical q values) and then the

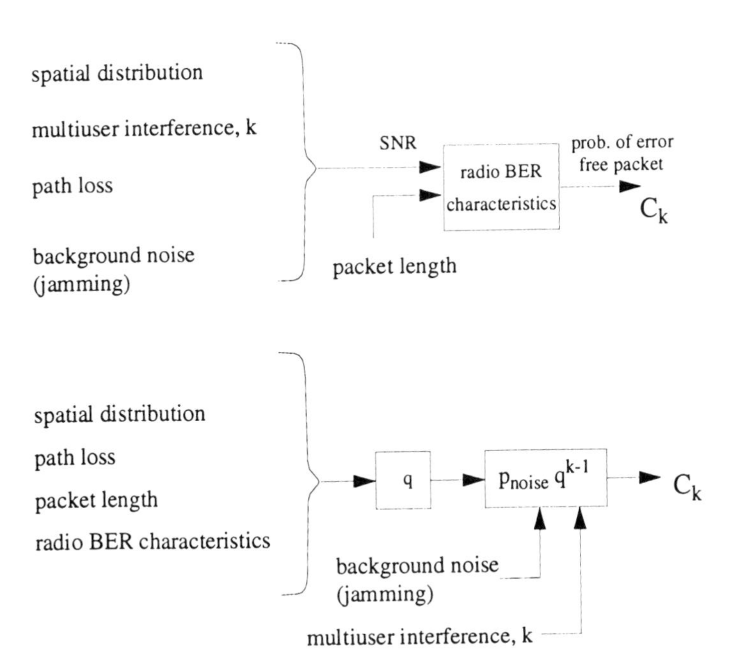

Figure 5.17 The two capture models considered and their input parameters

packet success drops fast with increasing k. Section 5.1 provided a preliminary discussion of this issue and the next Section addresses this in more detail for fully-connected networks.

5.3 Medium access control

This Section explores the many aspects of importance for implementing efficient packet-switching protocols over a radio channel by using a CSMA-based protocol. Although the term carrier sense is somewhat misleading with the use of spread spectrum, we continue to use the term because of its well established use within the communications community. With spread-spectrum signalling, channel-state sensing is based on detecting one or more particular codes, and therefore code-sense multiple-access would be a more adequate expression.

To date numerous CSMA protocols have been presented in the literature. Their difference stems primarily from the use of different random-access delay distributions when accessing the channel, behaviour under retransmissions and the fact that some are slotted while others are unslotted. The focus in this Section is tied to the combination of CSMA and spread spectrum, so only aspects of most importance to this combination are treated. We are going to use an unslotted protocol with uniformly-distributed scheduling delay. For a reasonably complete treatment of the many CSMA protocols the reader is referred to [28].

The network model serving as the basis for this section is shown in Figure 5.18. A number of mobile stations, say n, receive packets from the users according to some distribution. The packet arrival and the packet-length distributions are assumed to be identical for all users, the incoming traffic pattern is uniformly distributed and the path loss is assumed identical for all radio links. The benefits of these choices are that the offered traffic will then not be the cause of any non-homogeneous node behaviour and ease the modelling of the network operation as well as the interpretation of our case studies. For discussing the basic characteristics it is sufficient to assume that the nodes contain the MAC protocol only and that all nodes are within radio range of each other.

Communication between the nodes is achieved by sending and receiving packets over a radio interface. Therefore we have to make some assumptions about the radio services and characteristics before we can design and study a medium-access protocol. This is done in Section 5.3.1 where a reasonable generic radio interface is specified. Then we concentrate on the basic characteristics of CSMA protocols and in Section 5.3.3 one particular scheduling distribution is selected. Finally, we develop a

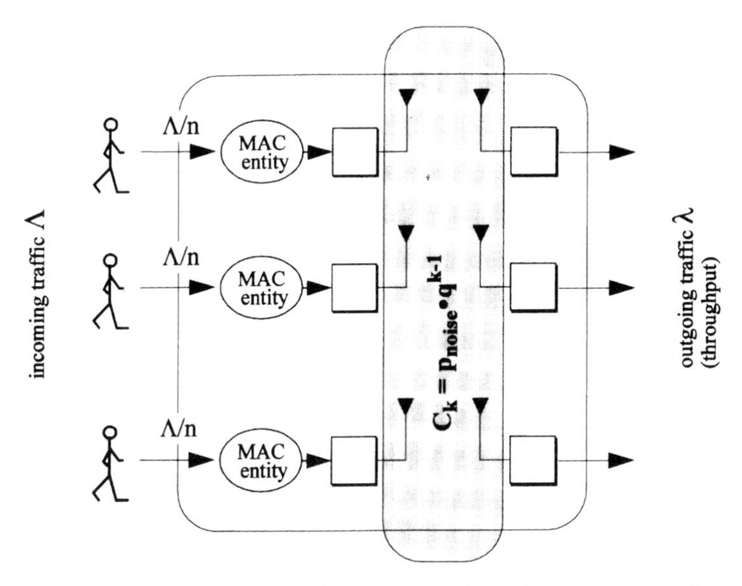

Figure 5.18 The incoming traffic (Λ) and the radio environment are configured to give identical operating conditions for each node

throughput model for a heavy-load network state and discuss network behaviour by means of specific examples.

5.3.1 Radio services and characteristics

The radio design for a packet-switched system involves much more than considering bit error rate as a function of E_b/N_0. Section 5.2 addressed just this issue by expressing an abstraction of the radio channel as a simple model, referred to as the capture model, taking background noise and multiuser interference as input and giving the packet-success probability as output. This section focuses on other radio aspects of importance to implementing efficient packet-switched data services under the CSMA scheme. The implementation of a CSMA protocol requires a carrier sense service from the radio. As described in Chapter 4, this can be implemented by measuring the channel energy level or by using particular codes. Seen by the network level, the carrier sense service must be appreciated according to: how fast it can track channel state transitions and the probability of signalling incorrect channel state. We open this discussion by considering Figure 5.19 illustrating the time diagram for a single packet transmission on the channel.

A radio has an idle-to-transmit switching time delay greater than zero. This means that some time passes from the point in time when a node

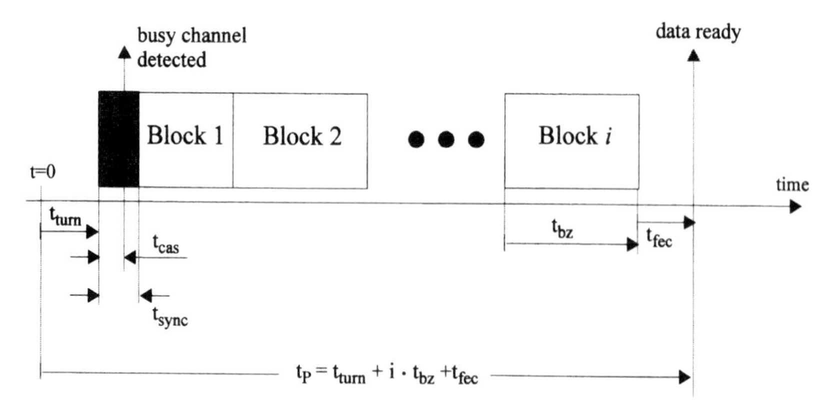

Figure 5.19 Time diagram for the transmission of a packet on the radio link. A radio switches to transmit mode at $t = 0$ and t_v time units later its neighbours detect a busy channel

decides to transmit to the time instance the signal is emitted from the antenna. Then the radio wave propagates through the air and reaches the receiver antenna after a delay determined by the speed of light. Let the sum of these delays be t_{turn}. Generally, the first part of a transmission is a dedicated bit pattern, called a preamble, with length t_{sync}. The preamble is needed to synchronise the receiver to the transmitter. A carrier-sense time, t_{cas}, is the time necessary to achieve an acceptable level of correct determination of the channel state: busy or idle. The first opportunity for a receiving radio to detect a transmission is therefore after the time delay $t_{turn} + t_{cas}$. The network receivers are unable to detect a transmission within this period and this period, defined as the vulnerable period, t_v, has great impact on the network performance. This issue will be illustrated later.

Imperfect carrier sensing causes two unwanted effects. First of all, an overlapping transmission occurs if the radio misjudges the channel state and signals idle when it is busy, denoted by the probability $p_{cas,I}$, while the network level prepares for a new transmission. The second effect addresses the cases where the radio signals busy while the channel is idle, denoted by the probability $p_{cas,B}$, leading to the abortion of an ongoing packet scheduling. However, the packet will be rescheduled for transmission at some time in the future. A high $p_{cas,I}$ does most harm because it increases the multiuser interference while a high $p_{cas,B}$ does not increase the interference[1] but only introduces an additional packet-service delay.

Practical use of radio demands implementation of a forward error correction (FEC) coder to reduce the channel bit error rate to a level suitable for data transmissions. A time delay, t_{fec}, will then pass from the last bit

[1]At least not directly, but possibly indirectly due to increased channel traffic.

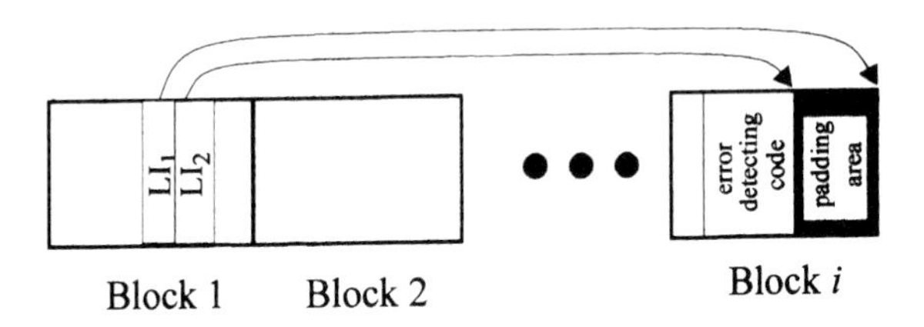

Figure 5.20 Block-oriented communication and padding

received to the time where the packet is available at the network level. We assume use of block-oriented FEC and the user packets have to be divided into an integral number of blocks. If a packet is not of correct length to fulfil this, dummy data must be filled in at the originating side and removed by the destination. A typical radio implementation provides no end-of-frame marker and then it is up to the network level to keep track of the frame length. Therefore, the network level generally needs the two length indicators (LI) in Figure 5.20. LI_1 keeps track of the number of blocks in the frame and LI_2 carries information about the information length (or equivalently, the length of the padding area). The procedure for the network level is to order the receiving radio to stop its demodulation process when the number of blocks given by LI_1 is received. If one or more bit errors have occurred in the LI_1 field, the packet length is misjudged. The LI_1 field may then take a value shorter or longer than the true length and further behaviour is determined by the radio implementation. When the entire frame length is received, an error-detection process is applied to the frame part instructed by LI_2 and with the frame error-detection code value (inserted into the frame by the source node) as an input. Corrupted frames are never delivered to the network level for further processing.

For simplicity of presentation, we define the packet length t_p to also include the radio-switching delays and the FEC delay.

The sections to follow present several examples with the main objective being to gain insight into network behaviour. To obtain meaningful comparisons it is important to use the same radio for all the cases and therefore we standardise on the particular radio parameters shown in Table 5.3. If we deviate from these, it will be explicitly stated in the text.

5.3.2 Basic characteristics

This section establishes the basis for exploration of CSMA and spread-spectrum signalling. We start by considering time instance t_1 in Figure 5.21 where a successful transmission ends. The nodes which have a data packet ready for service draw a random-access delay D and the node with the

Table 5.3 Default radio parameters

parameter	value
carrier sense delay (t_{cas})	6.4 msec
block length (t_{bz})	48 msec
synchronisation time (t_{sync})	6.4 msec
radio turn-on time (t_{turn})	3.6 msec
FEC processing delay (t_{fec})	0
CAS failure probability ($p_{cas,I}, p_{cas,B}$)	(0, 0)

shortest delay starts to transmit at t_2. The duration $(t_2 - t_1)$ is referred to as the channel-idle period, C_I. Within the vulnerable period $[t_2, t_2 + t_v]$ additional nodes may start to transmit because they have not yet detected a busy radio channel. The last transmission ends at t_3 and completes the channel-busy period, C_B. The sum of a channel-idle and a busy period is called a channel-access cycle. Let K_c be the random number of channel-access cycles needed before a successful packet delivery and X the time period between successful deliveries. From Figure 5.21 it is easy to realise that:

$$X = \sum_{i=1}^{K_c - 1} \{C_I + C_B(/failure)\} + C_I + C_B(/success) \tag{5.14}$$

If we define K_{Tx} to be the random number of simultaneous transmissions in a busy period and p_{net} as the probability of having a successful transmission then $p_{net} = P(K_{Tx} = 1)$ for a system without capture (i.e., $q = 0$). However, when $0 < q \leq 1$ the first out of k concurrent transmissions may succeed in a spread-spectrum system. If $K_{Tx} = k$, $(n - k)$ nodes are idle and the probability of addressing one of these is $(n - k)/(n - 1)$ under uniformly-distributed traffic pattern. The probability of capturing the first packet given the $(k - 1)$ overlaps, is directly given by the conditional capture probability C_k, see eqn. 5.5. Therefore, given the k transmissions,

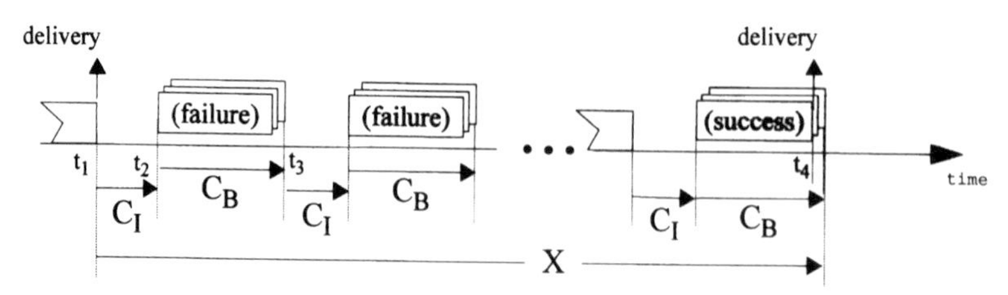

Figure 5.21 The delivery cycle X as seen by an outside observer: 'success' addresses the event that the destination receives the packet without any bit error, 'failure' is the opposite event

the probability of successful delivery is $C_k \cdot (n-k)/(n-1)$. By the total probability formula we then have:

$$p_{net} = \sum_{k=1}^{n} \frac{n-k}{n-1} \cdot C_k \cdot P(K_{Tx} = k) \tag{5.15}$$

If the transmission attempts are independent then K_c is geometrically distributed with the mean $1/p_{net}$. From eqn. 5.14 we find the first moment of X as:

$$\begin{aligned} E[X] &= (1/p_{net} - 1)(E[C_I] + E[C_B|failure]) + E[C_I] + E[C_B|success] \\ &= (E[C_I] + E[C_b])/p_{net} \end{aligned} \tag{5.16}$$

$E[C_B|failure] > E[C_B|success]$ because when failures occur the mean number of transmitting nodes will be higher. The number of packets delivered per unit time (throughput) is $1/E[X]$. We are, however, more interested in studying the MAC performance relative to the radio-transmission capacity and define the normalised throughput λ as:

$$\lambda = \frac{\textit{physical layer service time}}{\textit{MAC service time}} = \frac{t_p}{E[X]}, \quad 0 \leq \lambda \leq 1 \tag{5.17}$$

The random-access delay distribution affects p_{net}, C_I and C_B and we have to select a random-access distribution before continuing. This is discussed in the next Section.

5.3.3 Selecting the random access delay distribution

The best distribution is one which gives the nodes different access delays spaced t_v apart, giving a collision-free system without introducing channel-idle periods. With a central station to apportion the node access delays, such a system could be achieved simply by implementing an urn model having an integer number for each busy node. One number is drawn without replacement for each network node and the final result is broadcast by a single transmission to all the nodes in the fully connected net. However, network designers are likely to be challenged by fully-distributed randomly-changing network topologies.

The random-access distribution should also be fair, that is, on average all nodes should be dedicated the same share of the channel capacity. The exponential distribution is often used in the literature because its memoryless property eases the mathematical treatment. For practical use it has the disadvantage of introducing a channel-idle time which has no upper bound. This can be corrected by removing the upper tail of its density function. However, here we find it more convenient to use the uniform distribution defined by:

$$D(p_f, t_u) = p_f \cdot t_v + uniform\ distribution[0, t_u] \tag{5.18}$$

where p_f is the priority delay access factor and t_u is the upper bound for the uniform distribution. p_f is a natural number and t_u is a real number greater than zero. With this distribution, the channel-idle time cannot be longer than $p_f \cdot t_v + t_u$ if at least one node has a packet to send. Then the access delay D has the cumulative density function (c.d.f.):

$$F_D(t) = \begin{bmatrix} 0 & \text{for } t < p_f \cdot t_v \\ \frac{1}{t_u}(t - p_f \cdot t_v) & \text{for } t \in [p_f \cdot t_v, t_u + p_f \cdot t_v] \\ 1 & \text{for } t > t_u + p_f \cdot t_v \end{bmatrix} \tag{5.19}$$

With the access-delay distribution in place we can continue with the analysis, first by focusing on the channel-idle time C_I. Consider time instances t_1 in Figure 5.21. In a lightly-loaded network all nodes are usually idle at t_1 and the first node gets a packet at $t_1 + 1/\Lambda$, cf. Figure 5.18, draws a random-access delay D with an average $t_u/2 + p_f \cdot t_v$ and transmits at $t_1 + 1/\Lambda + t_u/2 + p_f \cdot t_v$. To analyse the network under heavy load is more interesting because it is in this state that the network designers meet the real challenges. This also simplifies the forthcoming analysis. We place the condition that all nodes have a packet under service and thus n nodes independently draw their random-access delay at time instance t_1. The smallest of these is the first node to transmit and therefore:

$$F_{C_I}(t) = P\left(Min\left[\underbrace{D, \ldots, D}_{n}\right]\right) = 1 - [1 - F_D(t)]^n \tag{5.20}$$

The average channel-idle period is then:

$$\begin{aligned} E[C_I] &= \int_0^\infty [1 - F_{C_I}(t)]dt \\ &= \int_0^\infty [1 - F_D(t)]^n dt \\ &= \frac{t_u}{n+1} + p_f \cdot t_v \end{aligned} \tag{5.21}$$

To find the length of the busy period is more complicated but [38] shows that its average length is:

$$E[C_B] = t_v \sum_{k=1}^{n} \frac{k-1}{k} P(K_{Tx} = k) + t_p \tag{5.22}$$

for fixed-sized packets of length t_p. Now consider the number of simultaneous transmissions K_{Tx}. Let $S_1 \leq S_2 \leq \ldots \leq S_n$ be the order statistics obtained by ordering the node-scheduling intervals (random-access delays). The node getting the smallest scheduling interval starts to transmit after the delay S_1, see Figure 5.22.

If $S_1 = t$ and $S_k \in [t, t + t_v]$ then at least k simultaneous transmissions have occurred. Exactly k simultaneous transmissions occur when:

(*a*) one of n nodes schedules a transmission at t;
(*b*) $k - 1$ nodes schedule a transmission within $[t, t + t_v]$; and
(*c*) the remaining $(n - 1) - (k - 1)$ nodes wait until after $t + t_v$.

All nodes behave independently up to $t + t_v$ and therefore the probability of event (b) is $[F_D(t + t_v) - F_D(t)]^{k-1}$ and the probability of event (c) is $[1 - F_D(t + t_v)]^{n-k}$ giving

$$P(K_{Tx} = k) = \binom{n-1}{k-1} \int_0^{\infty} [F_D(t + t_v) - F_D(t)]^{k-1} \times [1 - F_D(t + t_v)]^{n-k} n f_D(t) dt \tag{5.23}$$

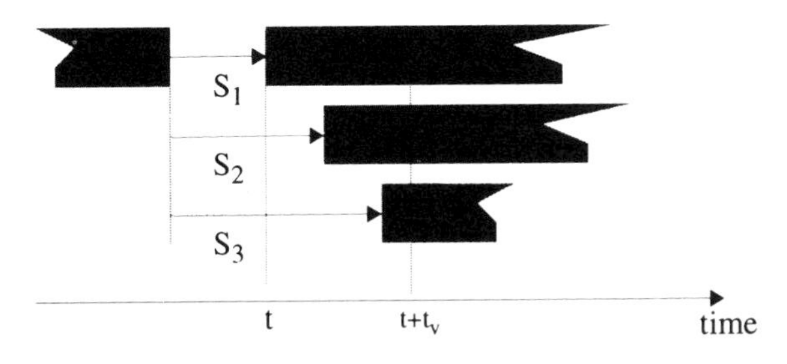

Figure 5.22 The first node transmits at t. At $t + t_v$ all the nontransmitting nodes detect a busy channel

From this we get:

$$P(K_{Tx}=k)=n\binom{n-1}{k-1}\int_0^{p_f t_v} 0dt$$

$$+n\binom{n-1}{k-1}\int_{p_f t_v}^{t_u+(p_f-1)t_v}\left(\frac{t_v}{t_u}\right)^{k-1}[1-(t-(p_f-1)t_v)/t_u]^{n-k}/t_u dt$$

$$+n\binom{n-1}{k-1}\int_{t_u+(p_f-1)t_v}^{t_u+p_f t_v}[1-(t-p_f t_v)/t_u]^{k-1}0^{n-k}/t_u dt$$

$$=\binom{n}{k-1}\left(\frac{t_v}{t_u}\right)^{k-1}\left(1-\frac{t_v}{t_u}\right)^{n-k+1}+\left(\frac{t_v}{t_u}\right)^{n}0^{n-k}\quad \text{for } 1\le k\le n \tag{5.24}$$

The probability of having a collision P_{coll} when $n > 1$ is simply:

$$P_{coll}=1-P(K_{Tx}=1)=1-\left(1-\frac{t_v}{t_u}\right)^n \tag{5.25}$$

This shows that the ratio t_v/t_u must be small to have few collisions, and as the radio designer increases t_v the network designer has to increase t_u. A useful result is the expected number of simultaneous transmissions which is simply given as $E[K_{Tx}]=\sum_{k=1}^{n} k\cdot P(K_{Tx}=k)$. This expression indirectly expresses the level of multiuser interference to be expected and will also serve other purposes in the forthcoming text. By inserting eqn. 5.24 into this expression, straightforward manipulation yields:

$$E[K_{Tx}]=1+n\cdot\frac{t_v}{t_u}-\left(\frac{t_v}{t_u}\right)^n \tag{5.26}$$

The remaining issue is how to select values for p_f and t_u. The only effect of p_f we can see so far is via eqn. 5.21 which shows that p_f has only the bad effect of increasing the channel-idle time. Therefore, we set $p_f = 0$ in the remainder of this Section. After having introduced the LLC layer in Section 5.4, the motivation for introducing p_f becomes clearer. Before closing this section, we give some examples illustrating the effect of different parameters.

Example 1

The objective of this example is to compare our protocol, hereafter referred to as the CSMA/CC (collision control) protocol, with the three classical access protocols: ALOHA, slotted ALOHA and nonpersistent CSMA, see [26]. The classical works consider a zero-capture system

($q = 0$), a radio with zero turn time ($t_{turn} = 0$) and an infinite-user population. The basic assumption taken by the classical works is that the generated channel traffic is Poisson distributed with the rate Λ and normalised to the fixed-sized user packet length t_p. In our context, the equivalent offered radio-channel traffic is $\Lambda = n \cdot t_p/(t_u/2 + t_p)$ because it is the uniformly-distributed access delay D with the average $t_u/2$ that controls the offered traffic. The network throughput is $\lambda = \Lambda \cdot p_{net}$ and p_{net} for the three classical protocols is given by:

$$ALOHA: \qquad p_{net} = e^{-2\Lambda}$$

$$slotted\ ALOHA: \qquad p_{net} = e^{-\Lambda}$$

$$nonpersistent\ CSMA: \qquad p_{net} = e^{-\Lambda \cdot \frac{t_v}{t_p}} \Big/ \left[\Lambda(1 + 2t_v/t_p) + e^{-\Lambda \cdot \frac{t_v}{t_p}}\right]$$

For the CSMA/CC protocol we have $\Lambda = t_p/(E[C_I] + E[C_B])$ and $p_{net} = (1 - t_v/t_u)^n$ (for $n > 1$ and $q = 0$), cf. eqns. 5.15 and 5.24. Throughput against the access-delay parameter t_u is plotted in Figure 5.23 for the packet length $t_p = 0.336$ s, the vulnerable period $t_v = 0.01$ s and the net size $n = 50$.

The effect of increasing the t_u-value is to force the nodes to have longer idle periods between the packet transmissions. Decreasing t_u leads to increased channel traffic and more frequent collisions. The Figure shows that ALOHA and slotted-ALOHA systems must operate with longer channel-idle periods because they have no carrier-sense function and thus

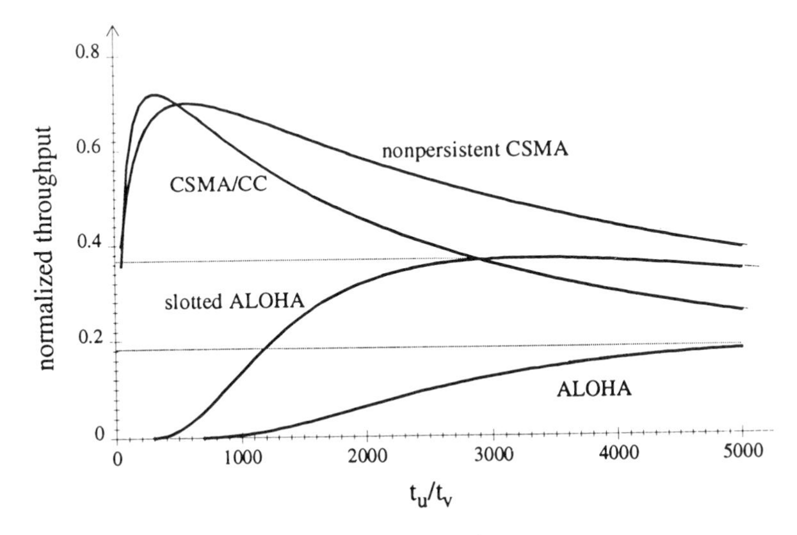

Figure 5.23 Network throughput as a function of t_u/t_v

the vulnerable period is determined by the packet length. Maximum throughputs for ALOHA and slotted ALOHA are $1/2e$ and $1/e$, respectively.

It is very important to be aware of the fact that the throughput peaks shown in this Figure are only attainable in a system exercising perfect control between t_u and the network load level. This is impossible in a real network for two reasons: the user traffic normally changes over time and thus the offered traffic is time variant, and hidden nodes are usually present in real networks. The latter point will be considered in later sections and we note for these reasons that the maximum throughput values shown represent optimistic results.

Example 2

The primary concern with the random-access technique is the high multiuser interference level that appears if the network-level protocols fail to regulate the use of the radio channel in an orderly manner. The starting point for discussing this issue is to consider the relationship between the spatial distribution, the radio BER characteristic and the network protocol combined into one single expression giving the probability of having a successful transmission, namely:

$$p_{net} = p_{noise} \sum_{k=1}^{n} \frac{n-k}{n-1} \cdot q^{k-1} \cdot P(K_{Tx} = k) \tag{5.27}$$

which is the result of inserting the capture model of eqn. 5.2 into eqn. 5.15. For a network case which can best be approximated by a zero capture system $q = 0$ where any overlapping transmission leads to bit errors, we have:

$$p_{net} = p_{noise} \cdot (1 - p_{coll}) = p_{noise} \cdot (1 - t_v/t_u)^n \tag{5.28}$$

which shows that the packet-success probability grows linearly with the probability of not having a collision. The MAC protocol can use a t_u value such that the average probability of collision, p_{coll} (eqn. 5.25), becomes arbitrarily small as illustrated by Figure 5.24. However, it is important to be aware that this does not give maximum throughput. This is an issue to be considered later.

Consider one particular node which schedules its packet for transmission at time instance t. This node transmits at t if and only if none of the other nodes transmit before $t - t_v$, because if they do, the carrier-sense signal will be high at t and the tagged node keeps off the channel. All nodes act independently in the interval $[t - t_v, t]$ and observed by the tagged node, the probability that it will be among the transmitting nodes is simply k/n

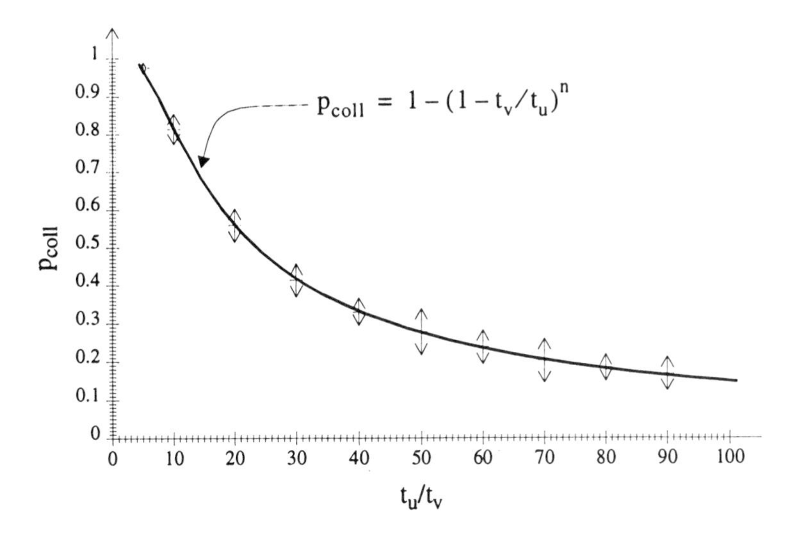

Figure 5.24 Probability of collision as a function of the MAC parameter t_u. The network size $n = 16$ and $t_v = 10$ ms. Simulated results are marked by arrows

when exactly k nodes transmit. Hence, the probability that one particular node transmits is $k \cdot P(K_{Tx} = k)/n$ and its expectation p_{Tx} becomes:

$$p_{Tx} = E[K_{Tx}]/n = \frac{1}{n} + \frac{t_v}{t_u} - \frac{1}{n}\left(\frac{t_v}{t_u}\right)^n \tag{5.29}$$

The last equality follows directly from eqn. 5.26. With $n = 1$, a single node net, we have $p_{Tx} = 1$ because the node cannot lose. If $t_v/t_u = 0$, $p_{Tx} = 1/n$ which is intuitively correct because exactly one node transmits at a time and each node gets the same share of the channel.

In an idealised system having perfect capture ($q = 1$) the first transmission always succeeds regardless of the multiuser interference level. For such a system eqn. 5.27 gives:

$$p_{net} = p_{noise} \cdot \frac{n - E[K_{Tx}]}{n-1} = p_{noise} \cdot \frac{n}{n-1} \cdot (1 - p_{Tx}) \tag{5.30}$$

For a system with perfect capture the packet-success probability decreases linearly with increasing $E[K_{Tx}]$. Figure 5.25 illustrates the progress of the average number of simultaneous transmissions as a function of the MAC parameter t_u.

Example 3

Although we have already commented that the countermeasure against a long radio synchronisation delay is to increase the t_u parameter, we have so

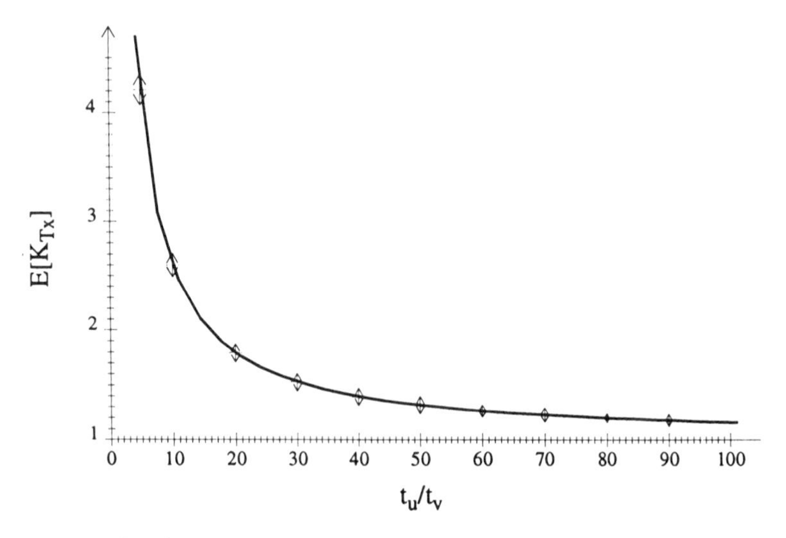

Figure 5.25 $E[K_{Tx}]$ *as a function of the MAC parameter* t_u*. The network size* $n = 16$ *and* $t_v = 10$ *ms. Simulated results are marked by arrows*

far not quantified this statement. This example considers throughput as a function of t_u/t_v. As t_u/t_v decreases, the number of simultaneous transmissions increases, cf. eqn. 5.26, the average network SNR level becomes low and the receivers fail to demodulate incoming packets. If $t_u < t_v$, all nodes transmit at any point in time under the heavy-load case and the network collapses. A 16-node network and a radio with zero capture forms the basis for the throughput curve in Figure 5.26. $t_{turn} = 0$ and $t_{fec} = 0$ are used such

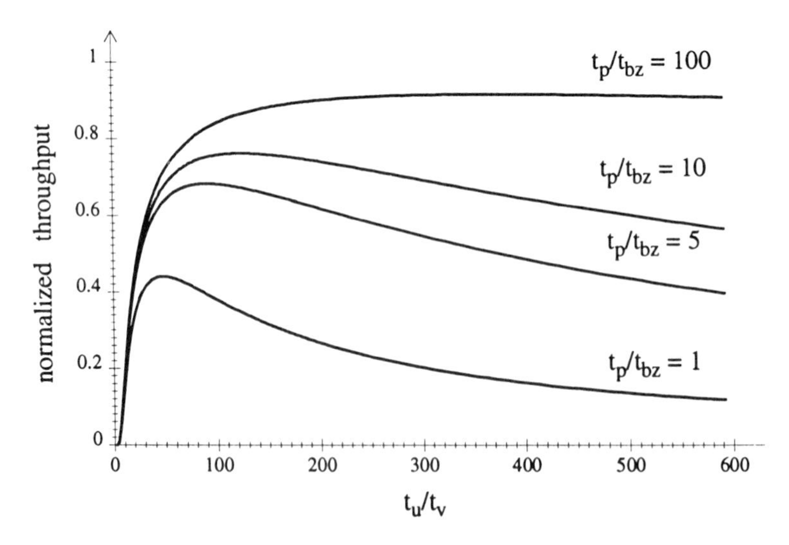

Figure 5.26 *Throughput as a function of* t_u/t_v *and the number of blocks per packet* t_p/t_{bz}

that t_p/t_{bz} expresses the number of blocks per packet. A t_u/t_v value giving maximum throughput for each packet length exists and this value sets the optimum balance between the channel-idle time and the number of simultaneous transmissions. A large t_u/t_v value gives few collisions but channel bandwidth is wasted owing to long channel-idle periods. On the other hand, small t_u/t_v values give short idle periods but more frequent collisions. With increasing packet lengths, the t_u/t_v value must be increased to maintain maximum throughput. The loss of a long packet represents more waste of channel bandwidth at each loss than short packets do, and therefore the network must be dimensioned to operate at a lower collision rate.

Example 4

With spread spectrum a question arises concerning the advantages of spreading to combat multiuser interference. On the asset side we have improved packet-capture probability which increases with the processing gain, G_p. The drawback is the longer time needed to transmit packets due to the reduction in the bit transfer rate if we are not allowed to extend the RF bandwidth accordingly. However, with our MAC protocol we have the option of increasing the access delay to reduce the multiuser interference. We initially neglect the increased transmission time and plot the network throughput as a function of t_u while selecting the processing gain G_p from the set $\{-\infty, -10, 0, 10, \infty\}$ [dB]. This set gives the corresponding q-value set $\{0, 0.010, 0.190, 0.617, 1\}$ for the nine-node net in example 2 of Section 5.2. Figure 5.27 presents the normalised network throughput for the packet length $t_p = 195.6$ ms. We have used $p_{noise} = 1$ because our MAC protocol has no ability to combat background noise.

The throughput remains nearly the same for G_p values below 0 dB and the net is best modelled as a zero-capture system. When the processing gain increases from 0 to 10 dB, the maximum throughput increases from 0.68 to 0.75. The system can be dimensioned to operate at a higher collision rate and a shorter channel-idle time by reducing t_u.

If we are granted a fixed RF bandwidth, the resulting packet length on the channel becomes $t_{turn} + 10^{G_p/10} \cdot t_{bz} \cdot E[n_P]$ (when G_p is given in dB and n_p is the number of blocks in the packet) because the block transmission time increases. Spread-spectrum signalling introduces longer channel-busy time periods, although each packet delivered contains the same amount of information (bytes). The dotted curve in Figure 5.27 shows the resulting throughput for $G_p = 10$ dB when this fact has been taken into account. The maximum throughput is only 0.0927 (occurs at $t_u = 0.93$ s). The optimum t_u value has increased because the packet transmission time

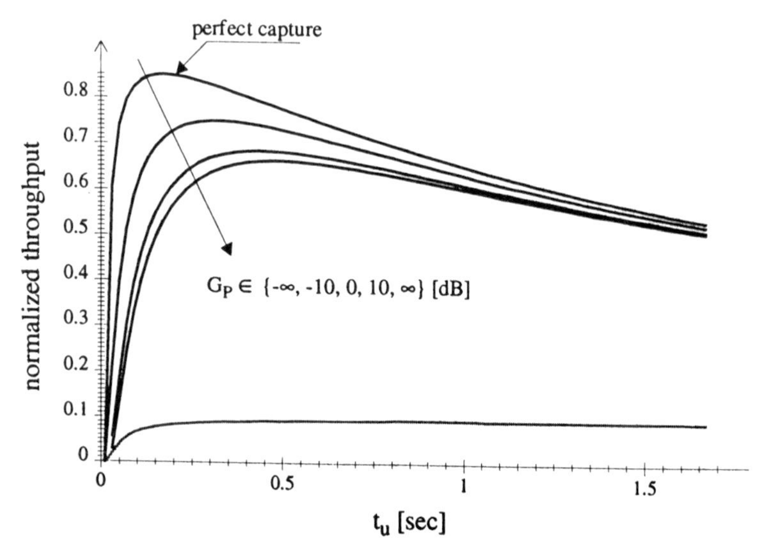

Figure 5.27 Network throughput for different processing-gain values

has increased from 0.1956 to 1.956 s and therefore each packet loss represents more waste of channel capacity. For the case considered we conclude that the best utilisation of the given spectrum is to use $G_p = 0$ dB because the capture probability does not grow fast enough to compensate for the enlarged block transmission time.

5.4 Logical link control

Most networks have to provide reliable delivery of data packets. This necessitates techniques which reduces the packet-loss probability across the radio links. There are basically two techniques for improving the radio-link success rate, the automatic-repeat request (ARQ) scheme and the forward-error correction (FEC) scheme. The ARQ scheme is used in conjunction with an error-detecting code. With the use of radio, both techniques are normally applied. An FEC scheme is implemented at radio level to bring the radio gross bit error rate to a level suitable for performing packet switching. Packets containing too many bit errors to be corrected by the FEC scheme will be retransmitted by the ARQ protocol at the link level until they are correctly received at the destination. It is intended in this Section to discuss alternative ARQ schemes for operation in a random-access network and to consider their relative performance. We look at the effects of multiuser interference and noise, with and without using a spread-spectrum technique. This section will use the same network model as in Section 5.3 and Figure 5.18, which results in homogeneous node behaviour.

5.4.1 ARQ protocols

This Section starts with a short introduction to ARQ protocols and then turns to a discussion of their relative performance. A more complete description of ARQ protocols can be found in e.g. [36] or [37]. Traditionally, ARQ protocols have been divided into three basic types:

(i) stop and wait (SW);
(ii) go back N (GBn);
(iii) selective repeat (SRn).

Under the SW scheme the sending entity is forced to wait for an acknowledgement for each data packet sent. The major advantage is simplicity and that no buffering is needed at the receiving end. Figure 5.28 illustrates the basic behaviour of this scheme where the LLC protocol entity A uses the MAC service to send data packets. To keep track of which packets have been acknowledged, each packet is labelled either 0 or 1. The receiver side B acknowledges a received error-free packet by sending an acknowledgement assigned a sequence number of the next expected data packet.

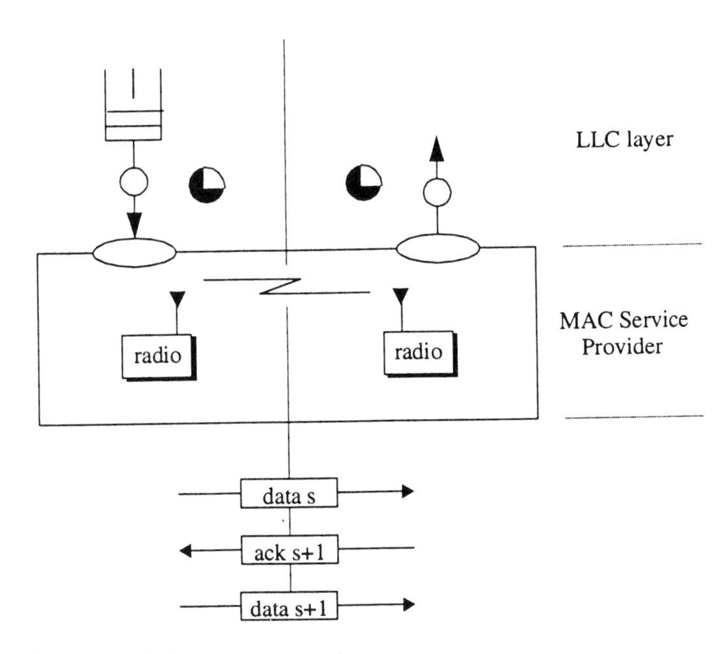

Figure 5.28 Model of the window protocol

Assume that A has just sent a data packet assigned the sequence number s. One of the following events may happen:

(i) data packet s is successful and ack $s + 1$ is successful;
(ii) data packet s is successful and ack $s + 1$ is corrupted;
(iii) data packet s is corrupted;
(iv) data packet s is successful and ack $s + 1$ is returned too late.

(i) is the preferred event where everything works nicely. Events (ii) and (iii) address the cases where bit errors occur and the received packet has to be discarded. In both cases A will wait forever for the acknowledgement from B. Whenever A sends a packet it has to start a timer. If this timer expires, A must retransmit the data packet previously transmitted and the deadlock situation is resolved.

When using a random-access scheme, access to the transmission channel is granted at random points in time, and once in a while the retransmission timer expires even if the data packet was received free from errors. Event (iv) addresses this case. Now A retransmits the data packet which already has been successfully delivered and the protocol efficiency decreases. Thus the retransmission timer has to be scaled according to the operating environment in such a way that event (iv) becomes rare, but not set so large that excessive delay is introduced when the events (ii) or (iii) happen.

A natural extension of the SW protocol is to allow transmission of more than one data packet before awaiting an acknowledgement. GBn and SRn facilitate up to n outstanding data packets and one single acknowledgement packet may even confirm successful delivery of more than one data packet. The difference is that the SRn scheme adds more control fields to data and acknowledgement packets, giving the receiving side the possibility of signalling which packet(s) are missing among the set of outstanding packets. Only these selected packets are repeated.

The MAC protocol analysis in Section 5.3 has clearly illustrated that packet collisions lead to performance degradation. Any mechanism that can reduce this collision rate has the potential of giving increased network performance. Loss of acknowledgement packets is particularly costly because losses introduce retransmission of data packets already received. The MAC layer priority service, described in Section 5.3.3, gives the opportunity of assigning extra access delays to data packets through the parameter p_f in eqn. 5.18. By using a p_f value which is slightly larger than the vulnerable period t_v for data packets and zero for acknowledgement packets, the acknowledgements are always sent first. In a system where only one data packet can be successfully received at a time, only one acknowledgement will also be scheduled for transmission and we have achieved collision-free transmission of acknowledgements. The operation

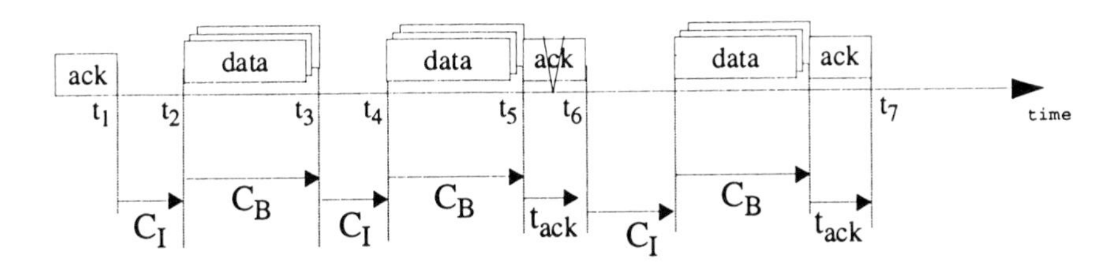

Figure 5.29 Illustration of the SWIA link protocol

of this scheme, to be referred to as the stop and wait with immediate acknowledgement (SWIA) protocol, is illustrated in Figure 5.29.

The system is entered at the completion of a successful acknowledgement packet where all the data packets in the system have been acknowledged at time t_1 in the Figure. The busy nodes draw a random-access delay D, see Section 5.3.3, and the winning node transmits at t_2. The busy period ends at t_3 but none of the data packets succeed. New attempts take place at t_4 and at t_5 one is successfully delivered. At t_5 an acknowledgement is pending and is given the access delay $D(p_f = 0, t_u = 0)$. The data packets are given the access delay $D(p_f = 1, t_u > t_v)$. The acknowledgement packet is therefore transmitted immediately. This transmission is completed at t_6 but a noisy radio channel leads to corruption. New scheduling attempts are repeated until one data packet and its corresponding error-free acknowledgement is returned, marked as t_7 in the Figure.

The question naturally arising is the relative performance of the schemes SWIA, SW, GBn and SRn. Which of them offers the highest throughput performance? Analytical treatment of the SWIA scheme needs only modest extensions to the throughput analysis in Section 5.3. This is, however, not the case for the other schemes and therefore a simulation model is used under the following performance comparison. The network configuration taken as the basis for this comparison is identical with the 16-node network used by example 3 in Section 5.2. The other fixed parameters of importance are shown in Table 5.4. The link-protocol alternatives also differ in the protocol-control information size needed but the differences are so small that this can be neglected.

The SWIA protocol can schedule data packets for retransmission immediately after the transmission has ended since the acknowledgement is given precedence. If the other protocols use the same strategy, many unnecessary retransmissions take place because the acknowledgement competes with the data packets and thus can generally first be returned after some time.[1]

[1]The average delivery cycle time delay $E[X]$ is given by eqn. 5.16 in a system without ARQ. If the retransmission delay is set higher than $2 \cdot E[X]$ then sufficient delay should have been introduced to prevent significant loss of capacity due to unnecessary retransmissions.

Table 5.4 Parameter values for the different ARQ protocols

	Parameter	LLC protocol variant	
		GB2, SR2, SW	SWIA
Network data	user traffic:		
	packet-arrival distribution	infinite	
	packet-length distribution	fixed	
		$t_{p,data} = t_{turn} + 4t_{bz} = 195.6$ ms	
	retransmission timer	8 s	0 s
	acknowledgement packet	$t_{p,ack} = t_{turn} + t_{bz} = 51.6$ ms	
Radio data		Table 5.3	

The difference in performance between the four link-protocol candidates is a function of the radio-link quality and for this reason both the background noise and the multiuser interference are taken as simulation variables. The simulation model implements the capture model of eqn. 5.12 and the background noise is simply controlled through the p_{noise} parameter while the multiuser interference level is set by changing the access-delay parameter t_u. Both a perfect-capture $(q = 1)$ and a zero-capture $(q = 0)$ system are included. By these variables we encompass a wide range of operating conditions.

The network traffic is scaled to give a heavy-load condition (i.e. every node has a packet to send at any instant), thus the data packets experience the same multiuser interference level in all the four cases, and the normalised LLC throughput is estimated over the wide range of simulation variables shown in Figure 5.30.[1] The simulated results show that the SWIA is superior.

The additional delay introduced by the SWIA protocol when serving data packets, is compensated for by having error-free acknowledgement packets. Increased performance as a result of permitting a larger number of outstanding packets must come from reduced acknowledgement traffic, caused by the event: one acknowledgement packet confirms the receipt of more than one data packet. As the radio-channel quality deteriorates, the probability of experiencing this event decreases and no gain is observed. Both GB2 and SR2 require more protocol-control information than the two others so their performance is actually poorer than stated. We register that the optimum t_u-values are as follows:

[1]Normalised LLC throughput is defined as being identical with the definition of the normalised MAC throughput of eqn. 5.17.

	$q=0$	$q=1$
SWIA	$(p_f = 1, t_u = 0.81$ s) for data	$(p_f = 1, t_u = 0.21$ s) for data
GB2, SR2, SW	$(p_f = 0, t_u = 1.0$ s) for data and acks	$(p_f = 0, t_u = 0.5$ s) for data and acks

Note the higher t_u values for the GB2, SR2 and SW protocols which mean that they should operate under a lower collision rate. A second observation is the decreasing optimum t_u value as q increases. A system with a strong capture effect should be operated at a lower average channel-idle period and at a higher collision rate because every overlapping transmission does not always lead to packet loss.

Figure 5.31 presents the performance results with background noise when using these t_u values. $q = 1$ represents a case where the radio can suppress any number of overlapping transmissions but is still affected by background noise, or foreign interfering sources (through p_{noise}). We conclude that the SWIA is the preferred link protocol and the next Section analyses this protocol in more detail.[1]

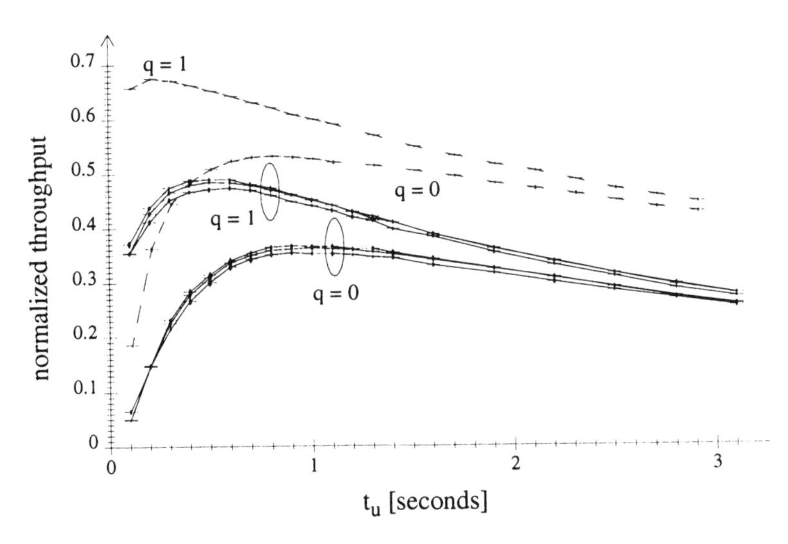

Figure 5.30 Throughput against the access-delay parameter t_u. SWIA is shown as dotted lines and SW, GBn and SRn as solid lines

[1]A real radio requires an additional time delay, the FEC processing delay, after a packet receipt prior to delivering the packet to the network level. In contrast to SW, GBn and SRn, the radio channel cannot be used by the other network nodes during this period of time under SWIA. The SWIA scheme will consequently deteriorate faster than the three other methods as the FEC processing delay increases.

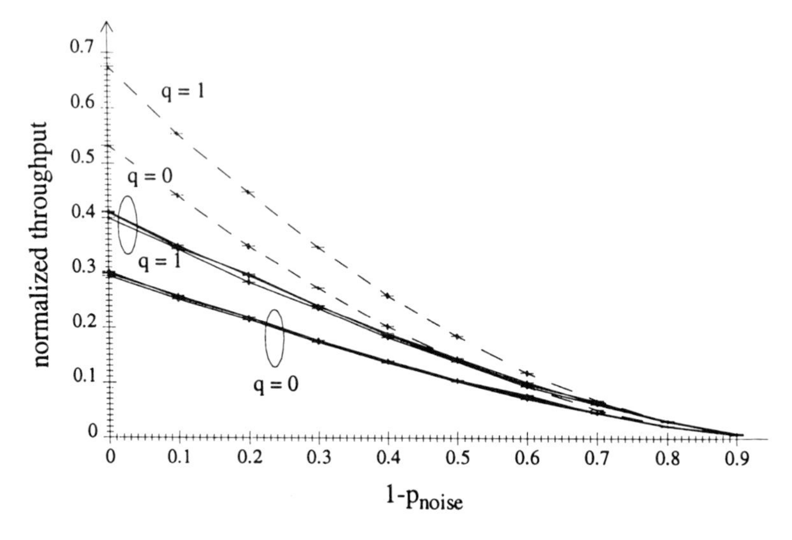

Figure 5.31 *Throughput against the background noise. SWIA is shown as dotted lines, and SW, GBn and SRn as solid lines*

5.4.2 The characteristics of the SWIA protocol

Section 2.1.4 on user traffic and performance measures introduced a general network model for packet-switched data traffic and discussed the network throughput and transit delay as a function of the offered traffic. It is important to take a closer look at these delay components to gain insight into the network behaviour since the network transit delay is the sum of other random delays. This Section takes this important issue one step further by performing analysis of networks in a saturated state.

The analysis in Section 5.3 exhibited an unstable MAC protocol. With the use of an ARQ scheme, a very strong feedback between the radio-channel quality and the offered channel traffic arises because lost packets lead to retransmissions. Even at small user-traffic levels high channel load can be observed due to the combination of a degraded radio channel and an ARQ protocol. Poor channel quality is not necessarily caused by any external source but may be due to the network protocol itself if it fails to regulate the multiuser interference to an acceptable level. From the network designer's point of view, the multiuser interference level can best be regarded through the probability of collision (cf. eqn. 5.25) given by:

$$p_{coll} = 1 - \left(1 - \frac{t_v}{t_u}\right)^n \tag{5.31}$$

This equation shows that t_u must be increased when the number of users n increases to maintain an acceptable interference level. At the time of network deployment, the network is initialised to some p_{coll} value by selecting an appropriate t_u value. A real net must allow new users to enter after the end of the network-deployment phase yet still maintain stability. However, a network function which regulates p_{coll} to the new situation is easy to implement because late entries usually require exchange of a management packet between this new node and the old nodes. Thereby the nodes become aware of the network state change. A much more difficult task is to track the events of sudden user-traffic changes. User traffic is commonly time variant, meaning that the packet-arrival distribution, packet-length distribution and the traffic-pattern distribution are functions of time. In a typical situation several nodes will be competing in each contention period. To deliver packets with low delays the network must use a short average-access delay (remember that the average channel-idle period is given by eqn. 5.21). With a sudden increase in the user traffic, the number of active nodes increases, leading to increasing p_{coll} if the access-delay parameter t_u is not increased accordingly.

Figure 5.32 depicts the components forming the network transit delay as the sum of the queuing delay Q and the LLC protocol entity serving time delay, X_{llc}.

An incoming user packet arriving while another is under service has to be queued for a time Q and a first-in-first-out (FIFO) queuing discipline is usually employed. A packet arriving at an empty node but during a channel-busy period, sees the time delay given by the residual channel-busy period (H). However under the heavy-load assumption, the queue is never empty and thus $H = 0$. The time delay X_{llc} represents the time delay that a packet experiences before it arrives successfully at the destination node and includes one or more scheduling attempts as well as any retransmissions. It is important to notice that the X_{llc} delay is not the time delay between packet removal from the queue. The latter will be higher. The reason for this will be explained in what follows. X_{mac} represents the time delay between the instances when a node grabs the channel and when it transmits.

The analysis starts by addressing the four major events in Figure 5.33 as seen by one particular node:

E_0: the tagged node completes its delivery cycle by receiving an error-free acknowledgement packet from the destination node; from this point in time a new packet can be taken under service;

E_1: the tagged node transmits the data packet but it is lost;

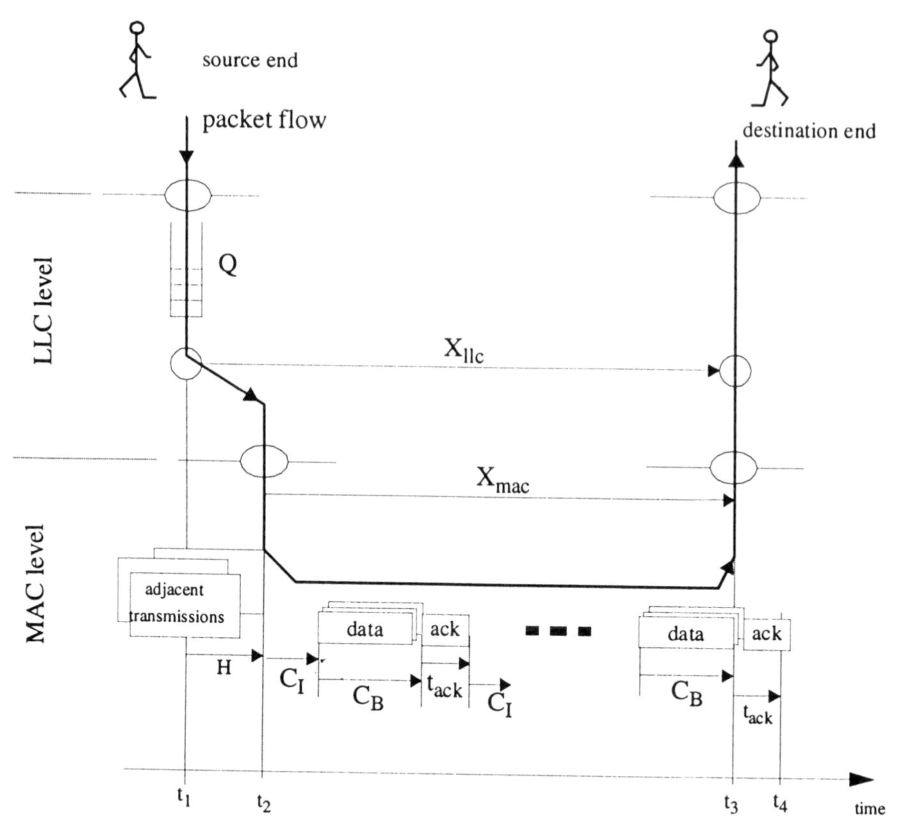

Figure 5.32 A hierarchical model for presenting the network transit delay components

E_2: the tagged node transmits the data packet which is successfully delivered to the destination;

E_3: the addressed node has received the data packet but the return of the acknowledgement fails.

First, we focus on the MAC service time from time t_0 in the Figure. Under the heavy-load assumption the tagged node has always a packet waiting for service and immediately takes it into service. The tagged node must compete for channel access and wins with the probability p_{tx}, see eqn. 5.29. An extra delay $\overline{C}_I + \overline{C}_B + p_{net} \cdot t_{p,ack}$ is added each time it loses.[1] When the node finally transmits, the MAC service time completes at t_5 such that:

$$\begin{aligned}\overline{X}_{mac} &= (1/p_{tx} - 1)(\overline{C}_I + \overline{C}_B + p_{net} \cdot t_{p,ack}) + \overline{C}_I + \overline{C}_B \\ &= (\overline{C}_I + \overline{C}_B)/p_{tx} + (1/p_{tx} - 1)p_{net} \cdot t_{p,ack}\end{aligned} \tag{5.32}$$

[1]This section contains longer equations than the previous section; therefore the previously used expression for averages (E[ ]) is substituted by the more compact notation.

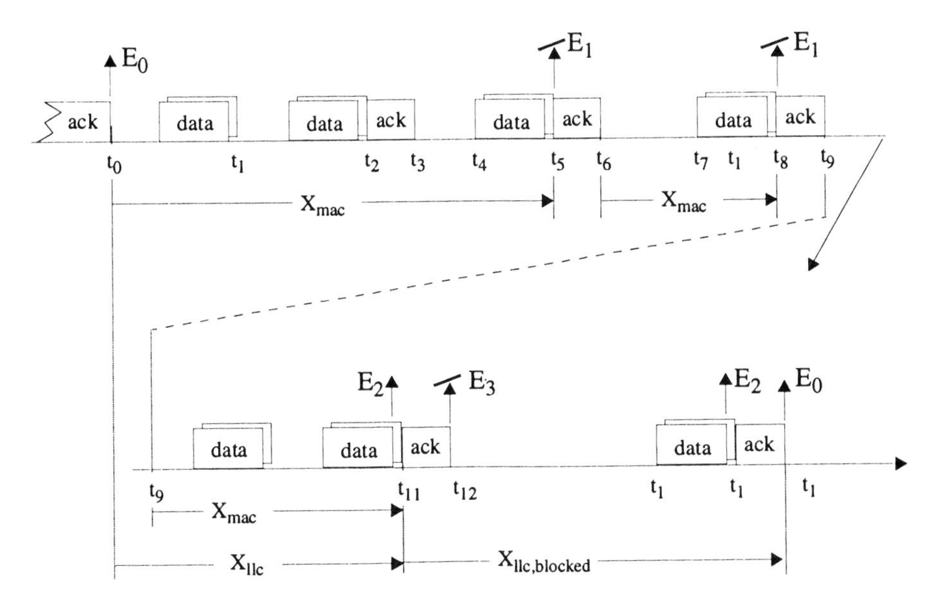

Figure 5.33 Detailed view of the link-layer-service time delays

This is based on the fact that the scheduling attempts are independent and that $1/p_{tx} - 1$ gives the average number of scheduling attempts needed before the tagged node transmits. After the time delay X_{mac}, the packet has been transmitted once on the link and the probability of successful delivery on a random link $i \to j$ is:

$$p_{succ,i\to j} = p_{net}/\overline{K}_{Tx} \tag{5.33}$$

since the probability of having a successful packet over the network is p_{net} from eqn. 5.30, and when the tagged node transmits it has the same probability of succeeding as the other transmitting nodes.

In the Figure, t_4 marks the point where the tagged node transmits the packet for the first time but unfortunately fails. However, another node succeeds in delivering a data packet and consequently an acknowledgement packet will always follow. The tagged node retransmits the packet for the first time at t_7 and hence the packet experiences two additional delay components, the acknowledgement packet ending at t_6 and the X_{mac} delay. This leads to the average LLC service time:

$$\overline{X}_{llc} = (\overline{X}_{mac} + p_{net} \cdot t_{p,ack})(1/p_{succ,i\to j} - 1) + \overline{X}_{mac} \tag{5.34}$$

By eqns. 5.29 and 5.33 this can be simplified to:

$$\overline{X}_{llc} = \frac{n \cdot p_{Tx}}{p_{net}} \overline{X}_{mac} + (n \cdot p_{Tx} - p_{net})t_{p,ack} \tag{5.35}$$

If the expression for X_{mac} (eqn. 5.32) is inserted, we have:

$$\overline{X}_{llc} = \frac{n}{p_{net}}(\overline{C}_I + \overline{C}_B) + (n - p_{net})t_{p,ack} \tag{5.36}$$

A complicating factor when determining the link-level throughput is caused by event E_3 for which the data packet succeeds but the acknowledgement fails. With the SWIA protocol, only background noise can cause this event and therefore the event E_3 does not occur if $p_{noise} = 1$. At the occurrence of event E_3, the node cannot take a new packet under service before it has received the acknowledgement (the transmitter window size is one). Let $\overline{X}_{llc,blocked}$ be the time period during which the tagged node is blocked. It is easy to realise from Figure 5.33 that history repeats itself from t_{12} after the acknowledgement is sent, and we can write:

$$\begin{aligned}\overline{X}_{llc,blocked} &= t_{p,ack} + (1/p_{noise} - 1)(\overline{X}_{llc} + t_{p,ack}) \\ &= (1/p_{noise} - 1) \cdot \overline{X}_{llc} + t_{p,ack}/p_{noise}\end{aligned} \tag{5.37}$$

The average time between packet removal from the queue is $(\overline{X}_{llc} + \overline{X}_{llc,blocked})$. By normalising the LLC throughput in the same way as for the MAC throughput in eqn. 5.17, the LLC normalised throughput for node i can be written as:

$$\begin{aligned}\lambda_i &= t_{p,data}/(\overline{X}_{llc} + \overline{X}_{llc,blocked}) \\ &= p_{noise} \cdot t_{p,data}/(\overline{X}_{llc} + t_{p,ack})\end{aligned} \tag{5.38}$$

This is the throughput per node and since the network throughput λ is simply the sum of the throughput for each node, we have:

$$\begin{aligned}\lambda &= n \cdot \lambda_i \\ &= p_{noise} \cdot p_{net} \cdot t_{p,data}/(\overline{C}_I + \overline{C}_B + p_{net}(n + 1 - p_{net})t_{p,ack}/n)\end{aligned} \tag{5.39}$$

A number of factors have been neglected in generating this result. For example, the tagged node sees a different $(C_I + C_B)$ distribution when it loses the access attempt and when it becomes a transmitting node. One objective of the first example below is to check the significance of the approximations used, and the following examples discuss other interesting performance issues.

Example 1

The objective of this example is twofold: first, to check the validity of the approximations used for generating the basic results in eqns. 5.32, 5.36 and 5.39 for the heavy-load state, and secondly, to widen our insight into the relationship between the access delay D and the layer-service time delays. Figure 5.34 presents simulated and theoretical results for the heavy-load case under a wide range of multiuser interference levels. The collision rate increases with decreasing t_u and many retransmissions are required for each packet delivered. The LLC service time therefore becomes longer

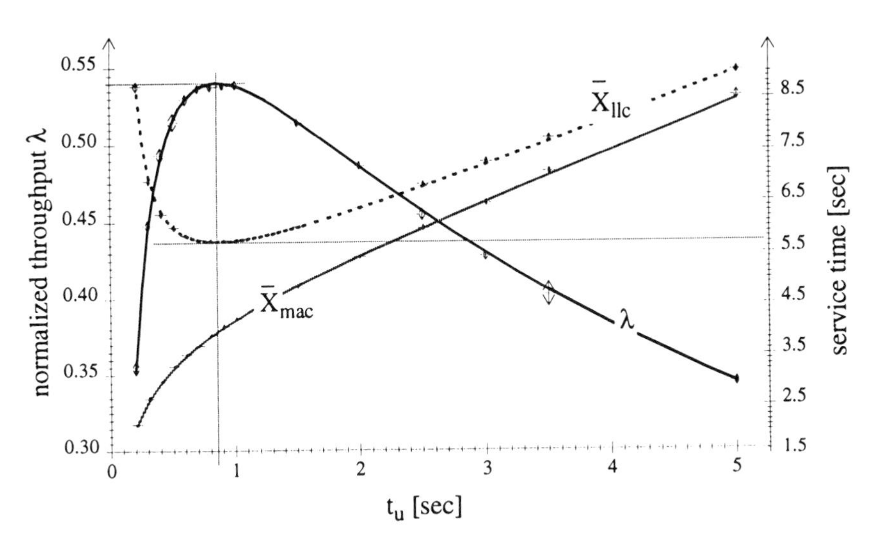

Figure 5.34a Throughput and service times against the access-delay parameter t_u for $n = 16$ and $p_{noise} = 1$. Simulated results at 90% confidence intervals marked by vertical arrows

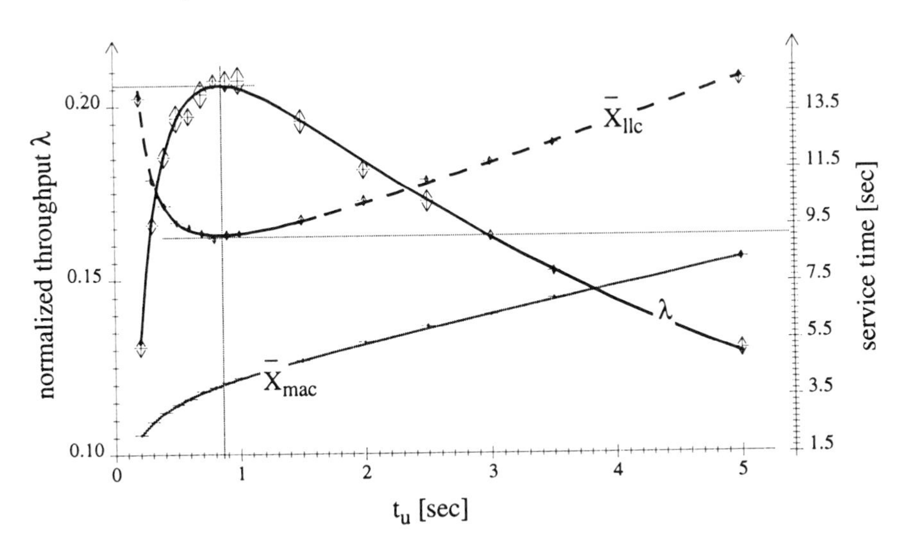

Figure 5.34b Throughput and service times against the access-delay parameter t_u for $n = 16$ and $p_{noise} = 0.6$

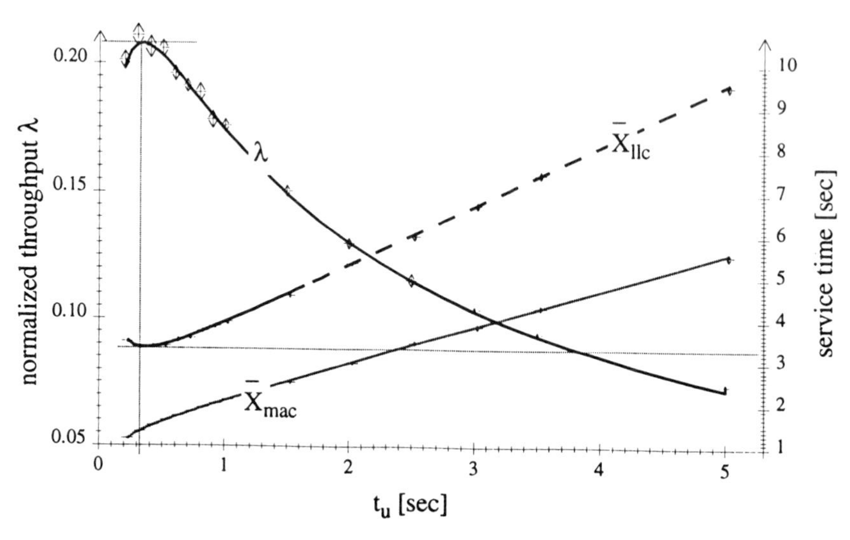

Figure 5.34c *Throughput and service times against the access-delay parameter t_u for $n = 6$ and $p_{noise} = 0.6$*

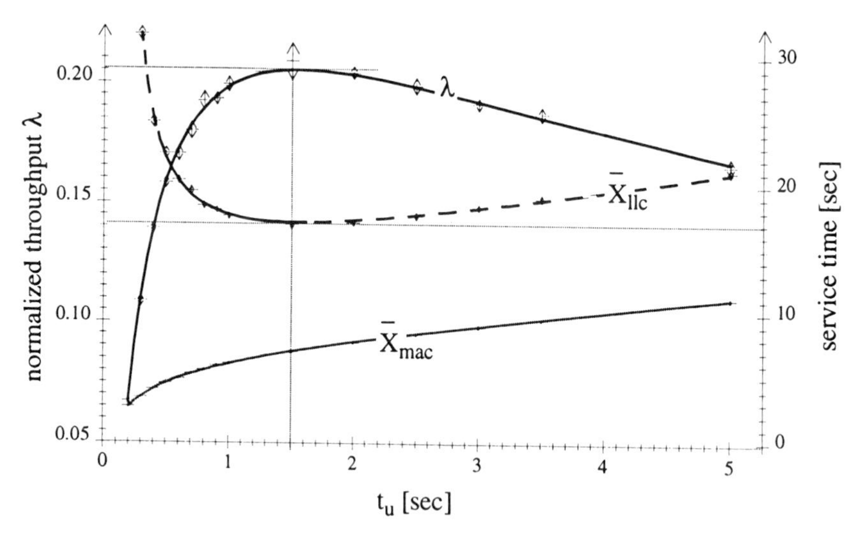

Figure 5.34d *Throughput and service times against the access-delay parameter t_u for $n = 30$ and $p_{noise} = 0.6$*

with decreasing t_u. On the other hand, large t_u values give few retransmissions but increased average MAC service time due to longer channel-idle periods, which again leads to increased LLC service time. $\overline{X}_{mac}$ approaches $\overline{X}_{llc}$ with increasing t_u in a noiseless system. $\overline{X}_{mac}$ decreases with decreasing t_u because fewer scheduling attempts are required. Roughly, we can say that the MAC service time is only indirectly affected by the multiuser interference, although the LLC service time is affected directly through p_{net}.

Example 2

In a real network the users generate packets at a rate which changes over time and the traffic distribution can only be regarded as stationary for a period of time. By traffic distribution we mean both the time between arrivals and the traffic pattern. This means that the number of nodes contending for channel access is time variant. Optimum performance at low-load level is attained at a low average access delay $\overline{D}$ because the stochastic fluctuation of the user-packet arrival distribution is sufficient to give a low collision rate. As the load level increases, $\overline{D}$ must also be increased. Loss of radio transmission capacity is either due to collisions or channel-idle periods and this calls for a dynamically-adjusted $\overline{D}$ according to the traffic conditions. Increasing $\overline{D}$ limits the traffic into the radio channel and has the potential of preventing network collapse.

We wish to quantify potential efficiency gained by adaptive flow control and take a fixed-sized network of 30 users as an example. During the busy hour all 30 nodes are active and as the packet-arrival intensity drops, the number of active nodes decreases. The throughput for the SWIA protocol as a function of the access-delay parameter t_u (since $D = F_D(t_u)$, see eqn. 5.19) and number of active nodes, is plotted in Figure 5.35.

With 30 active nodes the throughput peaks at ($t_u = 1.56$, $\lambda = 0.54$). If the t_u is kept fixed while the traffic drops to a level where only two nodes are active at a time, the resulting throughput reduces to $\lambda = 0.25$. The throughput peak for two active nodes is at ($t_u = 0.13$, $\lambda = 0.56$) and then the potential gain for an adaptive scheme is high. The throughput may be more than doubled in this example. The Figure shows that ($t_u = 0.13$, $\lambda = 0.09$) for 30 active nodes, so adjusting t_u from the optimum setting at low traffic to the optimum value for the busy hour has the potential of giving an even higher performance improvement. The final result of an adaptive flow-control scheme depends on its implementation loss, the packet-control information fields it needs for signalling purposes and its ability to correctly adjust t_u to the traffic conditions.

Example 3

The previous example explained the benefit of adjusting the access delay according to the number of active nodes. This example considers the throughput and service times as the access delay remains fixed when the

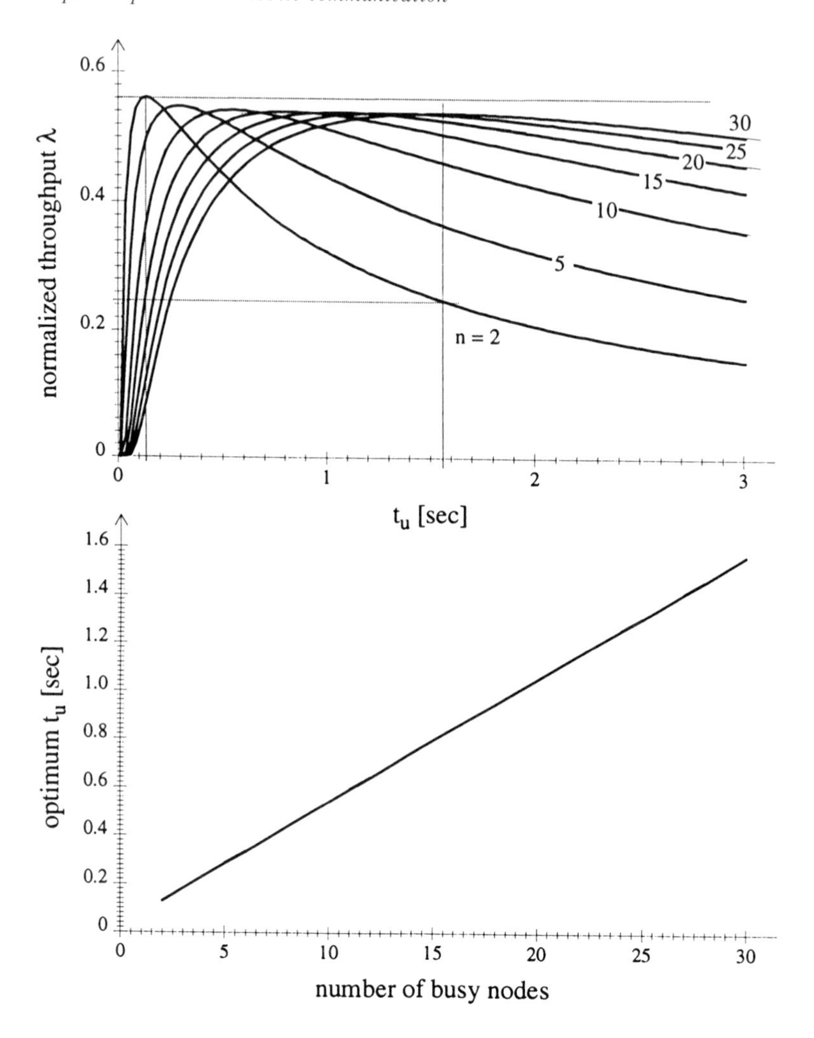

Figure 5.35 LLC throughput against the access-delay parameter t_u and the optimum t_u value as a function of the number of busy nodes

operating conditions alter. By taking the t_u values, maximising the throughput for the number of busy nodes in Figure 5.35 and reorganising the curve to present throughput as a function of the number of nodes, Figure 5.36 is produced. Here, the throughput drops slowly with increasing number of nodes when the network sets t_u correctly. We have also included optimum t_u values for $n = 2$ and $n = 30$ and and the Figure illustrates the loss in performance at nonoptimum operating conditions. The reader should remember that this example assumes fixed packet length $t_{p,data} = 196.6$ ms. Example 3 in Section 5.3 illustrated the dependency on

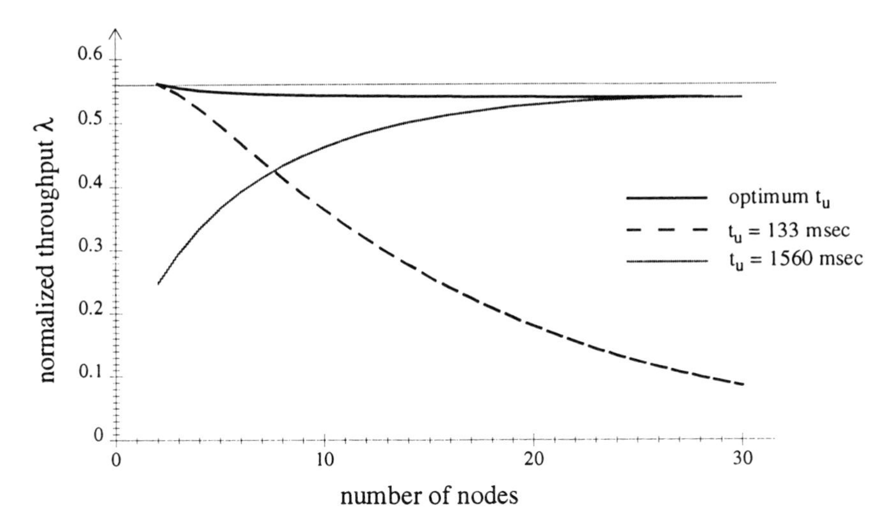

Figure 5.36 LLC throughput as a function of the number of nodes in a heavy-load state

the packet length. Real networks must handle different sized packets as well as different numbers of active nodes and therefore t_u must be set according to the packet-arrival distribution and the packet-length distribution. This is the subject for the next Section.

5.4.3 Flow control

Flow control comprises the operating rules for handling the traffic flow for controlling the use of the network resources. A network without flow-control functions can be compared to a car without brakes; it will sooner or later be involved in an accident (network collapse). The operating time until collapse becomes stochastic depending on the network users and the radio environment. The network resources of most interest are communication capacity and storage capacity. This functionality must be distributed among the nodes and each node takes its share by properly monitoring its surroundings and controlling the traffic which it injects into the radio channel. Today, computer memory is low priced and therefore the buffer size is commonly regarded as less important.

Flow control can be broken into two major components: local flow control and peer-to-peer flow control. Local flow control addresses local issues such as speed matching between a mobile station and its attached terminal, speed matching between the adjacent protocol layers inside a node, storage allocation and service ordering within the layers. Peer-to-peer flow control controls the traffic streams between peer-layer entities and often requires explicit control-information fields in the packets. We

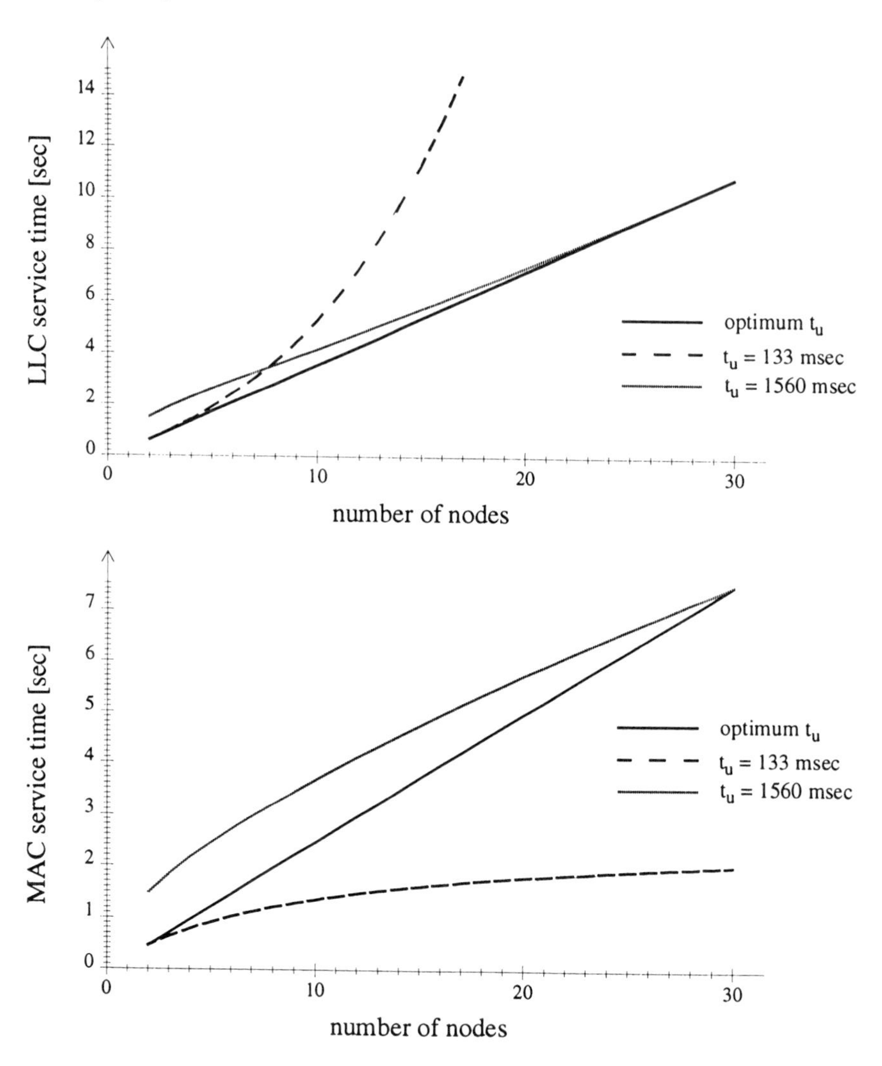

Figure 5.37 Service times against the number of nodes (n) in heavy-load state

are not so concerned about the exact implementation and therefore little will be said about local flow control.

This section looks at peer-to-peer flow control at the link level. The SWIA protocol restricts the LLC entity to having only one outstanding packet at a time to each of its peer entities, so the number of packets under service within the LLC service provider should be easily controllable. We are more concerned about the load on the radio channel.

Important performance metrics for flow control are to which degree it prevents network congestion and deadlock, enhances the throughput-delay characteristic and executes fairness. Congestion is the event where

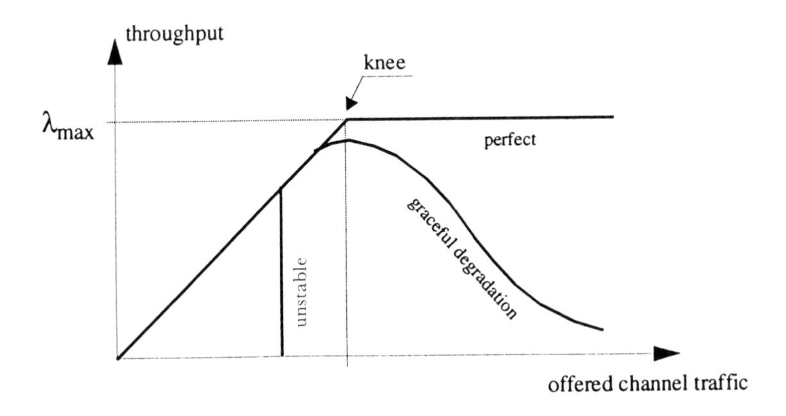

Figure 5.38 Flow control

some part of the network is unable to accept more packets; deadlock may occur as a consequence of packet re-assembly and/or resequencing [23]. Throughput degradation happens in the case of unbalance between the channel-idle period and the collision rate as exemplified by Figure 5.38. Every network configuration has a maximum throughput capacity, marked as the knee in the Figure. A perfect flow-control function shall let the throughput grow linearly with the offered traffic at levels below the knee and maintain maximum throughput when the offered traffic increases beyond the knee. At the knee the network queues grow indefinitely and the network must either reject new user packets or stop the user-input traffic. A flow-control function with instability problems can lead to network collapse, i.e. a fast drop in the throughput before the knee is reached. Even if this occurs at the knee it will make problems for practical use. The flow control must exhibit a smoother adaptation and it is better that the drop starts a little before the knee followed by a graceful degradation. With the SWIA protocol, the offered channel traffic increases very fast when the packet loss starts because lost packets are immediately scheduled for retransmission. Consequently, this case is more demanding than one without ARQ.

Fairness expresses the sharing of the network resources. A typical question will be if it is fair that user C in Figure 5.39 can send up to, say ten packets/s, while user E is limited to five packets/s? The Figure shows a network configuration where the nodes can be partitioned into the two fully-connected sets $\{A, B, C, D\}$ and $\{A, E, F\}$. Node A serves as a relay for traffic between the two sets. The operating topology is normally invisible to the users and user E is not aware of when node A relays its packets.

Using this example a short analysis of three possible flow control strategies is given:

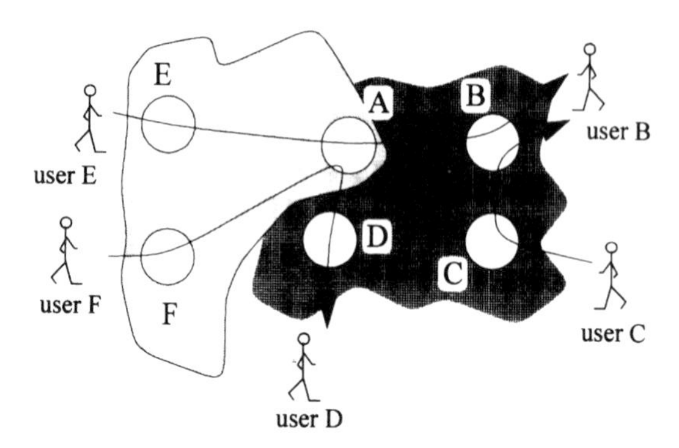

Figure 5.39 A network serving three identical user-packet streams

FS1: *Link-based throughput assignment*
With this strategy each network link has the same throughput and the nodes must be allocated channel capacity according to their radio channel qualities. For example, if the link $A \rightarrow D$ has a higher packet-loss rate, more capacity must be dedicated to node A so that A can compensate for the lost packets by more frequent transmissions on this link. This may lead to an excessive overall network throughput degradation.

FS2: *User-based throughput assignment*
With this strategy each user has the same throughput. Here the nodes must be allocated channel capacity according to the user-traffic distribution. For example, node A which handles two identical user streams must be given twice as much capacity as node C, leading to degradation from user C's point of view. If the users had been able to look at the topology, user C would probably express the opinion that this is unfair because user E injects multihop packets, burning more radio-channel capacity than single-hop packets.

FS3: *Node-based channel-capacity assignment*
With this strategy each node is allocated the same share of the radio-channel capacity. If the nodes have the same probability of winning channel access, they automatically get the same channel-capacity usage. The drawback with this scheme is that the users experience different throughput levels.

The designer's dilemma then becomes how much capacity shall be taken from the high quality links and dedicated to the low quality links? Despite the nonperfect property of FS3, we are going to use this strategy in the forthcoming discussions for two reasons: we believe that this is the only practical way to get a reasonable overall throughput (in a radio network the performance will always be location dependent), and this strategy is easily implemented by just letting the nodes use the same access-delay distribution. Although this provides fair access to the media, it does not optimise the network throughput.

The user-traffic distribution is generally time-variant and cannot be predicated in advance by the network nodes. Consequently, the number of busy nodes changes and, as illustrated by example 3, performance degradation might result. The challenge now is to find a distributed algorithm for estimating the number of contending nodes. If each node knows this number, the preferred t_u can be set by a simple table look-up.

In a random-access system we see two possible concepts for estimating the number of contending nodes. One is to extract the information from some of the network distributions by stochastic sampling, e.g. estimate the average channel-idle period and calculate the number of busy nodes via eqn. 5.21. The other approach is to let the stations explicitly signal their load level through dedicated control fields in outgoing packets. With the assistance of Figure 5.40 we first consider the former method. Examples of distributions which alter with changing load level are the channel-idle period C_I (eqn. 5.21), the channel-busy period C_B (eqn. 5.22), the number of simultaneous transmissions K_{Tx} (eqn. 5.24), the number of scheduling attempts before transmissions and the retransmission rate.

A node can never measure the K_{Tx}-distribution, but may sample the C_B-distribution in the periods in which it does not transmit. The delay difference to be sampled is $(14/15 - 15/16)t_v = 0.0042 \cdot t_v$ when the ambition is to differentiate between 15 and 16 active nodes, cf. eqn. 5.22, in a 16-node network.[1] If the average channel-idle time period (eqn. 5.21) is used instead, we get $(1/16 - 1/17)t_u = 0.0037 \cdot t_u$ and this is not so demanding (because t_u must always be set much larger than t_v). We realise

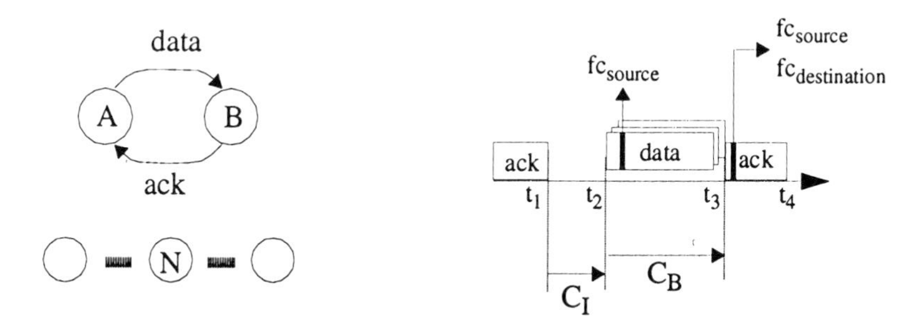

Figure 5.40 A successful delivery cycle involves three parties, the nodes A, B and N. The corresponding time-sequence diagram is shown to the right

[1]The equations used are conditioned on the understanding that at least one node has a packet under service and are only valid for an outside observer. However, the error introduced has no practical consequence.

that the sampling of the C_B-distribution may be unrealistic in real networks and a less demanding solution is to sample the C_I-distribution. For a system with varying packet lengths we have to precalculate t_u to maximise the throughput for the average number of blocks per packet and then let the nodes set their t_u based solely on the number of nodes.[1] Stochastic sampling also requires that a number of samples are available before a conclusion can be drawn.

To illustrate a serious pitfall with methods working under the assumption that the other nodes use the same access-delay distribution, consider the case where the nodes regulate their access-delay parameter (t_u) as a function of the packet loss experienced. A low-loss rate over some period of time is interpreted as low traffic and t_u is lowered accordingly. As the loss rate increases, t_u is enlarged. A node which does not experience a sudden increased loss rate continues to use the same t_u value, while the nodes detecting the enlarged loss rate increase their t_u. The former node gets a higher likelihood of winning channel access and experiences a lower collision probability and decreased loss rate. Thus, it optimistically concludes that there is a lower congestion level. If it continues to reduce its access delay, it may eventually take over the channel completely. The fairness principle is broken. The method with explicit signalling of the traffic level has the potential of overcoming this problem because here the concept is to base the regulation on explicit signalled information. By introducing a header field fc_{source} in the data and the acknowledgement packets, the nodes can signal their load level upon sending data packets. To further improve the reliability of the method an additional field $fc_{destination}$ can be introduced into the acknowledgement packet. This field informs the nodes which have missed the data packet about the load level at its source node. The semantics are:

$fc_x = empty$: x has an empty queue (the node holds the data packet under service only)

$fc_x = not\ empty$: x has a queue size strictly larger than zero

Consider the implementation of this method in the network in Figure 5.40. The behaviour of three actors A, B and N must be regarded. A sends a data packet to B which returns the acknowledgement. Node N represents a passive listening node. The $fc_{destination}$ is filled in by node B and informs the nodes which missed the data packet about A's load level. All the stations maintain a state list with an entry for every network member and update this according to the state diagram in Figure 5.41. This is simply a one-to-one mapping between the queue status $\{empty, notEmpty\}$ and the node state

[1] The statistical fluctuation caused by different packet lengths is easy to filter out because the length is the sum of a number of large blocks ($t_{b_z} \gg t_v$).

a

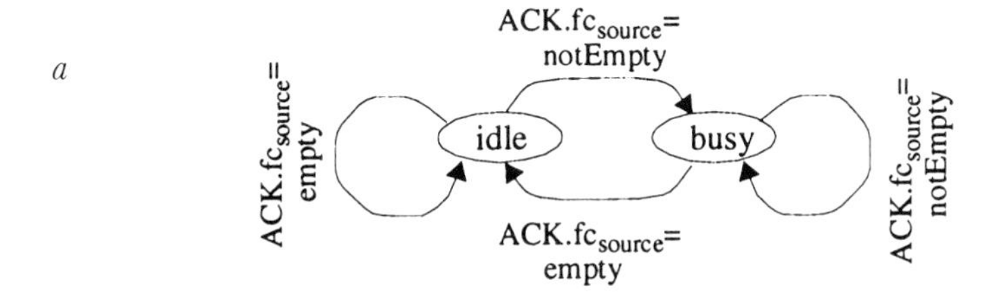

b

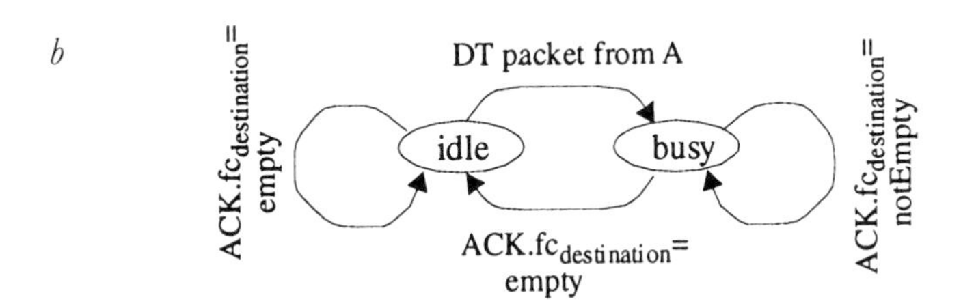

c

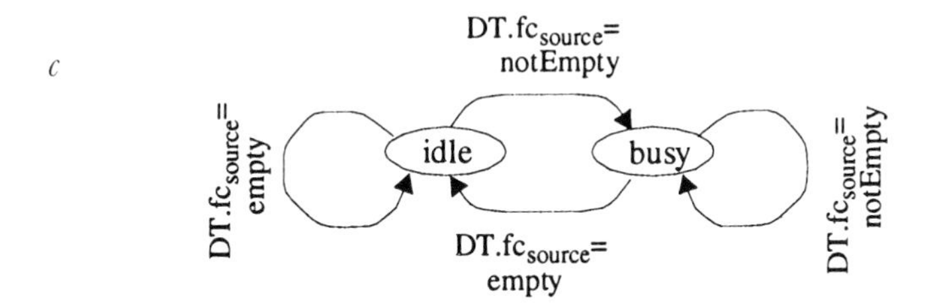

Figure 5.41 *Flow-control state diagram for the three actors A, B and N*
a Node *A* and *N* updating status at *B* (time instance t_4)
b Node *N* updating status at *A* (time instance t_3 and t_4)
c Node *B* updates status at *A* (time instance t_3)

{*idle, busy*} with an exception for node N at time instance t_3. Here, A is always registered as busy at N.

Let $\hat{n}_i$ be the number of busy nodes estimated by node i. $\hat{n}_i = 1$ means that node i has noted that all the other network nodes are idle, and $\hat{n}_i = n$ says that all the network nodes have a packet under service. At the time of scheduling, node i selects a value for the access-delay parameter $t_u = f_{t_u}(\hat{n}_i)$ prior to drawing the access delay D from the distribution given by eqn. 5.19.

An obvious imperfection of this flow-control method is seen by regarding a network under low load (which gives insignificant multiuser interference and few active nodes) and high background noise. A passive listening node receiving the data packet but missing the acknowledgment, registers the source node as busy. The tendency is then to overestimate the number of busy nodes. However, the method never sets $\hat{n}_i$ higher than the number of network members and the introduction of an upper limit for t_u is not needed.

Two possibilities exist for the traffic to build up without giving the flow-control procedure the possibility of adapting to the new situation. The first is a fast increase in the traffic during one or more packet-transmission times. The second is a full blocking of the radio channel for a period of time, e.g. caused by network jamming. All packets are lost and packets rescheduled for retransmission and fresh packets start to fill up the network buffers. When the blocking comes to an end, all the network nodes erroneously expect a low load state and then use a low t_u value. Consequently, they experience a high collision probability. The probability of reaching all nodes by this first transmission is $[P(K_{Tx} = 1) \cdot p_{noise}]^{m-1}$ for a system with zero capture, a very unlikely event for a network of some size. The contrary event, the probability that one or more nodes are able to demodulate, is much higher, namely:

$$(1 - (1 - p_{noise})^{n-1}) \cdot P(K_{Tx} = 1) \approx P(K_{Tx} = 1)$$

for reasonable background noise levels. The nodes reached will register the event of increased load and use a higher t_u-value. This leads to a decreased collision probability in the next busy period. For a short period of time the fairness principle is broken, but we expect the flow control to bring the system back into a new steady state where the fairness applies.

Having discussed the expected characteristics, the discussion will proceed with a case study presented through four examples. A crucial supposition for having an efficient flow-control function is its ability to estimate the number of busy nodes correctly. This will therefore be considered in the first example. Thereafter example 5 looks at the throughput and delay performance. The examples continue with the network configuration from Section 5.2 in example 3. At the completion of this case study we hope to have confirmed the flow-control presentation and validated the statements given above.

Example 4

The first test of the flow control is to check its capability to estimate the number of busy nodes under steady-state at different load levels. The preliminary discussion above pointed out that its performance degrades with increasing background noise (or jamming) and therefore we compare a noiseless system ($p_{noise} = 1$) with a very demanding scenario where $p_{noise} = 0.6$. In the latter case 40% of all transmissions are corrupted by background noise.

Figure 5.35 illustrated how the access-delay parameter should be set under different traffic conditions. The same scenario is used here, and the access-delay parameter should be regulated according to:[1]

$$t_u = f_{t_u}(\hat{n}_i) = \begin{bmatrix} 30.5 + 51.3 \cdot \hat{n}_i & \text{for } \hat{n}_i > 1 \\ 133 & \text{for } \hat{n}_i = 1 \end{bmatrix} \quad [\text{ms}] \tag{5.40}$$

What the designer should actually try to avoid is underestimating the number of busy nodes because this leads to network instability due to high multiuser interference. The previous examples have shown that it is better to have performance degradation caused by too long an idle period than high collision rates since the former gives more graceful degradation. Therefore, $t_u = 133$ is used when $\hat{n}_i = 1$ even though zero should have been used. With this lower limit the average packet-success probability is never less than $0.29 \cdot p_{noise}$ for the network under consideration ($p_{net}(t_u = 133,\ n = 16)$).

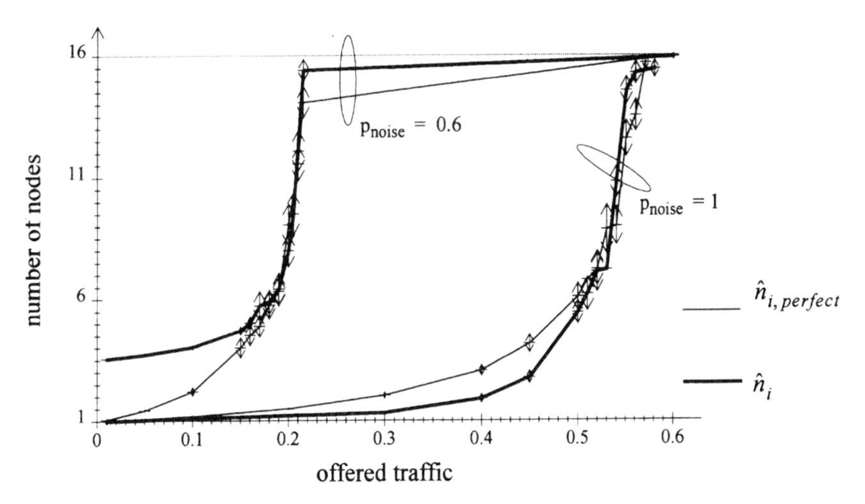

Figure 5.42 Estimated number of active nodes at the time of scheduling as a function of the normalised offered traffic. $\hat{n}_{i,perfect}$ is the exact number of busy nodes and cannot be estimated by other than an idealised node. $\hat{n}_i$ is the number of busy nodes as estimated by the network nodes

[1]This is an approximation because the theoretical model in Section 5.3 has two shortcomings: it does not take into account the fact that a node which is idle at the last busy period might become busy during the scheduling of the tagged node and the number of competing nodes changes from contention period to contention period. However, the error introduced has little effect in the steady-state.

The use of a simulation model facilitates undemanding implementation of an estimator for the exact number of busy nodes at the time of scheduling, $\hat{n}_{i,perfect}$. The plot in Figure 5.42 presents the simulation results for $\hat{n}_i$ and $\hat{n}_{i,perfect}$. The results confirm many of the assumed properties outlined in the above discussion: at low load level the flow control's ability to correctly measure $\hat{n}_i$ deteriorates with increasing background noise and $\hat{n}_i$ can never increase beyond the number of nodes in the network.

We find the error $\hat{n}_{i.perfect} - \hat{n}_i$ acceptable for all the load levels and also note that this difference gets smaller with increasing load level. This gives us an optimistic starting point for the next example where network throughput is the subject.

Example 5

The previous example focused on the flow control algorithm's ability to correctly estimate the number of busy nodes at different load levels. From the network user's point of view, the success of the algorithm is better judged via throughput and delay measurements. This example continues the study of flow control by considering just these estimates. A comparison between a set of possible strategies would probably strengthen the trustworthiness of the conclusion and the following additional strategies are therefore included:

- *fixed 1*: a network without adaptive flow control. t_u is fixed and set to 50 ms. This configuration suffers from high multiuser interference.
- *fixed 2*: a network without adaptive flow control. t_u is fixed and set to 133 ms.
- *fixed 3*: a network without adaptive flow control. t_u is fixed and set to 851 ms. This is the optimum t_u value for the heavy-load state given by the throughput model eqn. 5.39.
- *perfect*: a network with an idealised flow-control algorithm (i.e. cannot be implemented in real networks) where the nodes have the knowledge of the exact number of competing nodes at the time of scheduling (implemented by the estimator $\hat{n}_{i,perfect}$ used in the previous example). This represents the upper limit in performance.

The model in Section 5.4.2 gives the opportunity to precalculate some of the expected performance measures and this is done in Table 5.5.

The configuration giving the highest network capacity is represented by the strategy *fixed 3* and as shown by Figure 5.43, the knee (cf. Figure 5.38) occurs at the anticipated point $\lambda = 0.54$. The throughput performance fulfils our expectations by growing linearly up to the knee and then remaining constant thereafter. The three strategies *nonperfect*, *perfect* and

Table 5.5 Expected performance in heavy-load state

strategy	λ	X_{mac}	X_{llc}
perfect, nonperfect, fixed 3 ($t_u = 851$ ms)	0.540	3.99 s	5.74 s
fixed 2 ($t_u = 133$ ms)	0.240	1.67 s	12.97 s
fixed 1 ($t_u = 50$ ms)	0.025	0.83 s	123.42 s

fixed 3 are identical and throughput gains can be observed by introducing an adaptive scheme.

The *fixed 2* strategy illustrates the effect of setting the access delay too small for the number of competing nodes. The network never reaches its potential capacity and even exhibits a drop in throughput. However, no catastrophic instability problem is discovered. This is in contrast to the *fixed 1* strategy. Here the network collapses completely beyond the 0.2 level. An interesting observation is the sudden throughput drop at the offered traffic level of 0.15. What happens is that the stochastic fluctuation of user traffic brings the system into a state where all nodes become busy. The scheduling delay is the sum of packet interarrival time and access delay,

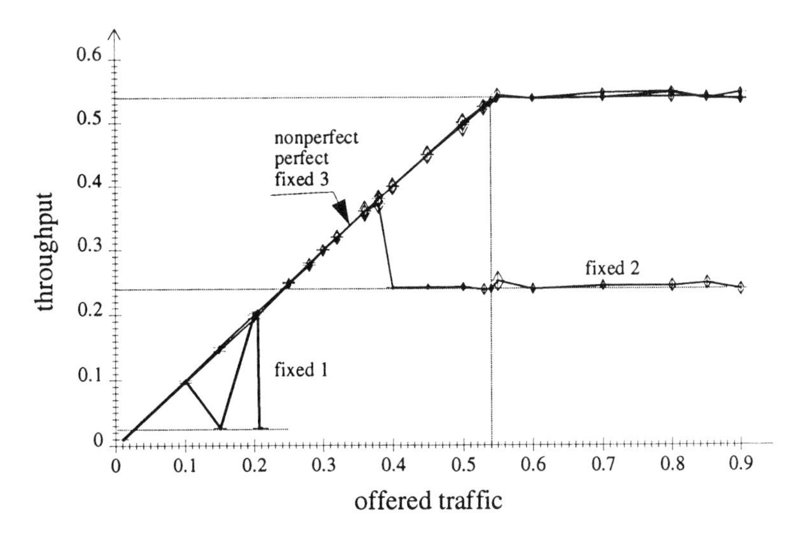

Figure 5.43 Normalised throughput as a function of the offered load in a noiseless radio environment ($p_{noise} = 1$)

but as saturation is reached, only the access delay contributes and the probability of collision reaches a high level ($p_{coll} = 0.97$ by eqn. 5.31). Many retransmissions are needed while the user packets continue to arrive and fill up the network buffers. The network is unable to recover from this state.[1] The simulations show that the nodes estimate the number of busy nodes to be 16, so they interpret the situation correctly, but they are forced to use this short access delay.

The tendency to erroneously estimate the number of busy nodes at low load levels implies that the access delay is set to a nonoptimal value. This cannot necessarily be observed as a degraded throughput plot since the network operating point stays in the linear region far below the system capacity.

From the network users' point of view, the most important stochastic variable is the network transit delay. Three factors contribute to an increasing transit delay as the load level rises:

(i) A larger number of nodes compete for access and consequently each of them must wait longer before getting access to the channel. $\overline{X}_{mac} \approx \overline{X}_{llc}$ and $\overline{Q} \approx 0$.
(ii) The retransmission rate increases due to collisions. $\overline{X}_{llc} > \overline{X}_{mac}$ and $\overline{Q} \approx 0$.
(iii) The network queues start to fill up. $\overline{X}_{llc} \gg \overline{X}_{mac}$ and $\overline{Q} > 0$.

Figure 5.44 presents the estimated time delays against the offered traffic for the different flow-control strategies. The *nonperfect* flow control performs close to the *perfect* and gives considerably lower delays than the *fixed 3* strategy. Recall that the throughput analysis was not able to distinguish between the *nonperfect* and *fixed 3* strategies.

The lesson to be learned is that steady-state analysis can hide long periods of misbehaviour because under this form of analysis the measurements are integrated over a long time scale. For example, a network providing throughput of 0.6 during 23 hours and 0 during 1 hour in a 24-hour cycle, gives a time average of 0.58; possibly an acceptable average over the 24-hour time period, but still not likely to be accepted for practical use due to the long blocking period. Consequently, the network designer must also encompass transient analysis of the network. This is exemplified in the next section on flow control and network jamming.

[1]The reasons why the next higher load level does not have the same degraded performance are that all the simulation runs are started with empty queues and that the next run did not experience the same sample path as the previous.

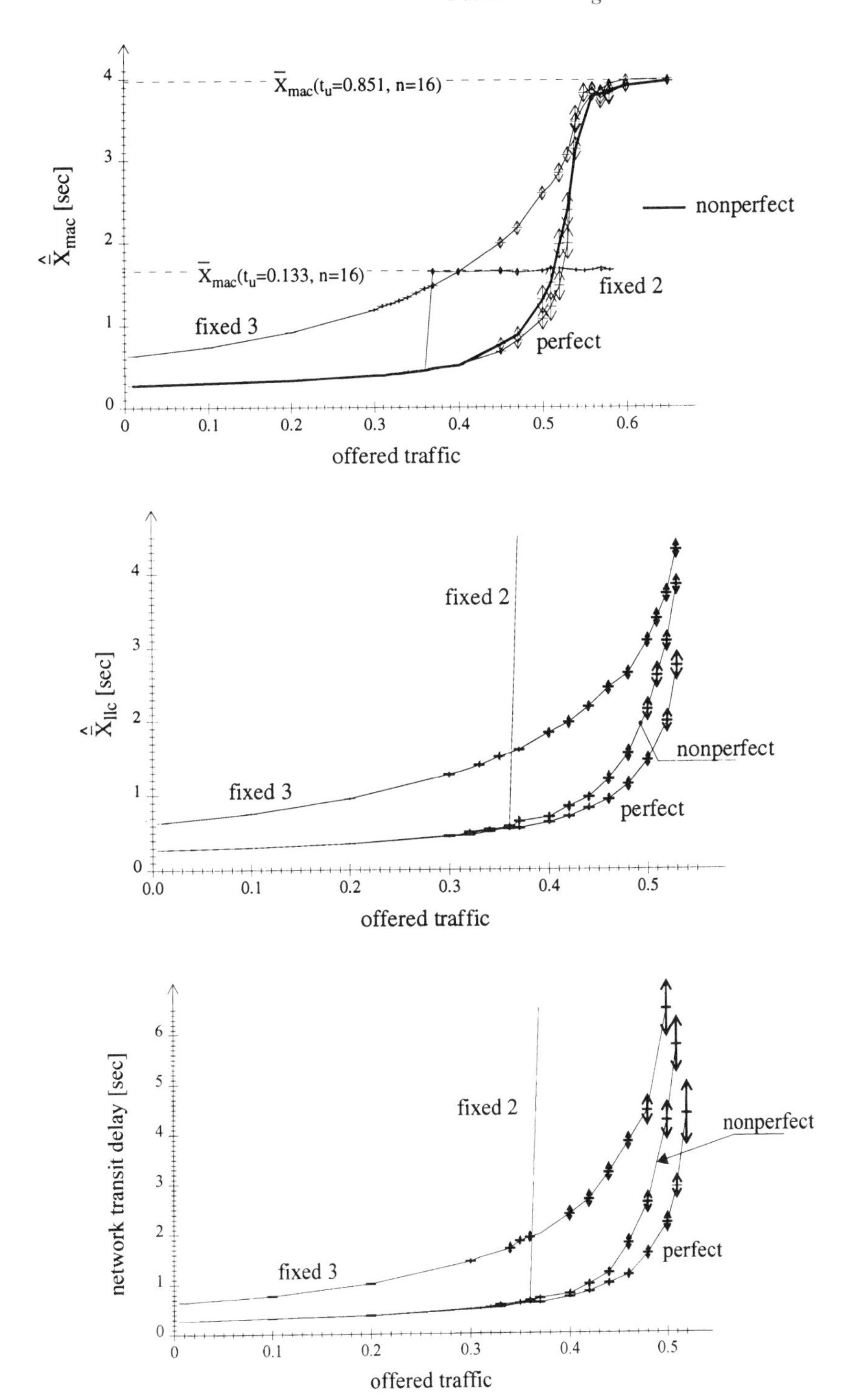

Figure 5.44 Estimated time delays against the offered traffic

5.4.4 Flow control and network jamming

The original motivation for performing a jamming evaluation is the military scenario where reliable communication services are of upmost importance. Communication systems based on radio are very vulnerable because no physical barrier exists between the jammer and the network receivers; only the path loss serves as a guard. Although spread-spectrum signalling gives additional protection, it is only a matter of distance or power before jamming becomes a problem. Therefore, the highest ranking question is to evaluate the shortest distance possible for the scenario under study. We have no preference for a particular scenario and model the effect of a jammer in the forthcoming by setting the p_{noise} parameter in the capture model (eqn. 5.12). The jamming problem is expected to be given more attention within the domestic and public domain as the dependency of reliable communication services increases, in the same way as security has become an important issue. Many security mechanisms developed for military networks have been adopted by commercial products.

Before turning to a specific example, a short overview of how jamming can affect network operation will be given. The first obstacle for a jammer to overcome is producing sufficient power to exceed the receiver-sensitivity level. Already at this point we see a problem by considering a simple tone jammer which is any transmitter simply injecting energy at the same frequency as the network receivers' bandpass filters. Consider a tone jammer attacking a CSMA network which uses a simple energy-detection device for implementing the carrier-sense function. The nodes detect a busy channel under the CSMA scheme and stay off the channel until the jamming signal vanishes. This could be partly solved by increasing the carrier-sense threshold level, but has the side effect of giving a higher collision rate in an ordinary operation mode. A better solution might be to introduce a special code to follow the carrier sense. If a legal code is not received, the channel is set to the idle state.

With spread-spectrum signalling, the channel-state transition from idle to busy is detected by the receipt of a certain code. A tone jammer is no longer sufficient for generating false channel states, but a code jammer is needed. However, a code jammer succeeds better than a tone jammer whenever it generates the correct code because then the network receivers detect an incoming signal and start demodulation. They stay locked onto this first signal up to the point in time where they can conclude that it is not legal. This time delay depends on the radio design. Certain radio solutions might be unable to detect the end-of-transmission on their own, and the network level has to attach a packet length indicator (LI) field to the head of each packet. The radio continues with the demodulation process until the network entity has received the number of bytes assigned by the LI field before ordering the radio to stop and enter the preamble search mode. The LI field will take a random value and the worst-case delay is determined by the largest number possible in the LI field. For a code jammer to work efficiently, it must know the code currently in use. To make it difficult to possess this knowledge, networks may shift the code periodically.

When the carrier-sense problem is solved, the next barrier is to produce sufficient energy to introduce a high packet-loss rate. To which degree this leads to performance degradation depends not only on its absolute value, but also on how it affects other network-control functions. If the protocols interpret the increased loss rate as increased traffic and then adjust the access delay accordingly, the jamming sets the network into a nonoptimal regulation state besides introducing extra packet loss. The best action for the network is to regulate the control variables according to its internal traffic level. The main obstacle to providing efficient jamming control is the difficulty of differentiating between packet losses caused by multiuser interference and those caused by jamming. Apart from increasing the internal control traffic such as acknowledgement traffic, jamming also increases the traffic generated by the routing function, as explained in Section 5.7.

In order to produce a case study, the jamming scenario depicted in Figure 5.45 is used. We evaluate the reaction of the 16-node network from example 4 under the influence of an airborne jammer.

The situation is analysed from time instance t_1 where the beam front reaches the first four nodes. Assume that these four nodes become busy while the radio channel is fully blocked. We freeze the time scale for a moment and look at what happens. The busy nodes are isolated and

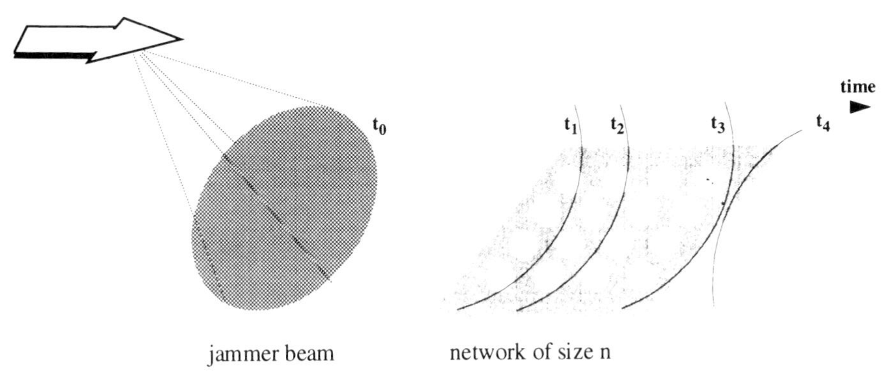

Figure 5.45 *Network jamming. The jammer covers the whole network area during the time period $t_3 \leq t \leq t_4$ and comes to an end after t_4*

retransmit the packets at a high rate using a short access delay because they are in a low-load state and are unable to detect the sudden increase in the activity level. These nodes amplify the effect of the jammer by increasing the collision rate in the network regions which it has not yet reached. The nodes not disposed of by the jammer signal detect the increased level and increase their average access delay.

Then the jammer moves further to the right and at t_3 the jammer overlaps the network completely. An airborne jammer generally has the benefit of having the path loss $1/r^2$ compared to a typical path loss of $1/r^4$ at ground level. Besides the lower path loss, an airborne jammer also has the advantage of reaching a large geographical area during jamming, so that backup routes cannot normally be found.

It should be remembered from Section 5.4.3 that the primary concern with adaptive flow control comes with the danger of using short access delays. By lowering the access-delay parameter t_u from its optimum setting, more multiuser interference is introduced and thus reduces the jamming margin. The throughput curves showed that the performance deteriorates faster with decreasing t_u and therefore it is better to use a larger value giving longer channel-idle periods but less multiuser interference.

This example makes it possible to envisage any misbehaviour of the adaptive flow control while focusing on the operating mode of a reasonably realistic network. The most demanding state occurs when the multiuser interference takes its highest value and this occurs when $\hat{n}_{i,perfect} - \hat{n}_i$ takes its maximum. This is achieved by applying the time-variant noise model:

$$p_{noise}(t) = \begin{cases} 0 & \text{for } t_1 \leq t \leq t_4 \\ 1 & \text{otherwise} \end{cases} \tag{5.41}$$

in conjunction with the time-variant Poisson packet generator:

$$\lambda(t) = \begin{cases} \infty & \text{for } t = t_3 \\ 0.1 & \text{otherwise} \end{cases} \tag{5.42}$$

where λ is the average arrival rate in packets/second. For the time period t_3 to t_4, the jamming beam covers the network area totally and fully blocks the radio channel. All nodes have full queues when the channel state goes from blocked to unblocked at t_4. The net operates in low traffic noiseless mode prior to jamming such that all nodes have $\hat{n}_i = 1$ at t_3, and because no signalling of the increased activity level is possible during jamming, $\hat{n}_{i,perfect} - \hat{n}_i$ is $n - 1$ at t_4. This is the maximum error possible and produces the highest multiuser interference. If some noise had been applied prior to jamming, the previous section showed that $\hat{n}_i$ would be greater than one and the flow control would have a better starting point at t_4.

This combination of channel blocking and traffic introduces a very demanding situation. We even use the pessimistic assumption that the time delay $t_3 - t_1$ is too short for signalling the increased load level. Similar but less hostile threats exist on the civilian scene. Examples are a sudden increase in the input traffic, or the deployment of a new network too close to an existing one. However, the probability that all the bad events occur simultaneously here is practically zero. This is in contrast to communication systems serving weapon-control applications where the user-input traffic is highly correlated with channel blocking.

We have yet to determine one network parameter before we can start the simulation study, the number of buffers in each node, say s_q. The high traffic generated at time t_3 fills all the network buffers and then the buffer size will have great impact on the time needed to expedite the packets after the jamming vanishes. A packet-lifetime control function is often used to restrict the age of the network packets. This function is especially useful for real-time traffic (e.g. weapon-control systems) where late deliveries will be deleted at the end destination anyway. By removing these packets at the network level, fewer network resources are wasted. We do not consider this alternative solution but set $s_q = 10$ which effectively restricts the maximum number of packets in the system. At t_4 each node holds $s_q + 1$ packets. One packet is now under service while s_q packets wait for service. Packets under service are persistently retransmitted until delivery is confirmed. The service time is now a function of time, and steady-state analysis gives an incorrect answer. However, if steady-state analysis is used as an approximation, we find the average time needed to serve these waiting packets to be $s_q \cdot (\overline{X}_{llc} + \overline{X}_{llc,blocked})$, but during this service period new packet arrivals occur at an average of $s_q \cdot (\overline{X}_{llc} + \overline{X}_{llc,blocked}) \cdot \lambda/n$. When the channel is released at t_4, the link protocol must not show the unstable behaviour seen in Figure 5.38 but reach a new steady-state operation after a recovery time $\overline{T}_{recovery}$ given by the approximation:

$$\overline{T}_{recovery} \approx (s_q + 1) \cdot (\overline{X}_{llc} + \overline{X}_{llc,blocked}) + (s_q + 1) \cdot (\overline{X}_{llc} + \overline{X}_{llc,blocked})^2 \cdot \lambda/n \tag{5.43}$$

The first term is due to the time needed to deliver the packets existing at t_4 and the last term represents new arrivals during their service. By using the heavy-load steady-state model with optimum t_u (which is 851 ms for this case, cf. example 5), we have $\overline{X}_{llc} + \overline{X}_{llc,blocked} = 5.8$ seconds, giving the optimistic recovery time of 63 seconds. How much longer the recovery time will actually be, depends on the time delay until the nodes use this optimum t_u value.

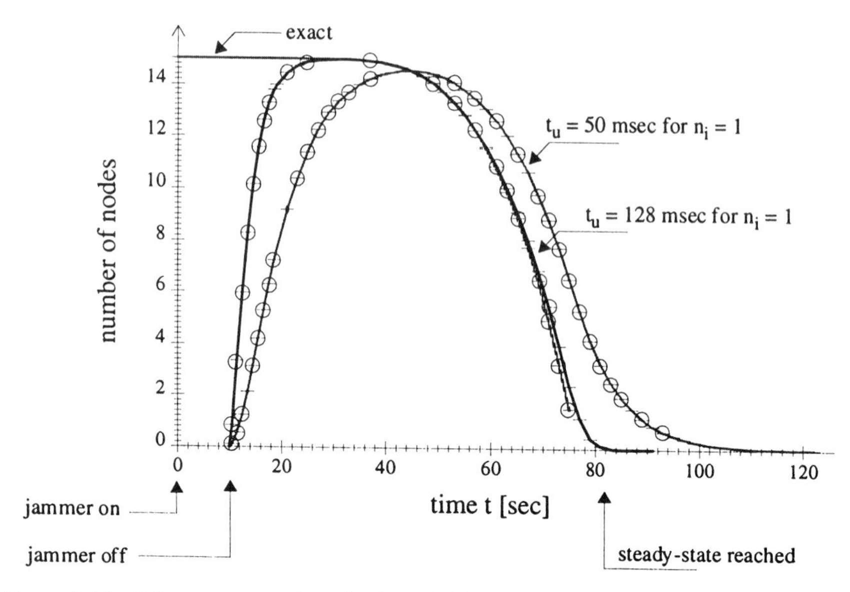

Figure 5.46 The average number of active neighbours as estimated by each node

The above description has given us an idea of the expected behaviour, and we can now perform transient simulations. We take t_3 as the reference point by setting $t_3 = 0$ and assume a jamming duration of ten seconds (i.e., $t_4 = 10\,\mathrm{s}$). The primary concern is how well the flow control estimates the activity level. For this reason the first simulations focus on the number of active neighbours that a random node measures as a function of time. By inspecting each node periodically during the simulation runs, the sample averages in Figure 5.46 have been formed.[1,2] The curve marked 'exact' shows the exact number of active nodes as a function of time. The estimation of this quantity is done by sampling each network node periodically during simulation.

Just before time zero, the network operates in the steady state far below its capacity where the average number of busy neighbours is practically zero. It operates without any background noise and we know from our earlier steady-state analysis that the nodes correctly measure the activity level. During jamming the radio channel is fully blocked, so the

[1]The double vertical arrow represents the accuracy of the corresponding time ensemble at a 90% confidence level. Ensembles subject to confidence control represent an average of at least 10 000 samples. Time ensembles sampled without a confidence control have been marked by circles. The same semantics apply for all the transient simulation results in this section.

[2]An ensemble is formed by collecting samples over a time window since sampling at an exact point in time generally implies infinite simulation run times. The time window size is Δt seconds such that an ensemble time average located at time instance t represents an average over the time period $t \pm \Delta t$.

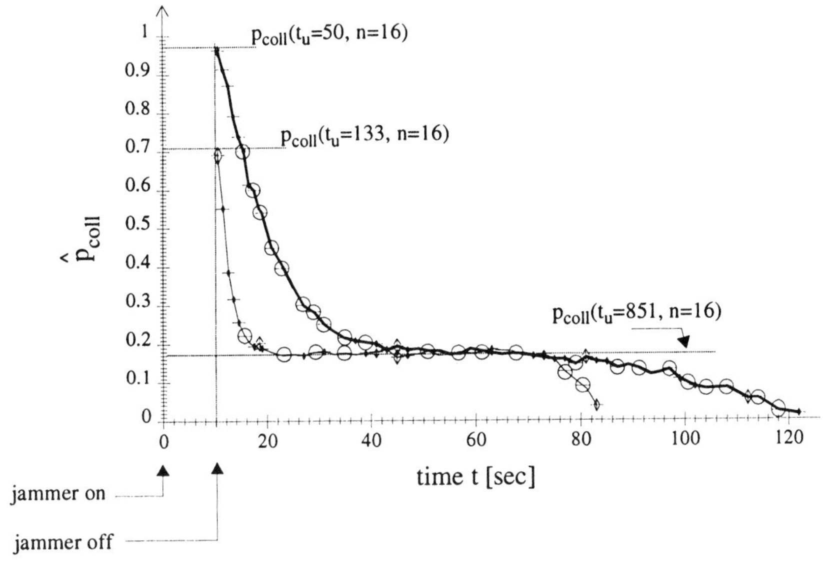

Figure 5.47 The probability of collision as a function of time

nodes cannot change their view during the first ten seconds. When the radio channel switches to unblocked at $t = 10$, flow-control signalling information can again propagate. After approximately 12 seconds all nodes have taken the correct view of their neighbours' state, namely that the fifteen neighbours are all busy. We note a recovery time of roughly 70 seconds after the end of jamming.

For curiosity we have also included the case where the lower t_u bound is reduced from 133 ms to 50 ms leading to a change in the worst-case collision probability from 0.71 to 0.97, and then a significant increase in the adaptation time is expected. This is confirmed by the Figure showing that the nodes use the correct access delay distribution first at $t = 40$.

We are happy to note the flow control's capability to track the number of busy nodes after a rapid shift in the load level. When the radio channel re-opens, the first notable effect is the high level of multiuser interference because the network nodes use short access delays. An impression of the multiuser interference can be attained by studying p_{coll} as a function of time. At $t = 10$ the nodes use $t_u = 133$ ms because they have not yet detected the high load level, p_{coll} is 0.71 and since $q = 0$, any overlapping transmission leads to packet loss and many retransmissions occur. Despite the many collisions, Figure 5.47 shows that p_{coll} drops quickly with time and confirms that the nodes regulate the access delay satisfactorily. After ten seconds p_{coll} reaches its steady-state heavy-load value, i.e. $p_{coll}(t_u = 185\,\text{ms}, n = 16) = 0.17$.

The network operates at a low load level before $t = 0$, all the network buffers are filled up at $t = 0$ and the link entities take the first packet into service at the same instance as the buffers are filled up. Since the radio channel re-opens at $t = 10$, the first delivery possible must be after this time instance and the first packet delivered has an LLC service time strictly longer than ten seconds. The LLC service time for a packet starts when it is removed from the queue and ends when it is successfully delivered to its destination. This packet also has zero queuing time because it is immediately taken under service upon arrival. The service time for the second packet starts when it is removed from the queue and the second packet contributes less to the average service time for two reasons; it does not contain the channel blocking time nor does it suffer the same high multiuser interference as the first packet. The latter is due to the fact that by the delivery of the first packet and the error-free acknowledgement packet to follow in the noiseless system, the nodes get the opportunity to take a step towards a more optimum access delay distribution by increasing their t_u. The average LLC service time as a function of time is shown in Figure 5.48.

The packet lifetime grows linearly with time as long as the packet resides in the queue. The network queues are instantaneously filled up at t_3 and when the radio channel switches to unblocked at t_4, they have the age $t_4 - t_3 = 10\,\mathrm{s}$ (except for the packet possibly under service just before t_3).

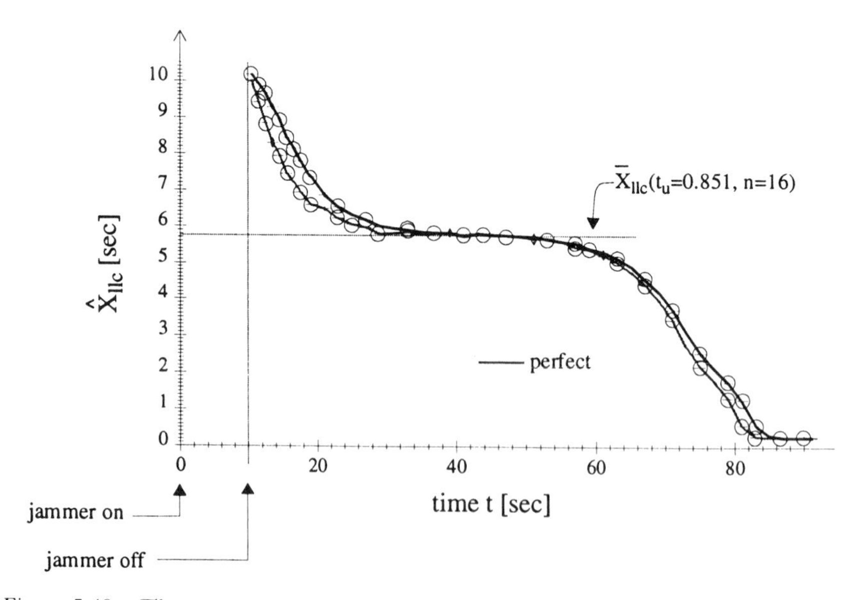

Figure 5.48 The average LLC service time as a function of time. The sampling of the X_{llc} distribution occurs at the same time instance as the delivery of the packet

Figure 5.49 shows how the queuing time grows linearly with time up to a certain point somewhere around 50 seconds from whence the delay increases less rapidly. Some of the nodes have completed the delivery of the oldest packets and taken the newer packets (those which arrived after t_3) under service. These packets contribute lower delay values but it should be remembered that new packet arrivals must be rejected until some buffer space becomes free. The steady-state packet arrival rate is only 0.1 packets/second and is the reason for the speedy reduction in the delay when the network recovers from the overload state.

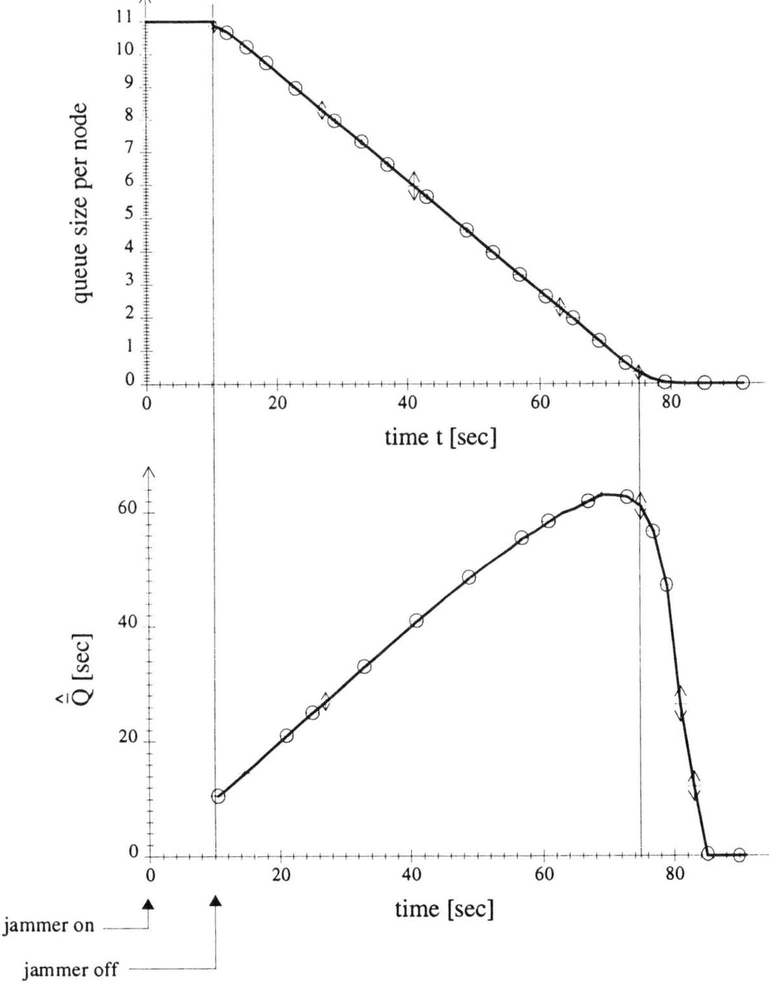

Figure 5.49 *The average queue size per node and the queuing delay as a function of time. Queuing delay distribution samples are collected each time a packet is removed from the queue. Queue-size sampling is performed periodically at a sampling rate of 10 Hz*

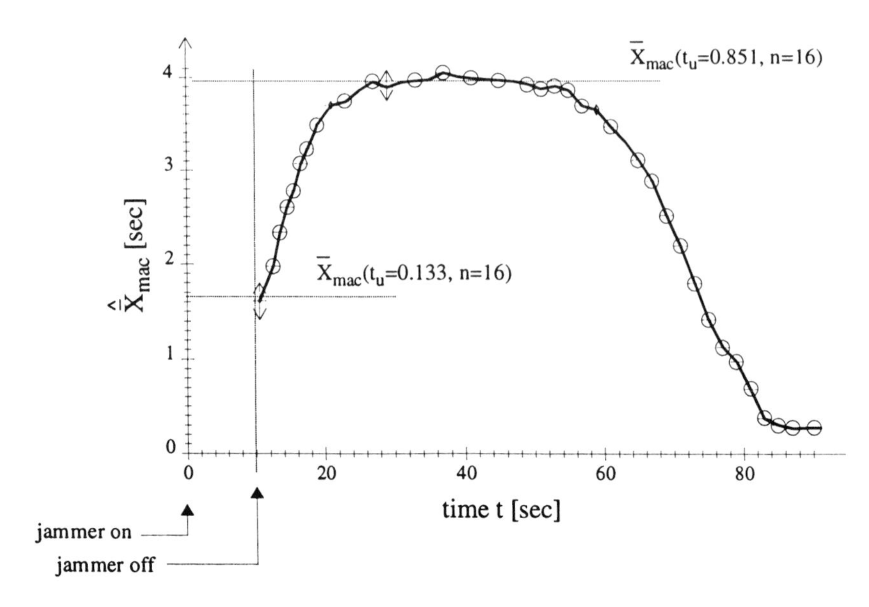

Figure 5.50 The average MAC service time

For completeness, the average MAC service time is plotted in Figure 5.50. At t_4 sixteen nodes compete for channel access, all using a low average-access delay. As they detect the high-load situation, they increase the access delay accordingly. Hence the growing service time up to the point in time where they reach the largest t_u value (851 ms). They keep on using this maximum value which gives a fixed service time as long as the heavy-load level remains. As the load level decreases, two factors contribute to the decreasing average MAC service time. The flow-control signalling system brings this information to all the nodes, which will stepwise lower their access delay until the lowest value possible is reached (133 ms). The second effect, which is the most important, is the fewer scheduling attempts needed owing to the smaller number of active nodes.

5.5 The hidden-node problem

The LLC layer provides delivery of packets across single links and the previous section considered the case where all the network nodes have direct radio contact, i.e. where the network constitutes a complete graph. Within this environment, the impact of multiuser interference was studied, including how it was possible to reduce its magnitude by adjusting the random-access delay distribution. Stable network operation could be achieved by the correct setting of the random-access distribution according to the network traffic and the network size.

Real networks must handle situations where pairs of radios are disconnected by providing the required functionality to serve traffic streams between the disconnected users, for example from node A to node D in Figure 5.51. The result is complex topologies and intricate traffic distributions.

This Section leaves the simple topology cases where all nodes are directly connected, and turns to more complex topologies. Whereas previously link-layer design and analysis were reasonably straightforward, at least by the modelling simplifications introduced in the previous Section, this Section will show that the design and analysis become much harder. This complexity is caused by many elements, but the most important is the fact that the network connectivity is a function of the user spatial distribution and the traffic on the radio channel. This traffic is again a function of the user traffic and the network-protocol functions. Thus, the connectivity between two nodes becomes a stochastic process and the network topology is not as clearly defined as it is with wired networks. This is seen by considering the relay path $A \rightarrow B \rightarrow C \rightarrow D$ where the link traffic $A \rightarrow B$ is affected by transmissions from node 4. Node 4 is a hidden node observed from A. When ARQ is used, a very strong dependency is introduced, for example between node A and node 4, even at low load levels. If the retransmission rate increases on link $4 \rightarrow B$, the congestion levels at nodes 3, 5 and 6 become higher because they share the channel with node 4. This local congestion will propagate through the network and, with the absence of a proper flow-control mechanism the network may collapse.

The link design problem is now, given the topology and the user traffic, a search for the solution which achieves the best performance. This is a considerable task and the best we can hope for is to give some guidelines. To give an illustrative example of the problem, consider a system

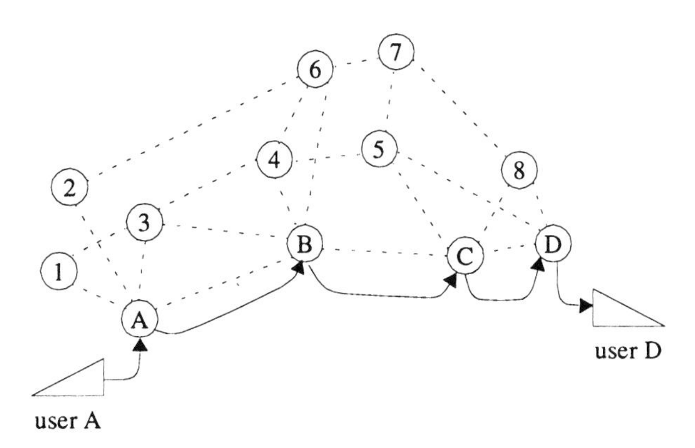

Figure 5.51 Example topology; radio connectivity marked by dotted lines

where all nodes can detect a transmission in any part of a network regardless of topology. We cannot be sure that this always will give the highest throughput due to spatial re-use. Spatial re-use means that two or more noninterfering concurrent transmissions may succeed. For example, in Figure 5.51 the link $1 \rightarrow 3$ and the link $7 \rightarrow 8$ do not interfere and this net has many disjunct links. If the user traffic is such that only disjunct links are activated, then it is easy to design a protocol and the network capacity becomes higher than the radio-transmission capacity. Another design rule could be to guarantee a minimum capacity at every location. Even with good radio connectivity, performance degradation can be experienced with the presence of hidden nodes and we expect that the price will be too high due to the large loss in the overall throughput.

The design process is composed of two main steps. The first is functional design where the goal is to describe the protocol functions needed to provide the layer services. The next step is to validate that the design meets the quality of service required. A performance analysis should always be conducted as an integral part of the design process. Much effort has been put into the performance modelling of multihop networks, but unfortunately all works rest on assumptions which are not normally fulfilled in practice. Consequently, the models end up as approximations to real systems and the magnitude of the error becomes a function of the actual network traffic. A review of the different models for multihop packet-radio networks is given in [21]. In [33] it is argued that it will never be possible to give a tractable expression for the link delay in general cases. It is difficult to imagine the possibility of having a simple explicit mathematical performance model describing a complex reality. Performance analysis must, on the other hand, be an integral part of the network-design process and this Section uses a simulation model to gain insight into network behaviour.

The example topology in Figure 5.51 is sketched based on radio connectivity. The two necessary conditions for a connection to exist are: the radio signal must be above the receiver sensitivity, and the radio signal must be strong enough to give sufficient SNR conditions. It is the second condition which introduces the dependency between traffic and connectivity. For instance, the following set of topologies might be experienced:

Topology A: two separate fully-connected networks $\{1, 3, A\}$ and $\{8, C, D\}$ if the traffic streams are restricted to staying within these two sets.

Topology B: a nearly complete network (cf. Figure 5.3 in Section 5.1) $\{B\} \cup \{C, D, 8\}$ if the traffic streams are restricted to staying within this set.

Topology C: a star network if the traffic streams are directed from the nodes in the set $\{B, 3, 6, 5\}$ toward node 4 while the other network nodes have no traffic.

Section 5.4 showed that it was possible to find efficient solutions for topology A and excellent performance can be achieved with the two noninterfering regions. Section 5.1 explained why CSMA becomes inefficient in topology B and topology C. The strategy to be used in the following is to address the topologies B and C under different protocol alternatives and from there turn to the more complex case. Before proceeding, remedies against the hidden-node problem must be considered in more detail.

Many solutions to the hidden-node problem have been presented in the literature and the two most interesting are the BTMA protocol and the RTS/CTS protocol. Section 5.1 presented BTMA as a promising technique for alleviating the hidden-node problem. The original study which concludes that BTMA with hidden nodes performs almost as well as CSMA without hidden nodes is presented in [26]. BTMA requires duplicated radio equipment and this makes it less popular for practical use. A dedicated radio must handle the data channel while the other radio serves the busy-tone channel. Besides the additional hardware, the busy tone must of course be assigned its own signalling frequency. The BTMA technique requires additional functions at the radio level and Appendix 5.A specifies the radio services and characteristics taken as the basis for this Section. Among others, the capture model used for the single-hop nets in Section 5.4 has to be revised when considering multihop cases.

Figure 5.52 specifies the three different protocol stacks to be discussed. The CSMA variant is still included since it is interesting to compare it to

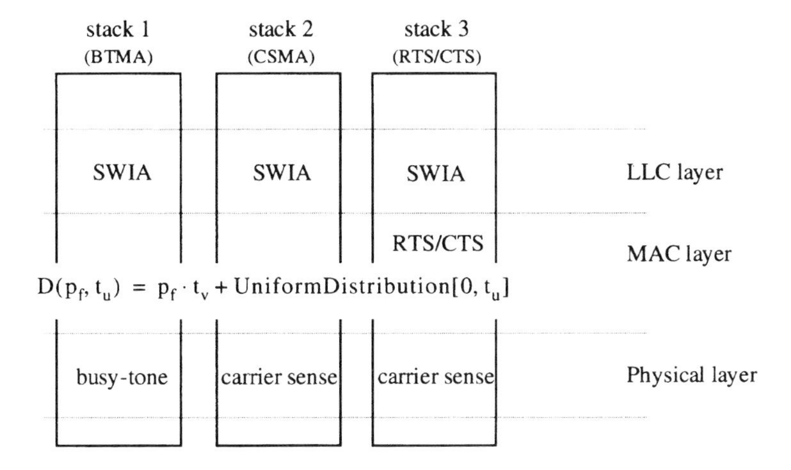

Figure 5.52 The protocol stacks

the other two techniques. All the three stacks have been implemented with the LLC stop-and-wait protocol with immediate acknowledgement (SWIA) from Section 5.4. Protocol stack 3 includes the RTS/CTS scheme presented in Section 5.1. To recapitulate, the intention of the RTS/CTS exchange is to inform the hidden nodes about the awaiting data packet such that they stay off the channel for a period of time. The stacks use the same random access delay distribution as presented in Section 5.3.

The method used to stabilise multihop networks is in principle identical to the method previously used for single-hop nets, which is to add a random-delay component prior to accessing the channel. The purpose is to reduce the collision rate by spreading the packet transmissions apart. The drawback is the longer idle periods leading to lower utilisation of the channel capacity. Section 5.3 showed that the average random delay needed to stabilise a single-hop network was a large multiple of the vulnerable period, t_v (the radio turn-time delay plus the carrier-sense time delay). With the introduction of hidden nodes this vulnerable period is increased from t_v to the entire packet length and thus we expect that large access delays must be added to stabilise multihop networks. Here we see the benefits of BTMA, which maintains the vulnerable period in order of t_v time units (this is further explained in Appendix 5.A).

Before proceeding with the performance analysis, the RTS/CTS must be extended with functionality beyond the description in Section 5.1. The RTS/CTS scheme has one passive and two active parties. The initiator which sends the RTS is denoted node A, and the addressed node which returns the CTS is designated B. The passive nodes are denoted C_x. Any node located in the neighbourhood of both A and B is classified as a C_{AB} node and is in a position where a full view of the interaction between A and B is possible. A $C_A(C_B)$ node is a node which can hear A (B) only. If a C_A node in Figure 5.53 does not detect any channel activity at $t_1 + t_{p,CTS} + t_v$, it can conclude that the RTS/CTS signalling has failed.[1] The presence of a signal does not mean that the signalling has succeeded because one of its neighbours might have started a transmission. A C_B node which receives the CTS packet at t_2, may first at $t_2 + t_{p,data} + t_v$ recognise failure if no activity is detected. This brings us to the question of for how long the different nodes should defer their transmissions? Clearly, nodes with pending CTS packets and ACK packets are always allowed to transmit (these packet types are scheduled for transmission with a zero access delay so the rule is always fulfilled). The same applies to data

[1]The packet length t_p is extended with a subscript denoting the packet type to which it applies.

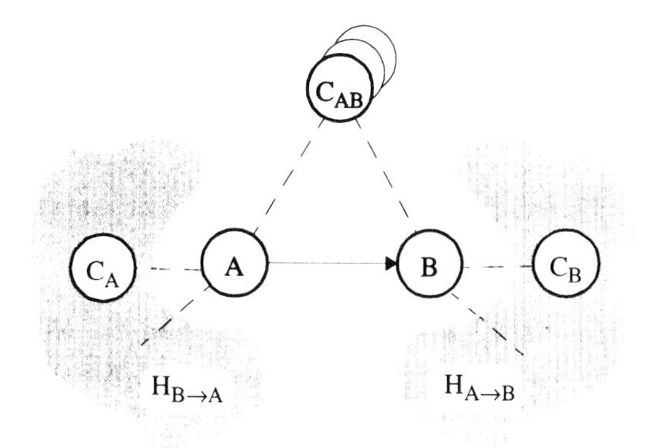

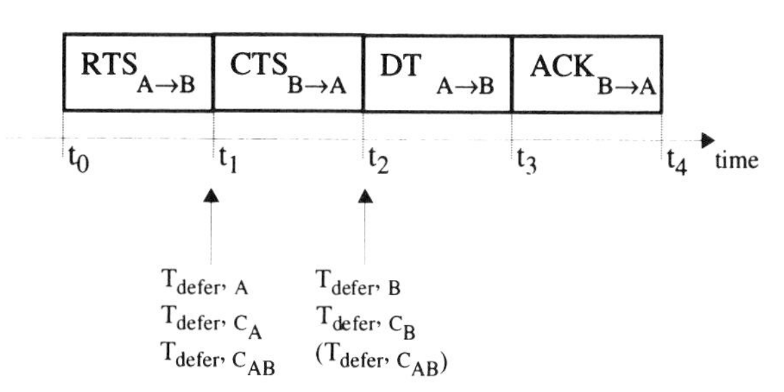

Figure 5.53 Categorisation of the node types according to their location and time-sequence diagram for the RTS/CTS protocol delivery cycle

packets to be sent as a response to CTS packets. The following five types of transmission-defer timers are needed:

$$
\begin{aligned}
T_{defer,C_A} &= t_{p,CTS} + t_{p,DT} + t_{p,ACK} \\
T_{defer,C_B} &= T_{defer,C_A} - t_{p,CTS} \\
T_{defer,C_{AB}} &= T_{defer,C_A} \\
T_{defer,A} &= T_{defer,C_A} \\
T_{defer,B} &= T_{defer,C_A} - t_{p,CTS}
\end{aligned}
\qquad (5.44)
$$

where the subscript indicates the node position in the network according to Figure 5.53.

A node may generally have one defer timer running while it gets a new RTS (or a CTS) from another neighbour. This new defer-timer period does not generally coincide with the remaining time period of the timer running and every node must therefore keep track of separate defer timers for each neighbour node.

Motivated by the performance improvements showed in the earlier sections with the introduction of a random-access delay, the same approach is used for the RTS/CTS protocol by defining a stochastic-delay component of the transmission-defer delay. In multihop nets the vulnerable period becomes the whole packet length and therefore we set:

$$T_{defer,C_A} = t_{p,CTS} + t_v + (t_{p,DT} + t_{p,ACK} - t_v) \cdot uniform[a, b] \qquad (5.45)$$

where a and b $(b \geq a)$ are two system parameters available for tuning the protocol to the operating environment. Note that the sum of the data-packet length and the acknowledgment-packet length is multiplied by a random weighting factor. This has the effect of spreading the transmissions more apart as $t_{p,DT}$ increases. Real networks must tackle randomly-distributed packet length and this strategy is expected to contribute to a more stable network. The length of the variable-sized data packet is carried by both the RTS and the CTS packets such that the nodes can learn about the reservation time period and become able to independently draw their random defer-time delay. A node is prohibited from sending a new RTS as long as one timer is active and can only respond to an incoming RTS after the fixed part of the transmission defer-time period has ended.

The originator of an RTS could learn about a failure much earlier than the other nodes and may be allowed to send a new RTS after a shorter time delay than that signalled by the RTS packet. The benefit is a potential reduction of the average channel-idle period but the method is rejected since it gives the originating node a higher precedence and consequently may introduce unfairness.

If the RTS/CTS is to give a performance gain of practical interest, six conditions must be fulfilled simultaneously:

(a) there must be a high probability for the RTS packets to reach node B;
(b) the RTS packets must effectively choke node B's hidden nodes (the nodes in the set $H_{A \rightarrow B}$);
(c) there must be a high probability for the CTS packets to reach node A;
(d) the CTS packets must effectively choke A's hidden nodes (the nodes in the set $H_{B \rightarrow A}$);
(e) the length of the data packets must be considerably longer than that of the control packets (RTS, CTS and ACK);
(f) losses must be caused by multiuser interference (and not due to other noise sources upon which the protocol has no effect).

A short packet has a higher chance of surviving than longer packets because of the reduced probability of colliding with a transmission from a hidden node. When the RTS/CTS operates in an environment where this scheme effectively prepares the way for the data packet, the longer data packet gets a better chance of reaching its destination. If the control packets and the data packets are of the same length, it would be better to skip the RTS/CTS/DT/ACK sequence and instead use the simpler DT/ACK sequence where the data packet is extended with the functionality of the RTS packet. Two other factors enhance the network's capacity. Listen-before-transmit reduces the collision rate between neighbours and added random-access delay has the potential of increasing the success rate of the RTS packets.

The performance of a radio network is related to both internal system parameters as well as environmental parameters. It is easier to give a unified presentation of the examples to follow and make comparison between them if the examples are based on a common parameter-value set. The set in Table 5.6 is therefore used as the default and any deviation from this is explicitly stated in the text.

Protocol stack 3 is the most demanding protocol to understand since the RTS/CTS protocol is definitely the most complex. The following example exemplifies its operation.

Example 1

This example focuses on the operation of the RTS/CTS protocol and the two parameters in the transmission defer timer in particular. The star network in Figure 5.54 should be an excellent candidate for studying the impact of these parameters when the traffic streams are directed toward the centre node.

As pointed out so many times before, the protocol designer should be most concerned about operation under heavy load and therefore the network is set in a heavy-load state. With a set to zero, the defer timer can expire $t_{p,CTS} + t_v$ time units after the start of the corresponding delivery

Table 5.6 Default parameter values

	parameter	value
Network data	user traffic:	
	packet arrival distribution	Poisson
	packet length distribution	fixed $t_{p,DT} = 483.6$ ms
	nodal buffer size	infinite
	$t_{p,RTS} = t_{p,CTS} = t_{p,ACK}$	51.6 ms
Radio data	Table 5.3	
	$q = 0$ (zero capture system)	
	$p_{noise} = 1$ (no background noise)	

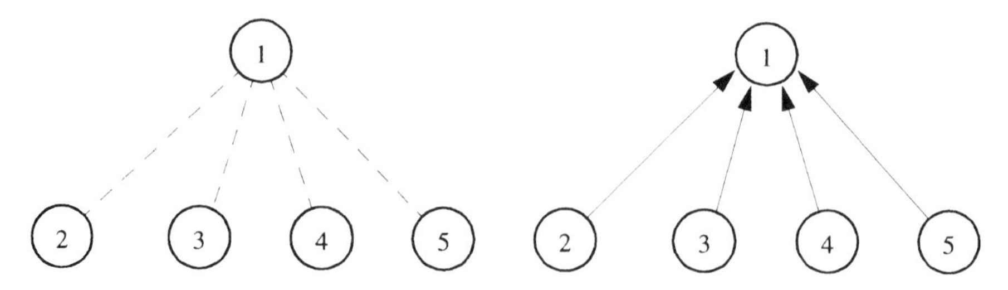

Figure 5.54 Network topology (left) and the traffic streams for example 1

cycle. Then an RTS will be sent by a hidden node within a short time limit and with a high likelihood of hitting the data packet (the acknowledgement packet is sent by the centre node so the carrier-sense function prevents a collision). The probability depends on b. The collision rate can be reduced in one of two ways, either by setting a to a larger value or by increasing b. If a is set larger than one, the transmission defer timer will never expire before the end of the corresponding delivery cycle.[1]

Figure 5.55 shows the simulated results for the star network using the default parameters in Table 5.6. The t_u parameter is set very small (i.e. 1 ms) because the carrier sensing has limited impact with this traffic configuration.

The most interesting findings are summarised in Table 5.7

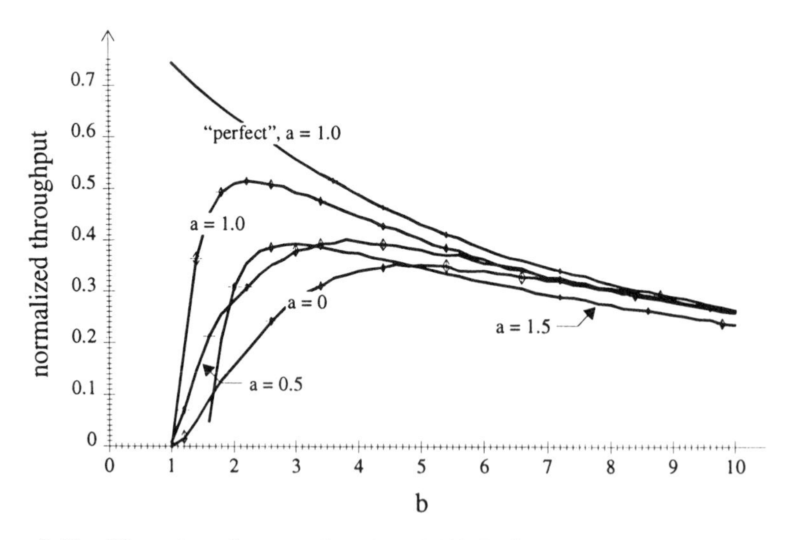

Figure 5.55 Throughput for protocol stack 3 (RTS/CTS) versus b

[1]This statement is only true if the nodes are perfectly synchronised. However, this will not be the case due to collisions.

Table 5.7 Summary of results for example 1

a	$b-a$	capacity
0	5.0	0.35
0.5	3.3	0.40
1.0	1.2	0.52
1.5	1.5	0.39

Here, $b-a$ expresses the range of the random delay at the highest throughput level and the Table shows that the largest range occurs for the smallest a. This is logical since $a = 0$, as just described, opens up the possibility of having transmissions hitting an ongoing delivery cycle. To which degree this occurs is fully determined by b and the largest b value must occur for the smallest a value.

For comparison purposes we have included the performance of an idealised RTS/CTS protocol with a set to one,[1] the curve marked by 'perfect', knowing the exact state of the hidden nodes. Under this protocol the nodes refrain from transmitting when one of their hidden nodes refrains, and thus only one RTS packet is required for each data packet since the CTS packet always succeeds in this noiseless environment. The optimum access delay with this protocol is zero. The access delay does not affect the collision rate but only degrades the performance by introducing longer channel-idle periods and maximum throughput is therefore reached at $b = 1$. At this point only the random-access delay contributes to a channel-idle period and since its contribution is practically zero, the normalised throughput becomes:

$$t_{p,DT}/(t_{p,RTS} + t_{p,CTS} + t_{p,DT} + t_{p,ACK} + t_v) = 483.6/648.4 = 0.745$$

The interesting outcomes of this example are that a should be set to one and b should be set to 2.2. However, the optimum b value is certainly a function of the number of hidden nodes and therefore it must be adjusted accordingly. Figure 5.56 gives the relationship between b values and the number of hidden nodes when the optimisation criterion is maximum throughput for the given traffic condition.

The network's capacity deteriorates slowly with increasing n, but since it is equally shared by the nodes each node has a different view. Observed by a single node its capacity decreases approximately as $1/n$.

[1] $a = b = 0$ gives the best performance for this idealised protocol.

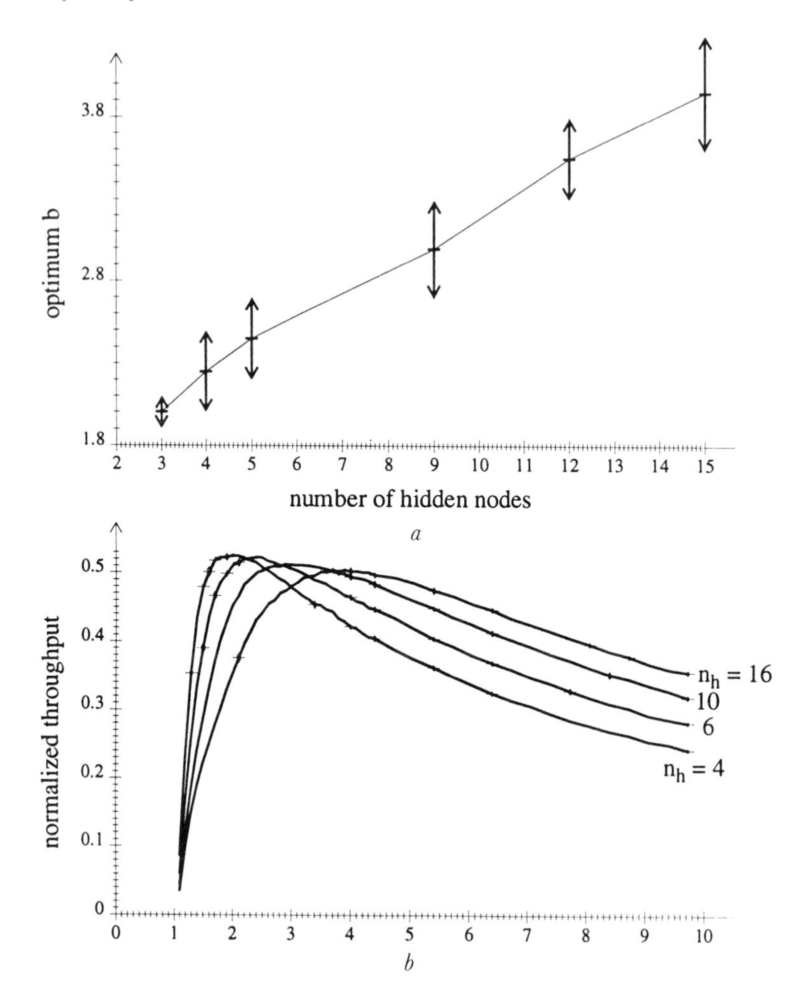

Figure 5.56 (a) Optimum b versus the number of hidden nodes in the star, and (b) throughput versus b for a different number of hidden nodes (n_h)

To start a performance comparison of the protocol stacks in Figure 5.52 from the complex topology in Figure 5.51 is not tractable because many effects interact making it impossible to analyse the significance of the different properties. We must start with a simpler topology and gradually move towards complex cases as experience is gained. The most multiuser interference-intensive topology is the star topology used in example 1, and therefore the next example considers just this topology.

Example 2

The objective of this example is to compare the performance for the various protocol stacks within an environment where the dominating challenge comes from the hidden nodes. The star network in the previous example should be an excellent candidate also for this example. Section 5.4 showed the need to add a random-access delay that was a large multiple of t_v which is the vulnerable period under CSMA for single-hop nets. When the vulnerable period becomes the packet length, we expect to need to add large delays to the packet service times for the protocol stacks 2 and 3. The benefit of stack 1 (BTMA) is that the vulnerable period remains at t_v for the case considered.

With the protocol stacks 1 and 2, the system is tuned to the traffic by adjusting the MAC parameter t_u. Since the vulnerable period of stack 1 remains at t_v while the vulnerable period of stack 2 becomes the packet length, we expect the throughput to peak at very different random delays. Figure 5.57 presents simulated throughput results for the three protocol stacks; the stacks required very different random-access delays. To be able to present the performance within the same plot, a factor β had to be introduced such that:

$$\begin{aligned} \beta &= t_u/(10 \cdot t_v) && \text{for stack 1} \\ \beta &= t_u/t_{p,DT} && \text{for stack 2} \\ \beta &= t_{ab}/t_{p,DT} && \text{for stack 3} \end{aligned} \tag{5.46}$$

where

$$t_{ab} = (t_{p,DT} + t_{p,ACK} - t_v) \cdot (b - a) \tag{5.47}$$

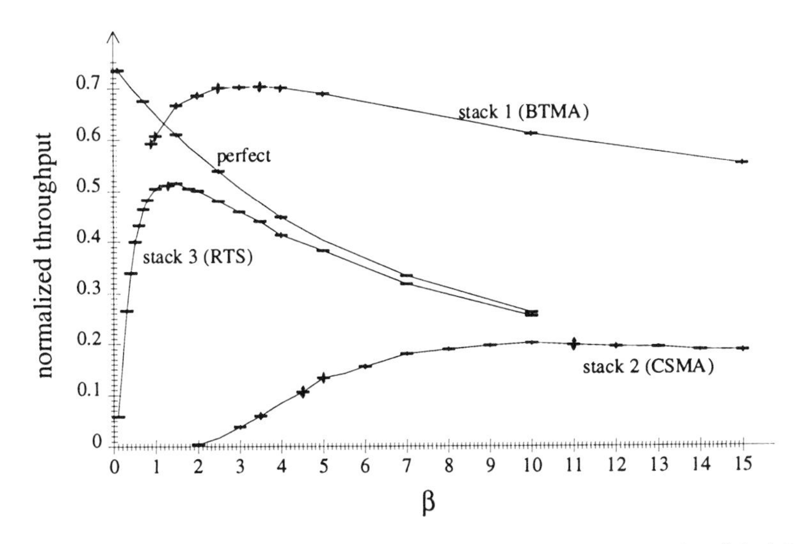

Figure 5.57 Normalised throughput versus β for the three protocol stacks; β is defined in the text

which is the range of the random part of the transmission defer timer. Example 1 explained why $a = 1$ is the best choice and with this value eqn. 5.45 can be rewritten as:

$$T_{defer,C_A} = t_{p,CTS} + t_{p,DT} + t_{p,ACK} + (t_{p,DT} + t_{p,ACK} - t_v) \cdot uniform[0, b-1] \tag{5.48}$$

The most interesting findings are summarised in Table 5.8.

The highest throughput level occurs at the point where the optimum balance between the channel-idle period and the collision rate is found. For the BTMA protocol this occurs at a much smaller access delay than for the CSMA protocol. This five-node star net turns into a fully-connected four-node CSMA net under the BTMA scheme and the imposed traffic. The throughput model in Section 5.4 confirms the estimated throughput but also the fact that a slightly higher throughput is reached by increasing t_u to 345 ms (the capacity improvement is from 0.703 to 0.704 which is a change outside the accuracy of the simulated results). As expected, the BTMA has the highest throughput, but the RTS/CTS protocol also has reasonable performance.

Note that the throughput deteriorates faster with decreasing access delay, so if the designer is uncertain of the access delay to use, he should select the largest. Any collision introduces additional channel noise thus degrading the average signal-to-noise ratio and giving less resistance against other interfering sources. Military network operators regard the jamming margin as a very important property and can sacrifice some capacity by introducing a larger access delay. This reduces the multiuser interference and gives improved resistance against jamming.

The centre node has a synchronisation effect on the leaf nodes because each time it transmits, the leaf nodes in receiving mode detect the channel-state change from idle to busy and will therefore defer from transmitting.

Table 5.8 Summary of results for example 2

Protocol stack	β	Capacity	Delay components
1 (BTMA)	3.0	0.70	access delay: 10 + uniform [0, 300] [ms]
2 (CSMA)	10.0	0.20	access delay: 10 + uniform [0, 4836] [ms]
3 (RTS/CTS)	1.5	0.52	access delay: 10 [ms] transmission defer timer: 569 + uniform [0, 725] [ms]

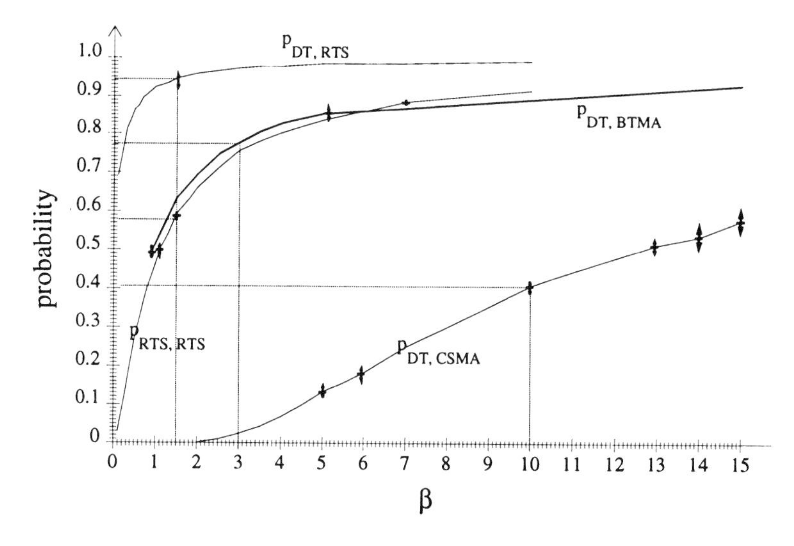

Figure 5.58 Probability of packet success against β *for the three protocol stacks*

Once the centre node stops transmitting all the leaf nodes will simultaneously start a new schedule.

To get an impression of the multiuser interference level, the packet-success probability ($p_{x,y}$) has been estimated as a function of β in Figure 5.58. The notation used is $p_{x,y}$ where x is the packet type and y is the protocol stack. The dotted lines indicate the points where maximum throughput is reached for each stack. Since $p_{DT,CSMA}(\beta_{optimum}) = 0.41$ while $p_{DT,RTS}(\beta_{optimum}) = 0.95$, data packets experience a higher interference level under CSMA.[1] As the curves show, data packets handled by the CSMA protocol can be given the same interference level simply by increasing t_u. The drawback is the reduction of the system capacity. $p_{DT,RTS}$ rises fast with increasing β which illustrates the ability of the RTS packet to prepare the way for the data packet.

The large access delay needed to reach maximum throughput for stack 2 indicates that it introduces much larger packet delays, even at a low load level, than the two other stacks. This is confirmed by Figure 5.59 which presents the throughput-delay characteristics.

The positive property with the network case in example 2 is that the nodes experience the same traffic situation. This makes it easy to draw conclusions about the best solution and the protocol parameter values to use. The next example considers a simple network topology yet a difficult case to analyse.

[1] $\beta_{optimum}$ is the β-value which gives the highest throughput.

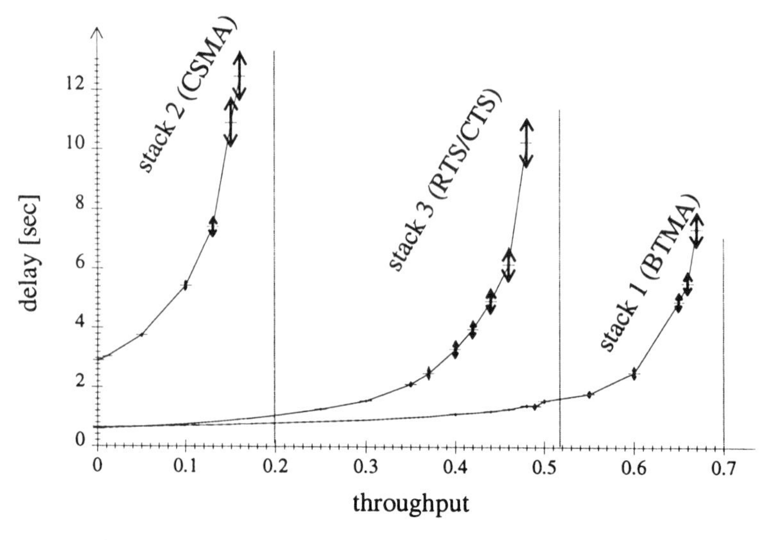

Figure 5.59 Throughput-delay characteristics for the three protocol stacks

Example 3

Consider the collection of nodes in Figure 5.60 where all the nodes $\{2, \ldots, n\}$ are in direct radio contact. One single node (node 1) has contact with one other node only (node 2).

This case gives a mixed traffic case where the set of the network links can be partitioned into five equivalent subsets:

$$S_1 : \{3, \ldots, n\} \rightarrow \{3, \ldots, n\}$$

$$S_2 : 2 \rightarrow \{3, \ldots, n\}$$

$$S_3 : \{3, \ldots, n\} \rightarrow 2$$

$$S_4 : 2 \rightarrow 1$$

$$S_5 : 1 \rightarrow 2$$

The link sets are ordered according to the challenges that a network designer meets. S_1 forms a fully-connected set of nodes without any hidden nodes. The best solution here is protocol stack 2 (CSMA). The other link sets operate with hidden nodes (a hidden node can hit the data packet or the acknowledgement packet). Link set S_3 is classified as more demanding than S_2 since the former meets the hidden node when sending the data packet. The same reasoning applies to the ordering of S_4 and S_5.

The worst case scenario arises when defining two traffic streams such that one stream is uniformly distributed among the nodes in the set

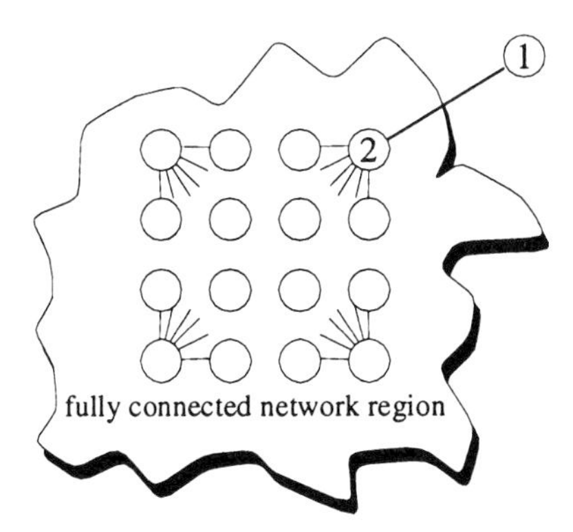

Figure 5.60 A network where node 1 is connected to one node only and the others are fully connected

$\{3, \ldots, n\}$ and the other stream is directed from node 1 to node 2. If node 2 had transmitted, node 1 would be synchronised to the next contention period in the main part of the network and thus it could compete more efficiently. Node 1 succeeds if it is the first node to transmit in the next busy period and if no other hit its packet (or eventually, the capture effect rescues it when $q > 0$). Node 2 will never inject packets into the main part of the network with this traffic pattern and thus node 2 does not inform node 1 about the state of the main part.

With the use of a BTMA protocol this network is turned into a fully-connected CSMA net, and if the bandwidth expansion needed to implement the busy tone is neglected BTMA would without doubt be the best solution.

CSMA gives excellent performance for the links in set S_1 since they do not suffer from the hidden-node problem, but gives fatal performance on the link $1 \rightarrow 2$. Example 2 showed how the RTS/CTS protocol can clear the way for the data packet, but the necessity for this is that the short RTS packet succeeded. Section 5.4 showed that the average channel-idle period for a fully-connected net under heavy load is approximately $t_u/(n+1)$ and therefore the average access delay has to be increased beyond the optimum value for the link set S_1. Since the set S_1 contains a large number of links, the overall network throughput is reduced by this action. The problem with this topology is that the highest overall network throughput does not coincide with the highest link throughputs and the network designers' dilemma is whether or not the overall network throughput should be sacrificed for the benefit of link $1 \rightarrow 2$.

Assume that the selected strategy is to implement both protocol stacks 2 and 3. The outgoing traffic from the nodes in the set $\{3, \ldots, 10\}$ uses the simple CSMA variant but the nodes react to incoming CTS packets (sent by node 2) in conformance with the RTS/CTS protocol. The RTS/CTS protocol is used on link 1→2 which suffers from the hidden-node problem.

Figure 5.61 presents simulated throughput for link 4→3 and link 1→2 against the network offered load. Node 4 is located in the well connected region. All the network nodes employ the same Poisson packet-arrival distribution but two factors contribute to different node throughput. Node 1 uses the RTS/CTS protocol which sends additional packets for signalling purposes and thus introduces more transmission overhead. Link 1→2 also operates under a higher collision rate than the links in the set S_1.

Prior to sending data node 1 must have successfully delivered the shorter RTS packet to node 2 (and received the CTS). This can only occur within a time window determined by the channel-idle period in the well connected region. The average time window is large at low load levels and node 1 seldom fails. Upon increasing traffic the window decreases and reaches its smallest size when every node in the set {3, . . . , 10} has a packet for service at all time instances. The average time window is approximately given by $t_u/9$ and stays fixed regardless of further increasing traffic. The success rate for node 1 is determined by the size of the time window and the Figure shows that its minimum size is sufficiently large to let some RTS packets pass through. Node 1 throughput stays constant because the time

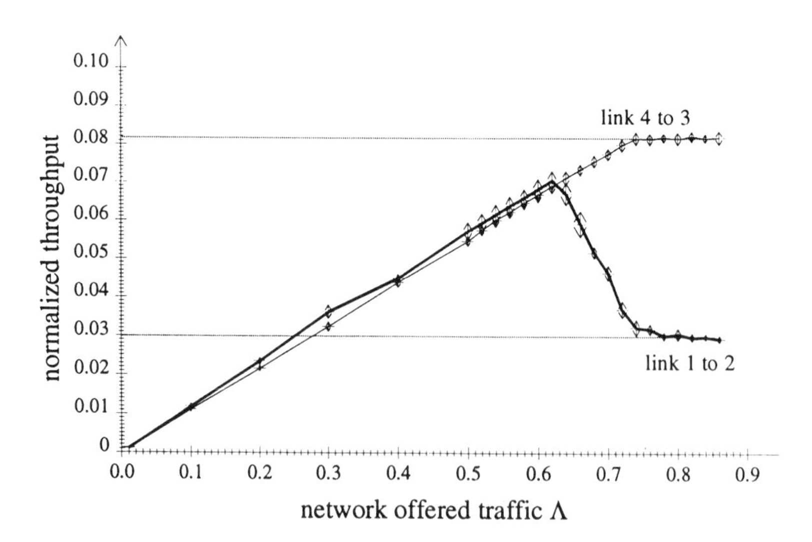

Figure 5.61 *Link throughput versus the network-offered traffic*
Network parameters used:
network size $n = 10$
stack 1 protocol parameters $t_u = 800$ ms
stack 3 protocol parameters $(t_u, a, b) = (800\text{ ms}, 1, 2)$

window is not modified after the well connected region has reached saturation point.

The network has ten nodes but one does not send data packets, and the initial node throughput is thus 1/9 of the network throughput. The throughput grows linearly up to the point where the first link(s) reach the saturation point. As expected link $1 \rightarrow 2$ reaches saturation first and this occurs at $\Lambda = 0.62$. After this point its throughput starts to drop while the link throughput for the nodes in the set $\{3, \ldots, 10\}$ continues to increase up to $\Lambda = 0.74$. Beyond this level the throughput remains constant.

If the network designer is not satisfied with the resulting performance, he is faced with three options:

(i) Restrict the network input traffic to $\Lambda = 0.62$.
(ii) Enlarge the channel-idle period in the well connected region by increasing t_u for the nodes in the set $\{3, \ldots, 10\}$.
(iii) Change the RTS/CTS protocol parameters to find better values seen from node 1's point of view.

In the preliminary discussion of the complex topology in Figure 5.51 it was concluded that the network could behave as a fully-connected net, a star or as a nearly-complete net depending on the imposed user traffic. All the network links would probably carry some traffic in a typical real case and the resulting characteristic becomes a mix of the different topologies. This leads to a very difficult case to analyse. For simplicity of the forthcoming discussions, all the 21 network links are offered the same user traffic, that is, 42 identical traffic generators generate the user packets independently for each link. The offered user traffic will then not be the cause of different link throughputs. Another problem is the huge number of links. 42 links are present and cannot be divided into a few equivalent sets as was done in example 3. Thus 42 curves are needed for displaying the link throughput. This is impractical and therefore only the overall network throughput is considered (the network throughput is the sum of the throughput of all the links since only single-hop traffic is applied).

To establish a parameter set that can be stated as the optimum set is not possible in this nonhomogeneous network, or at least not immediately, but examples 2 and 3 give some guidance on the initial values to use. Example 2 indicates that maximum throughput is achieved for $t_u = 300$ ms and $t_u = 4836$ ms for stack 1 (BTMA) and stack 2 (CSMA), respectively, for a five-node star. Since Figure 5.51 contains one five-node star, these values are also used even though it is certainly not the optimum choice for all the network regions. However, large access delays give better network stability and smoother degradation. The access delay for stack 3 (RTS/CTS) must be set higher than in example 2 if the carrier-sense function is to reduce the collision rate among competing neighbours. Therefore $t_u = 300$ ms is used for stack 3 and the defer timer parameters are $a = 1$ and $b = 2$.

To make intuitive judgements about the expected performance of the different network regions is hard, but a review of the topology indicates that node B has a worse starting point than node 1 since the former shares the radio spectrum with five adjacent nodes while node 1's number of neighbours is only two. To extend the reasoning to also cater for the effect of hidden nodes is impossible, e.g., which of the two links $B \rightarrow 3$ and $B \rightarrow A$ will have the highest throughput? Without considering the expected performance further on an intuitive basis, simulations are performed instead and Figure 5.62 presents the results.

The most surprising result is the much better capacity under CSMA than seen in example 2. A number of factors contribute to this. The average nodal degree is larger now and the prerequisite of a well operating carrier-sense function has been enhanced. Moreover, the number of non-interfering links has increased such that spatial re-use contributes to higher throughput. Although CSMA gives satisfying capacity, it still suffers from a large delay due to the long random-access delay required.

The second observation worth mentioning is the stability of the network. The throughput curves do not decrease for any of the protocol stacks as the offered traffic increases.

A user data packet arriving at an empty node is immediately scheduled for transmission if the node is allowed to transmit. Transmission is prohibited for various reasons with the three protocol stacks, e.g. stack 3

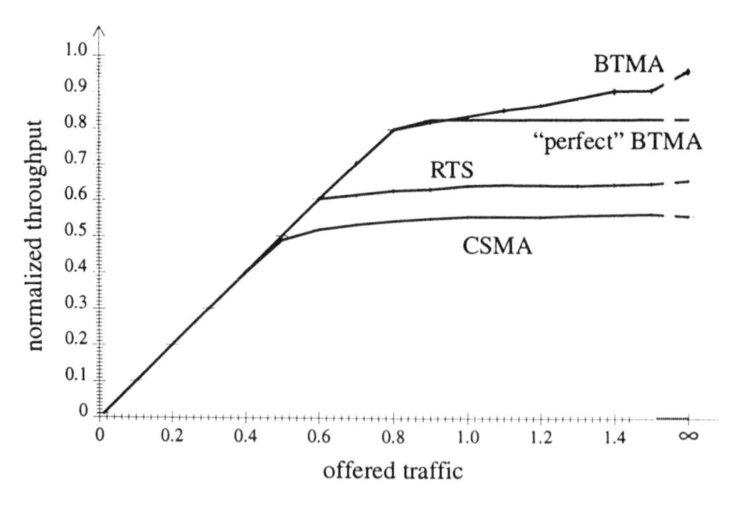

Figure 5.62 Simulated performance results for the complex topology; note the discontinuity of the offered traffic

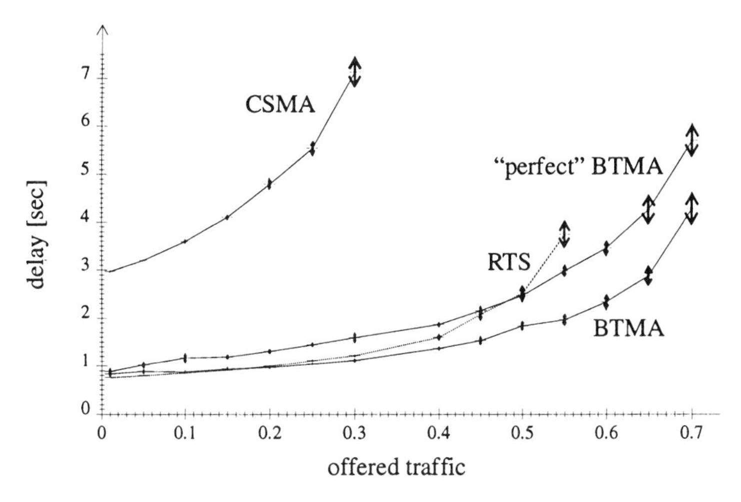

Figure 5.62 (continued)

(RTS/CTS) demands an idle radio channel and no transmission defer timers running, but none of the stacks introduces additional delays at low load levels. Only one packet transmission is needed in a noiseless system and therefore the average network transit delay in a noiseless lightly-loaded system is simply:

stack 1 (BTMA): $t_v + t_u/2 + t_{p,DT} = 0.6436\,\text{s}\ [0.643, 0.646]$

stack 2 (CSMA): $t_v + t_u/2 + t_{p,DT} = 2.9116\,\text{s}\ [2.887, 2.936]$

stack 3 (RTS/CTS):

$$t_v + t_u/2 + t_{p,RTS} + t_{p,CTS} + t_{p,DT} = 0.7468\ \text{s}\ [0.745, 0.749]$$

The numbers enclosed in brackets represent the corresponding simulated 90% confidence intervals. As the user traffic increases, the usage of the channel becomes more frequent and this affects the packet-service time in two ways. First, the number of nodes competing for channel access grows and from the point of view of a single node, the available channel capacity decreases since it must share the bandwidth with other nodes. Secondly, the probability of collision rises, primarily due to the hidden nodes, and the number of transmissions needed until an error-free acknowledgment is returned increases. This feedback between channel traffic and packet loss is the cause of the fast dropping throughput in a network without proper protocol parameter values. When the packet service time becomes longer, an arriving packet will more often find the node busy serving an older packet and it must be placed in the node queue. The queue starts to build up and as the user traffic continues to increase the queue length grows accordingly. The radio-channel saturation point is reached when all the network nodes have a packet under service at all time instances and then the node experiences the largest delivery times. The user traffic can only increase beyond this point for short periods of time. Otherwise, network buffer overflow occurs and the network has to discard the user packets.

A performance measure often of interest is the end-to-end delay seen by the users, for example the network-transit delay between user A and user B in Figure 5.51, which is a three-hop route. The estimated delay reflects the average delay over all the network links while the true delay along a particular route depends on the traffic through its links as well as the traffic on the adjacent nodes. Thus only true end-to-end delay measurements can give the correct picture, but these require as many as $12 \cdot 11 = 132$ delay values to be presented.

The performance engineering is not completed but the system must be investigated for other parameter and traffic conditions (i.e. different packet lengths and traffic pattern) before conclusions can be reached about the quality of the communication services. This is a tremendous task and the results are too numerous to be presented. Only a selected set is therefore given in Table 5.9. Figure 5.63 shows the resulting link throughput and clearly illustrates the region dependency.

Apart from multiuser interference, real systems must operate with foreign interfering sources, hereafter simply referred to as noise. The network protocols need to maintain communication over noisy radio channels but in contrast with the multiuser interference, the noise level remains fixed regardless of the network parameters. Noise leads to imperfect carrier/busy-tone sensing and packet corruption (by introducing bit errors). Packet corruption is the more severe of the two.

All the protocol stacks use an activity-sensing technique which becomes less reliable as the noise level increases. However, stack 3 meets an additional problem since the nodes do not send data packets unless they receive error-free CTS packets. This becomes evident when considering a lightly-loaded network where the multiuser interference can be neglected.

Table 5.9 Results after the optimisation phase under the heavy-load state

Protocol stack	Parameters	Normalised network throughput
1 (BTMA)	$t_u = 450$ ms	[1.019, 1.035]
2 (CSMA)	$t_u = 3400$ ms	[0.582, 0.600]
3 (RTS/CTS)	$(t_u, a, b) = (100\text{ ms}, 1, 4)$	[0.721, 0.733]

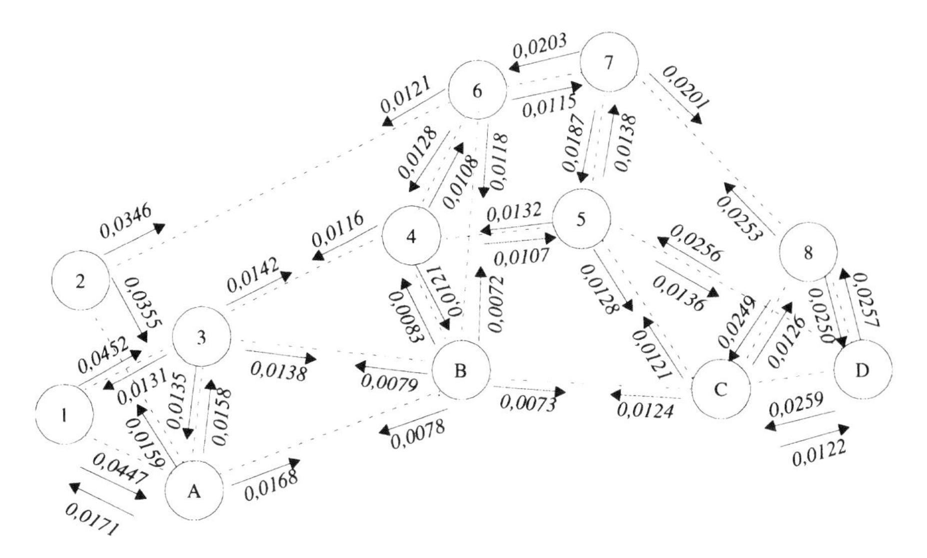

Figure 5.63 Final link throughput (estimated averages) for protocol stack 3 under the heavy-load state

The probability of having a successful RTS/CTS sequence is p^2_{noise} and the probability of having a successful DT/ACK sequence is the same for the three stacks. If the nodes refrain from transmission, the data packet can never reach the destination and contribute, and thus stack 3 degrades faster than the two others. This is confirmed by Figure 5.64 which shows estimated throughput against background noise. If the RTS and CTS packets serve their mission, namely choke the hidden nodes for a period of time, the packets must be received by the hidden nodes but the probability that one or more nodes misses rises quickly with increasing noise and the number of hidden nodes. The Figure shows that the final results are far from pleasant.

The capture model used gives a somewhat pessimistic view of the RTS/CTS scheme since it does not distinguish between different packet lengths; the shorter RTS/CTS packets have a better chance of surviving than the longer data packets. To reach more precise conclusions about the performance under background noise requires a more detailed modelling of its effects.

By this time the reader should be aware of the impossibility of finding one single solution giving the highest performance for all user-traffic distributions, and of the contradiction between maximum link-by-link throughput and the overall network throughput. The network topology is a function of the user spatial distribution and the user traffic. Since both generally are time variant in a real network, the network topology

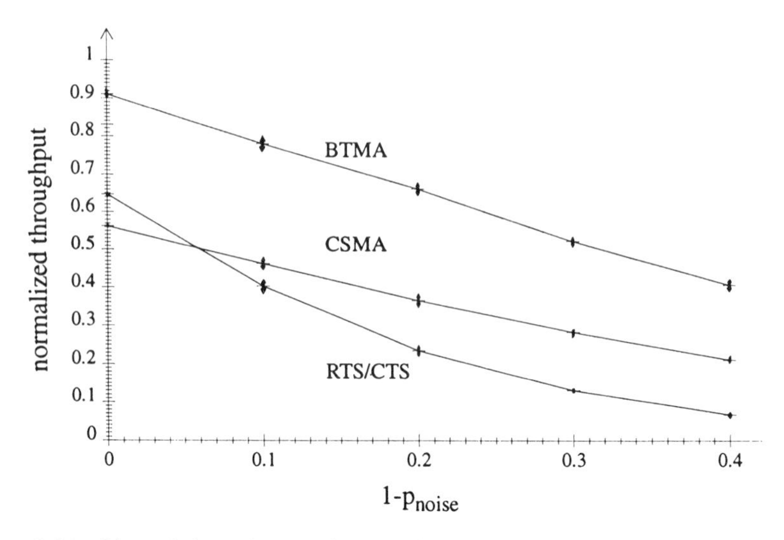

Figure 5.64 Network throughput against noise level

changes over time and the best solution will also vary over time. A network designer could make an attempt to adapt to changing traffic conditions by implementing a dynamic assignment technique. Each network link is carefully monitored by a process which makes a decision about the protocol to use by estimating traffic conditions. For example, it selects the RTS/CTS scheme for links with hidden nodes. The final benefits of this approach depend on a huge number of factors. The network-design process is therefore about finding the best compromise between a set of factors.

A dynamic assignment strategy is not considered here but to provide an impression of a more dynamic scene, the next example considers a network where the topology changes over time.

Example 4

Suppose we have a number of nodes distributed over some geographical area where all the nodes are within the radio-coverage area of each other. Then consider what happens as the network topology gradually moves toward lower nodal degree as for the network in Figure 5.65. This can be caused by mobility, or changes in the radio-channel quality. The impact of the hidden nodes will rise as the network fragmentation increases. At the start time t_1 the topology constitutes a fully-connected nine-node network and as the time passes, the radio link length changes such that:

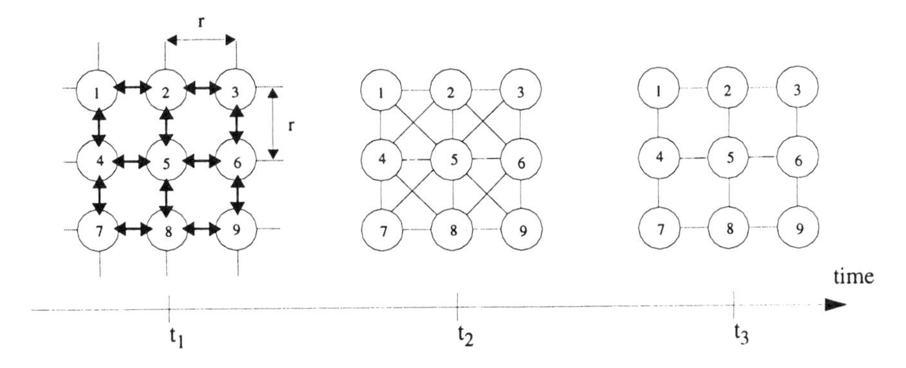

Figure 5.65 The network spatial distribution for example 4. The arrows show the traffic streams setup; the traffic streams remain fixed over time

time t_1 : *transmission range* $> 2\sqrt{2} \cdot r$ (fully connected net)

time t_2 : *transmission range* $= \sqrt{2} \cdot r$

time t_3 : *transmission range* $= r$

The user traffic pattern is marked by the double thickness arrows in Figure 5.65 and the packet arrival rates are set high enough to guarantee a heavy load state. The nodes will exhibit different behaviour even though the example strives to set up a homogeneous scenario by defining symmetric traffic generators. Another benefit of these traffic streams is that they do not invoke the routing function as the topology alters. This is important since the objective is only to study the effect of hidden nodes.

The lesson learned from the above is that stack 3 (RTS/CTS) is inefficient in fully-connected nets, CSMA is inefficient with hidden nodes and BTMA tackles both cases. The network designer has to choose a strategy for selecting protocol parameter values since one single set does not optimise the bandwidth usage for all time instances. This example presupposes the strategy: optimise protocol stacks 1 and 2 for maximum throughput at time instance t_1 and stack 3 for time instance t_3. The next step is then to find the best parameter set. Stacks 1 and 2 have one parameter, t_u, to be tuned to the traffic and since BTMA performs identically with the CSMA in a fully-connected net, an optimal t_u value can simply be found from the throughput model presented in Section 5.4. This model shows that the highest capacity is reached at $(t_u, \lambda) = (0.72\,\text{s}, 0.696)$.

Picking feasible parameter values for stack 3 is far more demanding but $t_u = 87$ ms and $a = 1$ should be safe. The motivation behind this low t_u

value is the low average nodal degree at time instance t_3. Thus the carrier-sense function has limited impact and the access protocol operates as a 1-persistent CSMA protocol (when serving the RTS packet). $t_u = 87$ ms gives the highest system capacity in a two-node net when serving packets of the same length as the RTS packet. The drawback of using a simulation methodology is that many simulation runs are needed since a large range of b values must be exercised before one value can be stated as the optimal value. This is time consuming and maximum throughput is observed at $(b, \lambda) = (3.8, 0.61)$.

With the system parameters in place, simulation can start and the estimated steady-state performance measures are given in Figure 5.66.

All three protocol stacks show capacity improvement as the network fragmentation increases due to spatial reuse with the exception of stack 1 (CSMA) which degrades quickly from t_1 to t_2. The parameter set reflects the network traffic at one particular time instance and will not be optimal for other time instances. In practice, one has to trade-off between the time instances.

This section has built up a picture of a packet-switching radio network as one which transmits independent data packets over single links. However, real networks must be prepared to handle streams of data that are set up on an end-to-end basis. The next section turns the focus on the procedures needed to move the packets through the network in a store-and-forward fashion under time-variant topology configurations.

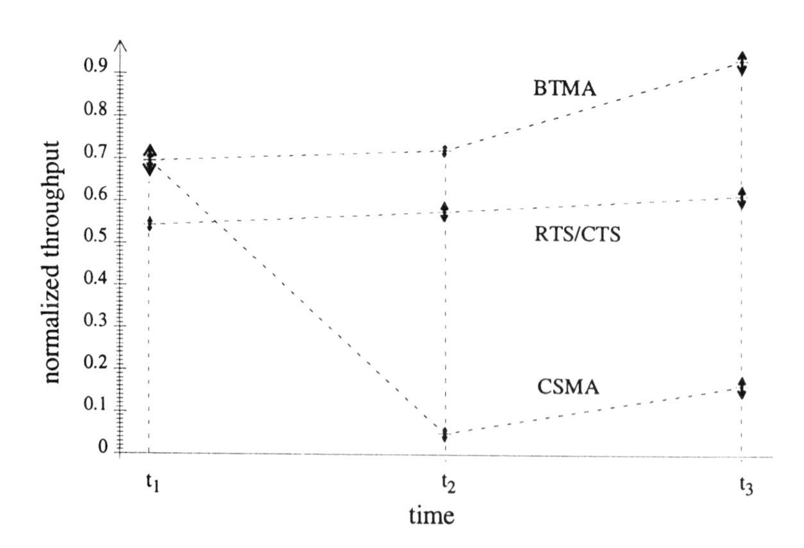

Figure 5.66 Steady-state throughput at time instances t_1, t_2 and t_3

5.6 Network layer

The previous section discussed methods for packet exchange between adjacent nodes within a general network topology. The quality of the communication service was enhanced by an automatic retransmission function. The next issue pertains to techniques for moving packets between end users through the network while supporting the quality of service requested by the users. The following lists some of the highest ranking functions for the network level design:

- *Relaying*: the procedure for store-and-forwarding packets between end nodes. Should the nodes increase their transmission power to reach the end destination in one hop? Should the network nodes send a relay packet to more than one other relay node for increasing the probability of successful delivery?
- *Retransmission*: should the network nodes retransmit packets on a hop-by-hop basis, or is it better to let the originating node take this responsibility?
- *Flow control*: if a link protocol can prevent network collapse, is additional traffic control needed at the network level?
- *Routing*: how shall the network find end-to-end paths between all pairs of users?

Routing in dynamic radio networks is a very comprehensive subject and is covered in detail in Section 5.7. The design decision which has the greatest impact on the final solution is relaying. Once this strategy has been selected, the degree of freedom for the two latter functions, retransmission and flow control, has also been reduced; therefore we will start by considering relaying.

5.6.1 Relaying

The relaying function has the overall responsibility for providing packet exchange between the end users. It extends the network service-coverage area beyond the radio-coverage area of single nodes by using one or more nodes as intermediate relays. The goal is to maintain continuous communication services with fixed quality under mobility. Earlier sections have shown that the available transmission capacity varies within the service area and to provide a constant quality over time is rarely possible.

Relaying has been addressed as a separate function since it is very important for providing packet-switched services. However, relaying uses and works in close co-operation with other functions, for example routing. Routing assigns the paths to be used between the end nodes but must be told about the quality experienced on the existing routes. Relaying possesses this information.

In the hypothetical example in Figure 5.51 routing states that the path $A \rightarrow B \rightarrow C$ shall be used towards node D. Assisted by the time-sequence diagram in Figure 5.67, drawn under the assumption of having the SWIA protocol described in Section 5.4 implemented at the link level, we take a closer look at what happens on the relay path $A \rightarrow B \rightarrow C \rightarrow D$:

t_0: the user attached to node A has just delivered a packet addressed to the user located at node D. Node A analyses the destination address, gets the next hop from routing and reliable delivery to the next node is provided by the ARQ protocol at the link level. Node A contends with the other busy nodes in the neighbourhood for access to the radio channel.

$t_1 - t_3$: node A gets access to the channel, transmits the packet and receives the acknowledgement. The transmission on the first hop is completed from node A's point of view and node B takes over.

$t_4 - t_6$: the same procedures as those just described for t_1 to t_3 are repeated.

Here we take a break in our passage of the time-sequence diagram and focus on the time instant t_5. Because node A is within radio range of node B, A is able to receive B's transmission, which is able to let this confirm receipt of the data packet sent in $[t_1, t_2]$ and therefore node B need not send an explicit acknowledgement. This type of acknowledgement, where no explicit acknowledgement packet is sent, is called passive acknowledgement. To implement passive acknowledgement each packet has to be assigned a global unit identifier (GUI)[1] which enables node A to distinguish this packet from the other packets sent by B. Figure 5.68 shows the time-sequence diagram for passive acknowledgement. On the last hop the end node D must confirm receipt by sending an explicit acknowledgement.

Here we see two design alternatives and an interesting challenge is to decide which of the two gives the highest performance. The two schemes are compared by breaking up the delay components. X_i denotes the MAC service time (see Section 5.3), Q_i denotes the queuing delay and subscript i is the node number. By inspecting Figure 5.68, the forwarding delay D_F for node A in the case of a passive acknowledgement can be written as:

$$D_{F,passive} = \sum_{i \in \{B, C\}} (Q_i + X_i) \tag{5.49}$$

[1]A GUI can be assigned by the node at which a packet enters the network. The three components' source address, destination address and sequence number are sufficient to form a GUI if the sequence-number range is large enough to prevent reuse within the lifetime of a packet.

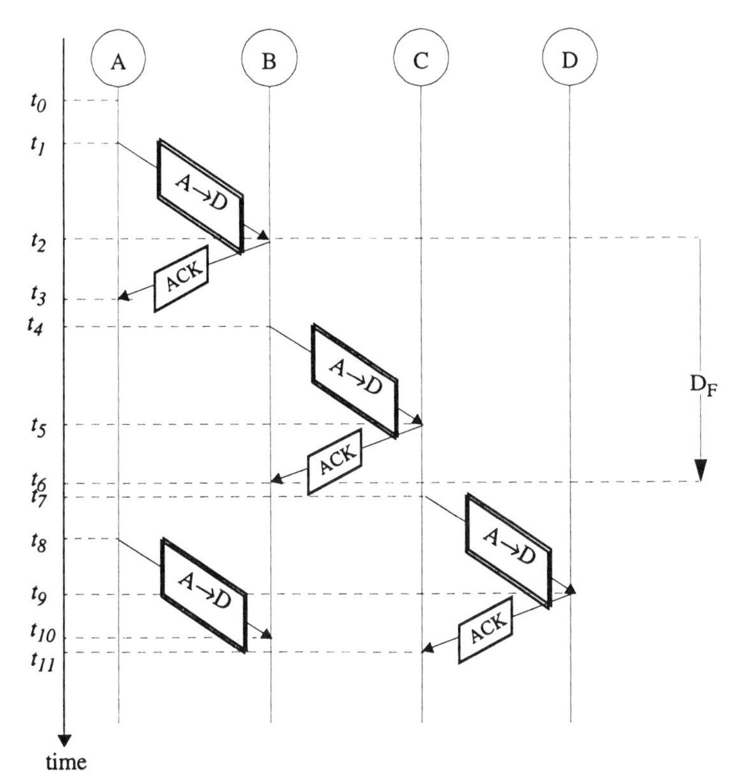

Figure 5.67 Time-sequence diagram for relaying with the use of explicit acknowledgement

Node A must stay off the air during this period so as not to interfere with its own packet forwarding. Figure 5.67 gives the result for the explicit acknowledgement method as:

$$D_{F,explicit} = Q_B + X_B + 2t_{p,ack} \tag{5.50}$$

Q_i and X_i are random variables and must therefore be estimated by A. It is possible to implement procedures to measure $Q_B + X_B$ but since C is hidden from A, node A cannot easily sample the distribution $Q_C + X_C$. The best A can do is to assume that the traffic condition at C is nearly the same as that at B, and use the same estimate for both. Also note that the average of eqn. 5.49 is greater than the average of eqn. 5.50 since the former contains twice the number of queue delays and service times.[1] Thus, node A must take longer breaks with passive acknowledgement but

[1]This statement is only valid at low traffic levels where $D_{F,passive} \approx 2t_{p,data}$ and $D_{F,explicit} \approx t_{p,data} + 2t_{p,ack}$ and when $t_{p,data} > 2t_{p,ack}$. To make a general statement is hardly possible owing to the strong dependency on topology and traffic.

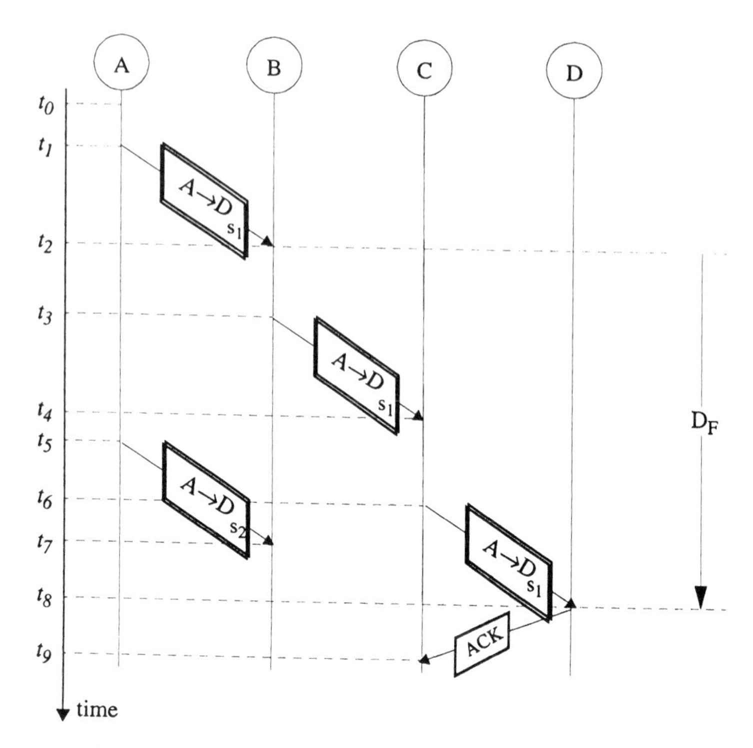

Figure 5.68 *Time-sequence diagram for relaying with the use of passive acknowledgement. The packet transmitted at t_5 interferes with B's passive acknowledgement. The last hop must use the explicit acknowledgement method*

this does not necessarily lead to worse performance since other effects may compensate for this (e.g. decreased number of retransmissions).

The above discusses the actions to be taken by the originating node *A* in order not to destroy its own store-and-forwarding operation only. Reduction of the network signalling traffic is achieved by letting the forwarded data packet serve as an acknowledgement, but since data packets normally are longer than an acknowledgement packet, they have a higher probability of being hit by a hidden node. The hidden node problem and possible countermeasures were discussed in Section 5.5 and will not be treated further here. Instead, methods for improving the relaying function are looked at.

A strategy for improving the data-packet forwarding success rate is to use point-to-multipoint relaying. With this method a relay packet is at each hop sent to more than one node for further relaying. For example, if node *B* in Figure 5.51 addresses 'all-hearing-this-packet' the nodes {3, 4, *C*} get the responsibility of serving this packet. Here some filtering mechanism is clearly needed to limit the number of new packets generated in the next hop. The two rules applied in [19] are: (i) the relay node must be at a smaller distance (in number of hops) from the end destination than the last transmitter and (ii) upon hearing a faster forwarding by one of its neighbours, the relay node removes the packet from its buffer. Duplicates may still be generated by hidden nodes, or by nodes that are not able to detect the next forwarding (exposed to an overlapping transmission or captured by another transmission). The effect of the point-to-multipoint system is to circumvent hidden nodes by using a path passing around difficult network regions. It does not directly improve the acknowledgement-success probability.

What about using power control for adaptive transmission ranges? High power gives increased network connectivity, contributes increased progress towards the destination and reduces the number of simultaneous transmissions under CSMA. A few studies [22] consider variable transmission ranges.

5.6.2 Retransmission

All packet-switching networks apply retransmission to enhance the quality of the communication service. The difference is the level of its application, hop-by-hop or end-to-end, and the protocol scheme used. Discussing retransmission strategy is actually a discussion of where to place the ARQ protocols. If retransmission is performed on an end-to-end basis, the ARQ protocol operates directly between the node where the data packet enters the net and its end-destination node and thus the acknowledgement confirms that the data packet has reached the exit node.[1] Retransmission hop-by-hop addresses the cases where the ARQ protocol operates between nodes in direct radio contact and the significance of a received acknowledgement is only that the adjacent node has received the packet.

For a moment, assume that retransmission of lost packets is done at the link level only. A route found to be excellent at the start of a user-communication session may become degraded during the session, for example due to user mobility, and the network must find an alternative route. Although this should be transparent for the users, some of the packets may experience a situation where they are discarded within the network (a node turns its power off, moves outside the service-coverage area, etc.). An analysis of the likelihood of this situation can only be done for a particular network under well defined conditions. If the user

[1] An equivalent solution in this context is to let the transport protocol in the end system perform retransmission of lost packets.

requirements are such that an upper limit of unsignalled packet loss shall always be provided, an ARQ protocol must operate with end-to-end significance.

The conclusion is, retransmission must generally be implemented at an end-to-end level, but then is it necessary to perform retransmissions at the hop-by-hop level? There are several points of importance here. A typical packet route contains radio links of varying quality, some bad but hopefully most with a high probability of packet success. If packets are not retransmitted on a hop-by-hop basis, a packet which is lost on an intermediate link will be retransmitted from the point where it entered the network and again traverse the high-quality links. Thus, when traffic streams are set up between high-quality network regions and low-quality regions, additional traffic is imposed on the high-quality regions. ARQ protocols should therefore be implemented on a hop-by-hop basis for better utilisation of the transmission capacity. The retransmission rate at the link level is also a direct measure of the link quality and might give valuable link-status information to the routing function which shall determine the best route to use. Without an ARQ protocol at the intranetwork level, measurement reports on the link qualities become less reliable. Slower adaptation to the changing environment is also expected. Moreover, a hop-by-hop ARQ protocol eases the detection of congestion at remote regions since only a small number of outstanding data packets is allowed.

5.6.3 Flow control

Section 5.4 discussed some aspects of flow control but most attention was paid to stabilising the medium-access protocol. Many of the problems encountered in hard-wired packet-switched networks such as store-and-forward deadlock and reassembly deadlock [23] are also met in radio networks and the flow-control schemes developed can serve as the basis for radio nets. This Section takes a broader view of flow control through Figure 5.69, identifying the level of flow control.

The node-to-node level embraces flow-control methods working between adjacent nodes. The classical example is a receiver which stops the sending node by using appropriate commands. An ARQ protocol restricts the output traffic between peer nodes but a receiver cannot delay the acknowledgement packets to perform flow control because this leads to time-out at the sender and unnecessary retransmissions are the result. But the acknowledgement packet gives a good opportunity to signal flow-control status information back to the

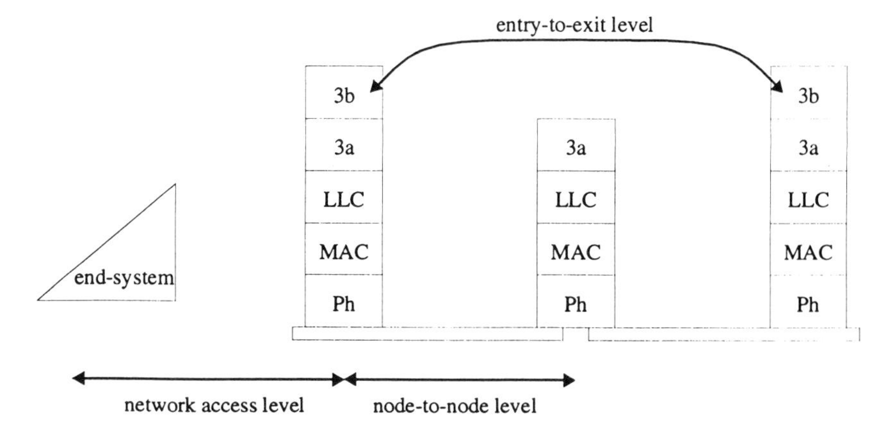

Figure 5.69 Flow-control levels

sender by extending its control field. Moreover, the broadcast channel in radio networks gives the possibility of reaching all adjacent nodes by one single transmission and thus many traffic sources can be turned off simultaneously.

The entry-to-exit-level flow control works directly between the end-source and end-destination node with the objective of protecting the end-destination node from being congested by a fast sender. If an appropriate node-to-node flow control is implemented, the congestion moves upstream towards the source, but it is not desirable to let congestion information propagate by backpressure. Instead a credit mechanism is often used by which the end-destination node tells the end source about the number of packets that can be accepted. This credit can also be applied for buffer reservation such that reassembly deadlocks are prevented.

Network-access-level flow control has the objective of regulating the input stream of new packets based on local and/or global load-level measurements. This function must define the precedence ordering between the new input traffic and the relay traffic depicted in Figure 5.70.

An ideal observer can monitor the capacity used along the path and acquire instantaneous and complete status information. However, a single node can only inspect the capacity used within its neighbourhood and perform flow control locally by placing selective restrictions on the capacity which each stream may acquire. If the nodes shall operate more like an ideal observer, real-time measurement of how much capacity each stream captures must be implemented in a dynamic and heterogeneous environment. This demands bandwidth for additional signalling and can even introduce unfairness.

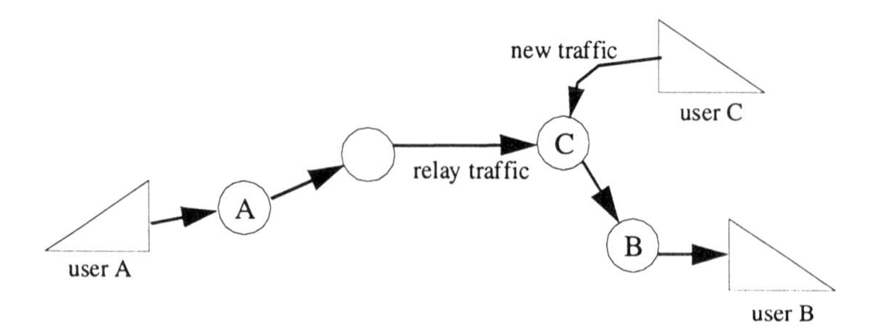

Figure 5.70 Network-access flow control and fairness. If node C favours transit traffic over new traffic then C's user packets are first sent when node C's relay queue becomes empty

5.7 Routing

One of the most important tasks of the 3a layer is relaying. This is closely related to routing which is the task of deciding where to relay packets submitted by users (attached to the local network node) or received from neighbouring nodes. Generally, in a packet-switched wireless network there is more than one alternative path from the sender to the receiver. And even if there is only one path, there is still the problem of how to find it. In the case of multiple paths, the problem is to choose the most cost-effective one.

The primary object of routing is to optimise the network performance in the sense of maximising the end-to-end throughput of user traffic. This is done by choosing routes which minimise the use of network resources. But this task is met by two challenges related to routing in dynamic (mobile) networks:

(i) routing consumes network capacity due to the need for information distribution caused by changes in the network topology;
(ii) the performance of the routing function is dependent on correct network information at any time; in principle, this is not possible in a dynamic network.

Routing in dynamic networks may be considered as a balancing act between the desire to update routing information frequently, and the need to minimise the network capacity used for this. Figure 5.71 gives a schematic presentation of this balancing act. High-mobility networks will generally experience a lower throughput. Low routing updating results in poor network performance due to incorrect routing. As routing updating becomes very frequent it eventually consumes all network capacity. The

en.

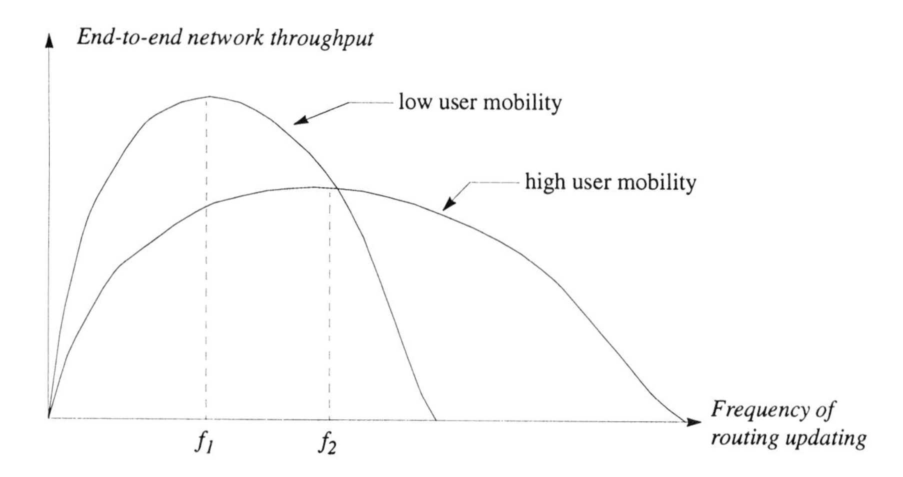

Figure 5.71 Schematic representation of the relationship between network dynamics (user mobility), required routing signalling and network performance

optimum performance (at f_1 and f_2, respectively) is found somewhere in between.

The process of routing is illustrated in Figure 5.72. The circles represent radio nodes, and the lines indicate all links connecting the radio nodes. Two nodes that are not connected by a line are out of each other's range, and cannot communicate directly. Node S(ource) has a packet for node D(estination). This packet can be forwarded either via node A, B, C or E, as alternative choices.

Some of the aspects that should be given a thorough consideration when choosing a routing strategy are listed below:

- stability (steady state) — all information (routing tables) should represent the status of the network as time reaches infinity, and for distributed strategies be consistent throughout the network (in all nodes);
- simplicity is important in order to be able to verify the algorithm;

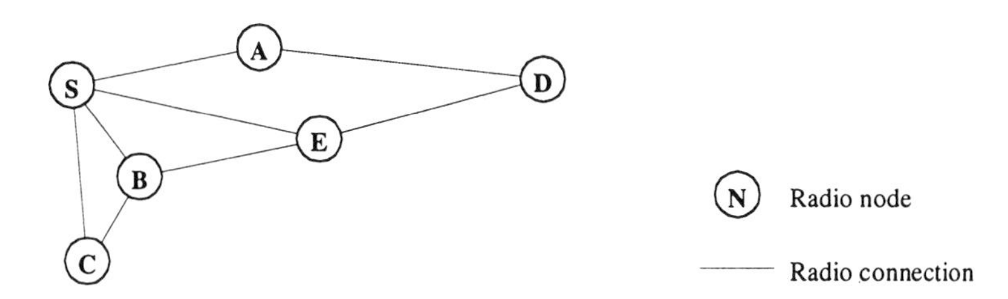

Figure 5.72 Routing—which node to choose for forwarding

- robustness is particularly important in mobile networks in order to withstand the dynamic changes (in the connectivity) of the network;
- transient behaviour — the routing-updating algorithm must adapt to dynamic changes of the network;
- fairness — the interface between flow/congestion control and routing; the flow/congestion control gives information to the routing about congested links and in this way the traffic load may be more evenly distributed, not straining centrally located nodes more than necessary;
- route metric — what should be the criteria for optimum routing? This is implementation and network dependent.

This Section will focus on routing strategies for land-mobile packet-switched wireless networks transmitting with omnidirectional antennas on a single common channel. This limitation excludes many of the possible routing strategies.

Let us consider two examples. The first of these will illustrate the necessity of a routing function by comparing the network performance under static routing to that under idealised routing. The second example will give an indication of the cost of routing (occupying network capacity) due to the need of information distribution.

Example 1—network performance as a function of routing

To illustrate the necessity of routing, consider the following example: a network of 16 nodes, transmitting with omnidirectional antennas and additional parameters given in Table 5.10. Consider two instances of time, t_0 and t_1. The network topologies at these instances are outlined in Figure 5.73. To evaluate the influence of routing on network performance the two following transitions may be studied:

Table 5.10 Parameters for the example simulations

User traffic	traffic pattern	end-to-end addressing uniformly distributed over all possible destinations
	packet-arrival distribution	Poisson at the link level
	packet-length distribution	fixed with length 483.6 ms (150 bytes)
Protocol stack		3 (RTS/CTS) with fixed parameters $(t_u, a, b) = (683\text{ ms}, 1, 1.5)$
Radio data		from Table 5.3
Routing algorithm		shortest path

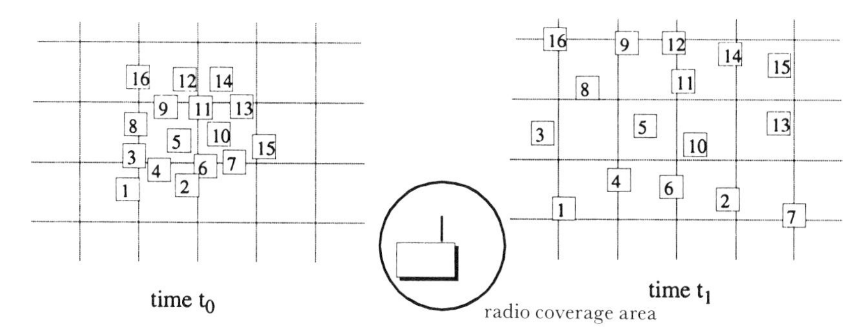

Figure 5.73 Mobility; performance is analysed for the time instances t_0 and t_1

Transition from time t_0 to t_1:
Some routes existing at time instance t_0 become invalid at time instance t_1. Performance degradation is caused by:

(i) many (infinite) retransmissions;
(ii) loss of end-to-end traffic streams until the routing can find alternative routes.

Transition from time t_1 to t_0:
All existing routes at t_1 are still valid at t_0 but they are not the shortest routes possible. The routing continues to use the old routing table for a period of time. Performance degradation is caused by:

(i) some packets traverse more hops than needed leading to a higher end-to-end delay;
(ii) unnecessary high link traffic increases the link delay.

The following two routing strategies will be compared for this transition:

(a) static routing with shortest path routing optimised for time instance t_1;
(b) idealised routing which can find the shortest path routes instantaneously.

The simulated network performance (throughput against delay) for this transition is plotted in Figure 5.74. The middle curve represents the initial state at t_1, where both routing strategies will perform equally. On the left is the performance of the static routing strategy after transition to t_0; on the right, the performance of the idealised routing strategy after transition to t_0. The difference in throughput-delay characteristics for *a* and *b* gives the estimate for the highest improvement possible for an adaptive routing algorithm.

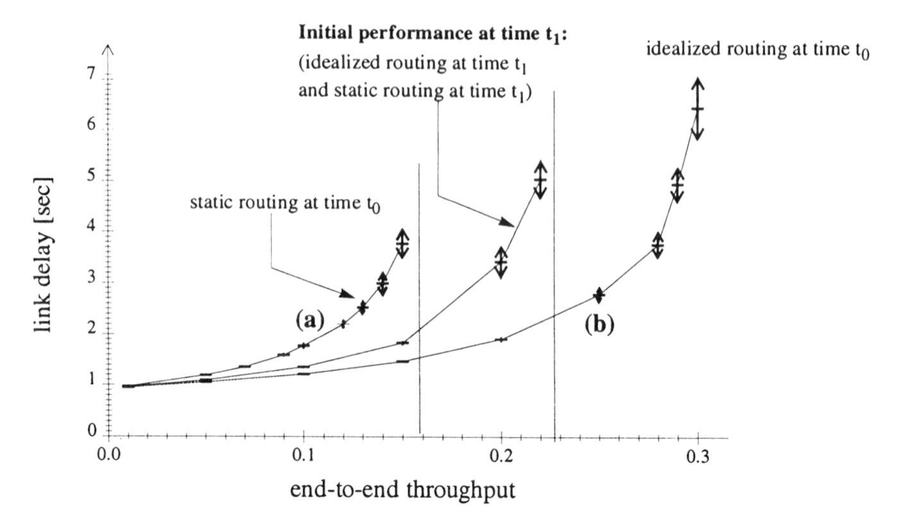

Figure 5.74 Average hop-by-hop delay against end-to-end network throughput

The network-hop counts for the two time instances are given in Table 5.11.

Example 2—channel capacity consumed by routing-information distribution

Assume a network of size M (M nodes in the network), and that all nodes share a common channel and are within radio range of each other. Further, assume that all nodes disseminate routing information (PROP, further discussed in Section 5.7.3) at regular time intervals, twice every minute. Each PROP will contain information about the route to all other nodes in the network ($= M - 1$). In the PROP, the length of each entry (a total of $M - 1$) will take 0.005 s to transmit (e.g. with a channel capacity of 5 kbit/s each entry will be 25 bits long). The fraction of total channel capacity consumed by PROPs then is given by eqn. 5.51, and plotted in Figure 5.75:

$$C_{PROP} = M\left((M - 1) \cdot \frac{1}{30} \cdot 0.005\right) \tag{5.51}$$

Table 5.11 Routing data

	Time instance	
	t_0	t_1
Average number of hops	1.35	2.65
Largest number of hops	3	6

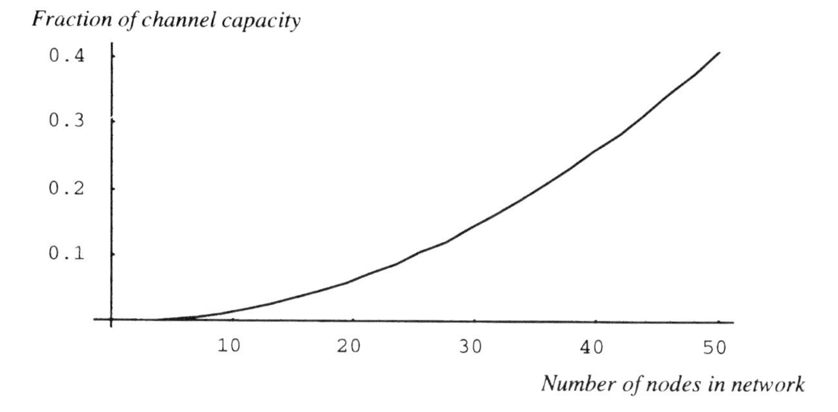

Figure 5.75 Fraction of channel capacity occupied by PROPs for timer-driven strategy

When applying an event-driven distribution strategy, assume that there is a significant change to a link on the average once every 30 minutes. The average number of link changes each second is given by:

$$N_{change} = M(M-1) \cdot \frac{1}{30 \cdot 60} \qquad (5.52)$$

Further, assume that a PROP is not sent until a node has registered five changes. Assuming (as a simplification) that on the average a link change results in a change in the routing table of 25% of the nodes, the fraction of total channel capacity occupied by PROPs is given by eqn. 5.53 and plotted in Figure 5.76:

$$C_{PROP} = M(M-1) \cdot \frac{N_{change} \cdot 0.25}{5} \cdot 0.005 \qquad (5.53)$$

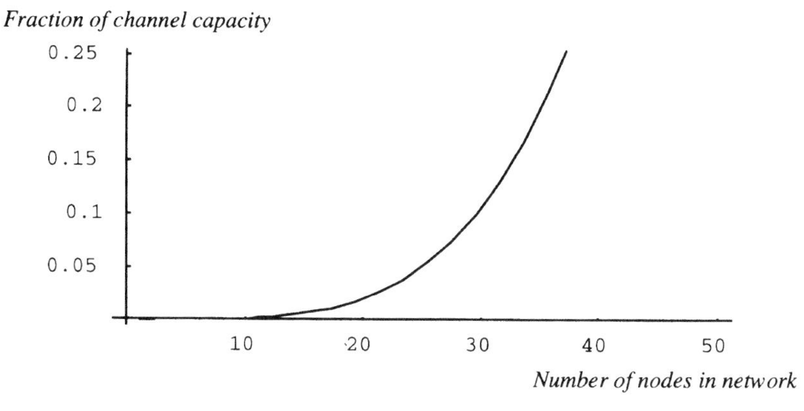

Figure 5.76 Fraction of channel capacity occupied by PROPs for event-driven strategy

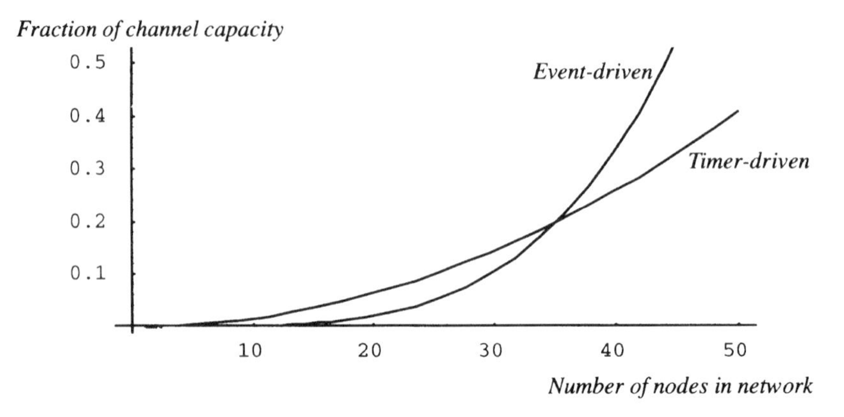

Figure 5.77 Comparing resource usage for event-driven against timer-driven PROP strategies

Plotting these two curves together, as in Figure 5.77, we see that they cross at a network size of 35 nodes. Of course, the problem of timer-driven PROP distribution is the slow updating of routing tables when the network size increases, as the increased dynamics is not compensated for. The event-driven strategy adapts to changes in network size and dynamics, but at the obvious consumption of network capacity.

5.7.1 How to perform routing

Optimal routing must be based on the best possible available information. There are different strategies for obtaining this information and where and how to perform the route calculation. Given a certain strategy, routing consists of minimising one or more of a number of different factors which somehow represent the cost associated with each route.

5.7.1.1 Routing strategies

There are three different basic ways of performing routing: centralised, isolated or distributed.

A more general introduction to routing can be found in many textbooks such as [39], or for packet-switched wireless networks in particular in [40].

Centralised routing requires the presence of a dedicated routing centre which receives information about all changes in the network, calculates the best routes to be used for all connections and returns the processed information back to the nodes. The advantage of this approach is the centralised control that assures a uniform route calculation and the best overall resource usage. The disadvantages are the vulnerability (due to the centralisation) and the adaptation time and the traffic generated by the necessary information distribution in dynamic networks.

Isolated routing is a distributed algorithm in its simplest form. Each network node performs routing based on own information only. No specific exchange of routing information between nodes takes place. One such method is called backward learning. Each packet is marked with the original sender's address. In addition, there is a hop counter that is updated for each new radio hop. Any node that eavesdrops on a packet can use the information to update its own routing table. One fatal drawback of backward learning, however, is the lack of a mechanism to handle the loss of a link connection

Truly distributed routing implies the exchange of routing information between network nodes, either regularly or event-driven (due to a change in the routing table) or a combination of these. Obviously, this information distribution is done at the sacrifice of network capacity available to the users. Information exchange generally increases with network size (number of nodes) and network dynamics.

Distributed routing algorithms may be static or dynamic. Static-routing determination implies that the source node determines the route to the ultimate destination, also deciding each relay node to be used. Information about relay nodes then must be included in the packet. In dynamic route determination a (relaying) node only determines the next node to be used on the way to the destination, then transfers the responsibility to the consecutive relay nodes along the path. Dynamic route determination obviously has an advantage in dynamic networks.

The rest of this Section will be dedicated to dynamic distributed routing algorithms only, as these have the greatest potential in packet-switched wireless networks.

5.7.1.2 Route calculation metric

First, it is necessary to establish a set of criteria on which the routing function can be based. Generally, routing is the task of finding the cheapest possible connection from source to destination, if such a connection exists. For a distributed algorithm, the task for each node is usually limited to finding the next node, not to establishing the complete route. As a packet traverses the network (hopefully) on its way to the destination, each relay node will perform a completely new routing calculation based on its own local knowledge about the network. Obviously, the same set of criteria should be used in all nodes. There are a number of different criteria available for calculating the cost of a given route:

- number of radio hops to destination;
- link quality (previous retransmission rate and SNR-measurement/ noise margin), either as poorest link quality or an average quality of the links involved;
- node congestion;
- node mobility rate (type of node, past history, etc.);
- delays related to previous transmissions.

In a single channel packet-switched wireless LMN, factors such as node congestion and delay relating to one route are not independent of other routes. The most vital parameter is probably the number of radio hops, as each hop occupies the common resource (the channel), resulting in reduced end-to-end network throughput. To distinguish between routes of equal length, it is possible to use such parameters as the quality of the poorest link (along the route) and the average link quality. Also, if there is some knowledge about the mobility or dynamics related to each node, nodes associated with a high degree of dynamics should be considered as being less reliable, and be the last choice.

5.7.2 Evaluation of routing strategies

In order to be able to make a decision between different routing-strategy candidates, some kind of quantitative measure for the performance is needed. Also, the performance must be related to a sensible set of boundary conditions such as geographic scenario, connectivity, rate and pattern of mobility/dynamics and traffic volume and pattern. It will not be possible, neither is it sensible, to measure the performance of alternative routing functions in a realistic network. Also, analytical approaches are impossible without making too many simplifications. The only practical approach to this problem is through a simulation model.

In order not to make routing more complex than necessary, we will not introduce output power optimisation, but use static transmission power.

As previously mentioned, a routing function must be based on a number of cost criteria to be minimised. Criteria frequently used in multi-channel networks, such as retransmission rates and queue lengths, are not as well suited for single-channel wireless PS networks, since the different links are not independent of each other. Also, we will not use flow- or saturation-control information as parameters for routing. The following prioritised list of criteria is chosen:

(i) minimum number of radio hops;
(ii) maximum value for the quality of the poorest link (trying to avoid routes using potentially unreliable links);
(iii) maximum average link quality.

The second criterion is used to choose between routes with the same number of radio hops, and the third to choose between routes that all have both the same number of hops and the same quality of the poorest link.

5.7.2.1 Scenario

The performance of a certain routing strategy is not independent of the network scenario. Ideally, the different strategies should be tested for all possible scenarios, but this is impossible to accomplish. The best approach is to select a few snapshots of scenarios which contain elements of real-life situations, where the routing function is thought to be stressed. The vital element of dynamics can be achieved by varying the radio range of nodes, altering their geographic positions or changing connectivity by varying the SNR values of the different radio links.

The choice of scenario should be based on any knowledge available about the network being designed. For instance, there is no use in evaluating a routing function based on its performance in a relatively static scenario if the network is going to be used in highly dynamic environments. If little or nothing is known about the application of the network in design, it may make sense to choose regular or random node structures and a random but evenly-distributed traffic pattern between nodes. Network dynamics could be introduced through randomly moving the nodes or altering their radio range. The performance can be studied for different network sizes and rates of mobility (or dynamics).

5.7.2.2 Network dynamics

By network dynamics we understand the variation of the quality of the links. This may be due to mobility (nodes moving), variation in signal strength due to changes in the environment, or changes in the noise level. In addition, nodes entering or leaving the network may introduce an extra element of dynamics.

Network dynamics introduces a need to update routing information. Correct routing at any time requires frequent distribution of routing information, which obviously consumes network capacity. On the other hand, inaccurate routing due to outdated routing tables is also expensive. The frequency of routing-information distribution ends up as a trade-off between the cost of distributing updates and the liability of using potentially obsolete information.

5.7.2.3 *Performance measures*

The performance of the routing function is very important in dynamic networks, whether the dynamics are caused by node mobility, nodes entering or leaving the network or the presence of interference sources. In order to evaluate different routing strategies we need some kind of quantitative measure other than network throughput and delay, since these are not available to the routing function.

Before this can be done a reference must be established. This reference is based on instantaneous access to all necessary information about the network, e.g. the quality and state of all links and the state of all nodes. Then, based on the chosen routing criteria, the best possible route, the reference route, can be established for all pairs of nodes.

We have chosen to define the following performance measures, using number of radio hops as the criterion of best route choice:

(*a*) Probability of reaching destination in a network, p_S:

$$p_S = \frac{1}{n_{ref}} \cdot \sum_{i=1,\, i \neq j}^{m} \sum_{j=1}^{m} \delta_{ij} \cdot I_{ij} \tag{5.54}$$

where m is number of nodes in the network, and

$$I_{ij} = \begin{cases} 1, & \text{if the routing function finds a route between nodes } i \text{ and } j \\ 0, & \text{if routing fails} \end{cases} \tag{5.55}$$

and

$$n_{ref} = \sum_{i=1,\, i \neq j}^{m} \sum_{j=1}^{m} \delta_{ij} \tag{5.56}$$

where

$$\delta_{ij} = \begin{cases} 1, & \text{if a reference route exists between nodes } i \text{ and } j \\ 0, & \text{if no reference route exists} \end{cases} \tag{5.57}$$

Routes that cannot be established are not counted (in the case of a disjointed network). If there is more than one route interconnecting two nodes, only the shortest one is counted. The reference route is the shortest possible route between a pair of nodes (not necessarily known to the network nodes). In a completely jointed network the number of reference routes is $m\,(m-1)$.

(*b*) Network excess hop rate, e_H:

$$e_H = \frac{1}{n} \sum_{i=1}^{n} \left(\frac{r}{h}\right)_i - 1$$

where n = number of routes that reach destination
h = number of hops for the reference route
r = number of hops for the chosen route

Excess hop rate represents the overuse of resources caused by imperfect routing, e.g. reaching a destination by using routes with more hops than necessary. Generally, this can be replaced by an excess resource usage factor, defined to include other cost factors in addition to the number of radio hops. This could be end-to-end delay, number of nodes disturbed by the needed number of transmissions along the route etc.

The upper limit of performance is represented by the reference routes, and perfect routing would give $p_S = 1$ and $e_H = 0$. The two network-performance parameters are instantaneous, but equivalent average measures can be defined, thus giving the performance of the routing function in a network over a certain time interval (e.g. the duration of the simulation).

5.7.3 Routing information updating

As mentioned, a truly-distributed routing strategy must be based on the best possible network knowledge available in any node. This is best achieved by exchanging routing information between the nodes. This exchange is often performed by each node transmitting specific packet-radio organisation packets (PROP) [18]. These packets are usually distributed using the semibroadcast service. The PROP contains an extract of information from the node's local routing table, which again is partly obtained from the PROPs received from other nodes. The nature of the semibroadcast means that a PROP is only received by the nodes within radio range, i.e. the neighbours of the transmitting node. Information from nodes further than one hop away is obtained through other nodes as they update their routing tables, which again is the basis of their own PROPs. This means that the propagation of network changes is associated with a delay that is a function of network width (in number of hops from end-to-end). Thus, in a dynamic network there are bound to be some discrepancies in routing information between the different nodes.

It may be preferable to distinguish between two different kinds of routing information, both in a node's own tables and in the PROPs. Information about the status of all links to a node's neighbours, i.e. one hop away, may be distributed as link-PROP information. Information about nodes that are more than one hop away is distributed as path-PROP. There may be different strategies used to distribute these two categories of information, e.g. it may be sensible to distribute link-PROP information more frequently than path-PROP information, depending on the network topology and connectivity.

It is possible to imagine at least two different strategies used to distribute routing information in PROPs. The first strategy that comes to mind is to distribute information whenever there is a change in the network. This event-driven strategy may be called the asynchronous update algorithm (AUA). On the other hand, all changes should not necessarily result in the distribution of a PROP, as a minor change in the quality of a link may not have any consequence to the choice of routes. In addition, this would probably generate much unnecessary distribution of information. The greatest disadvantage of this event-driven strategy (AUA) is the intense packet generation which may occur if the network is subject to a significant change. In such a case the network operation may be severely disturbed by this extra load of control packets, which results in additional network congestion and packet delays.

Another distribution strategy, trying to avoid potential network overload, is the timer-driven strategy, or periodic update algorithm (PUA). PROPs are distributed by all nodes at regular time intervals. In order to reduce the chance of all nodes transmitting almost simultaneously, causing a short but high network overload, a random defer-time delay may be added prior to each PROP. This time delay should be much larger than the random-access delay added by the MAC protocol. The drawback of the PUA strategy is obviously that network changes propagate slowly, depending on the length of the PROP interval.

A third strategy may be a combination of PUA and AUA, giving a combined update algorithm (CUA). Let PROP distribution be event driven, but only whenever there is a minimum number of changes, or a minimum time since a PROP was previously transmitted from this node. In addition, distribution of PROP should also take place at a maximum time interval, as there is always a risk of nodes missing a PROP. In addition, it is possible to include a few bits of link quality information in any ordinary data packet sent.

If information about nodes' instantaneous mobility is available, it is possible to use this information in the routing function. Mobile nodes introduce an element of uncertainty. If two or more approximately equal routes are available to the destination, the shortest or less resource-consuming route may turn out not to be the best choice if this involves a larger number of mobile nodes.

5.7.4 Flooding

The only routing strategy that is guaranteed to always find a route (if it exists) even in a dynamic network is called flooding. The source node issues a semibroadcast packet, questioning all its neighbours for the destination. All neighbours that receive this packet (and who are not the destination) will transmit a new semibroadcast to all their neighbours. This procedure is performed until the destination is reached. Obviously, the packet must contain a unique identifier to prevent nodes from transmitting the same packet more than once, as it is received from all the node's neighbours. An example of a flooding search is illustrated in Figure 5.78.

This routing strategy is very resource consuming, and several methods are devised in order to restrain the traffic. For instance, it is possible to let a node refrain from relaying a flooding packet if a certain number of its neighbours have already done so. This obviously reduces the generated traffic, but at the same time it introduces an element of uncertainty, as the search is no longer guaranteed to locate the destination. Anyway, flooding should be used with great care in a wireless network. Usually it is

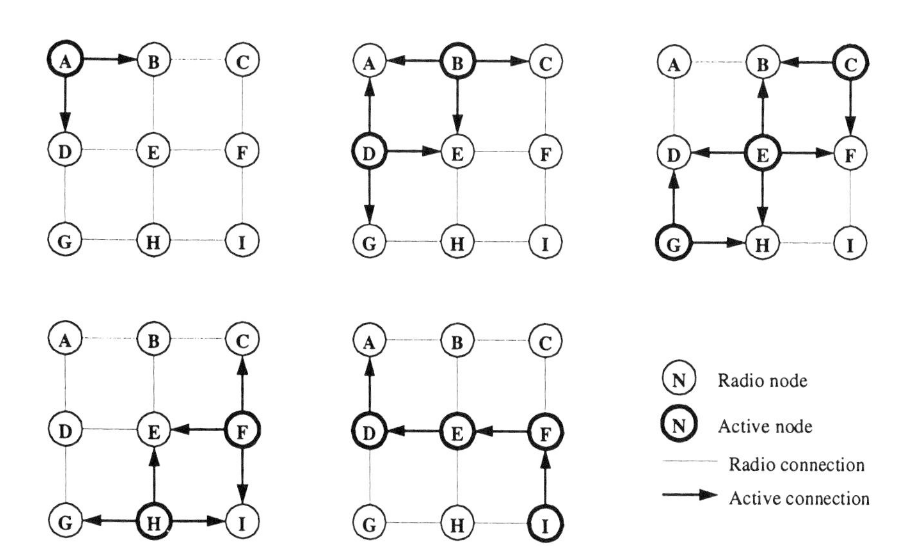

Figure 5.78 Illustrating the steps of a flooding search (node A trying to locate node I) in a small network, and the return of a selected route (one of a number)

combined with other routing strategies, and is used only as a last resort when other routing methods fail.

5.7.5 Hierarchical routing

The distributed-routing algorithm does not perform equally well as the network size increases, due to an increased fraction of obsolete information. In addition, the amount of routing information prevents frequent updating (or else the network will be overloaded with PROPs). One way of avoiding this problem is called hierarchical routing.

The network is divided into a number of groups (called clusters), each consisting of a limited number of nodes. Each of the groups is assigned a cluster head. Routing within the cluster is still distributed, all nodes exchanging routing information. But information about connections to nodes outside the cluster is only sent to the cluster head. Routing information about nodes belonging to different clusters is only exchanged between the cluster heads. This is illustrated in Figure 5.79, with two levels of routing. If the number of cluster heads grows too high, it is possible to introduce another level of hierarchy, and so on.

Hierarchical routing may result in a less optimal selection of routes. A combination of flat and hierarchical routing called semihierarchical is described in [41]. This method tries to obtain better (shorter) route selection while still benefiting from the reduction in routing information distribution caused by the purely hierarchical layering.

5.7.6 Robust routing

Military or other networks exposed to a threat of severe interference should have a backup mechanism in case the normal routing no longer

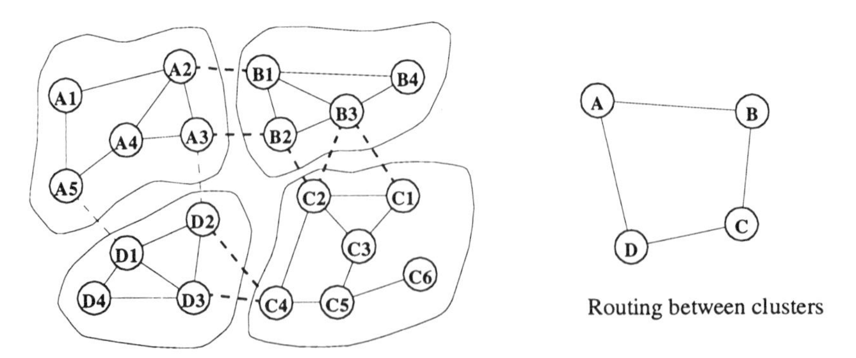

Figure 5.79 Hierarchical routing. The network (left) is separated into clusters. Routing at cluster level is shown on the right. There may be more than one physical route between two neighbour clusters

functions due to a rapid change in the noise level. Flooding is always available, but should, as mentioned, be used with great care due to the amount of traffic it generates.

Another alternative is to search for a more robust alternative route. This is performed continuously, in parallel to searching for the shortest or less resource-consuming route. Each node will maintain a set of alternative routes in addition to the normal routing table. Knowing the lowest noise margin of each route, the node may decide to use the robust route either if normal routing fails, or if it suspects that the noise has increased to such a level that the normal route is expected to fail. This may be performed by continuously measuring the local noise level, and assuming (not necessarily correctly) that changes in noise level are equal throughout the entire network.

References

1. PICKHOLTZ, R.L.: 'Theory of spread-spectrum communications — a tutorial', *IEEE Trans.*, 1982, **COM-30**, (5)
2. PAWLIKOWSKI, K.: 'Steady-state simulation of queueing processes: a survey of problems and solutions', *ACM Comput. Surv.*, 1990, **22**, (2)
3. RUSTAD, J.E., et al.: 'New radio networks for tactical communication', *IEEE J. Sel. Areas Commun.*, 1990, **8**, (5)
4. THORVALDSEN, T.: 'Narow band direct sequence/frequency hopping spread spectrum VHF radio'. MILCOM 92, *IEEE Military communications conference*, October 1992
5. KLEINROCK and LAM: 'Packet switching in a multiaccess broadcast channel: performance evaluation', *IEEE Trans.*, April 1975, **COM-23**
6. TOBAGI: 'Analysis of a two-hop centralized packet radio network–part II: carrier sense multiple access', *IEEE Trans.*, February 1980, **COM-28**
7. BOORSTYN, et al: 'Throughput analysis in multiple CSMA packet radio networks', *IEEE Trans.*, March 1987, **COM-35**
8. YU and HAMILTON: 'A buffered two-node packet radio network with product form solution', *IEEE Trans.*, January 1991, **COM-39**
9. SILVESTER and LEE: 'Performance modelling of buffered CSMA — an iterative approach'. Conference record, GLOBECOM 82, November 1982, pp F1.6.1–F1.6.5
10. TAKAGI, H.: 'Queueing analysis–vol 1' (North-Holland, 1991)
11. KLEINROCK: 'Queueing systems–vol. I' (John Wiley & Sons, 1975)
12. LEWIS, P.A. and ORAV, E.J.: 'Simulation methodology of statisticians, operations analysts, and engineers — vol. 1' (Wadsworth, 1989)
13. GLYNN, P.W. and IGLEHART, D.L.: 'Simulation methods for queues: an overview', *Queueing Syst. Theory Appl.*, 1988, **3**, (3)
14. LEE, I. and SILVESTER, J.: 'An iterative scheme for performance modelling of slotted ALOHA packet radio networks'. Proceedings of IEEE international conference on *Communications*, 1982

15. LAU, C.T. and LEUNG, C.: 'Capture models for mobile packet radio networks'. IEEE international conference on *Communications*, May 1992, **40**, (5)
16. LEINER, B.M., et al.: 'Issues in packet radio network design', *Proc. IEEE*, 1987, **75**, (1)
17. KAHN, R.E., et al.: 'Advances in packet radio technology', *Proc. IEEE*, November 1978, **66**, (11)
18. JUBIN, J. and TORNOW, J.D.: 'The DARPA packet radio network protocols', *Proc. IEEE*, 1987, **75**, (1)
19. DAVIES, B.H. and DAVIES, T.R.: 'The application of packet switching techniques to combat net radio', *Proc. IEEE*, 1987, **75**, (1)
20. PURSLEY, M.B.: 'The role of spread spectrum in packet radio networks', *Proc. IEEE*, 1987, **75**, (1)
21. TOBAGI, F.A.: 'Modeling and performance analysis of multihop packet radio networks', *Proc. IEEE*, 1987, **75**, (1)
22. KLEINROCK, L. and SILVESTER, J.: 'Spatial reuse in multihop packet radio networks', *Proc. IEEE*, 1987, **75**, (1)
23. GERLA, M. and KLEINROCK, L.: 'Flow control: a comparative survey', *IEEE Trans.*, 1987, **COM-35**, (3)
24. TOBAGI, F.A.: 'Multiaccess protocols in packet communication systems', *IEEE Trans.*, 1980, **COM-28**, (4)
25. KLEINROCK, L.: 'Packet switching in radio channels: part I — carrier sense multiple-access models and their throughput-delay characteristics', *IEEE Trans.*, 1975, **COM-23**, (12)
26. TOBAGI, F.A., and KLEINROCK, L.: 'Packet switching in radio channels: part II–the hidden terminal problem in carrier sense multiple-access and the busy-tone solution', *IEEE Trans.*, 1975, **COM-23**, (12)
27. SOUSA, E.S. and SILVESTER, J.A.: 'On multihop spread spectrum network modeling'. MILCOM 86: 1986 IEEE *Military communications conference* IEEE, New York, 1986
28. ROM, R. and SIDI, M.: 'Multiple access protocols — performance and analysis' (Springer-Verlag, 1990)
29. KOHNO, R. et al.: 'Spread spectrum access methods for wireless communications', *IEEE Commun. Mag.*, January 1995
30. SCHOLTZ, R.A.: 'The origins of spread-spectrum communications', *IEEE Trans.*, 1982, **COM-30**, (5)
31. IEEE P802.11D5: 'Wireless LAN medium access control (MAC) and physical layer (PHY) specifications'. IEEE standards department, July 1996
32. ETSI TC-RES: 'Radio equipment and systems (RES); high performance radio local area network (HIPERLAN); functional specification'. ETSI, July 1995
33. KLEINROCK, L.: 'On modeling and analysis of computer networks', *Proc. IEEE*, 1993, **81**, (8)
34. PROAKIS, J.G.: 'Digital communications' (McGraw-Hill, 3rd edn., 1995)
35. SOROUSHNEJAD, M. and GERANIOTIS, E.: 'Probability of capture and rejection of primary multiple-access interference in spread-spectrum networks', *IEEE Trans. Commun.*, 1991, **39**, (6)

36. ZHOU, X.Y. and KAMAL, A.E.: 'Automatic repeat-request protocols and their queueing analysis', *Comput. Commun.*, 1990, **13**, (5)
37. STALLINGS, W.: 'Data and computer communications' (Macmillan Publishing Company, 1985)
38. BERG, T.J. and EMSTAD, P.J.: 'Performance of packet switched services in spread-spectrum radio networks', *Telektronikk*, special issue on mobile telecommunication, 1995, **91**, (4)
39. TANENBAUM, A.S.: 'Computer networks' (Prentice-Hall, 1989, 2nd edn.)
40. LYNCH, C.A. and BROWNRIGG, E.B.: 'Packet radio networks. Architectures, protocols, technologies and applications' (Pergamon Press, 1987)
41. CALLON, R. and LAUER, G.: 'Hierarchical routing for PRnet'. BNN report 5945, SRNTN 31, June 1985
42. JAFFE, J.M.: 'A responsive distributed routing algorithm for computer networks', *IEEE Trans.*, 1982, **COM-30**, (7)
43. GARCIA-LUNA-ACEVES, J.J.: 'A fail safe routing algorithm for multihop packet radio networks'. Proceedings of IEEE INFOCOM'86

Appendix 5.A Radio characteristics for multihop networks

The radio services and characteristics specified in Section 5.3.1 are not complete enough to be valid for the new effects coming to light with the presence of hidden nodes. We have to reconsider the capture model and the carrier-sense detection sequence. The radio services for supporting the BTMA must also be specified. The objective of this appendix is to perform these tasks so that the main part of Section 5.5 can continue without any disruption.

In single-hop nets, we have seen that collisions occur only if two or more nodes transmit within the vulnerable period t_v of the first packet. Figure 5.80 illustrates how this vulnerable period is extended to the whole packet length when a hidden node is present.

Node 2 transmits at time t_1. Node 3 has no knowledge of the state of node 2 since they are not connected and node 3 starts to transmit near the middle of the first packet. When collisions occur in fully-connected CSMA nets, the entire information part of a packet gets the same degraded SNR. This is not true for multihop nets because the transmission from a hidden node can hit any part of a packet and therefore the noise is generally not uniformly distributed over the packet. Multihop nets require more complex capture models than single-hop nets. The FEC scheme in use has of course a great impact on the packet-success probability since their capability to correct bit errors differs. This Section will not consider

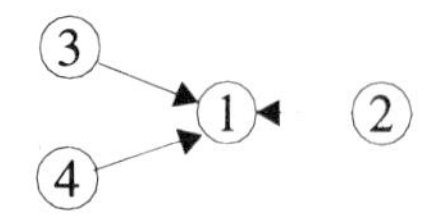

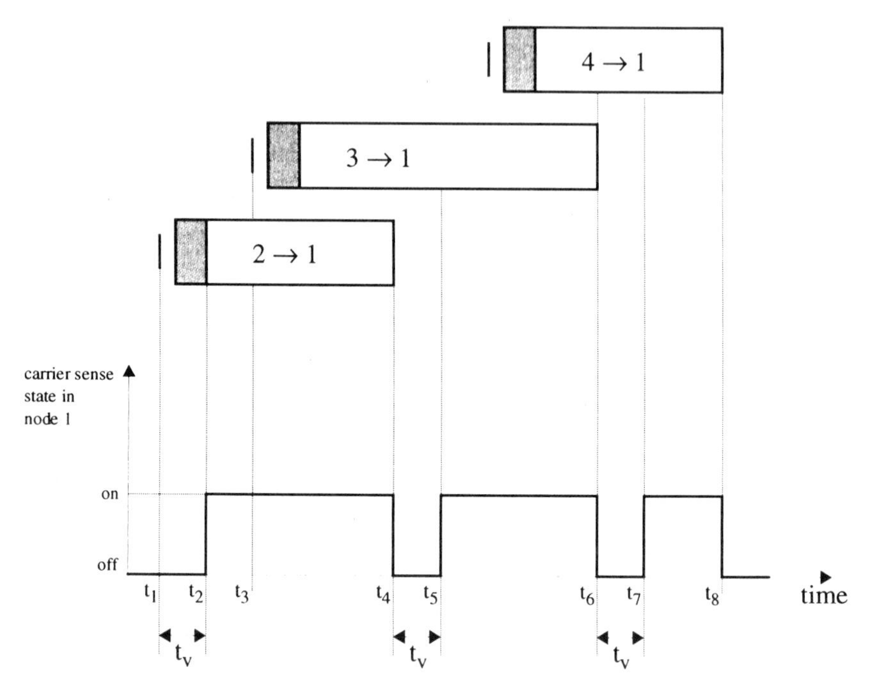

Figure 5.80 The resulting time-sequence diagram for the carrier-sense signal in node 1 for the three overlapping packets transmitted in the star network (at the top of the Figure)

different FEC methods and we want to continue with the simple capture model specified by eqn. 5.12 as:

$$C_k = p_{noise} q^{k-1} \tag{5.58}$$

Here k was defined as the number of simultaneous transmissions in the network. Since a multihop net may have two or more noninterfering transmissions, k must be redefined. The new definition of k is: the maximum number of simultaneous transmissions (i.e., $k-1$ overlaps) hitting the packet under consideration. Figure 5.81 exemplifies the meaning of this definition. Four packets are transmitted, partly overlapping and partly distributed over time. The values assigned to k are the

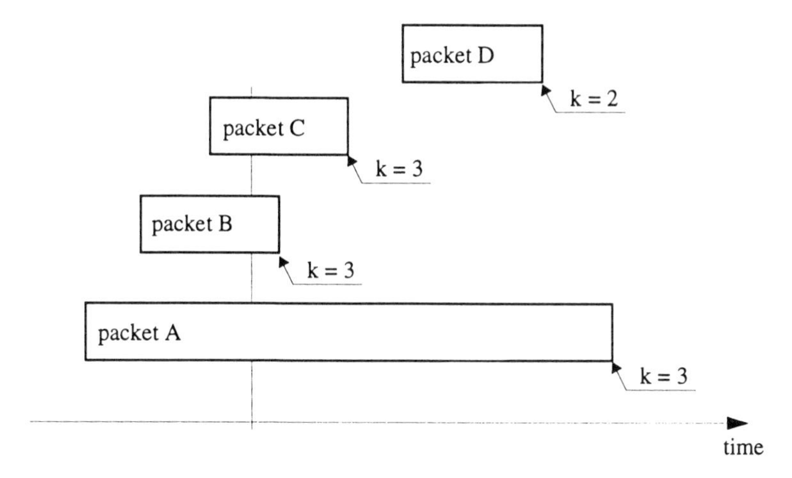

Figure 5.81 Examples of k-values for the redefined capture model

input values used when the probability of success is calculated by a receiving node.

With the capture model in place, we focus on the next item, the carrier sense. Only one small change to the previously-specified functionality is introduced; a node which completes its demodulation process or completes a transmission can first after a delay of t_v time units make a conclusion about the channel state. This is illustrated in Figure 5.80 where node 1 at time instance t_4 stops its demodulation, erroneously senses an idle channel but gets the true state t_v time units later. This models the effect of the correlation time delay introduced by spread-spectrum receivers.

Implementation of BTMA requires additional functionality at the radio level. A dedicated busy-tone channel must be provided. The process of detecting activity on the busy-tone channel is similar to the processes of detecting activity on the data channel under CSMA; a channel-sampling time delay is introduced and the probability of making a faulty conclusion is greater than zero.

Several variants of BTMA exist depending on the set of nodes emitting a busy tone. The BTMA protocol modelled in this section has most in common with the receiving destination BTMA (RD-BTMA) [21] and the following rules apply:

1. A transmitting node does not emit a busy tone.
2. A node acquiring synchronisation by receiving the preamble carried with each packet transmission shall immediately emit a busy tone and continue to emit the busy tone as it stays locked onto the packet.

3 From the point in time where a node changes state from transmitting or receiving, a channel-sampling time period is required before it can make a decision about the channel state.
4 A node which detects a busy tone but does not fulfil rule 2 or 3, shall not emit a busy tone.

Rules 1 and 2 reflect the fact that any conflicts occur at the receiver. Rule 3 models the effect of the correlation time needed by spread-spectrum receivers or the integration time needed by an energy-detecting device. Rule 4 prevents the busy tone from propagating more than two hops away from the transmitting node.

Nodes within the coverage area of the transmitter use the data channel to determine the channel state and for those nodes we assume the same functionality as for the CSMA protocol described above (i.e., the capture model is still valid as are the vulnerable periods). The main difference is the busy-tone channel which facilitates blocking of all nodes within the radio range of the receiving node (i.e., all nodes within a two-hop radius around the transmitter are blocked). Thus, the neighbours of the transmitter detect a busy channel after a delay t_v and since we neglect any delay components in the busy-tone channel, nodes also two hops away from the transmitter get the same vulnerable period.

With reference to Figure 5.80 this means that node 2 acquires synchronisation at time $t_1 + t_v$ and as the busy-tone channel has no time delays, node 3 will instantaneously detect the busy tone. Node 3 does not acquire synchronisation on the data channel and according to rule 2, it shall not emit a busy tone. If $t_3 - t_2 > t_v$, node 3 will defer the transmission of the packet. With these rules the star network in Figure 5.80 operates as a fully-connected three-node CSMA network both having the same vulnerable period t_v.

Chapter 6

A case study

The kinds of radio system treated in this book are likely to be developed in close co-operation between users and producers. On one side the users would like to have all possible requirements met or 'nice to have' features included; on the other hand they will meet the issues of cost and complexity forcing them to make certain choices. In addition, there might be several contradictions between the different wishes when transforming them into technical issues. Therefore, it will be necessary to make a kind of trade-off between all the requirements and issues involved before the final user requirements are used as inputs to the technical design.

It is not possible to establish a particular development method or approach that would be optimum for each and every radio system. It might not even be fair to consider something like a typical system. How to approach system development will largely depend on factors like volume, requirements with respect to functionality and performance, technology and so on. A system consisting of simple hardware and complex software will obviously be given another approach than a system with the opposite features. An important difference between commercial systems, and systems tailored to specific private or military needs is how cost influences solutions. In the commercial market place the 'right' price is set from market surveys and the system designer seeks solutions which reduce cost in order to increase profits. Traditionally, this has not been the case for military systems even though price will be of importance and might lead to trade-offs in functionality and performance. This means that price is a variable to consider for the designer.

In this Chapter we will present an example of a development process for a second-generation, high-complexity tactical radio system. However, the general approach would be applicable to a wide range of civil and military projects. If we were to refer to a known development method, our case would probably be seen as an example of the classical waterfall model. This model demands a systematic, sequential approach to development starting with definition of user requirements. However, as our case shows, a purely sequential approach is both impractical and impossible. As development proceeds, and for a case where not all of the initial user requirements can be met or achieved at a reasonable cost, it is usually necessary to go back and adjust assumptions or decisions made at an earlier stage to end up with the best solution.

6.1 The initial user requirements put forward

For the kind of system under discussion, initial user requirements are likely to say something about:

(i) some aggregated characteristics of the users and the scenario in which the radio system is to be used. For example, will the user require a hand-held radio and/or will there also be a use for the system integrated within a vehicle, and what will be the area to be serviced by the system?
(ii) whether the system will take a standalone form or shall the system be interoperable and/or compatible with other systems? What kind of environment will be challenging for the use of the system, for example will there be extreme levels of noise and interference or other foreseen difficulties?
(iii) a more detailed set of requirements related to the type of user services and performance figures (throughput, transit delay, priority, quality), use of standards, degree of autonomy, privacy and security and so on.
(iv) what the system can initially cost, and for how long is it likely to be in operation.

In the case used as an example in this Chapter, the user wants a mobile radio with a range of 10--15 km since that will be a kind of typical maximum distance between directly-communicating users, even though the total field of operations extends to some 40 km by 50 km. The radio should be hand held and battery powered with the smallest possible weight and volume. The radio environment could be characterised as rough with respect to climate and handling. Some sort of vehicle mounting is needed as the radio will also be used inside a vehicle. That vehicle might also contain other radio systems which might be in use at the same time. In fact, the new radio should somehow be interoperable with these infrastructure systems. The user will use the new mobile radio both for voice and data services. The system should offer both communication security and transmission security which means that no unauthorised user can introduce messages without detection or disturb accidentally or intentionally the transmission itself. The former problem is usually solved using some form of cryptography and the latter may be solved by using spread-spectrum modulation. The total area or scenario to be serviced by the mobile-radio system would exceed what could be covered by an 'everybody hear everybody' solution, and the need for some means of extending the service area beyond radio range is foreseen. Price is a major decision factor for a purchase.

6.2 The designer's first reflections on the user requirements

The designers will be working according to some form of plan which can secure necessary work progression and final match between user requirements and preferred solution. The designers realise that even though several work processes will be undertaken in parallel by a wide variety of specialists, and that several iterations will be needed in order to reach the desired goal, the overall process will benefit from some structural approach. A simplified illustration of the work approach is shown in Figure 6.1.

6.2.1 Choice of operating frequency

The choice of operating frequency will depend on a number of factors like obtainable radio range (propagation path loss, antenna efficiency and size), terrain features, frequency availability and regulations. Since the user has specified a scenario with no limit on where the radios may be located within a geographic area, a frequency not requiring strictly line-of-sight communication between transmitters and receivers will be advantageous. The expression 'radio range' is usually used to indicate the distance between transmitter and receiver where selected parameters like bit error rate satisfy certain threshold values. Usually the radio range is determined by the propagation loss between the transmitter and receiver, but as we have seen in Chapter 3 distortion of the signals may also limit the range for certain operating frequencies.

We may calculate the maximum propagation loss L_{lim} that any transmitter receiver link may experience and still be operational by:

$$L_{lim} = P_t + G_t + G_r - (E_b/N_o)_{threshold} - 10\log(N_o \cdot B)$$

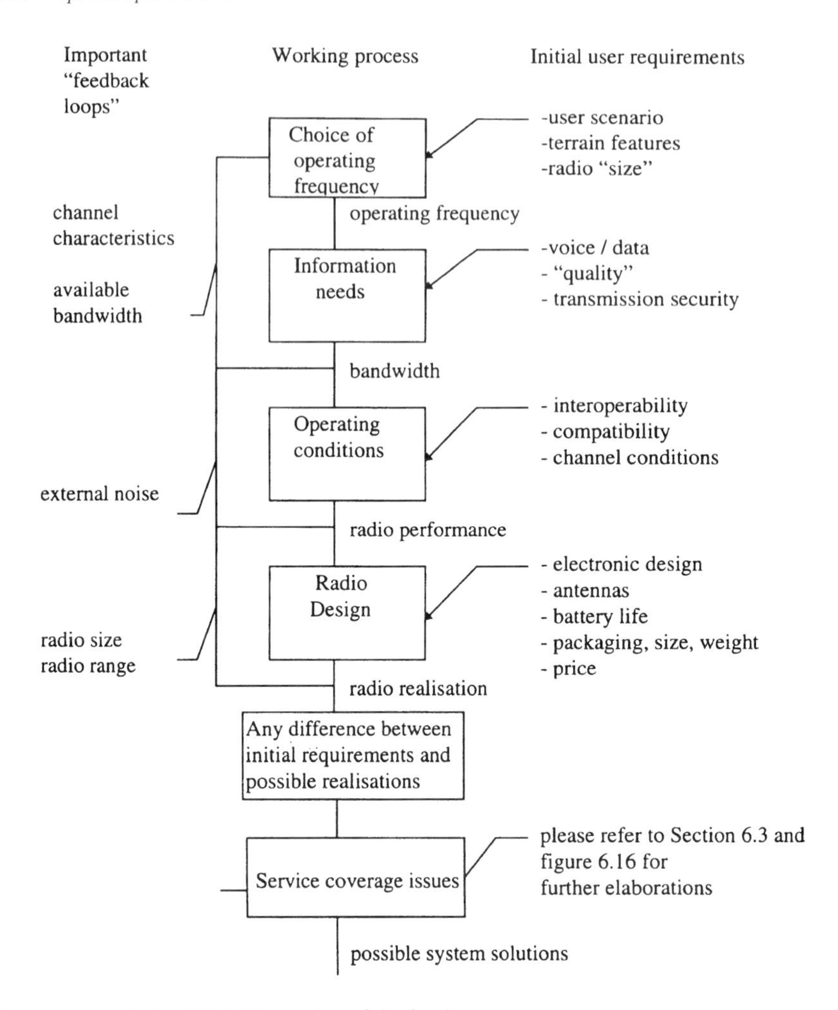

Figure 6.1 A simplified illustration of the development process

Such an equation contains many parameters specific to the system under consideration and it will be necessary to choose parameter values in order to obtain numeric values for maximum propagation loss.

For any given set of system parameters the choice of frequency range will give large variations in obtainable radio range. The most important influence on range will be related to a set of propagation effects like:

- free space attenuation;
- diffraction;
- losses due to vegetation;
- losses due to absorption.

In addition, reflections from the terrain and buildings may also influence the propagation loss.

As can be seen from eqn. 3.2 the free-space path loss increases with 20 dB/decade for increasing frequency.

Radio waves in the higher frequency range tend to propagate in straight lines and for any obstructions between transmitter and receiver the receiver will only be able to act on the part of the signal that bends around the obstruction. Such a bend may give considerable losses, as illustrated in Figure 3.5.

It will be difficult to give a general expression for the losses due to vegetation for higher frequencies, but some experimental data has been illustrated in Figure 3.7. Generally, the experimental data may give large variations due to the effect of vegetation. When the transmitter and receiver are placed outside vegetation (e.g. forest), but the vegetation will obstruct the line of sight the influence of the vegetation may become less important because the waves that propagate by diffraction above the vegetation may be stronger than those that propagate through the vegetation.

Gases and particles in the atmosphere may cause absorption of electromagnetic energy. This will, with respect to propagation, manifest itself as an added loss known as the absorption loss. The absorption loss measured in dB will under homogeneous conditions increase linearly with distance. However, this loss is not likely to be present for the frequency ranges of interest here.

As we have seen, perhaps the least meaningful radio-requirement parameter is range, since it depends to such a degree on the exact topography of the scenario, propagation characteristics of the particular path under consideration and local noise conditions. Nevertheless, it is most likely going to be used to indicate possible system solutions and choice of operating frequency. The designers on this occasion run some propagation simulation models to show how the operating frequency will influence radio range and positional freedom in establishing connections within the possible scenarios.

One of the results, used as an illustration for this case, is shown in Figure 6.2. The Figure shows the propagation loss for a set of possible operating frequencies for a scenario where the transmitter is situated within a wooded area extending for some 100 m and where an obstruction of 20 m height is introduced between the transmitter and receiver and where the antenna height is set to 2 m. In other words, the Figure illustrates a situation where losses due to both diffraction and vegetation will be present. The diffraction is modelled as simple knife-edge loss and the loss due to vegetation is estimated using an empirical expression. The influence of the reflectivity of the Earth is assumed to be approximately similar to plane Earth reflection.

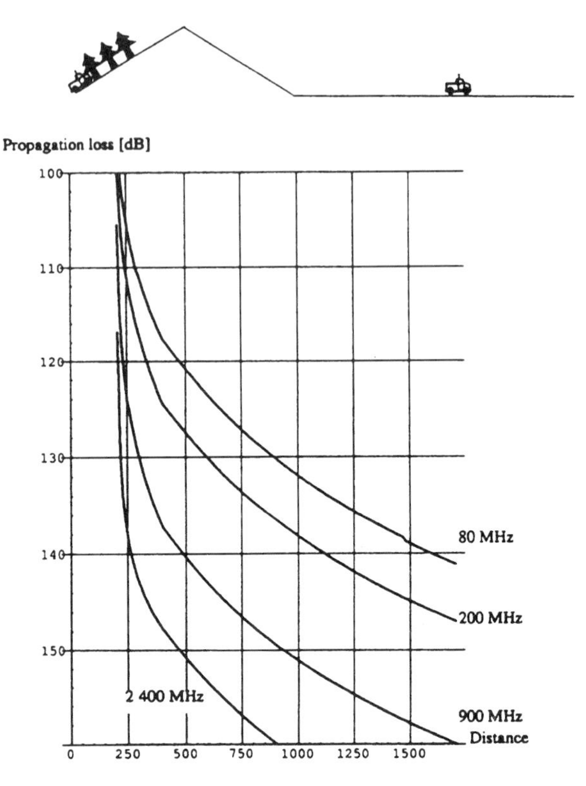

Figure 6.2 Simulated propagation loss for a selected range of operating frequencies, 100 m of wood (vegetation) between transmitter and receiver, a 20 m tall obstruction and antenna heights of 2 m

Based on such simulations the designers decide initially on the use of the VHF frequency band. The requirements for omnidirectional operation and reasonably small antennas, and at the same time for operation close to the human body, also support a VHF solution.

6.2.2 *Information needs*

The user has specified a system with the need for both voice and data services; the exact requirements are going to directly influence the bandwidth of the system. The required quality for the communication links is at this stage simply defined as a BER of minimum 1×10^{-1} for voice and a BER of 1×10^{-3} for data. It is assumed that these minimum requirements will be sufficient for possible use of error-correcting schemes and network protocols.

The user's demand for communication security and transmission security will complicate any solution, and in order to proceed further the designer would find it useful to discuss these demands in the context of bandwidth requirements.

The requirements are established, from needs for keeping the content of the transmission hidden from unauthorised users and for preventing degradation of the system in the case of heavy noise disturbances.

These demands are certainly going to make the radios more complex and costly than systems that do not have to comply with such requirements. The designers will be particularly anxious that such demands will increase power consumption to a level which makes it impossible to produce a hand-held solution at all. The designers will also be alerted by any bandwidth expansion which may make it impossible to make use of cost-effective and easily-available radio modules off-the-shelf.

So, the added complexity due to communication security and transmission security must be carefully controlled in order not to make the radio parameters too 'odd'.

To obtain transmission-security characteristics the designers consider some form of spread-spectrum modulation technique. In order to keep the bandwidth down, thus preventing the use of high-speed clock circuitry dissipating much power and specially developed RF modules, an orthogonal-coded direct-sequence spread-spectrum solution is considered. This particular coding, discussed in Chapter 4, gives a very efficient modulation which reduces bandwidth requirements and therefore increases the radio range for a given transmitter output power (not due to the code-spreading modulation but due to the encoding and reduced noise bandwidth).

The designers consider it possible to transmit both data at 2400 bit/s and digitised voice at 16 000 bit/s using 256-ary and 127-ary orthogonal signalling, respectively, within a bandwidth of approximately 50 kHz. At 2400 bit/s the transmitter sorts out the data in blocks of eight where the eight-bit binary combination points to a 0 to 255 numbered spreading sequence. At 16 000 bit/s voice the transmitter sorts out the data in blocks of seven where the seven-bit binary combination points to a 0 to 127 numbered spreading sequence.

If the bandwidth after filtering is restricted to approximately 50 kHz the spread-spectrum sequence length for data and voice can be:

code sequence length 2400 bit/s data: 50.000: $(2400/8) \approx 255$
code sequence length 16 000 bit/s voice: 50.000: $(16\ 000/7) \approx 31$

which gives an immunity to noise corresponding to a processing gain of 24 dB and 15 dB, respectively.

By using a short preamble, the radio will automatically be able to receive voice and data depending on the set up given by the preamble. The designers can thus obtain an integrated voice and data solution, and because of the narrow bandwidth it will probably not be necessary to design special RF circuitry. Such a bandwidth should also be easily available in the VHF frequency band.

For a data rate of 2.4 kbit/s and 16 kbit/s digitised voice (e.g. CVSD), the final radio range may vary by some 25% between voice and data. It may therefore be of interest to look for some narrowband voice-coding technique.

6.2.3 *Operating conditions*

Interoperability

The user has specified an infrastructure with which the new radio system should be both interoperable and compatible. The interoperability demands could preferably be solved by on the air interoperability, which means that the infrastructure systems and the new system should operate on the same frequencies and use similar modulation. The amount of direct communication between the infrastructure and the new system under design is, however, supposed to be rather limited but it is required that a subscriber to the new radio system may contact a subscriber in the infrastructure.

Compatibility

The main issue to be discussed at this point is the problem with collocation. It will be of interest to see how well radio systems (infrastructure + new system) operating at the same frequency and operating with several transceivers placed in the vicinity of each other, without any overall time management for when the different systems may transmit or receive, perform. Later discussion of collocation will be divided into two parts: (i) a mixture of radio systems operated in the vicinity of each other but outside the near field of the antennas in the same frequency range; (ii) the radio systems are collocated within 1--3 m of each other within or close to the near field of the antennas.

For both scenarios it is possible that the transmitter output powers may vary by a factor of 10--100, and since no overall time multiplexing exists, reception and transmission may interfere.

6.2.4 *Channel conditions*

This book has previously stressed the need to check theory against field measurements. For radio propagation in the VHF and UHF bands this is

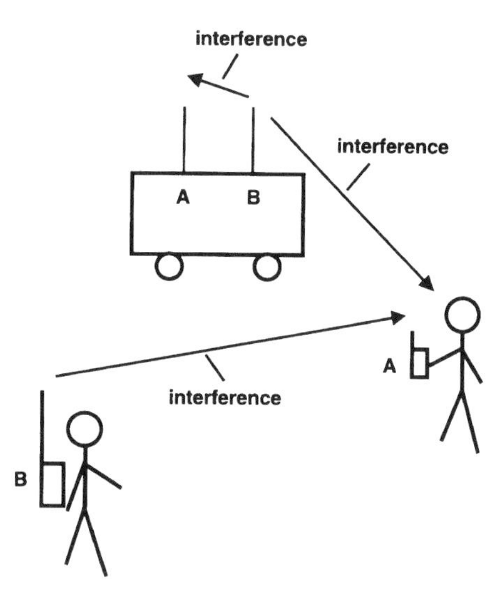

Figure 6.3 *The cositing scenario with the hand-held radio A used both in the field and inside a vehicle experiencing interference from two kinds of infrastructure radio system B (manpack and vehicular)*

essential. Not least, the existence and consequences of multipath may only be properly taken care of during the design process if there exist some experimental data.

The measurements shown in Figure 6.4 show propagation path losses in the VHF band for a scenario of mostly flat terrain with some forest. The results certainly do not follow any theoretical path-loss model, and must

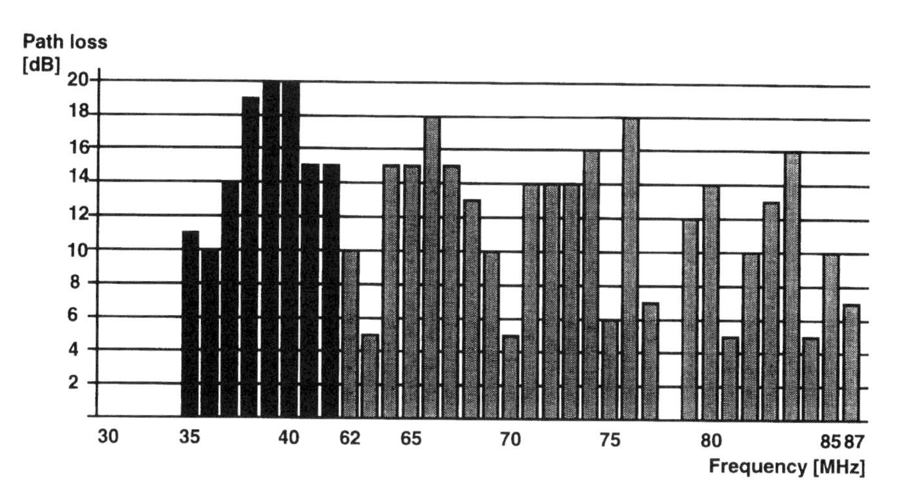

Figure 6.4 *A simple plot of path loss against frequency*

be explained by pointing to multipath effects. By just changing the frequency slightly, large variations in field strength are experienced.

In order to take the effect of multipath into account, more detailed measurements should be performed. A typical set up for multipath measurements is shown in Figure 6.5. The set up will continuously be able to measure the channel impulse response at a rate of typically 10--20 responses every second. These measurements make it possible to see how the impulse response is changing as a function of time or distance, depending on whether the receiver is at a standstill or is moving around.

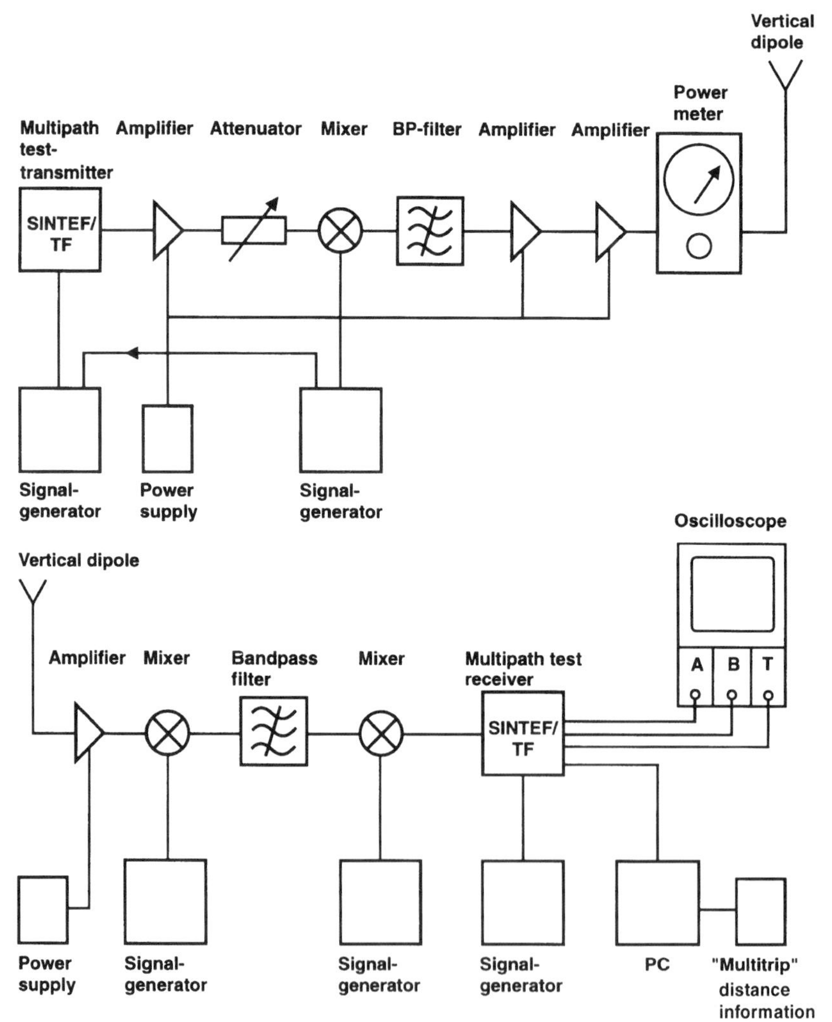

Figure 6.5 A simplified block diagram of equipment able to measure multipath characteristics

It is also possible to measure the total power received for each impulse response and in such a way get to know how the power is changing or how stable the channel remains. The radio paths may then be characterised by the length of the impulse response, the distribution of the signal components' mean path-length differences and variance and how stationary the different signal components making up the impulse response are.

With a transceiver bandwidth of 8 MHz the equipment will be able to resolve multipaths down to 125 ns which corresponds to a path-length difference of 37.5 m. In practice this will make it difficult to see the influence of reflectors close to the receiver, but merely show the influence of distant reflectors. At the same time a registration of 15 impulse responses every second will limit the possibilities for seeing changes in the impulse response during periods of less than 67 ms.

Figure 6.6 shows two follow-on impulse responses measured in the UHF band and recorded with a time difference of 67 ms. The receiver location has changed by some 0.13 m between the measurements. It looks like the first response is based on one major reflector and the second response is made up of two dominant reflections.

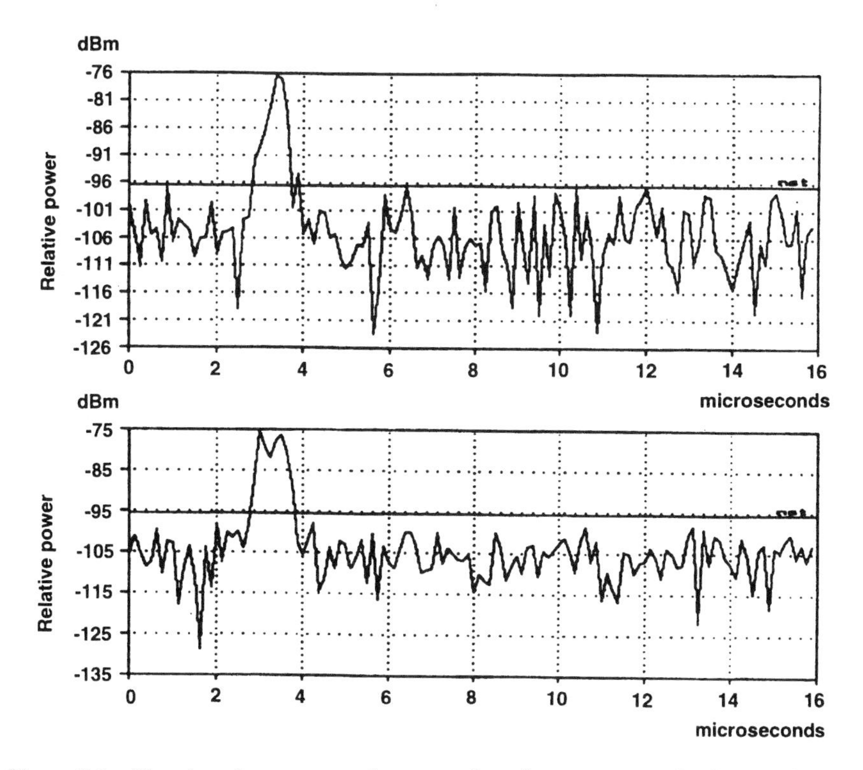

Figure 6.6 Two impulse responses from a series of measurements in flat terrain with some forest

The impulse responses can be analysed further to see how much the received signal will fluctuate (the cumulative distribution of received energy for different time delays). For any time delay where only noise is received the distribution will be Rayleigh. The same distribution will show up if the received signal is composed of signals from several paths with independent phase. If, on the other hand, one of the signals dominates, the distribution will be Rice.

Figure 6.7 shows *a* the average impulse response for the corresponding series of measurements and *b* the cumulative distribution of signal amplitude at different delays. Looking at Figure 6.7*a* where the Rayleigh distribution is shown as a reference, it can be seen that a dominant reflector exists causing different Rice distributions (more vertical lines than the reference Rayleigh).

The radio designer will have to make decisions on how the hand-held radio should handle situations with multipath, since these are likely to influence the radio range significantly. If possible it might be advantageous to let the spread-spectrum modulation resolve the multipath structure and then integrate the total energy (diversity reception).

6.2.5 Radio design

For a radio designer a good radio is most likely a radio with: a large radio range and good sensitivity, low sideband noise (low disturbance on other radio systems) and good collocation and large signal characteristics (low disturbance from other radio systems).

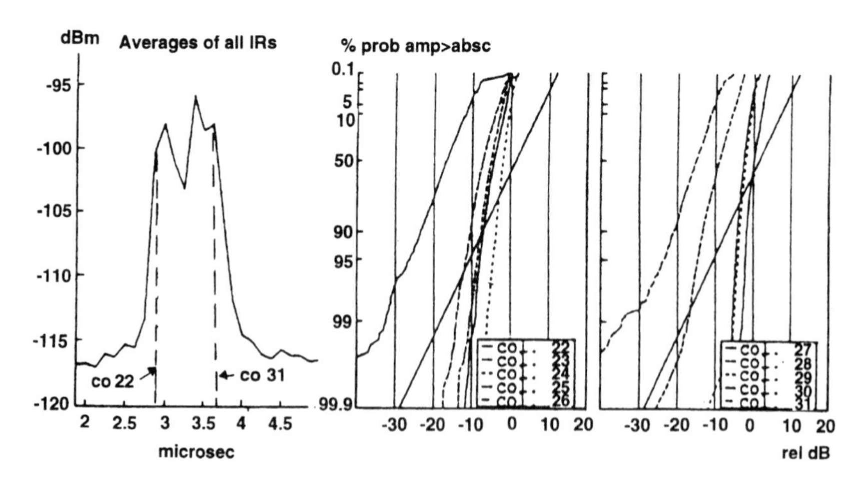

Figure 6.7 (*a*) *Average impulse response measured at certain propagation delays, and* (*b*) *cumulative distribution of the signal amplitude*

Such characteristics will be discussed in more detail and with regard to initial user requirements in what follows.

Transmitted power

The effective radiated transmitted power will depend on the available power and the antenna efficiency.

For a mobile radio the power supply is most likely to be some sort of rechargeable dry battery and the radio must be designed with regard to minimum power dissipation both for hardware and for software. A typical requirement may be a lifetime of some 12 to 24 hours for a duty cycle, transmit/receive/standby of 1/1/ 8 at a temperature range of −15°C to 40°C. The nominal battery supply will typically be limited to some 7.5 V with the internal voltage typically limited to some 3.5 V. Such a solution should preferably not exceed a total weight (radio + battery) of typically 500 g. The possible transmitter power within this framework is likely to be limited to the order of 1 W.

Antenna

The antenna gain (which for the small antennas required here will be a loss) will be an important parameter for radio range. The initial requirement could be for an omnidirectional antenna with a large bandwidth and an acceptable efficiency, able to operate close to the human body.

The choice of antenna will be a balance between simplicity in use and a wish for high antenna gain. However, several of the antenna requirements may be in direct opposition. When the size of the antenna is reduced, the bandwidth is also reduced and it becomes difficult to match the receiver to the antenna efficiently. Figure 6.8 illustrates the relationship between the antenna input compared to the antenna output power for electrically-small antennas. The electrical dimensions play in this regard

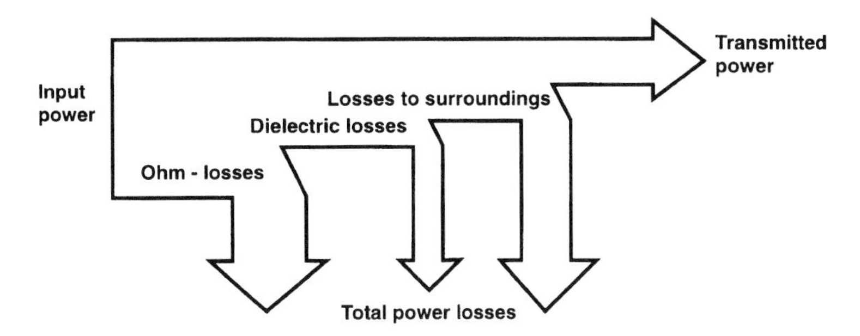

Figure 6.8 Important sources for losses in electrically-small antennas

an important role. The transmission efficiency will dramatically drop when the dimensions are decreased.

The requirement for an omnidirectional antenna and at the same time for operations close to the human body, favours a VHF solution. But at these frequencies there may be resonance between antenna and body and severe losses can be expected. Increasing the frequency on the other hand may be very unfavourable from a propagation standpoint. For vehicular use one solution may be to arrange for a radio frame which enables connection to a larger antenna and an extra power amplifier. In addition, such a frame may contain bulky filters which can suppress unwanted noise and interference.

The problem with short antennas (compared to the radio wavelength) combined with a large frequency range is the extremely low antenna efficiency. Two types of antenna solution for a VHF radio system are shown in Figure 6.9. The lower curve represents a 30 cm long helix antenna, and the upper curve represents a 130 cm long blade antenna.

Effective antenna height may not neccessarily be the same as the physical antenna height. For small antennas, experiments show that the ground characteristics will strongly influence the effective antenna height. This is particularly true for low operating frequencies, where the ground effect may change the radio range by as much as a factor of two. The effect of the ground condition on mean radio range based on a theoretical propagation model is shown in Table 6.1.

Generally it can be said that it is difficult to model electrically-small antennas in a reliable way, mainly because the antenna is so dependent on its close surroundings which tend to become a part of the antenna.

Finally, we may then plot the different noises for different antennas to see whether the radio range is going to be limited by the external noise (not including interfering signals) or the internal noise. The results for the two antenna designs described are shown in Figure 6.10.

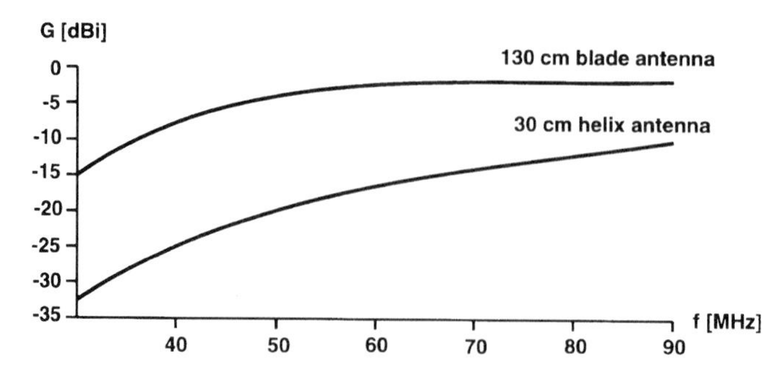

Figure 6.9 Antenna gain for two alternative antennas as a function of frequency

Table 6.1 Theoretically-calculated mean radio range (in km) between two hand-held mobile radios with 1 W output power and a 30 cm helix antenna

Frequency	30 MHz		60 MHz		88 MHz	
Ant.gain Tx/Rx	−32 dB/−32 dB		−16 dB/−16 dB		−11 dB/−11 dB	
	Voice	Data	Voice	Data	Voice	Data
Good ground	1.0	1.3	2.6	3.3	2.6	3.3
Poor ground	0.4	0.5	1.0	1.3	1.3	1.6
Mean ground	0.7	0.8	1.7	2.1	1.9	2.3

The Figure shows that for radios using the short helix antenna it would be very desirable to reduce the internal noise by a careful technical design of the radio circuitry.

Important transmitter characteristics

It is important to design a radio transmitter with a certain spectral purity in order not to disturb other radio systems. The sideband noise (selectivity on the transmitter side) depends on spurious noise, harmonics, sidelobe level, thermal noise and phase noise from the local oscillator. It is probably more important to keep the wideband noise low than to keep noise that will only occur at certain frequencies (e.g. harmonics) low. It is also more important to achieve a few spurious emissions at a relatively high level than many spurious emissions at a lower level. It is important that any sideband noise interferes with the fewest possible frequencies in order to keep the utilisation of the frequency band as high as possible.

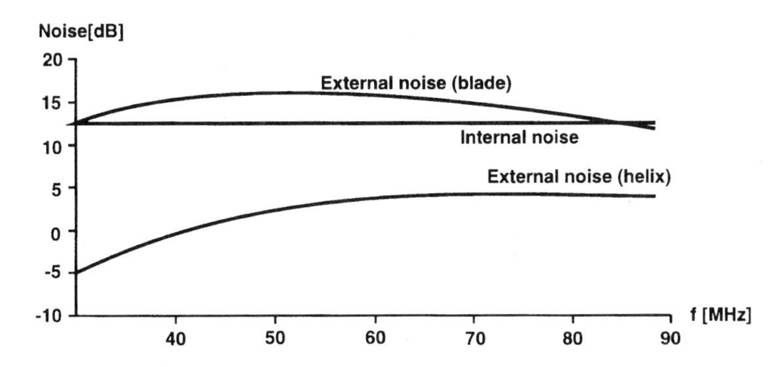

Figure 6.10 Contribution from external and internal noise for a VHF mobile radio with a helix or blade antenna

Typical figures may look like:

demand on suppression of harmonics: >80 dB
demand on suppression of a few (<10) spurious emissions: >90 dB

A typical transmitter characteristic for a mobile radio with a 50 kHz channel spacing is shown in Figure 6.11.

Background noise

External noise is going to influence the received signal, and both the wanted signal and the noise are going to depend on the antenna gain before they enter the receiver system. The external noise will, of course, be dependent on the location of the radio and the operating frequency, as the man-made noise changes. In addition to external noise, the radio receiver will be made up of components which will generate frequency-independent noise within the radio. This noise also interferes with the required signal (and the external noise for that matter) and will limit the receiver sensitivity.

Power output — electronic design — battery life

In order to reach the best possible compromise between the wish for a large radio range and the wish for long battery life, it is necessary to minimise the power overhead to run the radio circuitry.

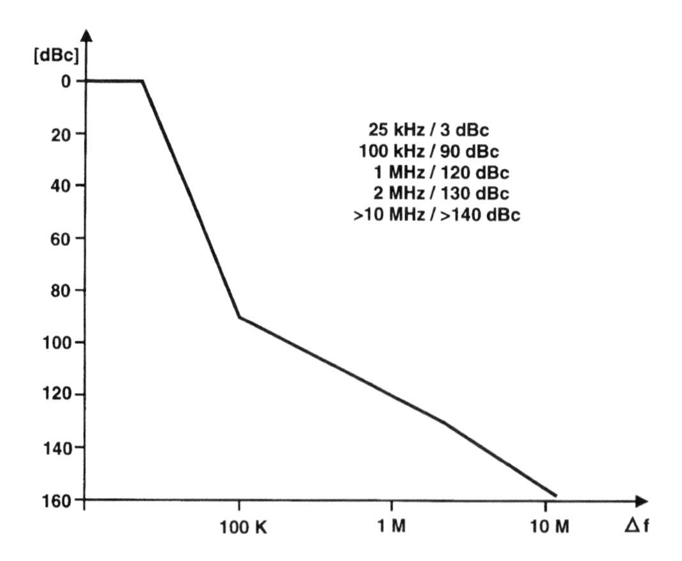

Figure 6.11 Spectral purity from a mobile radio transmitter

This will primarily be a discussion involving what functionality should be implemented in software and what should be implemented in hardware, not the least the duty cycle of the different radio functions. This kind of work is usually called partitioning.

If the radio as in Figure 6.12 is divided into an RF part, a digital part, a central processing unit (CPU) and a man--machine interface (MMI) unit, then certain observations can be made.

For the RF part there is a direct connection between performance, power dissipation and cost. The performance will, as discussed previously, be decided by the receiver selectivity, receiver sensitivity and what is usually called large signal characteristics. For large input signals it is important that the amplifier characteristics give the necessary linearity, that the radio chain stays symmetric and that the circuitry is able to dissipate the necessary power (or it will be damaged).

For the digital part, there is an inverse proportionality between flexibility and power dissipation. The most flexible module is the CPU but this also dissipates the most power. A bit less flexibility is available through the use of a digital signal processor (DSP), which uses less power than the CPU. The least flexible module will be an application-specific integrated circuit (ASIC), but this also uses the least power.

With such a general characteristic for the radio modules it will generally be good design to carry out those functions which will be in continuous operation with an ASIC, and the DSP can be used for the most processing-demanding functions which are also used most of the time. The CPU is used for functions which are only needed every now and then.

For the MMI unit it will be important to find a solution which allows the display to be turned off when not needed but which can respond with a short turn-on time when required.

Power dissipation is thus still, even after the development of a well established CMOS VLSI technology and highly automated design methodology, the largest challenge for the realisation of mobile radio equipment.

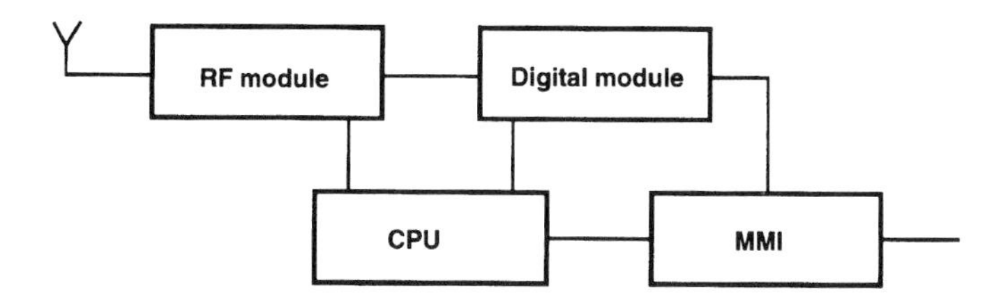

Figure 6.12 A simple decomposition of the mobile radio

Power use is an important trade-off between processing power (performance and functionality), battery life and the size and weight of the equipment. Foreseeing a fairly steady capacity from batteries in the years to come, increased battery life can only be achieved by finding new solutions which require a smaller number of operations per unit of time. In other words, it is necessary to develop more efficient signal-processing algorithms and architectures and to develop techniques to reduce the power dissipation for each operation.

The dynamic power dissipation in digital CMOS circuitry can be achieved through: a reduction in supply voltage, reducing capacitative load and/or reducing the switching activity.

Lately, methods for lowering the supply voltage as much as possible to lower power dissipation and compensate the resulting dropping performance (increased time delays) by making use of parallel operations and pipelining, have been established. However, as the number of processing elements (in a parallel solution) will increase the area of the circuitry, a trade-off between power consumption and the area of the circuitry has to be made. A simplified illustration of this trade-off is shown in Figure 6.13.

Using an ASIC in the development of a mobile radio will give the designer a great opportunity to make use of different schemes for reducing power dissipation at both system, architectural and algorithmic level.

At system level power dissipation is reduced by turning off those parts which are not in use. One may introduce several levels of on and off where the system or part of the system is brought to different dormant states.

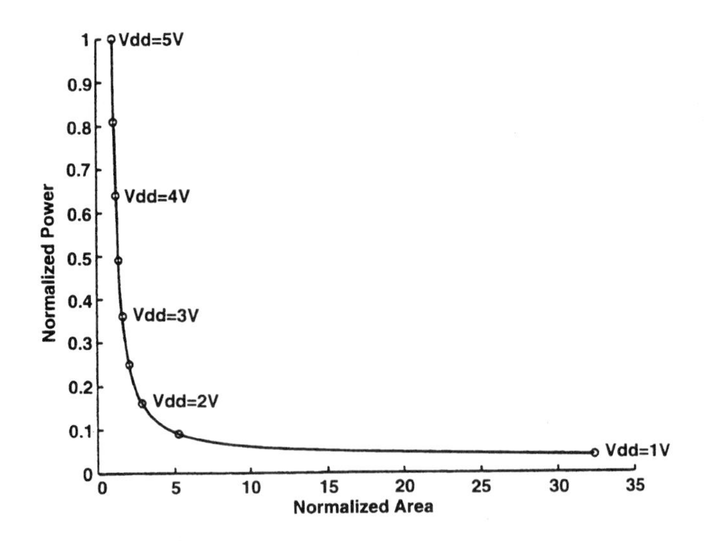

Figure 6.13 Power dissipation against area of circuitry

On an architectural level there will mainly be two factors of major importance. One of these is related to a change in architecture to compensate for lower speed and the other is a change of architecture that is technology independent. That is, if the system is not pushed speed-wise, one may choose an architecture which combines a small area and power with adequate performace.

In addition, the architecture should be optimised for power-reducing algorithms. Optimising the algorithms may give large power reductions. Using modern design tools, algorithms and architecture can be optimised and matched to each other. For DSP or general CPU the algorithm must use power-reducing operations and instructions.

Radio parameters important for collocation scenarios

In the following two main effects which will influence the collocation situation will be considered:

(i) interference from one system transmitting when another system is receiving in the same frequency band (in Chapter 5 referred to as multiuser interference);
(ii) an interfering signal that is mixed into the bandwidth of the received signal (reciprocal mixing).

The transmitted signal can be pictured as in Figure 6.14.

The modulator (phase noise) and the cosite filter determine the noise from the transmitter. The degree of interference depends on the frequency offset between the wanted frequency, to which the receiver is tuned, and the interference level at that frequency. The frequency offset can be specified by a specific value, for example 5 MHz or by a percentage, for example 5--12%. In the following an offset of 7% and 12% will be used for the worked examples. From Figure 6.14 it can be seen that for a frequency offset of 12% the transmitted spectrum can be expected to be approximately flat.

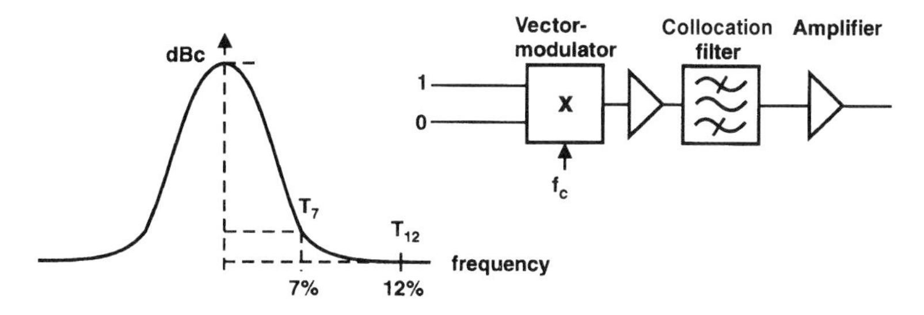

Figure 6.14 Illustration of transmitted signal

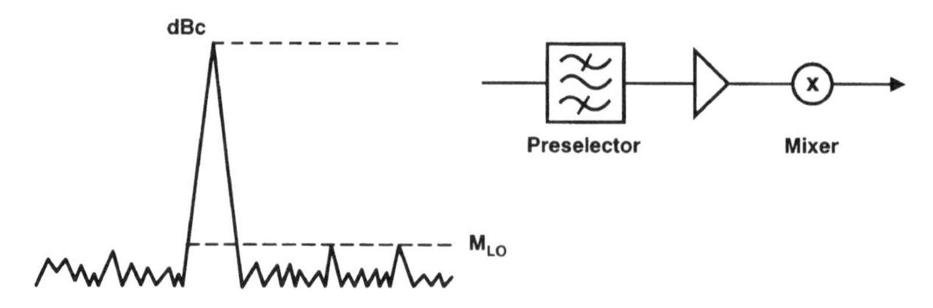

Figure 6.15 The receiver end with an illustration of a typical signal spectrum for the local oscillator (LO)

Looking at the receiver end, the signals from the antenna pass through some preselector before frequency changing. The situation is illustrated in Figure 6.15.

The interferer's spectrum 7% and 12% away from its transmitter centre frequency will first be attenuated by the preselector filters in the receiver and then mixed into the wanted signal bandwidth to a level being determined by the noise floor of the local oscillator. This effect will thus depend on the performance of a filter.

The major radio parameters influencing collocation problems are noise from the transmitter at a distance of 7% and 12% from the centre of the transmitted spectrum as well as the attenuation in the receiver's preselection and the noise floor of the receiver local oscillator.

The lower the noise floor of the local oscillator the more power dissipation (current drawn). Improving the preselector filters will reduce receiver sensitivity (important to radio range) and is likely to be costly. Thus, selectivity (important for rejection of interference) and sensitivity (important for radio range when the antennas are short and the frequency is high) will be opposing parameters.

For illustrative purposes some typical radio parameters are exhibited in Table 6.2.

Table 6.2 Central radio parameters important for collocation performance

	Noise from interferer dBc/50kHz		Preselector dB		Noise floor	Noise
	7%	12%	7%	12%	dBc/50kHz	factor
Hand-held radio	−99	−104	12	18	−85	11
Man-pack radio	−120	−123	32	44	−100	12
Vehicular radio	−120	−123	32	44	−100	12

Express then the noise from the interfering transmitter as:

$$N_{Transmitter} = N_T - R$$

where N_T is the noise level at relevant frequency offset and R is the path loss between the antennas, and the noise due to the mixing process as:

$$N_{Mixer} = N_M - R$$

where N_M is the noise from the mixing process excluding path loss between antennas.

The total reference noise without interference present is given as:

$$N_{Ref} = N_{Internal} + N_{Background}$$

$$= 10\log(kTB/10^{-3}(10^{F_1/10} + 10^{(F_2-G_R)/10} + 10^{(F_3-G_R)/10}))$$

where F_1, F_2, and F_3 are the internal, galactic and man-made noises, respectively, and G_R is the receiver antenna gain.

The noise from an interfering transmitter which raises the total noise (reduces the sensitivity) by Δ dB is given by:

$$N_E = N_{\mathrm{Re}f} + 10\log(10^{\frac{\Delta}{10}} - 1) \tag{6.1}$$

The required path loss between the antennas is given by:

$$N = 10\log(10^{(N_T-R)/10} + 10^{(N_M-R)/10}) \tag{6.2}$$

$$R = -N_E + 10\log(10^{N_T/10} + 10^{N_M/10}) \tag{6.3}$$

If, for simplicity, the path loss is considered to follow a $1/r^4$ propagation law, any increase in the total noise level by Δ dB will give a reduction in radio range of p% when:

$$p = (1 - 10^{-\frac{\Delta}{40}})100\%$$

Thus a 3 dB reduction in sensitivity will give a 20%–25% reduction in radio range, and a 12 dB reduction in sensitivity will halve the radio range.

The total loss between the transmitter and the receiver is given by:

$$L = R - G_R - G_S$$

where G_R and G_S are the antenna gain of the receiver and transmitter, respectively.

The necessary separation (for a given sensitivity reduction and hence reduction in radio range) between the interferer and the receiver can then be calculated using some form of preferred propagation model.

Table 6.3 Antenna gains used for the worked example

	30 MHz	60 MHz	88 MHz
Helix antenna	−32 dB	−16 dB	−11 dB
Blade antenna	−15 dB	−3 dB	−3 dB

Table 6.4 Necessary separation for a collocation scenario

	Necessary separation (m) 7%/12%		
	30 MHz	60 MHz	88 MHz
Hand held–Man pack (5 W)	<10	22/17	25/15
Hand held–Vehicular (50 W)	24/15	60/40	67/45

The gains for the antennas used throughout this Chapter are illustrated in Table 6.3.

The separation between the hand-held radio and the man-pack system as well as the hand-held radio and the vehicular radio, can then be calculated to give a 3 dB reduction in sensitivity which gives a 20%--25% reduction in radio range.

From Table 6.4 it is obvious that a considerable separation between the different radio systems is needed if the radio range is not going to be considerably reduced.

The final collocation situation refers to the situation where the hand-held system is used inside a vehicle also operating another interfering radio system. This time the antennas will be very close and it will be necessary to perform antenna isolation measurements since the near fields may introduce large errors to any calculations. If we let the background noise be represented by CCIR noise and assume as before a bandwidth of 50 kHz we have:

$$\begin{aligned}\text{CCIR noise dBm/50 kHz} &= -103 \text{ at } 30\text{ MHz}\\ &= -111 \text{ at } 60\text{ MHz}\\ &= -115 \text{ at } 88\text{ MHz}\end{aligned}$$

The internal noise may be represented by a noise figure of 12.5 dB which gives a total noise of:

$$\begin{aligned}\text{CCIR noise + internal noise dBm/50 kHz} &= -102.7 \text{ at } 30\text{ MHz}\\ &= -109.4 \text{ at } 60\text{ MHz}\\ &= -111.7 \text{ at } 88\text{ MHz}\end{aligned}$$

Typical attenuation between the antennas is measured to give:

	helix–helix	helix–blade
30 MHz	75 dB (5 m)	47 dB (1m)
60 MHz	49 dB (5 m)	23 dB (1m)
88 MHz	43 dB (5 m)	22 dB (1m)

The noise in the mobile radio using a helix antenna from a collocated radio 1 m away, offset by 7%, using a blade antenna and having an output power of 37 dBm and total noise at 7% of −120 dBc/50 kHz will then be:

collocation noise $f_c \pm 7\%$, dBm/50 kHz = −130 at 30 MHz
= −106 at 60 MHz
= −105 at 88 MHz

For a local oscillator noise floor of −85 dBc/50 kHz for a 7% frequency offset the reciprocal mixer noise becomes:

reciprocal mixer noise $f_c \pm 7\%$, dBm/50 kHz = −107 at 30 MHz
= −83 at 60 MHz
= −82 at 88 MHz

So the sum of the total noise from the collocated radio including reciprocal mixer noise gives:

−107 at 30 MHz
−82.9 at 60 MHz
−81.9 at 88 MHz

The sensitivity degradation due to the collocated transmitter will mainly be due to the reciprocal mixer noise (low receiver selectivity and high local oscillator noise floor) and will be:

1.4 dB at 30 MHz
26.5 dB at 60 MHz
29.8 dB at 88 MHz

Signal-to-noise ratio

The required signal-to-noise ratio will depend on the required error rate. For a data rate of 2.4 kbit/s and 16 kbit/s digitised voice, the final radio range may as we have seen vary by some 25% between voice and data.

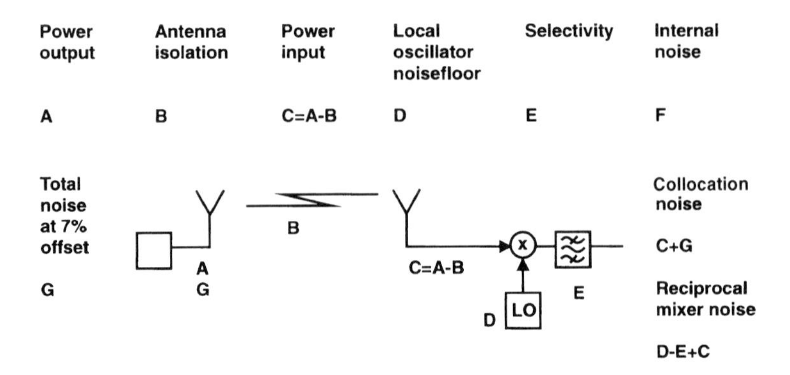

Figure 6.16 Relationships between the parameters discussed

For application of some narrowband voice-coding techniques in the hand-held radio, it will in particular be necessary to look into power dissipation, circuit area and cost.

There exists a large selection of voice-coding algorithms for 2.4 kbit/s and upwards. There will, however, be a number of factors influencing the choice for any particular application. From the decoding specifications one may obtain bit rate and algorithmic time delays. Similarly, it is possible to obtain estimates of complexity and memory use for a particular implementation. Based on such estimates it is possible to find the power dissipation.

The quality of the voice signal after coding/decoding is most likely validated by real experiments. In particular, background noise and channel quality for the application at hand will strongly influence the choice. Any use of objective benchmarking has up to now not been realistic in order to guarantee performance.

The designers will have to make a major choice between an ASIC (application-specific integrated circuit) or a general DSP (digital signal processor) solution. There will be four major factors that influence the use of a DSP:

(i) low power;
(ii) small size;
(iii) low cost;
(iv) high performance.

The power consumption is given by $P = C \cdot f \cdot U^2$ where C is the capacitance, f is the switching frequency and U the supply voltage. Thus it is necessary to look for reduced supply voltage, which reduces speed and thus allows reduced clock frequency. Or the design may keep the clock rate constant by changing technology to narrower line widths. Reducing the line width reduces the capacitance and thus power dissipation. In order to be able to reduce the clock rate one will need an increase of instructions within a clock cycle.

The general performance for DSPs has been short of what is needed for modern voice encoders. A lot of effort is being put into increased performance with a forecast of some ten times as many instructions per second within the next three to four years.

Electronic packaging technology

Since power consumption, size and weight are important parameters for a hand-held radio, new and advanced packaging technologies have to be assessed against production cost. The most effective technology, ASICs, has to be used for packing most of the digital electronics into one or two integrated circuits. Device sizes on silicon are rapidly downsizing (0.18 μm technology has been commercially available since 1997), packing density will be extremely high and power consumption per gate is at a level unthinkable a few years ago and still falling. Not all the digital electronics fits together in ASICs, and some has to be realised with standard components. These components may be packed very close together by use of MCM (multi-chip module) technology where uncapsulated IC chips are mounted on some kind of a substrate and protected by an encapsulant. There is a wide variety of technologies available, but as production cost is a limiting factor the MCM technology searched for should be in the area of laminated or ceramic substrates and nonhermetic encapsulation. For both ASICs and MCMs the initial or nonrecurring cost is in the \$60 000 to \$100 000 range, and a volume of say 25 000 is needed to get the per piece cost down to an acceptable level. For different reasons the production of modules should not be spread over a longer time than that indicated by the producer.

A new technology which has shown a very rapid growth makes use of packages which are slightly larger than the chip itself (up to 1.2 × chip size). The packages may be mounted on ordinary FR-4 boards, with standard pick-and-place machines. For large pin counts, special build-up boards with microvias and very fine lines have to be used. There is a large variety of these packages, which are called chip size packages (CSP), and they have been in use in hand-held electronics such as camcorders since 1995 and the first components have been commercially available since 1997. CSPs will be competitive with most MCMs both in terms of price and area savings in the future. Their greatest competitive advantage is that they fit into the infrastructure already in place in manufacturers' production lines.

To minimise production costs and to be able to test very closely packed electronics built-in or self-test is a must.

Analogue electronics may be packed in MCMs, but may be too expensive because special precautions have to be taken to get a high enough yield, due to the fact that component parameters have to be controlled to avoid module repair.

Other factors

One radio characteristic important to data services will be the transition-time delays (time between transmit--receive--standby). This is particularly time critical for receiving data in order not to lose information at the start of a message. Some typical requirements for transition times would be:

from transmit to standby: 500--700 μs
from receive to standby: 500--700 μs
from receive to transmit: 500--700 μs

If the radio user takes the hand-held radio with him down into ditches, or lies down behind hills, measurements show that the propagation loss is going to be increased by some 8.5 dB to some 19.5 dB over the frequency range of 30 MHz to 88 MHz.

A summary of findings

This section has shown that antenna design, sideband noise, collocation characteristics and power supplies are going to be important (and perhaps costly) parameters for the mobile radio design and will greatly influence the radio range.

6.3 Can the initial requirements be met?

System design is an iterative process as illustrated by Figure 6.17. Each iteration reflects a trade-off between many factors. This Figure sets the focus on the system's ability to fulfil the quality of the communication services, but other aspects may be of equal importance. For example, for the hand-held radio under study, the requirements for radio range which demand high output power and efficient (read large) antennas are in direct conflict to low weight (small power dissipation and small battery) and a small physical size (for example, a short antenna). It has also been shown how difficult it will be to make a radio with a maximum radio range (good sensitivity) which can operate in a collocation scenario (good selectivity).

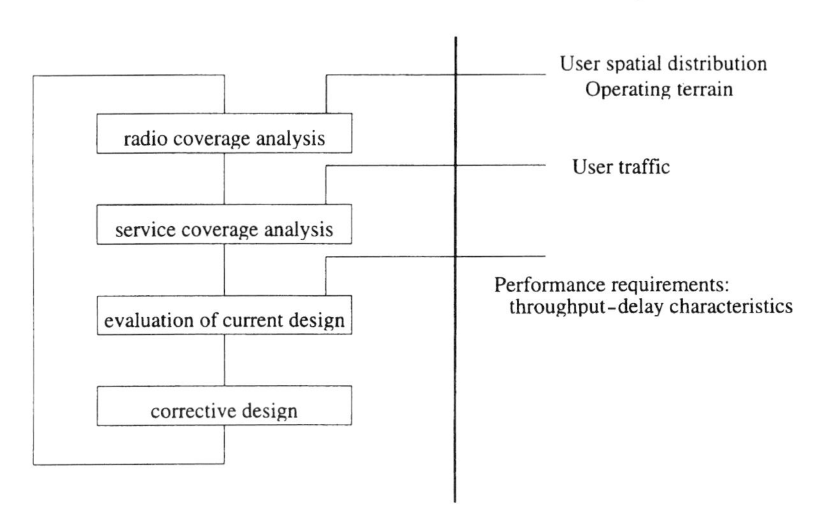

Figure 6.17 System design is an iterative process which is strongly related to the user requirements

Obviously, a necessary condition for establishing a communication channel is to have radio contact. However, if the transmission power is increased beyond the level needed to reach more than one adjacent user, this may introduce a higher interference level in other parts of a network and thus may deteriorate the quality of service in those network regions. The initial radio-coverage analysis should therefore consider the minimum power needed to avoid a situation with isolated users. The user spatial distribution and the operating terrain are the main input parameters to this step. The radio coverage predictions are usually performed by means of computer-assisted tools which present the results on graphical displays as illustrated by Figure 6.18.

Simulation tools can perform advanced modelling of the radio channel and its environment, and to a varying degree incorporate the many environmental parameters described in Chapter 3. Modern tools give the designer an opportunity to study the radio channel quality at different operating frequencies and the radio-environment parameters can be altered by changing environment parameters such as the presence of multipath and fading.

In this case, the designer concludes that the hand-held radio will not be able to operate with a typical radio range of more than 2–5 km. The question then becomes whether it is possible to design some form of function which may allow the radios to reach receivers outside the radio

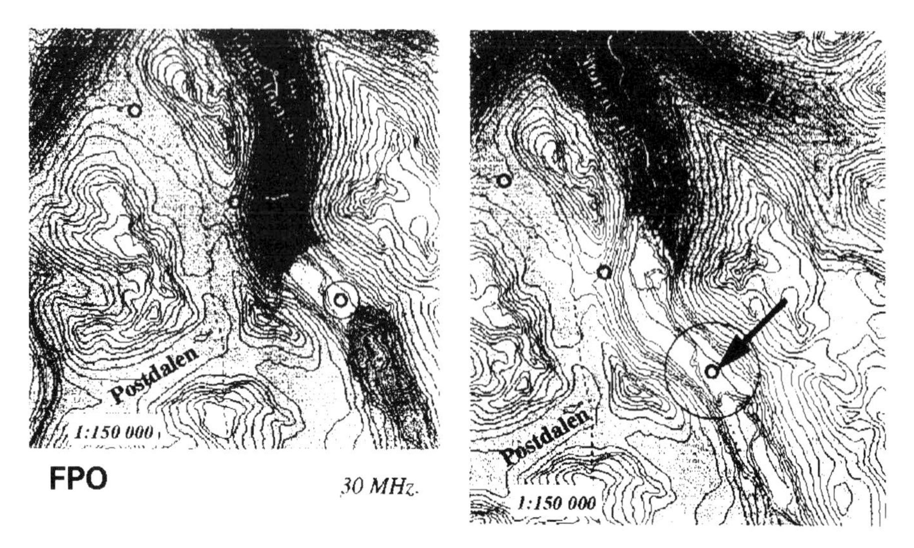

Figure 6.18 Example of radio-coverage area predictions at two frequencies
a 30MHz
b 60MHz

range. Such a function will complicate the radio, and the designer is likely initially to look for the simplest possible scheme.

6.3.1 From radio range to service area

How to extend coverage beyond radio range

Since the realistic radio range for the hand-held radio (the designer will more easily reach the range goal for a vehicular solution) is only 20--30% of what the user has specified, the designer will be looking for some simple (not increasing complexity, weight and volume and not increasing price) ways to establish a multihop system. In other words, the basic assumption of all-hear-all connectivity must be left.

A relay solution formed by two radios configured for this purpose can be introduced. The relay station must be established in a position which allows near real-time rebroadcasting to radios out of direct radio range.

Some form of via relaying may also be possible, making use of the network level as defined in the OSI interconnection model (see Section 2.2.3). To keep the system as simple as possible, such a system cannot be based on full packet-radio protocols, but rather some minimum of functionality. To obtain a multihop function in a simple system of hand-held radios it is particularly important to establish a mechanism which decides the path to be used for a particular end-to-end connection.

When source routing is used the addressing information contains the routing path that is the address of the hand-held radios through which the communication path shall be established. The routing path is determined at the source radio. A field is thus introduced in the packet overhead which identifies the source routing path to be used in the packet transmission.

The use of relay stations and the via relay function can be illustrated as in Figure 6.19.

Radio *A* wants to transmit to *B* which is well outside radio range for *A*. The following mechanisms exist for the two different solutions:

(*a*) the source address : *A*
the end address : *B*

A radio relay station receives and forwards a message on a near real-time basis on a different subnet. The packet data entities in both radios making up the relay station will regard the relaying to be performed on the physical level of the protocol. Since the relay function is performed on the physical layer, only the delay introduced by the relay station is considered in the estimation of access delays. This is valid for both data and voice.

Thus, all traffic in one radio subnet is transmitted in another radio subnet on a near real-time basis via a radio relay station.

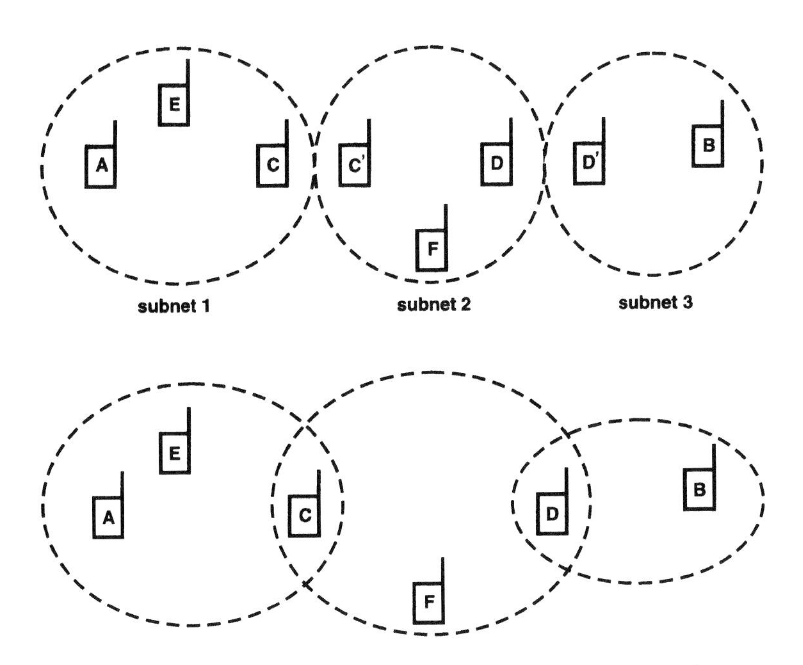

Figure 6.19 Communication between radios in an autonomous radio network
a use of relay station
b use of via relay

When A transmits then C will receive (being within radio range of A). Radio C sends a packet to C' which transmits the packet on the air. D will receive the packet from C' since it is within the radio range of C'. D sends the packet to D' which transmits the packet. The packet from D' is then received by B which is within radio range of D'.

In other words, by defining relay stations the only address information is the end address. This gives a minimum of overhead. If A wants E and F to receive and keep the message, E and F must be on the address list.

(*b*) Alternatively, all the radios that can hear the remote radios may retransmit the messages. This is called via relay.

The packet-data protocol for the hand-held radios can be based on the packet-radio protocol (PRP) as defined by OSI, but with an additional source-routing protocol (SRP) on the network layer. This protocol can allow multihop to take place without the full complexity (e.g. routing algorithm) of a PRP net. Source routing is performed by operating on the source-routing information transferred in 3a protocol control information (PCI), substituting the link destination address for each hop along the route.

The number of hops is likely to be limited due to the amount of addressing overhead introduced for routing purposes by each via relay. A large overhead reduces the efficiency of the protocol.

The packet-data protocol receives a service primitive from the service user in A with the following contents relevant for source routing:

source address : A
destination address : B
called address extension : B
calling address extension : A, C, D, B

The 3a PDU transmitted from A contains:

source address : A
destination address : B
3a PCI : hop counter $= A|A, C, D; B$

On this link the address extension facilities are conveyed transparently.

The 3a-PDU transmitted from C contains:

source address : A
destination address : B
3a PCI : HC $= C|A, C, D; B$

The 3a-PDU transmitted from D contains:

source address : A
destination address : B
3a PCI : HC = $D|A, C, D; B$

The service primitive received from PDP by the service user in B contains:

source address : A
destination address : B
called address extension : B
calling address extension : A, C, D, B

In designing the packet-data protocol it is assumed that no voice or unrestricted data is present on the radio network when packet data is to be transmitted. This condition is obtained by manually commanding silence due to data transfer. With no competing traffic on the air, the packet-data protocol does not experience collisions in complete networks. In topologies where via relays are used, collisions may occur when packets are retransmitted to obtain multihop. If, for some reason, a voice user wants to access the radio channel during a data-transfer period, the voice transmission can be given priority over the packet data on the access level. The voice may, for example, have shorter access delay (for example, none at all) than packet data. However, during transmission of packet data, voice will probably not be given priority to override the current packet-data transmission.

The medium access control (MAC) sublayer as defined in the OSI model can be replaced by a new MAC for the simple packet radio protocol discussed giving the following main functions:

- separation of voice, transparent and packet data traffic;
- prohibit the transmission of data packets when the radio channel is occupied;
- detect 'end of message' for voice and transparent data;
- detect 'end of message' for packet data (using request from the logical link control).

Even though the designers see a possibility for extending the service-coverage area beyond radio range, the exact performance and capacity of the possible solutions discussed here will depend on the parameter selections for the protocols and at this point remain undefined.

The designers therefore return to the user and try to sort out if there is any difference in operational needs between the users of the hand-held and the vehicular radios. They conclude that the hand-held radio is mostly used for voice and short data messages, while the vehicular radio represents the heavy-duty data requirements. Hand-held radio users are, however, by far the largest in number. They, therefore, conclude that it is important to stress the simplicity of the hand-held radio in order to keep the price down and rather choose a modular design where the vehicular radio is built upon the hand-held radio with add-on modules to give sufficient output power, collocation performance and full packet-radio network functionality. To make interoperability between the hand-held and the vehicular radio the designers go for the same modulation and RF format. The user scenario may then be sketched as shown in Figure 6.20.

The scenario is then defined by the designer as:

'The users of the packet-radio network are equipped with vehicular radios with a nominal radio range of 10 km; they operate within a geographic area of some 40 km × 50 km with typically some 20 users.

'They do primarily use formatted messages and receive tasks (where, kind of task, plan for performing task) consisting of approximately 30 kbytes and giving feedback (report from task) of approximately 30 kbytes. The traffic pattern is assumed uniformly distributed over all possible destinations. In addition, the users from time to time ask for background information regarding the task (a kind of handbook) of up to approximately 300 kbytes.

'Every user may be on the air typically once in an hour. There exists no typical user pattern with regard to geographical position, and the messages are generally not time critical, but it would be an advantage if the system could put a priority, if needed, on important messages. It will, however, be very important that all messages are received so that no tasks remain unattended.

'The load from the hand-held nets on the vehicular net is negligible.'

Radio-coverage predictions give reasonable accuracy for low-traffic regions where the background noise (or other variables not affected by

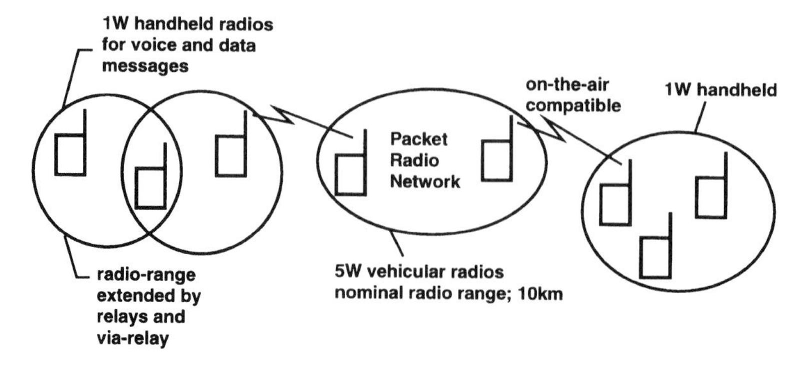

Figure 6.20 The user scenario

radio transmissions within the same cell) determines the link quality. Multiuser interference is, however, the dominant component in high-traffic regions. The radio-coverage analysis must therefore be followed by a step which takes multiuser interference into account. This is the task of the service-coverage area analysis using user traffic and radio-coverage predictions as the main input. The relationship between multiuser interference and user traffic has been given a detailed treatment in Chapter 5.

User mobility complicates the radio coverage prediction task since the radio path length becomes a function of time. The designer cannot overlook this important issue, nor can the designer analyse the situation for all time instances.[1] A practical approach is to analyse a concrete user scenario at particular time instances as shown in Figure 6.21. The example will be used as the concrete case to study in this subsection.

This example assumes that the designer has selected to consider the user locations at three different time instances for the given scenario. The radio-coverage analysis has predicted the radio-coverage areas to be perfect circles with no necessity to differentiate between the geographical locations.[2]

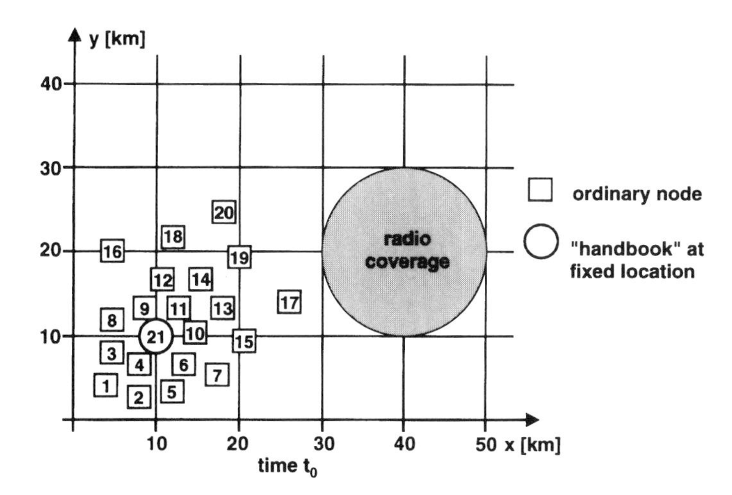

Figure 6.21 Mobility; performance is analysed for the time instances t_0, t_1 and t_2

[1]If a transient simulation model is used as in Section 5.4.4, the designer can get a reasonably accurate picture of a time-variant system. However, a system design would most likely use steady-state simulations in the initial phase since steady-state simulation models are generally less costly to implement and require shorter simulation run times. The same approach is used in this case study by performing steady-state simulations at a few selected time instances.

[2]Of course, this is not true for a real case but simplifies the presentation. It could be a reasonable approximation for flat terrain with identical radios equipped with omnidirectional antennas.

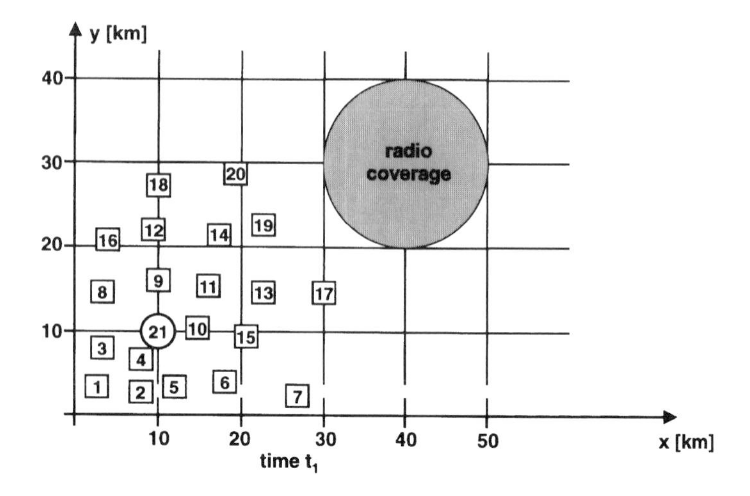

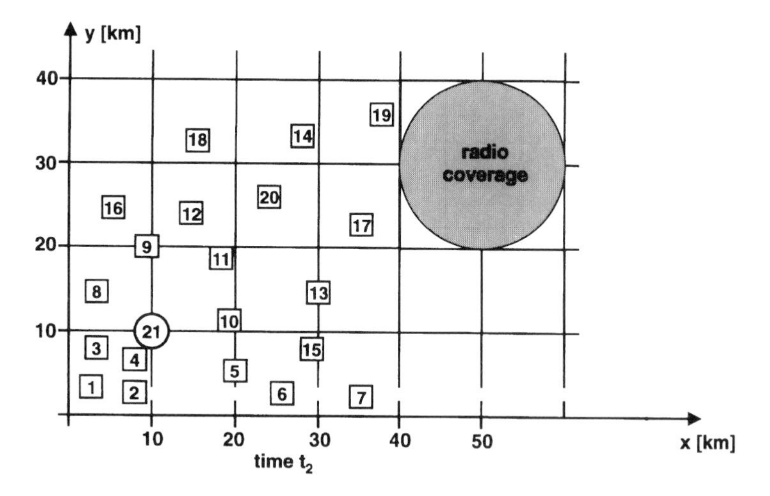

Figure 6.21 *(continued)*

The largest radio-coverage area required is determined by user number 19 at time instance t_2. The radiated power from his radio must be sufficiently high to reach user 14, otherwise he becomes isolated. With the lack of adaptive power control, all the radios must emit the same power and the network gets well connected at time instance t_0 and becomes more fragmented as time passes.

The radio-coverage predictions concluded that all the radios are not in direct radio contact (e.g., node 1 and node 20 never have overlapping radio-coverage areas). If the user-traffic distribution defines end-to-end traffic streams between disconnected users, communication cannot be established for all combinations of geographical locations and destinations. This is usually demanded and a network-relay function must be implemented. Techniques for extending the network-service coverage area for a packet-radio network beyond the radio-coverage area are outlined in Chapter 5. The designer starts with an initial design and we assume that the initial choice is the protocol stack 3 (RTS/CTS) using the radio characteristics specified by Table 5.3 in that Chapter. The initial RTS/CTS protocol parameter values are $(t_u, a, b) = (683 \text{ ms}, 1, 1.5)$.

The user traffic must be specified before the service-coverage analysis can be started, but the user traffic is again determined by the user applications. This case is set up with twenty identical users numbered 1 to 20. One special user exists, assigned the number 21, and is a kind of handbook from which the ordinary users ask for background information regarding the task. The ordinary users primarily use formatted messages and receive tasks consisting of 30 000 bytes and give feedback reports of 30 000 bytes. This is the traffic generated by the ordinary users during a typical busy hour. During the same period of time, the special user generates 300 000 bytes of data. This user is the user which requires the highest transmission capacity and this user remains fixed in one single position. The traffic pattern is uniformly distributed over all possible destinations and is identical for all the users. This completes the specification of the end-to-end user throughput requirements for the two groups of users. Their relative share (β) of the end-to-end network throughput become:

$$\beta_o = \frac{2 \cdot 30\,000}{2 \cdot 30\,000 + 300\,000} = 1/6 \qquad \text{for the ordinary users} \tag{6.4}$$

$$\beta_s = \frac{300\,000}{2 \cdot 30\,000 + 300\,000} = 5/6 \qquad \text{for the special user} \tag{6.5}$$

If λ is the end-to-end network throughput, the throughput required by an ordinary user is:

$$\lambda_o = \beta_o \cdot \lambda/20 \tag{6.6}$$

and the special user requires:

$$\lambda_s = \beta_s \cdot \lambda \tag{6.7}$$

A communication network is a complex stochastic process which must be underlaid by a structured analysis if the designer is to be able to understand its nature. A commonly used method is to apply computer simulations where new network functions are gradually introduced as'the designer gains insight into its behaviour by running simulation experi-

Table 6.5 Routing data

	Time instance		
	t_0	t_1	t_2
Average number of hops	1.37	1.86	2.59
Largest number of hops	3	4	6

ments.[1] Dynamic routing is an example of a network function which will probably be introduced at a later stage since it complicates the interpretation of the results. Therefore, the initial iteration applies static routing where the shortest paths are precalculated for all possible pairs of source and destination addresses. Table 6.5 summarises the routing data and clearly illustrates the increasing need for more transmission capacity as the nodes spread apart—from time instance t_0 to t_2 the packets must traverse nearly twice the number of hops.

With the input data in place, the simulation runs can be started. The service-coverage analysis is a very comprehensive study since the results vary from location to location as well as over all end destinations. A sensible approach is to consider the overall network throughput before going into the very detailed level. Figure 6.22 exemplifies this procedure by showing the 90% confidence intervals for the average hop-by-hop

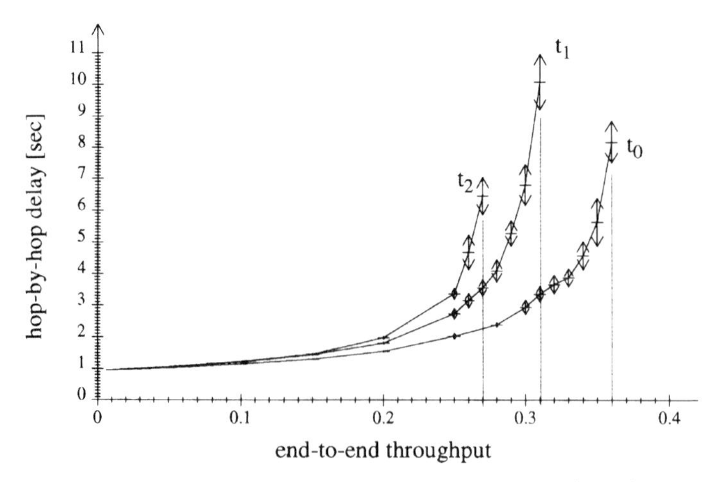

Figure 6.22 Average hop-by-hop delay against end-to-end network throughput

[1]The same approach is used in Chapter 5.

delays against the end-to-end user throughput. The estimates are formed by averaging over all the network links.[1]

The results show a fast growing delay as the load approaches the network capacity. To have reasonable delays for multihop packets, the network must be dimensioned to operate somewhat below its capacity level.

The evaluation of the current design regards the user throughput requirement which is specified by the number of bytes to be transferred during the busy hour, but the user information is carried as packets within the network. If a packet size of 150 bytes is used, the end-to-end network throughput required can by eqn. 6.7 be written as:

$$\lambda = (\lambda_s/\beta_s)/(150\ bytes) = \frac{6}{5} \cdot \frac{300\,000}{3600} \cdot \frac{bytes}{seconds} \cdot \frac{1}{150} \cdot \frac{packets}{bytes} \qquad (6.8)$$

$$= 0.67\ packets/s$$

The conclusion is that the user requirements cannot be fulfilled for time instances other than t_0. This fact is further emphasised by Table 6.6 where the throughput is converted to packets per second.

The Table also illustrates the capacity consumed by relaying the link-level throughput required. A presentation of the estimated end-to-end delays for all the users would demand too much space. However, the reader can acquire an impression of the end-to-end delays by combining data from Table 6.5 with the estimated link delays. Moreover, Figure 6.23 demonstrates how the network performance varies over the service coverage area by showing estimated link delays for a few links.

Evaluation of the current design concludes that three notable effects interplay to reduce the network capacity as time passes:

(i) the increasing number of hidden nodes causes a higher retransmission rate;

Table 6.6 Network steady-state capacity in packets per second. The bottom row is the link-level throughput at the corresponding end-to-end throughput

	Time instance		
	t_0	t_1	t_2
End-to-end user throughput required		0.67	
End-to-end network throughput capacity	0.74	0.64	0.56
Link level throughput	0.89	1.00	1.20

[1]This is the normalised throughput based on the physical packet transmission time 0.4836 seconds (one packet contains 150 bytes). Conversion to packets per second is simply normalised throughput/0.4836 s.

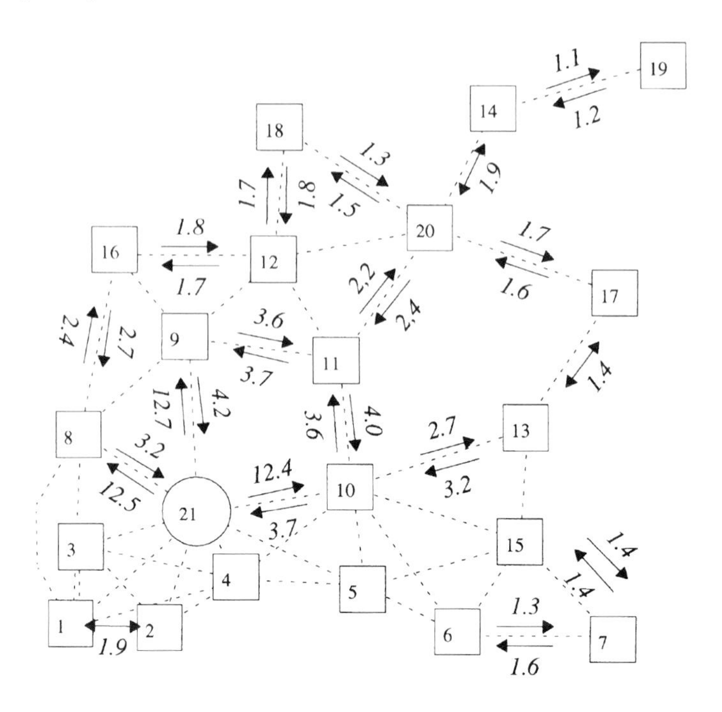

Figure 6.23 *Average link delays in seconds. The highest link delays occur at the outgoing links from node 21. These delays vary from 12 to 12.7 seconds and the large values reflect the high outgoing traffic rate from node 21. The nodes within the neighbourhood of node 21 serve as relays and are therefore forced to carry some additional traffic. The edges of the network area are within the low traffic regions and have the lowest link delays*

(ii) the increasing routing-path length consumes more network capacity (a higher link-level throughput is required without improving the end-to-end network throughput);

(iii) the increasing number of hops needed to reach the end destination introduces a larger end-to-end delay.

The last step in the initial iteration is to carry out a corrective design (i.e. to propose a new design that hopefully performs better) to fulfil the user requirements. For the case under study one attempt may be to find a more optimum protocol parameter set, or to extend the transmitting power to reduce the average number of radio hops between the end users. The new design serves as the input to the next iteration in the system-design process.

6.4 The need for a design exercise

The designers will, after being fairly confident that it will be possible to match the final user requirements and the technical solutions, finally enter into a design study or even exercise. The task at this stage will be to estimate and reduce risk by forcing the design team to make specific choices with respect to technical solutions and technology. From the previous discussion in this chapter the study is likely to focus on:

- functionality;
- performance;
- power dissipation;
- size;
- cost;
- development time.

The functionality must be studied in the context of possible implementation techniques. Of particular importance will be to see if it is possible to implement some form of low-rate voice coding in order to make the radio range for voice and data comparable.

Performance studies will include clarification of the radio range in the context of antenna and battery technology as well as low-power design possibilities. As seen from the discussion of PR networks in Section 6.3, these also put pressure on radio-range. The functionality and power dissipation will be connected through any use of partitioning techniques.

The power dissipation viewed in the context of functionality and technology costs must therefore also be clarified.

The size of the radio will, at large, be related to miniaturisation and technical production issues.

Index